Animal Movement Across Scales

AF729857

Animal Movement Across Scales

EDITED BY

Lars-Anders Hansson and Susanne Åkesson

OXFORD
UNIVERSITY PRESS

Animal Movement Across Scales. Edited by Lars-Anders Hansson and Susanne Åkesson.
© Oxford University Press 2014. Published 2014 by Oxford University Press.

OXFORD
UNIVERSITY PRESS

Great Clarendon Street, Oxford, OX2 6DP,
United Kingdom

Oxford University Press is a department of the University of Oxford.
It furthers the University's objective of excellence in research, scholarship,
and education by publishing worldwide. Oxford is a registered trade mark of
Oxford University Press in the UK and in certain other countries

© Oxford University Press 2014

The moral rights of the authors have been asserted

First Edition published 2014

Corrected 2015

All rights reserved. No part of this publication may be reproduced, stored in
a retrieval system, or transmitted, in any form or by any means, without the
prior permission in writing of Oxford University Press, or as expressly permitted
by law, by licence or under terms agreed with the appropriate reprographics
rights organization. Enquiries concerning reproduction outside the scope of the
above should be sent to the Rights Department, Oxford University Press, at the
address above

You must not circulate this work in any other form
and you must impose this same condition on any acquirer

Published in the United States of America by Oxford University Press
198 Madison Avenue, New York, NY 10016, United States of America

British Library Cataloguing in Publication Data
Data available

Library of Congress Control Number: 2014933931

ISBN 978–0–19–967719–1

Links to third party websites are provided by Oxford in good faith and
for information only. Oxford disclaims any responsibility for the materials
contained in any third party website referenced in this work.

Preface

The evolution of a book

Animal movement has been a core research field at the Department of Biology at Lund University for decades, although studies of different taxa have traditionally been performed at different departments, using different approaches. When we received a joint, long-term Linnaeus grant from the Swedish Research Council (Vetenskapsrådet) in 2008, we decided to not only continue our individual research paths, but also aim at opening up a broader avenue by utilizing our different scientific and technical approaches, taxa, and research networks. In so doing we decided that a book would be a suitable way to synthesize, but also give our own view of, the scientific field of animal movement. Our first step was to open up for suggestions of interesting chapters and twelve of us volunteered as lead authors, and we then met at a 3-day workshop where the outline of each chapter was discussed. After the inspiring workshop, the lead authors invited co-authors for their specific chapters and an almost twelve-month writing process was initiated. A major aim with this book project has also been to involve younger researchers in the writing and discussion process; a strategy that turned out to be very successful. Following the writing process, each chapter was sent out for international review and based on those comments the final version of each chapter was submitted. Then all interested lead authors were engaged in providing input to the synthesis chapter (Chapter 14), which not only constitutes the end of the book you now hold in your hand, but also of our scientific process of synthesizing the research field of animal movement. The evolutionary process of producing this book has been very creative and inspiring for us and we hope that you as a reader will find it useful and stimulating for your own studies and understanding of the fascinating field of animal movement across all scales! In order to guide you through the book we will below give some hints which may assist you as a reader.

To you as reader

During the writing process we decided that the target audience should be advanced undergraduate and graduate students, but a major aim has also been to provide an accessible and updated research-level text for professional researchers on animal movement, as well as established researchers from other fields entering the diverse area of animal movement. Moreover, we also wanted interested laymen to find the book engaging. In order to reach these categories of readers the chapters are written and edited in a way that makes the text as appealing and easily accessible as possible with many cross references among the chapters to make it easy to follow up intriguing subjects. To further highlight stimulating and noteworthy subjects, we have used ample illustrations to emphasize important concepts and processes. Moreover, at the end of the book we provide a glossary which hopefully will be useful for you who is new in the field of animal movement. Each chapter constitutes a unit and stands by itself, meaning that some illustrative examples or aspects of movement are taken up and discussed from different viewpoints in several chapters. This may be regarded as repetitive, although we hope you as reader will be indulgent towards this. An overall aim with this book has been to compile novel and interesting findings on movement ecology in order to stimulate and encourage the reader, rather than to provide a complete review of the research field. We, further, had no ambition to repeat or cover identical fields as other recent and excellent reviews or books (for example Dickinson et al. 2000, Nathan et al. 2008, Milner-Gulland et al. 2011, Clobert et al. 2012; Bauer and Hoye 2014). Instead this book has emerged through a creative process and thereby mainly reflects subjective cherry-picking by each author of especially absorbing, timely, and enlightening research. We indeed hope that you as a reader will find the subjects raised, and the viewpoints taken, as fascinating as we have found them during the writing process.

Thanks!

Just like animal movements, a book project requires a lot of energy and subsidies and this project would not have been possible without the support through our Linnaeus research platform CAnMove, funded by the Swedish Research Council (Vetenskaprådet) and Lund University. The Royal Physiographic Society and Stiftelsen Villa San Michele provided much appreciated logistics to L-AH during the editing of the book. All chapters have been reviewed and we would like to warmly thank all reviewers for very professional inputs, which have indeed improved the quality of the book. Since some of the reviewers would like to remain anonymous, reviewer names will not be provided here. Ben B. Chapman made a linguistic check of all chapters and Maria Sol Souza drew most of the illustrations, as well as the beautiful and thought-provoking cover. Anna Persson made the illustrations to Chapter 4. Emil Assarsson made a very professional job with the layout of the figures. Moreover, Thomas Alerstam has been supervisor or mentor for many of us and his enormous impact within the field of bird migration and navigation is clearly mirrored in the amount of citations of his work in this book. Finally, I would like to thank my wife, Ann-Christin, for fruitful discussions and ideas that would almost merit a co-editorship, and Linn, Sigrid, and Yrsa for being wonderful. The chapters of the book were written during fall 2012 and spring 2013 and edited in Askustorp, Hästveda, Sweden, and Villa San Michele, Capri, Italy, during summer 2013.

L-AH

References

Bauer, S., and Hoye, B. J. (2014). Migratory animals couple biodiversity and ecosystem functioning worldwide. *Science*, **344**, 1242552.

Clobert, J., Baguette, M., Benton, T., and Bullock, J. M. (eds) (2012). *Dispersal Ecology and Evolution*. Oxford University Press, Oxford.

Dickinson, M. H., Farley, C. T., Full, R. J., Koehl, M. A. R., Kram, R., and Lehman, S. (2000). How animals move: an integrative view. *Science*, **288**, 100–6.

Milner-Gulland, E. J., Fryxell, J. M., and Sinclair, A. R. E. (eds) (2011). *Animal Migration—A Synthesis*, Oxford University Press, Oxford.

Nathan, R., Getz, W. M., Revilla, E., Holyoak, M., Kadmon, R., Saltz, D., and Smouse, P. E. (2008). A movement ecology paradigm for unifying organismal movement research. *Proceedings of the National Academy of Sciences USA*, **105**, 19052–9.

Contents

List of Contributors xiii

1 An introduction to animal movement **1**

1.1 To move or not to move 2
1.2 Overview of the book 4
1.2.1 Large-scale patterns of movement 4
1.2.2 Movement strategies and adaptations 4
1.2.3 The mechanisms and codes of navigation and movement 4
1.3 Costs and benefits of movement 5

Part I Large-Scale Patterns of Movement **9**

2 Patterns of animal migration **11**

2.1 The evolution of animal migration 11
2.1.1 Ancient migrations 12
2.2 Taxonomic patterns of migration 13
2.2.1 Mammals 13
2.2.2 Birds 14
2.2.3 Fishes 14
2.2.4 Amphibians 14
2.2.5 Reptiles 15
2.2.6 Mollusca 15
2.2.7 Arthropods 15
2.3 Population patterns of migration 16
2.3.1 Between-population variation in migratory propensity and destination 16
2.3.2 Intra-population variation in migratory behaviour 17
2.3.3 Partial migration 18
2.3.4 Differential migration 19
2.3.5 Temporal patterns of migration 19
2.3.6 Daily migration 19
2.3.7 Temporal distortion 21
2.3.8 Seasonal migration 21

2.3.9 Semelparous migration 22
2.3.10 Transgenerational migration 22
2.4 Timing of the migratory journey 23
2.5 Spatial patterns of migration 24
2.5.1 Diadromous migrations in fishes 24
2.5.2 Latitudinal trends of bird migration 25
2.5.3 Altitudinal migration 26
2.5.4 Spatial scale and migratory routes 26
2.5.5 Landscape barriers to movement 26
2.6 Patterns of migration in a changing world 28
2.7 Future perspectives 30

3 Movement and migration in a changing world **36**

3.1 Habitat change 36
3.1.1 Destruction of stopover sites 36
3.1.2 Aquatic habitat change 39
3.1.3 Terrestrial habitat change 40
3.2 Climate change 40
3.3 Changing biotic interactions 44
3.3.1 Coping with change 45

4 Beyond dispersal: the role of animal movement in modern agricultural landscapes **51**

4.1 Community disassembly and movement 52
4.2 Effects of landscape fragmentation on movement and dispersal 54
4.2.1 Specialization and mobility mediates the impacts of habitat loss and fragmentation 55
4.2.2 Differential impact of habitat loss and fragmentation-implications for conservation 56
4.3 Mobility and spillover in mosaic landscapes 57
4.3.1 Spillover between habitats in heterogeneous agricultural landscapes 58
4.3.2 Dispersal ability and the differential response of fragmentation and disturbance frequency 59
4.3.3 Differential effects of fragmentation and disturbance frequency 59
4.3.4 The importance of source habitats for landscape-scale biological control 60
4.4 Mobility and multiple patch use 60
4.4.1 Agricultural change and resource separation 60
4.4.2 Landscape complementation and animal mobility 62
4.4.3 Differential mobility, structural change, and pollination 62
4.4.4 Landscape complementation and the matrix 63
4.5 Future perspectives 63

Part II Movement Strategies and Adaptations 71

5 Migration and flight strategies in animals: new insights from tracking migratory journeys 73

5.1 Migration strategies 73
5.2 Examples from tracking studies of migratory journeys 75
5.2.1 The giant leaps 75
5.2.2 Repeated flights and multi-day stops 76
5.2.3 Fly-and-forage 79
5.2.4 Soaring 80
5.2.5 The unexpected journeys 82
5.2.6 Feeding for flight 83
5.3 Migration in bats and insects 84
5.3.1 Bats 84
5.3.2 Insects 85
5.4 Future perspectives 86

6 Individuality in movement: the role of animal personality 90

6.1 Animal personality 90
6.2 Behavioural syndromes and behavioural types 91
6.2.1 Are behavioural syndromes adaptive? 91
6.2.2 Heritability of personality traits and behavioural syndromes 92
6.3 Plasticity as a trait 93
6.4 Animal personality and local movements 93
6.4.1 Movement variables as components of animal personality 93
6.4.2 Personality, consistency, and local movements in the wild 94
6.4.3 Fitness consequences of personality-dependent local movement 95
6.5 Personality and dispersal 95
6.5.1 Dispersal and behavioural traits 96
6.5.2 Aggression 96
6.5.3 Sociality 97
6.6 Behavioural syndromes and dispersal 98
6.7 Consequences of dispersal behavioural syndromes at the population level 98
6.7.1 Metapopulation dynamics 98
6.7.2 Range expansion 99
6.7.3 Invasive species and reintroductions 99
6.8 Individuality, personality, and migration 101
6.9 Partial migration and animal personality 101
6.9.1 To boldly go: partial migration and personality in roach 101
6.9.2 Boldness and blue tit migration 103
6.9.3 Comparing species-level behaviours 103
6.10 Individuality and migration 104
6.11 Future perspectives 104

7 Dispersal and phenotypic plasticity **110**

7.1 Phenotypic plasticity 110
7.2 Phenotypic plasticity and environmental heterogeneity 111
7.2.1 Agents of disturbance 112
7.3 Plasticity and costs related to habitat selection 113
7.3.1 Timing of dispersal 114
7.4 The dynamics and evolution of phenotypic plasticity and genetic adaptation 114
7.5 Phenotypic plasticity, dispersal rate, and the complexity of gene flow 115
7.6 The complex relation between dispersal and gene flow 116
7.6.1 Case study: dispersal in the terrestrial environment 117
7.6.2 Case study: dispersal in the marine environment 118
7.7 Plasticity and species invasion 118
7.8 Plasticity and the potential of speciation 120

8 Pathogens and hosts on the move **126**

8.1 Host movements and the risk of infections 128
8.1.1 Expansion of range limits 128
8.1.2 Increased vulnerability to parasites as a result of host movement 129
8.2 Moving host adaptations to reduce the impact of increased pathogen exposure 130
8.2.1 'Genetic footprints' in host immunity 130
8.2.2 Highly polymorphic host immunity genes 130
8.2.3 Tolerance instead of resistance to infection 132
8.2.4 Adjustments of immune function to meet infection risk 132
8.2.5 Spatial adaptations to local variation in pathogen distribution 133
8.3 Movement strategies as a mean to avoid infections 134
8.3.1 Mechanisms by which movement may reduce pressure from pathogens 135
8.3.2 The role of pathogen avoidance in the habitat-selection of migratory birds 135
8.4 Spreading of pathogens through movement 137
8.4.1 Emerging diseases in wild animals 137
8.5 Micro-organism strategies to take advantage of moving hosts 139
8.5.1 Parasites that influence and 'master-mind' the movement of their hosts 139
8.5.2 Passive dispersal of micro-organisms 140
8.6 Future perspectives 141

Part III The Mechanisms and Codes of Navigation and Movement **149**

9 Animal navigation **151**

9.1 Concepts and terminology 152
9.1.1 Orientation and navigation 153
9.1.2 Maps for navigation 154

9.2 The map and compass concepts 156
9.2.1 Biological compasses 156
9.2.2 Evidence for sun compass orientation in insects 156
9.2.3 Evidence for sun compass orientation in birds 158
9.2.4 Skylight polarization as a directional compass cue 159
9.2.5 The use of information from skylight polarization for orientation 160
9.2.6 Use of sun board and skylight polarization by Viking navigators 160
9.3 The star compass 161
9.4 The magnetic compass 163
9.5 Vector navigation in young migratory naïve birds 166
9.5.1 Fuelling before migration 167
9.6 Bi-coordinate maps—a global perspective of navigation 169
9.7 Path integration—an egocentric view of navigation 170
9.8 Landmark navigation 171
9.9 Future perspectives 172

10 Sensory mechanisms of animal orientation and navigation 179

10.1 Magnetic sense 179
10.1.1 Magnetic sense based on ferromagnetic particles 180
10.1.2 Biophysical and molecular mechanisms of ferromineral-based magnetoreception 180
10.1.3 Behavioural and physiological evidence for ferromineral-based magnetoreception 180
10.1.4 Electrophysiological recordings and lesion experiments 184
10.1.5 Conditioning experiments 184
10.1.6 Chemical magnetoreception based on a radical pair mechanism 185
10.2 Celestial compasses–sun, polarized light, and star compasses 188
10.2.1 Physiological evidence for sun and star compass orientation 188
10.2.2 Behavioural and physiological evidence for polarized light sensitivity 189
10.3 Future perspectives 190

11 Movements in the olfactory landscape 195

11.1 The olfactory landscape 196
11.2 Structure of odour plumes 196
11.3 Chemical features of odour plumes and trails 198
11.4 The evolution of olfaction and olfactory-guided movement 199
11.5 Odour tracking 200
11.5.1 Odour tracking in insects 200
11.5.2 Odour tracking in crustaceans 205
11.5.3 Odour tracking in fish 206
11.5.4 Odour tracking in birds 209
11.5.5 Odour tracking in mammals 210
11.6 Future perspectives 212

12 The genetics of migration **219**

12.1 How do we know that migratory traits are innate? 220
12.2 Quantitative genetic assessment of migratory traits 220
12.3 The genetics of migratory direction 223
12.4 The genetics of migratory timing 223
12.4.1 The threshold model of migration 224
12.5 Morphological and behavioural adaptations for migration 226
12.6 Genetics of migration: the molecular toolbox 226
12.7 Limitations and future perspectives in identifying 'the migratory gene' 228

13 The physics of animal locomotion **232**

13.1 Scaling of speed and cost of transport 232
13.1.1 Cost of transport (COT) 232
13.1.2 Speed of migration 234
13.2 Moving across a surface 234
13.2.1 Factors influencing speed on land 235
13.2.2 Factors influencing COT on surfaces 237
13.3 Moving through fluids 239
13.3.1 Flagella and cilia 239
13.3.2 Jet propulsion 241
13.3.3 Wings and fins 241
13.3.4 Factors influencing speed when moving in air and water 242
13.3.5 Factors influencing COT when moving in air and water 247
13.3.6 Morphology of wings and fins 250
13.4 Future perspectives 252

14 A synthesis of animal movement across scales **259**

14.1 Evolutionary compromises along the movement–sedentary gradient 259
14.2 Dispersal, gene flow, and assisted movements 260
14.3 Effects of environmental change on movement and migration 261
14.4 Finding their way 262
14.5 Patterns 263
14.6 Technical developments 265
14.7 Future perspectives on animal movement research 266

Glossary 269
About the Centre for Animal Movement Research (CAnMove) 274
Index 275

List of Contributors

Department of Biology, Lund University, Sweden
Susanne Åkesson
Staffan Bensch
Klaus Birkhofer
Jannika Boström
Christer Brönmark
Ben B. Chapman
Johan Ekroos
Lars-Anders Hansson
Dennis Hasselquist
Anders Hedenström
Johan Hollander
Kaj Hulthén
Christoffer L. Johansson
Niclas Jonzén
Marcel Klaassen
Miriam Liedvogel
Åke Lindström
Christer Löfstedt
Max Lundberg
Rachel Muheim
Florian T. Muijres
Jan-Åke Nilsson
Emily O'Connor
Ola Olsson
Lars Råberg
Maj Rundlöf
Ravinder Sehgal
Henrik G. Smith
Maria Strandh
Erik Svensson
Glenn P. Svensson
Sylvie Tesson
Machteld Verzijden
Maren Wellenreuther
Helena Westerdahl

Center for Environmental and Climate Change, Lund University, Sweden
Yann Clough
Henrik G. Smith

CHAPTER 1

An introduction to animal movement

Lars-Anders Hansson and Susanne Åkesson

All organisms on our planet move at some stage of their life cycle and animal movements on land, in air, and in water are therefore pervading features of animal life. Movements covering just a few, to many thousands of body lengths are performed by a variety of organisms—from the smallest protozoans to the largest whales, and can extend over widely different spatial and temporal scales, from the millimetre to global scale and from seconds to months. Animal movement is also a general feature of our everyday life, since even people living a modern, urban lifestyle cannot avoid seeing amazing examples of animal movement, such as bumblebees finding a flower with a surplus of nectar in a park, mosquitoes efficiently finding blood in our veins, and swifts (*Apus apus*) targeting their nest entrances under rooves with an astonishing precision and speed after an endurance flight lasting several days. Despite all those astounding achievements taking place around us, we rarely give a thought to how these animals manage, and how and why they move. Consider, for example, the never-ending row of ants walking the same path 24 h a day on the wall of the little house in Italy as shown in Fig. 1.1. Irrespective of whether the sun is burning or if it is completely dark, they just continue moving along their invisible track. Obviously they are not using the sun or the Earth's magnetic field to navigate, but rather their track is laid out with odour, allowing all individuals to follow the same, safe way (Fig. 1.1). Apart from a sophisticated odour detection system they also need morphological adaptations for locomotion to be able to walk along the wall without falling down. Hence, even those tiny organisms have unbelievable adaptations to not only handle chemical information for finding their way, but also to secure attachment and proprioceptors (stretch receptors) to successfully fight physical laws for moving up a wall!

Given such astonishing examples, it is not surprising that animal movement has, throughout our history (Nussbaum 1978), fascinated laymen as well as researchers. Similarly, it is not surprising that research on animal movement ecology is now entering a new era with the development of novel molecular, electronic, and other technical methods, which allows analysis of the movements of individual animals in the context of the complex variety of ecological trade-offs that determine the evolution of movement habits. In this book we aim at addressing how and why animals move and in what ways they differ in their locomotion and navigation performance. Our ambition is that the book will provide considerable synergistic connections both between studies performed on different taxa and between different spatial and temporal scales. Throughout the book we suggest that optimization is a useful approach for understanding the evolution of movement patterns among different animals, as well as their travelling performance, movement strategies, and paths followed. Taking foraging movements and dispersal into account as well as migration (see Box 1.1) is crucial for a detailed understanding of the spatial scale of adaptation, and also for analysing the consequences of movement at the population and community level, in the context of landscape and climate change as regarding the spread of invasive species.

Animal Movement Across Scales. Edited by Lars-Anders Hansson and Susanne Åkesson.
© Oxford University Press 2014. Published 2014 by Oxford University Press.

Figure 1.1 Ants on a wall in southern Italy are an illustrative example of locomotion on vertical surfaces, endurance movements, and the use of chemical senses in navigation. Artwork by Maria Sol Souza.

1.1 To move or not to move

To minimize energy expenditure, and thereby optimize its performance, an organism might reduce its movement, which is the case in, for example, terrestrial plants and also many marine invertebrates. Although the seeds or progeny can be highly mobile, adult stages of plants and sedentary animals generally live their lives at the place where they are born or settled and make the best of that situation by using nutrients available in the soil where they stand, and fighting enemies and grazers without moving from that site. Actually, by just considering trees, for example those beech trees, *Fagus* sp., that have been standing for more than 70 years outside the window of the room where this text is being written without moving at all, an immobile strategy seems rather successful. In contrast, the pied flycatcher, *Ficedula hypoleuca,* is constantly moving while vigorously defending his nestbox hanging in the closest beech tree, simultaneously singing and catching mosquitoes. On the other hand, the lake we can glimpse between the beech trees contains cyanobacteria (or blue-green algae, as they are also called), which are evolutionarily older than most other life on Earth, but are, just as the pied

Box 1.1 Movement terminology

The reasons for animal movements differ widely, from daily foraging movements to long distance migrations or dispersals over unknown lands and waters. In order to set the frame for the chapters in this book, we have used the following very broad definitions:

Movement. Individuals or populations (or parts of populations) that change position at any temporal or spatial scale. Movement includes all other ways of displacement.

Dispersal. Individuals or populations (or parts of populations) that move to reach new areas, but do not return.

Migration. What constitutes animal migration is a matter of some debate, with many definitions in the literature (Baker 1978, Dingle and Drake 2007, Brönmark et al. 2014). We have here used the following, rather broad, definition: Individuals or populations (or parts of populations) that move between two well-defined habitats on a temporally (reasonably) predictable basis. Migration includes, e.g., the seasonal migrations of birds between wintering and reproduction areas, fish migrations from lakes to streams (see Chapman et al., Chapter 2), but also the once-a-lifetime migration of eels from freshwaters to their natal marine habitat to spawn and then die (Aarestrup et al. 2009).

Homing. Refers to when an animal returns to a known goal, e.g. its home.

Foraging movements. Individuals that move between resting places, nest sites, etc., and feeding grounds in a temporally reasonably predictable way, e.g. bees moving from flower fields to the hive or bird parents feeding their nestlings. These types of movements are difficult to distinguish from migration, but generally occur at a shorter time scale. The most striking difficulty when distinguishing between migration and feeding movements is the very well known diel vertical migrations (DVM) of aquatic zooplankton. This type of movements may actually fit better as 'feeding movement' or as diel vertical movements, than as actual 'migration'. However, the term diel vertical migration is so established as a research area that any attempts to re-categorize it would lead to a revolution. Hence, in order to avoid such responses, we will here adopt the traditional view while noting the difficulties with such distinctions.

flycatcher, able to move in order to gather resources. Hence, mobility may not necessarily be a trait that has evolved *from* sedentary living, but instead may be one type of life strategy; that is, sedentary life is not necessarily inferior to being able to move! Actually, based on the number of beech trees compared to pied flycatchers outside the window suggests that the fitness of the sedentary life strategy is by no means worse than the mobile way of living! Hence, we may conclude that sedentary and mobile life strategies are both successful, suggesting that mobility comes with costs in comparison to a sedentary strategy, and vice versa. Such costs of movement include, for example, the machinery behind morphological adaptations for movement, such as legs, wings, and fins, but also knowing where to go, i.e. having sensory equipment to sense resources, threats, or mates. Sedentary organisms can instead spend the energy required for such movements on growth and reproduction.

Although most plants are in their adult stage sedentary or, with respect to aquatic habitats, free-floating without movement organelles, most animals are able to move as adults. Actually, the earliest invertebrate animals that evolved in the sea more than 500 million years ago appeared as two kinds: those that stayed put and waited for food to come to them, and those who moved to get it. In aquatic systems a sessile way of animal living is generally more fruitful than in terrestrial systems as a result of its three-dimensional food space and currents transporting suitable food-packages for consumption. Hence, not unexpectedly, sessile animal life is more common in aquatic than in terrestrial systems. However, in contrast to the generally sessile life-forms of plants, movement is the norm in the animal kingdom in both terrestrial and aquatic systems. Animals spend a considerable amount of time and energy on movement directed towards resources, such as food, shelter from predation, or sites suitable for reproduction. If these resources are of higher quality and more easily available elsewhere at certain times of the year or day, movement, dispersal, or regular migrations might be favoured by natural selection.

1.2 Overview of the book

In order to facilitate reading, the chapters of this book are separated into three major sections: In the first, which we have called 'Large-Scale Patterns of Movement', we address temporal and spatial patterns of movement at different scales, as well as how these might change in the future. Within the four chapters constituting the section 'Movement Strategies and Adaptations' we discuss alternative strategies to optimize movements, and, finally, 'The Mechanisms and Codes of Navigation and Movement', focusing on navigation, senses, locomotory energetics, and the genetics behind movement and migration.

1.2.1 Large-scale patterns of movement

In this section we address how migration has evolved many times independently and is an extremely widespread strategy in nature to combat predictable adverse changes in environmental conditions in time and space. This suggests that although there are no strong taxonomic constraints to the evolution of migration, migratory behaviour is to some degree evolutionarily unstable. A frequent observation, which is, among other aspects, discussed in Chapter 2 by Chapman et al., is therefore that animals move within relatively narrow time windows and in steps with circannual (seasonal) and circadian (daily) rhythms. Although temporal patterns of movement are diverse among life-forms there are many common themes which apply across temporal scales. Moreover, partial migration, i.e. when some but not the whole population migrate, may affect large-scale movements and migratory patterns in most species (Chapman et al., Chapter 2). Animals on the move are often vulnerable to abiotic as well as biotic threats, including human-induced phenomena, such as climate change and habitat fragmentation. Obviously, species with specific ecological traits are more prone to such threats than others. In the chapters by Lindström et al. and Smith et al. (Chapters 3 and 4, respectively), we highlight how different types of human-induced changes may alter the movements and migratory patterns we are used to seeing in, e.g., birds and insects. Finally, based on long-term data sets, we also provide a future perspective on how patterns of animal migration may change over time together with the potential evolutionary consequences thereof (Chapman et al., Lindström et al., Smith et al.; Chapters 2–4).

1.2.2 Movement strategies and adaptations

In this section we focus on what long journeys may teach us about life strategies in general and optimal movement and migration theory in particular. Different migratory strategies have most likely evolved because they conveyed improved survivorship to individuals that migrated in relation to those that stayed all year around at one site. The individuals migrating or dispersing may in this way have reached sites more suitable for reproduction, or providing higher food quantity and quality. When a moving strategy is adopted, animals migrate or disperse according to some alternative rules, for example minimizing the cost or duration of transport, or the total risk of predation during the journey (Hedenström and Lindström, Chapter 5). Although there might be some general rules for optimizing a dispersal or migratory journey, there are considerable differences between species, and even within species and populations at an individual level, e.g. between animals with differing personalities. The consequences of such personalities are discussed in Nilsson et al. (Chapter 6). Moreover, in Westerdahl et al.'s chapter (Chapter 8), we discuss the often neglected aspect that the moving animal may bring pathogens from one place to another and thereby spread diseases. From the viewpoint of the pathogenic micro-organism hitch-hiking on larger animals, this movement constitutes an efficient way of dispersal, whereas from the viewpoint of the host, migrating or dispersing may be a strategy to avoid pathogenic infections (Westerdahl et al., Chapter 8).

1.2.3 The mechanisms and codes of navigation and movement

Within the five chapters constituting this section, we discuss that all types of self-generated movements come with costs and the animal must produce forces to overcome resistance and gravity (Johansson et al., Chapter 13). For terrestrial locomotion, this is achieved by generating ground reaction forces, while in air and water this is done by generating fluid dynamic forces. Moving across scales thus has consequences for our expectations regarding the animal movement, including the occurrence of seasonal migrations, which may be limited by the speed and cost of locomotion (Johansson et al., Chapter 13). Here we discuss how animals move and address adaptations to generate forces and mechanisms to reduce the cost of transport. However, to overcome resistance and produce enough energy to actually move forward is not enough; an animal must also properly navigate in order to fulfil the actual reason for movement: to reach desired resources. Our understanding of the sensory systems behind animal movement has improved considerably during the past decades, including, for example, a better understanding of alternative compass mechanisms, including the magnetic compass, as well as strategies used for navigation, which are described in Åkesson et al. (Chapter 9) and Muheim et al. (Chapter 10). In addition to magnetic senses, moving animals may also use polarized light, the sun, and celestial navigation to find their way (Åkesson et al., Chapter 9, and Muheim et al., Chapter 10) and a considerable amount of taxa also utilize odour for navigation (Svensson et al., Chapter 11). For example, insects use odours to find mates or food, but even some birds use 'the smellscape' to find their nests or patchy food (Svensson et al., Chapter 11). Actually, the olfactory system was probably one of the first senses that evolved among animals on our planet; i.e. olfaction is possibly the oldest and most widespread sensory modality. However, despite that chemical communication may be the most potent and widespread form of information transfer in nature, the research field of chemical ecology is historically young. Although there are several reasons for this, such as the problems in analysing odour chemicals, a major reason is likely that we as humans have a poorly developed sense of olfaction and therefore find it more familiar to study how animal behaviour is influenced by our own major senses: visual and auditory cues. Hence, in this section of the book we aim at illustrating that moving animals are not relying on only one navigation instrument, but have an array of, more or less, reliable 'compasses'. In particular we describe how an animal needs a combination of at least one compass sense and a map to find their way back to a known site, or how they may rely on path integration and an ego-centric navigation system, measuring their own movements relative to the starting point (Åkesson et al., Chapter 9).

Another proximate cornerstone of animal movement research is the quest to understand the genetics behind movement and migration, including gene flow and local adaptations (Liedvogel and Lundberg, Chapter 12, and Hollander et al., Chapter 7). Movement and migratory strategies vary considerably between major groups of organisms, but also between closely related species or even within populations of the same species. From quantitative genetic analyses of cross-breeding and selection experiments, we know that a considerable proportion of the variance in a migratory phenotype has a genetic basis, but the number and identity of genes involved in modulating the migratory phenotypes are still largely unknown. In Chapter 12 by Liedvogel and Lundberg, we outline our current understanding of the genetic architecture of migratory phenotypes. We introduce molecular genetic tools and highlight how recent achievements in genomics potentially provide previously inaccessible ways to generate fully sequenced and annotated genomes of migratory species in the near future. Moreover, within the frame of the genetic code, an organism may be rather plastic and adjust its movement performance to the current environment. Such phenotypic plasticity may affect the ability for, e.g. local adaptations and dispersal (Hollander et al., Chapter 7).

1.3 Costs and benefits of movement

Despite the advantages of being mobile, any kind of active movement, whether in terms of running, swimming, or flying, comes with energy expenditure costs (Johansson et al., Chapter 13) and few organisms have adaptations that make them experts in more than one of the forms of movement. Hence, to become the best flyer, runner, or swimmer requires costly, specific adaptations, and having evolved such a specific morphology might not allow for further specializations from an energetic point of view. Moreover, there may also be trade-offs between adaptive traits so that the organisms specialized in one trait may be really bad at the others. Actually, this seems to be the general case and, e.g. the common swift, *A. apus*, lives the major part of its life flying (see, e.g., Hedenström and Lindström, Chapter 5), but can effectively not walk or swim. The cheetah, *Acinonyx jubatus*, is one of the fastest runners, but rarely tries to fly or swim, and a blue-finned tuna, *Thunnus thynnus*, is a very fast swimmer, but few people have seen it walk or fly. In contrast to these specialists, most organisms are more of generalists in their movement strategy and have spent their 'evolutionary energy' on developing capabilities for more than one way of movement. Hence, by graphically describing this thought experiment in a three-dimensional way, we observe most organisms being able to do, but not exceptionally good at, two, or even three ways of moving, whereas the real specialists mastering only one, but doing it really well, are few, represented by symbols tightly fitted to either the swim, run, or flying axes (Fig. 1.2). Hence, we find the symbols for most organisms in a swarm in the space between the axes in such a conceptual graph, whereas relatively few, the specialists, reach far out along one single axis (Fig. 1.2). The reason for this pattern is likely that the cost of evolving and managing specific motoric adaptations for, e.g., excellent flying involves trade-offs which are suboptimal for excellence in swimming and running, putting limits to the evolution of a super-organism mastering all means of movement to perfection. However, one way to approach being a super-organism is to use a specific morphological

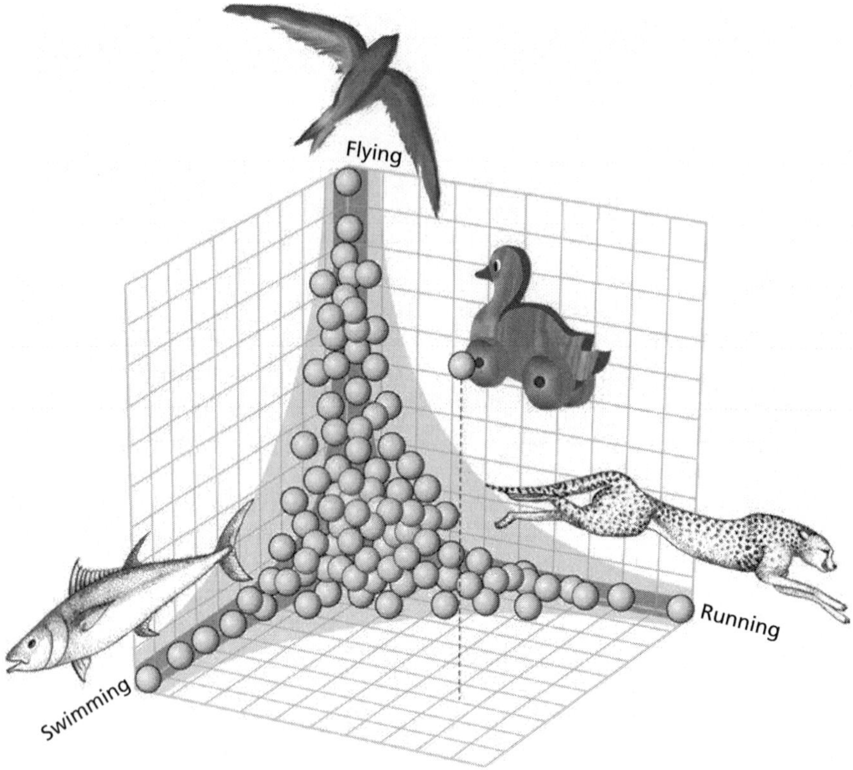

Figure 1.2 A conceptual illustration of different movement modes (running, swimming, flying), showing that few organisms are specialists in only one mode, and those that are generally have a weak performance for the other modes. In contrast, the majority of organisms are generalists and are reasonably good at several modes, as illustrated by the swarm of dots in the volume between the axes. A potential specialist on all modes may be the imaginary toy duck which is fast in air, water and on land. Note that this is a conceptual figure not based on data, but meant as a thought experiment.

adaptation to enhance movement efficiency when moving in different ways, for example as in the common guillemot (*Uria aalge*; Åkesson et al., Chapter 9), which is a reasonably good flyer, is a very good swimmer and diver, and is able to walk, albeit not very well. The trick here is that they use their feet and wings not only for walking and flying, respectively, but also for diving, i.e. using them as fins. In this way they reduce the cost (and space!) of mastering movement in several ways. Hence, it might be difficult to find, and even to imagine, a super-organism that could master all ways of moving. One possible imaginary super-organism may actually be the little yellow toy-duck on the back cover of this book, and also inserted in Fig. 1.2. We can see that it has probably long, pointed wings, which ensures rapid flight (Johansson et al., Chapter 13), it has large wheels which, for sure, will allow it to reach a high speed on flat ground, and it can obviously swim and even dive. Finally, it has a compact and strong body that can harbour a metabolism providing energy for the different adaptations. However, evolution has not yet combined all these features within the same taxa and we are therefore left to our imagination.

References

Aarestrup, K., Okland, F., Hansen, M. M., et al. (2009). Oceanic spawning migration of the European eel (*Anguilla anguilla*). *Science*, **325**, 1660.

Baker, R. (1978). Demystifying vertebrate migration. *New Scientist*, **80**, 526–8.

Brönmark, C., Chapman, B., Hansson, L.-A., Hulthén, K., Skov, C., and Nilsson, P. A. (2014). There and back again: migration in freshwater fishes. *Canadian Journal of Zoology*, **92**, 467–479.

Dingle, H. and Drake, V. A. (2007). What is migration? *Bioscience*, **57**, 113–21.

Nussbaum, M. C. (1978). *Aristotele´s de motu animalum*. Princeton University Press, Princeton, NJ.

PART I

Large-Scale Patterns of Movement

Chapter 2. Patterns of animal migration
Chapter 3. Movement and migration in a changing world
Chapter 4. Beyond dispersal: the role of animal movement in modern agricultural landscapes

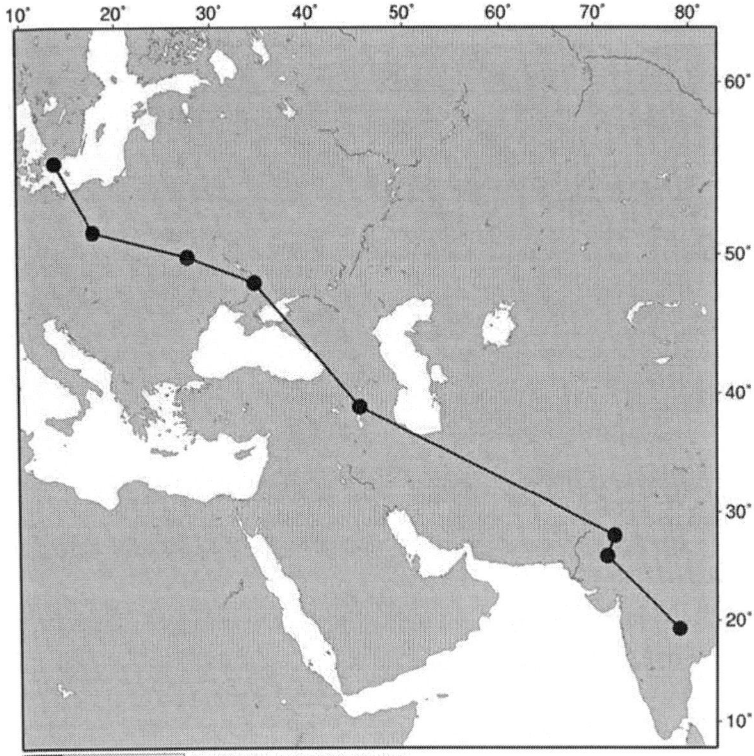

A migratory path from Scandinavia to India of little ringed plover, *Charadrius dubius*. Courtesy of Anders Hedenström.

CHAPTER 2

Patterns of animal migration

Ben B. Chapman, Kaj Hulthén, Maren Wellenreuther, Lars-Anders Hansson, Jan-Åke Nilsson, and Christer Brönmark

Animal migration is a fabulously rich and varied tapestry of behaviour, encompassing a diverse range of movements. Migration often evokes images of the spectacular seasonal journeys of birds from their European breeding grounds to the food-rich African plains and forests; or the leaping of Pacific salmon as they fight the river currents to return to their natal streams to reproduce and then die; or the teeming herds of wildebeest as they migrate across the Serengeti. Yet these iconic epitomes of animal migration tend to focus upon the seasonal, long-distance migration of charismatic species. The true diversity of migratory patterns in nature is far broader than these examples would suggest, with migration occurring at an astonishing range of spatial and temporal scales. From the seasonal journeys of Arctic terns, *Sterna paradisa*, that fly thousands of kilometres from the high Arctic to the South Pole (Fig. 2.1; Egevang et al. 2010), one can scale down to short distance migratory movements of blue tits, *Cyanistes caerulus*, that migrate less than 100 km (Nilsson et al. 2008), and even further to the daily migrations of micro-organisms such as zooplankton which move mere metres vertically in the water column in response to daily environmental fluctuations in predation risk and light (Hansson and Hylander 2009). One can also scale up, in time, to migrations that occur just once in a lifetime, such as in many semelparous fishes that migrate from the ocean to breed in freshwater habitats and then perish. Or we can consider the transgenerational migrations of insects such as the painted lady butterfly, *Vanessa cardui*, where it takes six generations to complete the migratory cycle from Europe to Africa and back again (Stefanescu et al. 2012).

In order to make sense of the bewildering complexity and heterogeneity of migratory patterns in nature it is critical that we classify some of the variation observed in the wild. In this way we can begin to conjecture and test hypotheses about the evolution of different migratory patterns. Furthermore, a clear understanding of the sheer variety of migratory behaviour will assist us to conserve migratory species and the habitats they rely upon in an ever-changing world (Smith et al., Chapter 4; Lindström et al., Chapter 3). In this chapter our aim is to highlight the diversity of migratory patterns in nature and categorize some of these patterns. To achieve this we discuss taxonomic, population-level, temporal and spatial patterns of migration. Throughout the chapter we will also discuss the evolutionary processes postulated to have shaped the different forms of migration. By doing this we hope to highlight that an understanding of the patterns of migration that can be found in the natural world requires an evolutionary approach, assessing the costs and benefits of different strategies through the collection of field and experimental data, and the testing of theoretical models. Finally we discuss evidence of changing patterns of migration, in response to both anthropogenic and natural processes (see also Lindström et al., Chapter 3).

2.1 The evolution of animal migration

For migration to evolve via natural selection the fitness benefits of migration must outweigh those of

Animal Movement Across Scales. Edited by Lars-Anders Hansson and Susanne Åkesson.
© Oxford University Press 2014. Published 2014 by Oxford University Press.

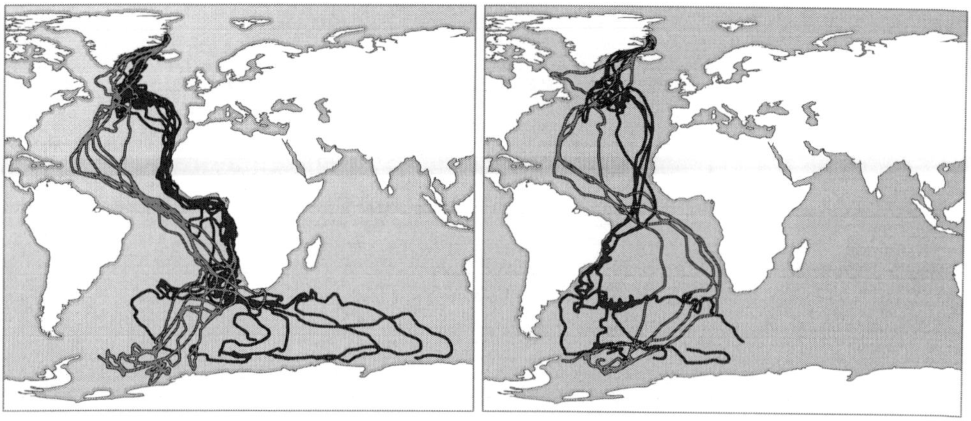

Figure 2.1 Yearly migratory journeys of Arctic tern, *Sterna paradisa*, between Arctic regions and Antarctic regions. Interpolated geolocation tracks of 11 Arctic terns tracked from breeding colonies in Greenland (n = 10 birds) and Iceland (n = 1 bird). Black tracks indicate autumn (postbreeding) migration (August–November), and grey tracks indicate spring (return) migration (April–May). Reproduced from Egevang et al. (2010) with permission from PNAS.

residency. Evolutionary fitness is essentially the contribution of an individual to the gene pool of following generations (i.e. lifetime reproductive success, which is a composite of two variables: survival and reproductive success). A great deal of research has investigated the costs and benefits of migratory versus resident lifestyles across a range of animals. For migration to be beneficial it should allow animals to maximize their survival and/or reproductive success in the face of a seasonally changing environment. These benefits can include refuge from predation (Hebblewhite and Merrill 2007, Skov et al. 2013), escape from adverse climatic conditions such as low winter or high summer temperatures (Newton 2007, Alonso et al. 2009), the avoidance of seasonal tropical storms (Boyle et al. 2010), and access to higher quality (or more) food. The costs of migration have been assessed in a number of species, and can include transport costs (Wikelski et al. 2003; Johansson et al., Chapter 13), heightened risk of predation (Beguer-Pon et al. 2012), and in some cases reduced feeding (Chapman B. B. et al. 2013). Below we illustrate some of the different forms of animal migration with reflections upon how these patterns have evolved. However, first it may be suitable to contemplate somewhat around ancient migrations.

2.1.1 Ancient migrations

It seems highly likely that now extinct creatures made migratory journeys analogous to those made by extant species. However, due to the obvious difficulties in studying the migratory behaviour of long-dead organisms, data upon their movements are scarce. Whether dinosaurs themselves made migratory journeys has been hotly debated since this idea was first proposed in 1928 by Friedrich von Huene to explain the European distribution of the prosauropod, *Plateosaurus* (von Huene 1928). Until recently, however, the evidence has always been circumstantial, with inferences from biomechanics and locomotive energetics, mammalian analogies, distribution maps gleaned from the fossil record, and the presence of enormous 'bonebeds' of dinosaurs (Bell and Snively 2008). However, recent advances in stable isotope analysis provide direct evidence for ancient migrations, in dinosaurs and also in extinct fish. Comparisons of oxygen isotope ratios ($\delta^{18}O$) of tooth enamel carbonate from the sauropod, *Camarasaurus*, with ratios analysed from ancient soil, lake, and wetland revealed that these dinosaurs made seasonal migrations of several hundred kilometres. Sauropods journeyed from lowland to upland environments, most likely leaving the basin in the dry season when plant growth may have been limited and drought more common (Fricke et al. 2011). Stable isotope analysis has also been used to reconstruct the migratory movements of a late Cretaceous fish, *Vorhisia vulpes*, which showed behaviour similar to modern anadromy. They migrated from brackish waters to marine habitats where they remained for 3 years before migrating back to the estuary, most likely to spawn and die (Carpenter et al. 2003). However, the idea that dinosaurs (especially polar dinosaurs) made extensive long-distance north–south migrations is still highly contentious, and for now at least, the jury is out.

2.2 Taxonomic patterns of migration

Migration is ubiquitous within all the major vertebrate groups, and extremely widespread among the invertebrates. There seems to be no clear taxonomic pattern of migration, which suggests that migratory behaviour is a commonly evolved response to periodic and predictable shifts in environmental conditions (Box 1.1). Examples of migration proliferate throughout the animal kingdom, in charismatic mammalian fauna such as ungulates and whales (Hebblewhite and Merrill 2007, Rasmussen et al. 2007), in a plethora of avian species (Newton 2007), and among fishes (Brönmark et al. 2014), crustacea (Groeneveld and Branch 2002), amphibians (Grayson and Wilbur 2009), and reptiles (Hays et al. 2002). There is also a growing recognition of the scale and importance of insect migrations (Holland et al. 2006, Chapman J. W. et al. 2010). Migratory behaviour has even been documented outside the animal kingdom, among algal protists that make daily migrations vertically in the water column (Jephson and Carlsson 2009). Although widespread among animals, migration is not equally prevalent within all families or genera: for examples among the fishes, migration is common among species within the salmonidae family (>80% of Canadian salmonid species are migratory), but thought to be almost absent from the Cobitidae loaches and the speciose Cyprinidontiforme killifishes (Lucas and Baras 2001). Similarly among the Jacanidae, a family composed of tropical wading birds, only one species, the pheasant-tailed Jacana, *Hydrophasianus chirurgus*, exhibits migratory behaviour. This suggests that although there are no strong taxonomic constraints to the evolution of migration, migratory behaviour is to some degree evolutionarily labile. It is clear that migration has evolved many times independently, and that it is an extremely widespread strategy in nature to combat predictable adverse changes in environmental conditions in time and space. In the following sections we briefly introduce some migratory behaviours of the different taxa, to emphasize the sheer breadth of this captivating phenomenon.

2.2.1 Mammals

Among the mammals, the ungulates perform the most visible migrations, and these are extremely well documented. Perhaps the most famous ungulate migrations are the annual movements of species such as zebra and wildebeest that migrate in vast, mixed-species herds across the plains of Africa. However, ungulates at more northerly latitudes also migrate, following the waves of freshly grown vegetation that appear after the winter in the lowlands and gradually spread northwards and to higher elevations through spring and summer. Young, rapidly growing shoots provide more digestible forage than older vegetation, and following the green wave of spring means that migratory herbivores access better food than their sedentary relatives (Hebblewhite et al. 2008). Carnivorous predators that hunt ungulates are much less likely to show migratory behaviour (with some exceptions, e.g. spotted hyena: Trinkel et al. 2004), likely due to the necessities of maintaining social structure and defending territories.

A new and exciting taxonomic frontier in migration biology is the study of bat migration (Hedenström and Lindström, Chapter 5). Bats are mostly nocturnal flyers, and relatively small, which makes them difficult to tag and track. However, a renewed interest in patterns of bat migration coupled with technological advances leading to reductions in tag size and weight has led to a number of fascinating studies revealing the prevalence of bat migration. Bats can make incredible migrations, in terms of both the numbers of animals that migrate and the distances travelled. For example, an estimated 5–10 million straw-coloured fruit bats, *Eidolon helvum*, migrate annually into Kasanka National Park in Zambia (Richter and Cumming 2005). Satellite telemetry showed that individual bats travelled great distances, with the longest recorded migration spanning >2,500 km in 149 days (Richter and Cumming 2008). Many other bat species also migrate

(Bisson et al. 2009), but migration is still thought to be a relatively rare strategy in this group, with estimates currently at around 3% of extant bat species (Fleming et al. 2003).

Finally among the mammals, the sea-dwelling cetaceans also make great seasonal migrations. The record for the longest mammalian migration is currently held by a female humpack whale, *Megaptera novaeangliae*, which migrated over 9,800 km from Brazil to Madagascar (Stevick et al. 2011).

2.2.2 Birds

Birds are the best studied and arguably the most iconic migrants, with a spectacularly diverse array of migratory behaviours (Newton 2007). Our understanding of the causes and consequences of avian migration is continuing at a rapid pace, fuelled by technical progression and innovation. One way in which bird migration is diverse is in respect to scale. At one end of the spectrum there are endurance migrants that travel great distances during their seasonal flights. Recent tracking of great snipe, *Gallinago media*, revealed that they make extremely long and fast endurance flights (4,300–6,800 km in 48–96 h), not only over deserts and seas, but also over large expanses of suitable habitat (Klaassen et al. 2011). However, the world record for endurance flight is held by the Alaskan bar-tailed godwit, *Limosa lapponica baueri*. This wader migrates 11,000 km from the Alaskan breeding grounds to wintering grounds on New Zealand. Moreover, they fly non-stop flight without rest for 8 days (Gill et al. 2009). Other extreme migratory behaviours include the high altitude journeys of bar-tailed geese, *Anser indicus*, over the Himalayas. These vertiginous flights have been recently tracked in great detail to shed light on this unusual strategy (Hawkes and Balachandran 2013). At the other end of the spectrum, birds can migrate over very short distances, latitudinally or altitudinally, requiring no specific adaptations to survive the migratory journey. As we will see in this chapter (and indeed throughout the book), research into bird migration has provided some of the most powerful insights into the evolution of this fascinating phenomenon (Liedvogel and Lundberg, Chapter 12).

2.2.3 Fishes

Migration is extremely widespread among fishes, and occurs in a diverse array of forms and for a number of functions, including feeding, breeding, and refuging from predators (Lucas and Baras 2001, Chapman B. B. et al. 2012a). Migratory behaviour in this group is often classified into typologies based on the habitats between which individuals migrate. Oceanodromy concerns migrations that occur solely in the marine environment, such as the vast seasonal migrations of Atlantic bluefin tuna, *Thunnus thynnus*, and the daily vertical movements of herring, *Clupea harnegus*. Conversely, migrations limited to freshwater are known as potamodromy, and are widespread across many groups (Brönmark et al. 2014), and include seasonal migrations from lakes into streams (Skov et al. 2008), or lateral migrations of fish into floodplains, which are well described from tropical and subtropical regions with dry and wet seasons. Further, many fish species perform a diel horizontal migration, where they migrate from the littoral zone of lakes into offshore areas at dusk and then return to the littoral at dawn (e.g. Muska et al. 2013). Perhaps the best-studied type of fish migration is diadromy, where individuals migrate across the salt–freshwater barrier. Migration in the fishes has traditionally received less research attention than in other taxa such as birds, most likely due to the relative difficulty in tracking individual movements over potentially great distances beneath the water. However, as we will see in this chapter, many classic examples of migration come from the fishes, and as a group their patterns of migration are as diverse as the birds, and perhaps even more so.

2.2.4 Amphibians

Amphibian migrations are mostly breeding migrations to suitable ponds and wetlands from overwintering sites, although foraging migrations can also occur (Semlitsch 2008). Amphibian migrations generally take place over a limited spatial scale: mean estimates of the distance travelled from the pond edge for 32 species were between 142 and 289 m (Semlitsch and Bodie 2003), although longer distance migrations have been recorded for certain amphibians (e.g. 2.4 km for western toads, *Bufo boreas* (Bartelt et al. 2004)). However, even these relatively long-distance migrations for amphibians would be extremely local movements for other migratory taxa. Many amphibians are prone to desiccation and their movements are usually constrained by weather, with migrations being triggered in many cases by nocturnal rainfall (Todd and Winne 2006). Sex differences in migratory behaviour are also very common in this group, with males typically arriving earlier to maximize reproductive opportunities at the breeding grounds (Semlitsch 2008).

2.2.5 Reptiles

Reptile migration is largely limited to the oceanic migrations of turtles and the short-distance migrations of snakes and tortoises. Lizards, for the most part, adopt hibernation as an alternative strategy to overcome challenges associated with predictable seasonal changes to the environment. The best documented (and most spectacular) migrations of reptiles are the great marine journeys of sea turtles. In many cases turtles migrate between foraging and nesting grounds, often over exceptional distances (Åkesson et al., Chapter 9). Leatherback turtles, *Dermochelys coriacea*, can range over thousands of kilometres to forage upon jellyfish in distant waters before returning to nest on their natal beaches, whereas terrestrial migrations in reptiles are typically on a much smaller scale. For example, giant land tortoises on the Galapagos Islands make elevational migrations driven by seasonal changes in food availability (Blake et al. 2012), and water pythons in Australia make migratory movements of up to 12 km to follow prey in the wet season (Madsen and Shine 1996).

2.2.6 Mollusca

The study of migration among the molluscs is still in its infancy, but is speculated to be quite widespread in this group (Semmens et al. 2007). Cephalopod migrations occur at all phases of their life history, from the passive drifting of eggs and larvae to diel vertical migrations, to long-distance (i.e. thousands of kilometres) adult migrations of animals to feeding and spawning grounds (O'Dor 1992, Semmens et al. 2007). These long-distance migrations are mostly undertaken by ommastrephid squids in the pelagic zone, with coastal cephelapods such as octopods or loliginid squids migrating over smaller scales (tens to hundreds of kilometres).

2.2.7 Arthropods

Many crustaceans make migratory journeys in the traditional sense (i.e. migrations which involve a return journey). For example, many zooplankton have been observed making daily migrations vertically in the water column (Gliwicz 1986), a phenomenon we will return to later in this chapter. Other crustaceans that migrate include Christmas Island red crabs, *Gecarcoidea natalis*, which make a highly synchronized mass breeding migration to the coast from the inland forest habitats where they forage for the rest of the year. Males migrate in a first wave, with females following shortly after. Social migrations are also common among lobsters. The spiny lobster, *Panulirus argus*, makes autumnal migrations into the deep water, potentially to avoid seasonal storms, and sometimes thousands of individuals line up to form long queues during migratory movements (Herrnkind 1985). The function of queue-forming is speculated to be defensive, as lobsters must travel across shelter-less substrate during migration. Individuals in queues can reduce their predation risk via increased vigilance, cooperative defence, and risk dilution (Herrnkind 1985). There may be additional benefits, e.g. drag reduction during movement (Bill and Herrnkind 1976; Johansson et al., Chapter 13).

Insect migration is also thought to be prevalent (Holland et al. 2006). Many arthropods such as insects are very short-lived, and hence in most cases traditional definitions of seasonal migration cannot be applied to these organisms. If we relax the requirement of a return journey in seasonal migration (Holland et al. 2006, Dingle and Drake 2007), billions of insects make migratory journeys each year. For example, individual tracking of green darner, *Anax junius*, dragonflies revealed that their migratory strategy was similar to songbird migrations, with individuals flying during the day and only on days with low wind velocity (Wikelski et al. 2006). Studies of noctuid moth migration have also highlighted that wind is an important factor in their migratory movements. Silver Y moths, *Autographa gamma*, migrate at high altitude (425 m) and are able to exert some degree of control over their flight direction via both choosing favourable winds and also adjusting their direction using an internal compass (Chapman J. W. et al. 2010). Migration has adaptive benefits in this species, with migrants having increased reproductive success (Chapman J. W. et al. 2012). Some insects (lepidopterans) have evolved a fascinating form of migratory behaviour, which encompasses multiple generations, such that the individuals that begin the migratory journeys do not complete them. Instead, their descendants complete different stages of the migratory journey (which can take up to six generations). We return to this phenomenon, known as transgenerational migration, later in this book.

2.3 Population patterns of migration

As we have seen, there is great diversity between species in migratory behaviour. However, also between individuals within a species, patterns of migratory behaviour are usually not uniform. Variation between—and within—populations has been widely observed, and such intraspecific diversity of movement strategies is fascinating both from an evolutionary perspective (i.e. what forces create and maintain this variation) and from a more applied point of view (i.e. how will population variation in migratory behaviour influence a species's response to global change (Lindström et al., Chapter 3)). Insights into the degree of within-species variation have been assisted by the advent of new tagging and tracking technologies that allow us to follow the migrations and fate of individual animals. In another chapter in this book we will probe further into individuality in migratory behaviour in more detail (Nilsson et al., Chapter 6); here we detail some of the population level consequences of these individual differences in movement behaviour.

2.3.1 Between-population variation in migratory propensity and destination

For species with broad distributions, different populations can experience radically different environments and thereby selective pressures. This can result in population differences in life-history strategies such as migratory tendency, and so comparing populations can illuminate some of the adaptations linked to the evolution of migratory behaviour. Monarch butterflies, *Danas plexippus*, have a wide distribution and include populations that differ in migratory behaviour, with some resident and some migratory. Morphological trait analysis has shown that migratory butterflies have large and elongated wings compared with individuals from resident populations (Altizer and Davis 2010), which is suggestive that wing shape may be an important adaptation to promote successful migratory behaviour (Johansson et al., Chapter 13). Comparisons between salmonid populations (sea migratory versus resident) have revealed the many life-history adaptations (such as fast growth rate and high investment in reproductive output) that can balance some of the putative costs of sea migration, such as increased predation pressure (Hendry et al. 2004). Similarly, many passerine birds with a broad distribution have both migratory and resident populations, for example European blackbirds, *Turdus merula*, and blackcaps, *Sylvia atricapilla*. In these cases, the migratory strategy of different populations is linked with latitude, with northerly populations tending towards obligate migration, central populations being partially migratory and southern populations being fully sedentary, highlighting the importance of winter climate.

Within a species, populations can differ in their movements and distribution during the migratory journey, which can produce interesting patterns of migration. There are at least three main patterns of latitudinal population movements: the first is known as *chain migration*, and occurs when populations maintain their relative latitudinal position following migration (Fig. 2.2). In this scenario, the more northerly populations in the breeding season remain more northerly in the non-breeding season following migration. This kind of migration pattern has been documented in many birds, including species of waders, passerines, and gulls (Newton 2007). More recently the post-breeding season movements of northern gannets, *Morus bassanus*, a large pelagic seabird, were tracked using geolocator tags and revealed evidence of chain migration also in this species (Fort et al. 2012). The second pattern is known

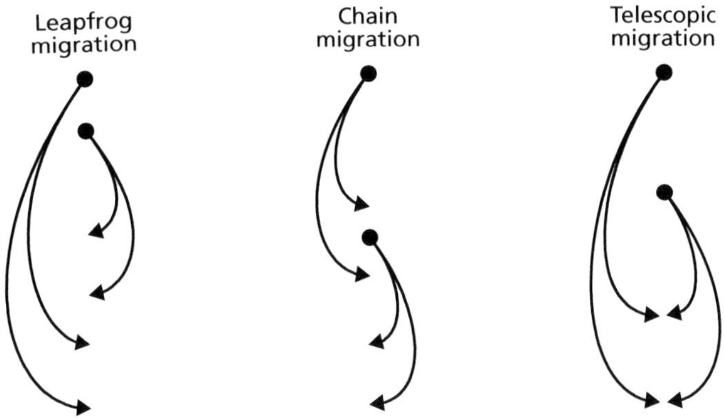

Figure 2.2 Conceptual illustration of different types of migrations, including leapfrog, chain and telescopic migration.

as *leapfrog migration* (Fig. 2.2), and is quite common among a variety of migratory bird species. Here (as the name suggests), migratory populations 'leapfrog' other populations during the migratory period, leading to a switch in the relative latitudinal position of the populations between breeding and non-breeding seasons. Analysis of 946 ring recoveries of bar-tailed godwits, *Limosa lapponica* (with data going back to 1935), showed clear evidence of a leapfrog migration pattern (Duijns et al. 2012). Finally, *telescopic migration* (Fig. 2.2) occurs when two populations coexist during one season and migrate to different latitudes during the migratory period.

Populations may also differ in their migratory trajectories and destinations. An example of this is the between-population variation found in Scandinavian willow warblers, *Phylloscopus trochilus*. More northerly populations spend the winter in East and South Africa while more southerly populations make a radically different journey to tropical West Africa (Chamberlain et al. 2000). The boundary between populations that migrate to alternative locations is known as a *migratory divide*. In many cases, such as in the willow warbler example, the reason why we see variation in flyways between different populations is still a mystery. One hypothesis here is that the migratory divide is a zone of secondary contact following post-glacial reinvasion of northern European habitats by populations of warblers with distinct migratory pathways (directions and destinations) from different refugia; i.e. there are historical reasons for the divergent migration patterns between these populations (Hedenström and Pettersson 1987, Bensch et al. 1999).

2.3.2 Intra-population variation in migratory behaviour

Many animals thus show great variation in migratory strategy among populations, but perhaps even more surprising is that individuals from the same population can radically differ from one another in their migratory behaviour. One of the most extreme forms of intra-population variation is known as partial migration, where populations are composed of both migratory and entirely resident individuals (Chapman B. B. et al. 2011). Partial migration is extremely common, and highly taxonomically diverse. Examples of partial migrants can be found among birds (Boyle et al. 2010), fishes (Chapman B. B. et al. 2012b), mammals (Mysterud et al. 2011), amphibians (Grayson and Wilbur 2009), crustacea (Hansson and Hylander 2009), and insects (Roff and Fairbairn 1991), and some speculate that partial migration is the most common form of migration in the animal kingdom. We can distinguish between three types of seasonal partial migration (Fig. 2.3): (1) migrants and residents are together during the breeding season and overwinter in separate locations; (2) migrants and residents breed separately and overwinter together; and (3) animals migrate to spawn or breed, but in some years skip spawning, and hence remain resident. Partial migration also occurs at timescales finer than seasonally—for example, this variation has been documented among tiny zooplankton that migrate (or not) up and down in the water column each day (Hansson and Hylander 2009).

Perhaps the most intriguing question when faced with the phenomenon of partial migration is why

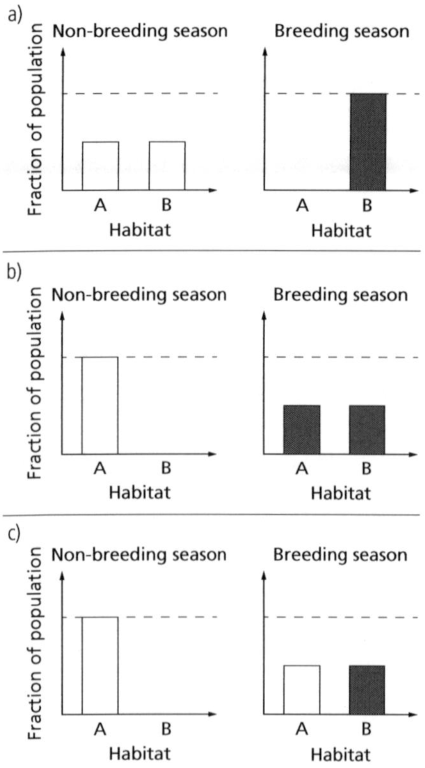

Figure 2.3 Schematic of three different types of seasonal partial migration: (a) non-breeding partial migration: residents and migrants share a breeding habitat but spend the non-breeding season apart, (b) breeding partial migration: residents and migrants share a non-breeding habitat and breed apart, and (c) skipped breeding partial migration where residents and migrants are apart during the breeding season, but as migration is required for reproduction only migrant individuals reproduce. Each panel shows the fraction of the population in each of the two habitats (A and B) during each of two seasons (non-breeding and breeding). Shaded bars indicate individuals that are reproducing. Reproduced from Shaw and Levin (2011) with permission from John Wiley & Sons.

some individuals stay and some go? Many hypotheses have been proposed to try and answer this question (Chapman B. B. et al. 2011, 2012a). Here, we present some of the ideas that have the best evidential support; for a more detailed and broader examination of theories of partial migration, see Chapman B. B. et al. (2011).

2.3.3. Partial migration

Competition for limiting resources during winter may be a key process in partial migration in some cases. For many species, food availability is reduced during the winter, and hence there is increased competition between individuals for access to the remaining resources. Some evidence (mostly from birds) suggests that poor competitors are forced to migrate to gain access to food during the winter months, for example among blue tits (Smith and Nilsson 1987). Further, in cases where migrants and residents breed separately, migrants may be out-competed for good nest sites, as is thought to be the case for American dippers, *Cinclus mexicanus* (Gillis et al. 2008). The idea that during seasons when resources are limited (whether these are food or nest sites), poor competitors are forced to migrate is known as the *competitive release* hypothesis. Individual animals may also differ physiologically in their ability to cope with the environmental challenges of certain seasons. The *fasting endurance* and *thermal tolerance* hypotheses postulate that individuals that cannot cope with either reduced food or changes in temperature adopt a migratory strategy, while those that can remain resident (Chapman B. B. et al. 2011). Another idea that was developed with birds in mind is known as the *arrival time* hypothesis (Ketterson and Nolan 1976). This is based upon the well-documented observation that, in many migratory species, those that arrive early have the best chance of securing a good breeding site. As long as it is possible to survive an entire winter at the breeding site, and the prior residency effect is strong enough, some individuals may take an early arrival strategy to the extreme and never leave the breeding grounds. Furthermore, an idea which has some support from fishes, zooplankton, and some ungulates is that the costs and benefits of a migratory versus resident strategy are dependent on the relative strength of trade-offs between predation risk and growth (the *p/g* hypothesis (Chapman B. B. et al. 2011)). Individuals can adopt either a high risk, high gain strategy (e.g. residency in cyprinid fishes, where both growth and predation risk are high (Brönmark et al. 2008, Chapman B. B. et al. 2013, Skov et al. 2013) or a low risk, low gain strategy (e.g. migration in cyprinid fishes to low risk and food-poor streams during winter). In addition to the earlier-mentioned causes, also the level of pathogens may affect individual decisions whether to move, which is discussed further in Westerdahl et al., Chapter 8.

2.3.4 Differential migration

Partial migration is an extreme form of within-population migratory variation, where individuals either migrate or do not. Yet animals can differ from one another in many migratory traits than just the simple propensity to migrate, for example in their migratory timing or destination. This kind of within-population variation is known as differential migration, and is exceptionally widespread among migratory species (Cristol et al. 1999). Explanations for differential migration are similar to those for partial migration. For example, the arrival time hypothesis has been cited as a potential reason for males migrating earlier than females during breeding migrations.

2.3.5 Temporal patterns of migration

Animals inhabiting the natural world are subject to environmental conditions that fluctuate over a wide temporal range. What is currently a high quality habitat may be suboptimal, poor, or even uninhabitable within months, days, or hours. However, many of these changes are cyclic and occur at predictable time intervals and organisms can adapt to these changes by migrating away during periods of low quality, returning once conditions have ameliorated, and in this way increase their fitness. A frequent observation is therefore that animals move within relatively narrow time windows and in step with circannual (seasonal) and circadian (daily) rhythms. Although temporal patterns of movement are diverse among life-forms there are many common themes which apply across temporal scales. For example, it has become increasingly evident that the ratio between predation risk and growth potential (p/g) may act as a fundamental selective force driving the evolution of animal movement patterns in many species. However, the p/g ratio can change at very different timescales, within a single day (e.g. when the capture success of visually oriented predator differs between day and night conditions: McMahon and Holanov 1995), or on a seasonal scale (e.g. metabolic demands of predators may differ considerably between summer and winter: Brönmark et al. 2008). Hence, similar ecological forces can act to shape the evolution of migratory behaviour on very different timescales. In this section we discuss different temporal patterns of migration. We begin by addressing the daily migrations of aquatic organisms, and zoom out to seasonal movements, and finally reflect upon the fascinating phenomena of transgenerational migration, before closing with some remarks on the temporal aspect of the migratory journey itself.

2.3.6 Daily migration

Habitats can change in various ecological properties over very short timescales. Differences in light intensity between the day and night can shift the cost–benefit ratio within a 24-h period, and some animals have evolved behaviours to track these regular changes by making daily migrations between different habitats. It has been documented that arthropods, arachnids, such as spiders (Bonte et al. 2000) and mites (Onzo et al. 2003), make predictable daily movements that can be classified as diurnal migrations. Perhaps the best studied of daily migrations is diel vertical migration (DVM) in aquatic systems, where organisms occur at different water depths during the diurnal cycle. This phenomenon has been reported for many species of phytoplankton, invertebrates, and fish (Gliwicz 1986, Jephson and Carlsson 2009, Mehner 2012). Some of the earliest work on DVM was with zooplankton, and identified sunlight as a key driver of such migratory movements. Sunlight can act as both a proximate cue (to induce negative phototaxis) and an ultimate factor (beneficial in terms of avoiding damaging UV radiation; Moore 1912, Kikuchi 1930). The ultimate causes driving the evolution of DVM have since remained a hotly debated subject with the majority of hypotheses focusing on bioenergetic efficiency, feeding opportunities, UV radiation, and predator avoidance. Of these hypotheses, predator avoidance has received the most attention and is now widely accepted as a powerful driving force of DVM in both invertebrates and fish (Gliwicz 1986, Mehner 2012), and also potentially in dinoflagellates (Bollens et al. 2012). However, DVM has also been reported in lakes lacking predators, and the magnitude of *Daphnia* vertical migration is often most pronounced in very transparent lakes (Dodson 1990), suggesting that ultraviolet radiation (UV) may indeed be a key player in the evolution of DVM (Williamson et al. 2001). In support of this idea, an experimental study found that UV strongly affected the depth distribution of all size classes of *Daphnia* (Fig. 2.4), whereas predation risk in combination with visible light (UV screened out) affected the depth distribution of larger, but not smaller, size classes (Hansson and Hylander 2009). The reason why small size classes (<0.9 mm body length) remained in surface waters during daytime is likely that small individuals are less vulnerable to predation than larger ones (Hansson and Hylander 2009), again suggesting that threats from both UV and predation are important in driving this migratory pattern. Also in fishes, interindividual variation in predation vulnerability may explain why some, but not all fishes within a population participate in DVM (Mehner 2012). The majority of piscivorous predators in aquatic ecosystems are gape-size limited and the outcomes of their interactions with prey are therefore dependent on the size ratio between the two. Given that predator avoidance is a key driver of DVM and that individuals seek to minimize the ratio between predation and growth (p/g), within-population differences in individual vulnerability related to size should, at least partly, predict patterns of DVM (Mehner 2012). This is an interesting analogy to population patterns of partial migration in bream, *Abramis brama*, where smaller fish that are vulnerable to pike predation make seasonal refuge migrations from lakes into streams (Skov et al. 2011). Pelagic marine organisms have also been recorded as combining longer migrations with patterns of DVM, for example European eels, *Anguilla anguilla* (Aarestrup et al. 2009). Pelagic organisms are thought to use DVM as a means to avoid predator exposure during the day and maximize feeding time at night, and potentially as a method for reducing transportion costs in oceanic fishes. DVM may also act as a way of maximizing the metabolic efficiency of juvenile fishes, who may migrate to warmer surface waters in order to speed up digestive processes and hence maximize food intake (Wurtsbaugh and Neverman 1988). A similar daily pattern of migration can also occur laterally in aquatic systems (known as diel horizontal migration, DHM), particularly in fishes and crustaceans. Zooplankton such as *Daphnia* spp. have been observed making daily movements from the vegetated littoral zone into the open, pelagic waters, particularly in shallow lakes (Burks et al. 2002, Romare and Hansson 2003). Diel horizontal migrations have also been studied in a variety of

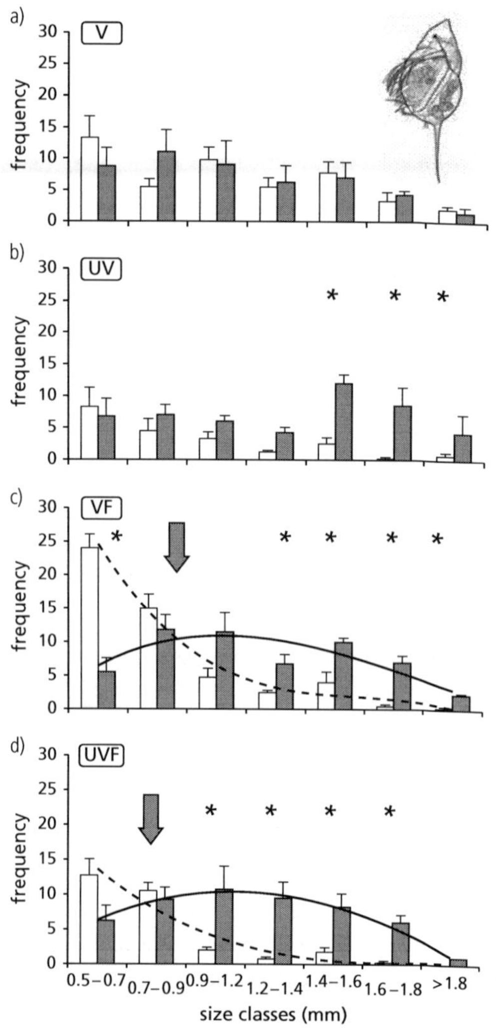

Figure 2.4 Predator-induced daytime vertical distribution among size classes of *Daphnia* in each size class (0.5 to greater than 1.8 mm) at surface (open bars) and bottom (grey bars) waters. Experimental treatments are: (a) visible light only (V), (b) ultraviolet radiation (UV), (c) visible light and predator cue (VF), and (d) UV plus predator cue (UVF). For clarity, best-fitted curve models are shown for surface (dashed lines) and bottom (solid lines) frequencies. Stars show significant (p<0.05; MANOVA; general linear model) differences between frequencies of each size class in surface and bottom waters. Arrows indicate where frequency curves cross, i.e. at which size the majority of *Daphnia* chose to avoid surface waters. Reproduced with permission from Hansson and Hylander (2009).

freshwater fishes (Muska et al. 2013). In many cases, DHM is thought to be a strategy for maximizing food intake while minimizing predation risk (i.e. driven by the p/g ratio).

2.3.7 Temporal distortion

Daily rhythms in movements and activity triggered by light cycles is the rule in many animals, but such movements require regular light/dark fluctuations and are lost when such stimuli disappears. For example, reindeer (*Rangifer* spp.) generally show circadian rhythms in movements and activity, but during periods of constant light (midnight sun) or constant darkness (mid-winter), they lose their diel rhythm (van Oort et al. 2005). Further, a latitudinal experiment showed that freshwater zooplankton perform strong diel migrations in Spain, where the day:night length is 15:9 h, but that this behaviour eventually ceases as the day length increases with latitude, and completely disappears close to the polar circle, where the daylight is almost constant in summer (Hansson et al. 2007). Similarly, DVM in fishes have been shown to cease when the difference between night and day is reduced, for example under ice and midnight-sun conditions (Steinhart and Wurtsbaugh 1999, Gjelland et al. 2009). However, some evidence shows that marine zooplankton in the Arctic winter continue to perform synchronized DVM in the polar night, which is potentially regulated by diel variations in solar and lunar illumination (Berge et al. 2009).

2.3.8 Seasonal migration

The quality of habitats also changes over longer, seasonal timescales. Seasonality in some form is ubiquitous across the globe, irrespective of geographical location. In temperate northern and southern hemisphere environments there are considered to be four seasons (spring, summer, autumn, winter), while in tropical habitats there are only two (dry and wet season). Animals have evolved mechanisms including hibernation and migration to cope with differences between the seasons in habitat quality and resource availability. Of the various kinds of migration, seasonal migration has historically received the most attention and in a sense constitutes the 'classical form' of animal migration, and typically involves repeat journeys over consecutive years for individuals. With the changing seasons many environmental and ecological variables also fluctuate, affecting the habitat's quality for many species. In many cases animals are thought to migrate simply to maximize resource intake by following seasonal shifts in food/water availability, as is probably the case with many of the great African ungulate migrations. Similarly, seasonal changes in the intensity and frequency of storms can affect feeding opportunities for tropical birds at higher altitudes, leading to migration downhill to drier sites (Boyle et al. 2010).

Alternatively, migration can be influenced by seasonal changes in the predation risk/growth potential trade-off (p/g). In cyprinid fishes that inhabit lakes in the summer, food availability dwindles in the autumn, and individuals migrate to streams to avoid predators during the food poor winter months (Skov et al. 2011). Migration can also be seasonal as it is strongly tied to breeding, which is seasonal in many species. In breeding migrations the optimal conditions for offspring differ from those of adults, and hence individuals migrate to spawning or breeding grounds to mate. This has been documented in a wide range of organisms, from amphibians to fishes to insects.

Seasonal migration should evolve when the fitness benefits of seasonal movement outweigh the costs of year-round residency. Quantifying the costs and benefits of seasonal (and often long-distance) migration is not trivial, however. A powerful study of migratory ground-nesting shorebirds provided strong evidence of a reproductive benefit to their Arctic migration (McKinnon et al. 2010). Monitoring the 'survival' of >1,500 artificial nests along a latitudinal gradient of approximately 3,350 km, researchers showed that by migrating to the Arctic, individual shorebirds had a twofold increase in nest success. Until recently, the seasonal journeys of billions of insects to utilize temperate areas during summer were thought to be an evolutionary dead-end, a sink where the progeny of migrants seldom returned (the 'Pied Piper' effect). However, evidence now shows that annual migrations in the noctuid moth *Autographia gamma* (Silver Y) can lead to a fourfold increase in reproductive success (Chapman J. W. et al. 2012).

2.3.9 Semelparous migration

Not all round trips are necessarily repeated on an annual basis and for some animals migration and subsequent breeding actually coincides with death, a phenomenon known as semelparous migration. Many diadromous fishes that migrate across the saline–freshwater interface, including migrants such as pacific salmon, *Oncorhynchus* spp., and eel, *Anguilla* spp., are textbook examples of semelparous species with a life history characterized by death after first reproduction. These migrants cross the interface between freshwater and saline habitats in two major migrations, each of which will never be repeated within the lifetime of a given individual. In the case of pacific salmon, migratory juveniles undergo a remarkable physiological transformation while still in their natal river to adapt to a saltwater environment, a process called smoltification. Following this transition smolts migrate downstream towards rich oceanic feeding grounds. After a couple of years at sea they have increased in size tremendously and become sexually mature. They migrate back to freshwater and spawn, often in their natal stream after epic journeys of thousands of miles, and die within days (Fig. 2.5). Their carcasses provide nutrients to the ecosystem, along with their sperm, faeces, and eggs. In particular phosphorus and nitrogen from the carcasses of successful migrants increases the growth of phytoplankton and zooplankton in the rivers, which provide a plentiful resource for juvenile fish, thus supporting the next generation of migrants.

2.3.10 Transgenerational migration

Most insects do not survive long enough to make a return journey during a seasonal migration cycle. A number of lepidopterans have evolved an unusual response to this problem, known as transgenerational migration. Transgenerational migration can be considered a special case of migratory semelparity, where individual butterflies or moths migrate southwards in the winter (Fig. 2.6), but do not survive to complete the migratory journey. Rather it is

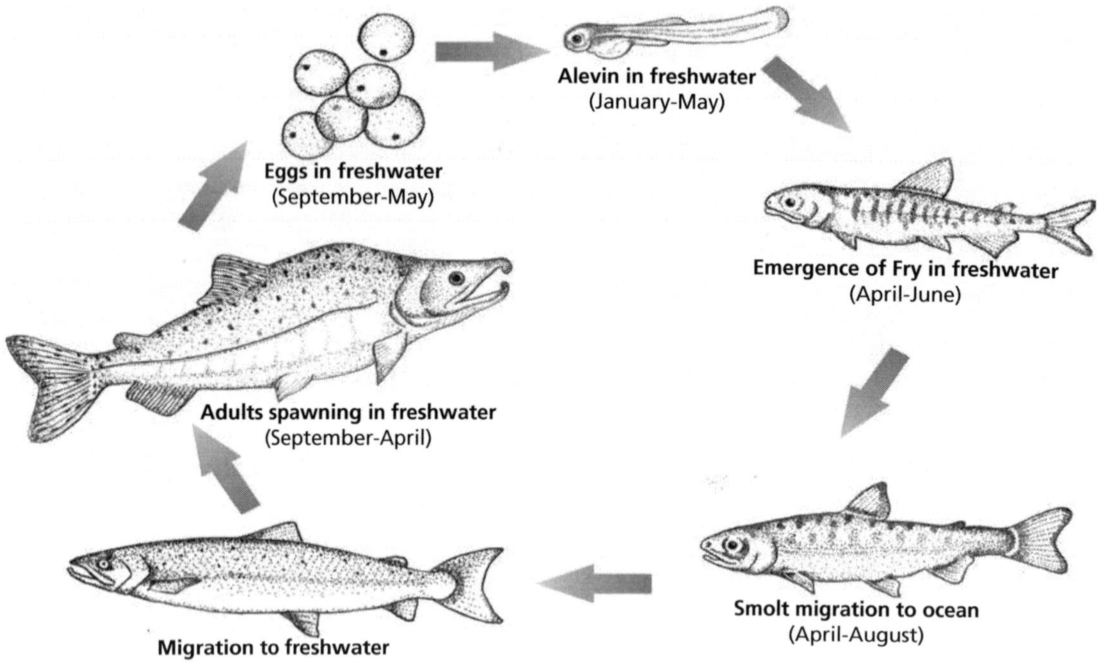

Figure 2.5 Life cycle of semelparous migratory salmon showing hatching from eggs and the first stages in freshwater and later the migration of smolt to the ocean. Adult salmon then return to freshwaters to spawn and then die, thereby closing the life-cycle loop.

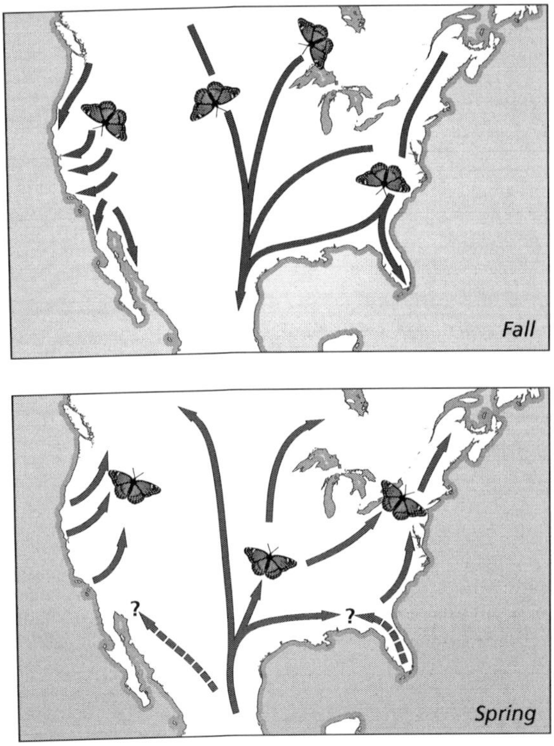

Figure 2.6 Migration routes of monarch butterflies, *Danaus plexippus*. Eastern North American monarch butterflies carry out a long-distance autumn migration to a restricted destination in central Mexico (upper plate). The population of monarch butterflies west of the Rocky Mountains also undergo a fall migration, overwintering in protected roosts along the Pacific Coast, but the distances travelled are far less than those of the eastern population, and the concentration of overwintering areas is lower. The lower plate shows the return migration of the monarch. The eastern North American migrants remain at the overwintering areas in Mexico until spring, when the same butterflies reproduce and disperse northward to lay fertilized eggs on newly emerged milkweed in the southern United States. Successive generations of spring and summer monarchs in the east and the west then migrate to repopulate the home range. Redrawn from Reppert et al. (2010) with permission from Elsevier.

an individual from a fourth (or fifth, or even sixth) generation that completes the migratory circuit. The most well-known transgenerational migrant is the monarch butterfly, *Danaus plexippus*. Monarchs migrate from northern American oviposition sites southwards to Mexico on lengthy migratory journeys as the temperature drops in autumn each year. How butterflies manage to return to the same overwintering sites as their ancestors (i.e. a site they have never been to) is still a matter of some debate, although monarchs have been shown to use a variety of navigational tools, including a time-compensated sun compass (Reppert et al. 2010: see also Åkesson et al., Chapter 9, and Muheim et al., Chapter 10). The genetics behind this transgenerational journey is further discussed in Liedvogel and Lundberg (Chapter 12).

2.4 Timing of the migratory journey

There is a great diversity of patterns of migratory timing in the animal kingdom. Migratory timing may be highly synchronized, such as is the case with the mass migration of Christmas Island red crabs. The degree of synchrony can vary depending on season; for example, the spring migrations of the European roe deer, *Capreolus capreolus*, are generally synchronized among individuals, whereas the timing of the autumn migration from summer to winter habitats varies considerably between regions, individuals, and even sexes (Cagnacci et al. 2011). Also in other species, such as migratory woodthrushes, *Hylocichla mustelina*, tracking has revealed surprising consistency in timing in spring, but not in autumn (Stanley et al. 2012). Synchronization may be active, i.e. due to benefits that can be accrued due to migrating en masse such as a dilution effect versus predators (Krause and Ruxton 2002), or in order to maximize reproductive success. Alternatively, it can be passive, and driven by exogenous cues shared by many individuals of the same species, such as photoperiod.

There is also variability in the diurnal timing of migratory journeys. A number of bird species migrate at night (nocturnal migrants), and use stars to navigate during their migratory sojourns (Emlen 1967; Åkesson et al., Chapter 9). Others migrate solely during daytime, and still more during both day and night. Migrating by night may have a number of benefits, including increased foraging opportunities during the day, predator avoidance during the journey, and atmospheric structure (i.e. a relative reduction in turbulence and strong winds during night (Alerstam 2009)). However, potential costs to nocturnal migration can arise due to sleep loss. Diurnal migration may be adopted by birds using thermal soaring migration, significantly reducing the costs of flight by daytime migration (Alerstam 2000; Hedenström and Lindström, Chapter 5; Johansson et al., Chapter 13).

2.5 Spatial patterns of migration

The environment is heterogeneous across space, and is composed of a wide range of habitats, opportunities, and risks for animals. As discussed earlier, this spatial diversity can promote migration in circumstances where different habitats are optimal for different life stages. Clearly, spatial and temporal heterogeneity are strongly linked in many cases to shape the patterns of migration documented in animal populations. Environments change over both time and space. However, in some examples the spatial variation is key in the evolution of migratory behaviour. Different optimal conditions for different ontogenetic stages can drive the evolution of breeding migrations in many species. This is especially evident in fishes, where many species have specific nursery grounds that provide the best habitats for fry to feed and grow. This results in fish commonly making breeding migrations to specific sites. For example, northern pike, *Esox lucius*, make yearly migrations from their adult feeding grounds in the Baltic Sea to spawn in flooded wetland habitats rich in zooplankton with relatively low numbers of predators (Engstedt et al. 2010). This kind of migratory shift between distinct aquatic habitats (i.e. freshwater and saltwater) is known as diadromy.

2.5.1 Diadromous migrations in fishes

Diadromy encompasses all migratory movements between salt- and freshwater habitats, and is prevalent among fishes, with examples also from crustaceans and molluscs. Diadromous migrations can be divided into categories based on the order and direction of habitat switches. Anadromy occurs when animals migrate from freshwater to saltwater as adults, returning to freshwater to spawn (Fig. 2.7, upper panel). Anadromous fishes include many species from the salmon, stickleback, herring, lamprey, and sturgeon families. The reverse situation, known as catadromy, whereby adults migrate to freshwater to feed and return to the ocean to spawn, has also been documented in fishes from the anguiliid eel, galaxiid, and scorpionfish families (Fig. 2.7, middle panel). Amphidromy is a third form of diadromy whereby spawning can occur in either fresh- or saltwater, and where juveniles migrate to the alternative habitat to feed and grow and then return to their natal habitat to continue to feed and spawn (Fig. 2.7, lower panel). Amphidromy is considered to be less widespread than either anadromy or catadromy, but has been documented in a range of families, and is particularly prevalent among the sicydiine gobies (McDowall 2007). Diadromy has been reported in just less than 1% of all fish species, but is of particular interest as many of these species are of commercial importance, such as the many anadromous salmonids (Helfman et al. 2009).

These forms of migration require a dazzling suite of adaptations to cope with the challenges associated with shifting from marine to freshwater environments (or vice versa). Diadromous fish undergo major physiological transitions, and in addition to this the journeys are often fraught with danger from predators and parasites, and energetically costly. Hence, how could diadromy have evolved? An elegant explanation lies with the notion that adult and juvenile habitat optima differ, as we discussed briefly earlier. For such long-distance migratory behaviour to evolve, clearly the costs associated with the migration must be outweighed by the benefits. Some evidence suggests that there are significant rewards for sea-migrating individuals. For example, juvenile anadromous Pacific salmon can increase their growth rate by up to 50% in just the first week of life in the ocean. As body size in fishes is often tightly correlated with fecundity, an additional benefit of high growth rate is potentially high reproductive success. Comparisons of anadromous and non-anadromous populations from the same species shows that sea-dwelling adults can produce up to three times the number of eggs than land-locked adults. Quantifying the costs of migrating is challenging, but estimates suggest that predation risk is significantly increased for oceanic migrants, and the energetic cost of migration itself can be very high, which may explain why anadromy is not more widespread. Given the apparently substantial costs of sea migration for freshwater fishes, one would predict that this behaviour should only

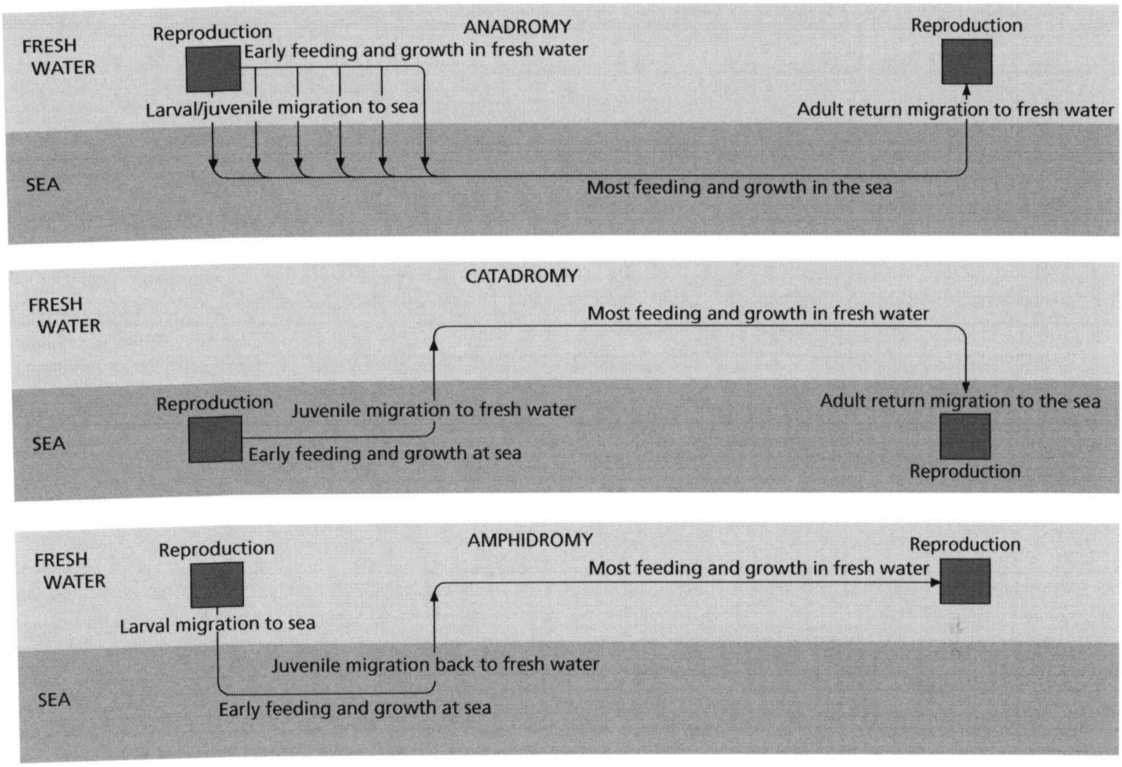

Figure 2.7 Different forms of diadromous migration. From top to bottom, the diagrams show anadromy, catadromy, and amphidromy. In each type, the light grey indicates time in freshwater and dark grey time in the sea. The *x*-axis from left to right shows part of the life cycle of an individual, from birth to first reproduction. Lines indicate the type of habitat occupied, and migration occurs when a line shifts from one habitat to the next.

evolve where the benefits to growth and thus reproductive output are significant. Following this logic one could then predict that the biogeography of diadromy would show that anadromous species were more prevalent at northerly latitudes where oceans tend to be more productive than freshwater habitats. Furthermore, catadromy should be more common in tropical regions where freshwater habitats tend to be more productive than oceanic habitats. An analysis of the proportion of diadromous species that are anadromous in relation to the relative productivity of the world's oceans and freshwaters at different latitudes shows this to be the case (Gross et al. 1988), making the evolutionary logic of this hypothesis quite compelling, and the circumstantial evidence strong. However, this analysis has been criticized on the grounds that primary productivity does not necessarily translate directly to food benefits for fishes (Dodson 1997), and also that other, e.g. historical factors, may be important (McDowall 2008). Discussions on the origin and biogeography of diadromy continue.

2.5.2 Latitudinal trends of bird migration

Just as latitude can predict patterns of diadromy among fishes, it can also predict the prevalence of migration as a strategy in other taxa, such as birds. It has long been recognized that a greater proportion of birds migrate as one moves from the equatorial to the polar regions. For example, 29% of North African (30°N) species migrate compared to 83% of bird species in Svalbard (80°N), a mean increase of 1.3% of breeding bird species being migrants for each degree of latitude (Newton and Dale 1996a). This pattern also holds in a similar analysis of eastern North American bird migration, with 12% of breeding birds migrating at 25°N compared to 87%

at 80°N (Newton and Dale 1996b). This relationship is probably driven by the increasing severity of winters at more northerly latitudes, and the challenges associated with surviving extreme conditions of cold weather and low food availability. Evidence in support of this hypothesis can be found by comparing the severity of winters in Western Europe and eastern North America (i.e. the east and west sides of the Atlantic). At corresponding latitudes, around 17% more breeding bird species migrate from eastern North America, where winters are more severe and resident birds encounter more adverse winter conditions (Newton and Dale 1996b).

This pattern continues to hold as one moves south towards the equator, with only 5 out of 807 breeding bird species (0.6%) migrating in Panama (8°N). Migrants here are insectivorous birds such as the common swallowtailed kite, *Elanoides forficatus*, which migrates south in the winter dry season. However, there is a growing realization that in tropical regions migratory movements still occur, but at different spatial scales, and in different directions.

2.5.3 Altitudinal migration

Altitudinal migration, where individuals migrate along elevational gradients whilst remaining in a restricted geographic area, occurs in a number of species, and it has been speculated that this kind of migration is much more widespread than previously documented, especially in the tropics (Boyle 2008a). Altitudinal migrations are often breeding migrations, with breeding sites being at higher altitudes than non-breeding sites (Burgess and Mlingwa 2000). The ecological reasons for these migrations are obscure as evidence for variation in food availability playing a role in altitudinal movements is inconclusive. Instead, recent experiments provide support for the role of nest predation. Boyle placed artificial white-ruffed manakin, *Corapipo altera* (a frugivorous altitudinal migrant), nests along an elevational gradient in the montane regions of Costa Rica, and showed that the probability of predation decreased as elevation increased (Boyle 2008b). Other factors also come into play to promote the evolution of altitudinal migration. Birds that breed in montane or moorland habitats, e.g. wallcreeper, *Tichodroma muraria*, and dipper, *Cinclus* spp., migrate to lower altitudes to escape cold weather conditions at higher altitudes during winter. In addition to birds, bats and some insects make altitudinal migrations, particularly in tropical forests.

2.5.4 Spatial scale and migratory routes

The scale of animal migration is highly variable, and taxon-dependent (Alerstam et al. 2003). Associated with the spectacular journeys of endurance migrants such as Artic terns, *Sterna paradisaea*, bar-tailed godwits, and European eels are adaptations to cope with the intense physiological demands and great hardships that these journeys impose upon migrating individuals. These extreme endurance migrants have the capacity to migrate more than 80,000 km annually, fly 11,000 km non-stop, or undertake a transatlantic migration in a fasting state (Gill et al. 2005, van Ginneken et al. 2005, Egevang et al. 2010). At the other end of the migratory distance spectrum lie short-distance migrants such as blue tits that migrate less than 100 km, or potamodromous fishes, and snakes, whose migrations can be less than 10 km. The spatial scale of migration may also influence the types of cues animals rely upon to navigate during their journeys (Åkesson et al., Chapter 9; Muheim et al., Chapter 10). Aside from scale, many spatial characteristics of the journey vary between species, populations, and even individuals (differential migration, see earlier). For the most part selection should act to optimize migratory journeys in terms of maximizing benefits while minimizing costs of transport, and this kind of framework has great explanatory power when applied to questions of migratory trajectories and routes. However, one must also consider historical exigencies and the context of past movements, as these can provide explanations of less intuitive routes. Perhaps the most obvious constraint on migratory routes for animal migrations is topography, or landscape barriers to movement. Below we discuss types of landscape barriers and their implications for migration, and movement more generally.

2.5.5 Landscape barriers to movement

Landscape barriers such as mountain ranges, rivers, and deserts can have major impacts upon ecological processes due to their disruptive effect upon animal movements such as migration and dispersal (Box 2.1; Lindström et al., Chapter 3). Many terrestrial migrants migrate along relatively predictable routes, for example along valleys or ridges in mountainous regions. Anthropogenic developments such as pipelines, train tracks, roads, and urbanization have changed the landscape topography and this has had impacts upon migration and migratory routes for many species (Lindström et al., Chapter 3; Smith et al., Chapter 4). In particular, many studies have assessed the impact of barriers arising from the rapid expansion of transport infrastructure that has fragmented terrestrial ecosystems worldwide. These studies have documented that roads, railroads, and their traffic disrupt animal migration and ecological processes, and also increase mortality and lead to habitat degradation, loss, and isolation (Forman and Alexander 1998, Southwood and Henderson 2000). Roads have a number of ecological effects, but perhaps the most damaging is the disruption they cause to animal movement (Forman and Alexander 1998), which can be severe enough to cause genetic differentiation between populations on either side of the barrier (e.g. Robinson et al. 2012). Roads tend to especially inhibit the movement of small mammals (Rondinini and Doncaster 2002) and the severity of this effect is influenced by several factors including traffic volume, physical obstacles such as fences, the level of human activity, road surface type, and road width (McDonald and Clair 2004). Even relatively narrow and rarely used roads can affect animal movement, but much variation seems to exist among species (McDonald and Clair 2004). The ability of animals to cross anthropogenic barriers like roads and to move through poor-quality habitats or places where human activity is concentrated may determine whether habitat fragmentation constrains animal movement, dispersal, and migration. Anthropogenic constraints on movement can prevent a species from occupying perfectly suitable habitats (Holt 2003) and can threaten species survival if the total available habitat size for a species falls below a specific threshold, as seen in some forest birds, such as the scarlet tanagers, *Piranga olivacea* (Rosenberg et al. 1999). The establishment of connectivity measures such as stepping-stone habitats or corridors among isolated populations can mitigate the negative effects of fragmentation; i.e. structural connectivity enhances or restores functional connectivity among populations (Lindenmayer and Fischer 2006), and can

Box 2.1 Implications of landscape barriers to movement

The direct effect that landscape barriers exert on animal movement means that many ecological processes are affected. Among the most important are migration and dispersal dynamics, disease transmission, invasion dynamics, gene flow, range expansions and shifts. The type of ecological process affected and the strength of the effect upon movement strongly depends on the nature of the barrier, the type of habitat on either side of the barrier and whether the barrier presents an impermeable or semipermeable barrier (Smith et al., Chapter 4). A direct consequence of any barrier to animal movement, regardless whether the barrier is impermeable or semipermeable, is that it alters the immigration and emigration rates between interconnected populations, which has been documented extensively using estimates of gene flow (Root et al. 2009, Zalewski et al. 2009, Pavlova et al. 2012). If the barrier between populations is impermeable, then populations might get entirely cut off from the remaining species pool, or if the barrier is semipermeable, then the isolated population might suffer only reduced rates of genetic exchange with populations on the other side of the barrier. In general however, isolated populations are often smaller, and so are likely to have greater temporal fluctuations and are affected more severely by any genetic and ecological stochastic effects that increase variation in individual reproductive success. These dynamics can lead to frequent population bottlenecks (Keller et al. 2010), which over time can pose a serious threat to population persistence (Wilcox and Murphy 1985). It thus becomes clear that landscape barriers may not only reduce the genetic exchange of some populations with others, but that this reduction in gene flow can also decrease the effective population size and increase risk of extinction of the isolated population (Lammi et al. 1999). The risk of extinction is particularly increased when these small populations experience random demographic and genetic processes that further reduce population size and variation in traits, and these processes also reduce the general potential for re-colonization and re-population following disturbance (Deutsch et al. 2008, Lindsay et al. 2008, Olden et al. 2011, Loxterman and Keeley 2012).

promote successful migration in many, but not all, cases (Smith et al., Chapter 4).

Habitat fragmentation in freshwater habitats has always occurred in form of natural features, such as sandbars, landslides, waterfalls, and boulder cascades. However, the number of anthropogenic barriers to animal movement has also risen steeply in recent decades and further truncated range of animal movements with an astounding variety of in-stream features and effects such as dams, culverts that create velocity, jump, or exhaustion barriers for aquatic organisms, thermal pollution discharge, channelization and hardening of rivers and streams, and stream enclosures. Many dams provide no mechanism for adult fish to pass above the dam, thus blocking off spawning habitat upstream. When fish passages do exist, many migratory fish have trouble finding the fish ladders on dams or die when exposed to higher water temperatures inside the ladders. Even if the fish manage to find the ladders and survive the arduous journey to their spawning grounds, individuals are often too exhausted from the journey to spawn successfully (Baisez et al. 2011). As a result, the number of adults returning to spawn is often far below the number needed to ensure the survival of many migratory species. The ability of freshwater fish to move in response to environmental change is further constrained by the dendritic arrangement of riverine ecosystems, as well as a variety of natural and anthropogenic barriers.

Marine systems are generally considered more homogenous than terrestrial and freshwater environments since fewer barriers to the movement of organisms are present. Major natural barriers to movement in marine ecosystems are ocean currents, which differ significantly from many terrestrial and freshwater barriers in that they are not stable. Ocean currents can be circuitous and oceanographic features like eddies and fronts can prevent mixing and diffusion of pelagic larvae, decoupling pelagic larval dispersal from Euclidean distance (Weersing and Toonen 2009). Ocean currents and environmental gradients do not pose a barrier to all animals, however, and some might even utilize these for long-distance travel. For example, adult saltwater crocodiles, *Crocodylus porosus*, adopt behavioural strategies to use surface water currents during long-distance travel, enabling them to move quickly and efficiently over considerable distances (Campbell et al. 2010). Climate change has been predicted to alter ocean circulation patterns and environmental gradients at a range of scales in both space and time. Some of the changes can already be seen, and their effect on movement measured (for a review see Doney et al. 2012).

As we have discussed, landscape effects especially apply to terrestrial and aquatic organisms and their movements; however, topography can also influence aerial migrants such as birds. For example, few migrants migrate over high altitude mountain ranges.

Analogous in many ways to terrestrial topography are weather systems. Aerial migrants pay significant energetic costs associated with extended journeys, and so detours are surprising and require explanation. One potential reason for detours is that maximizing aerodynamic efficiency may lead to longer (but less costly) migratory journeys (Alerstam 2001; Hedenström and Lindström, Chapter 5). Alternatively, detours may be driven by historical factors. Swainson's thrushes, *Catharus ustulatus*, make circuitous migrations, apparently following non-optimal migratory routes. Recently it has been proposed that these counterintuitive patterns of migration can be explained by a Late Pleistocene range expansion in this species, with the retention of an ancestral route, leading to the unusual migratory behaviour in this passerine (Ruegg and Smith 2002).

Finally, ecological constraints can come into play. Amphibians and terrestrial crabs are prone to dessication and require access to water or wet conditions during migration, and so environmental variables can impact upon routes. Similarly, many ungulates may track the 'green wave' of forage and, hence, have routes constrained by food availability along the migratory journey.

2.6 Patterns of migration in a changing world

The patterns of migratory behaviour we have discussed are not static. Over evolutionary time migratory behaviour has changed in step with environmental changes as new environmental configurations lead to different selective pressures (Lindström et al., Chapter 3). Migratory behaviour can also be plastic and responsive to more rapid change, and in longer-lived species components of migratory behaviour can be learnt. Perhaps a more worrying trend is that human activities can have a profound influence upon patterns of animal migration, either directly via, for example, overfishing, and hence extirpating migratory individuals, or indirectly via the effects of our behaviour upon global patterns of climatic variation (Lindström et al., Chapter 3). These impacts are mostly adverse, and many of the great migrations have vanished or are in decline (Wilcove and Wikelski 2008). In the recent past flocks of millions of passenger pigeons, *Ectopistes migratorius*, would migrate across North America in spectacular and vast migrations, until hunting drove them to extinction, and many migratory species are now in decline due to habitat loss, exploitation, damming, and other factors (Wilcove and Wikelski 2008).

Subtler change to migratory patterns can be driven by anthropogenic climate change. For example, some birds have shifted the timing of migration in step with changes in temperature (Cotton 2003, Stervander et al. 2005; Fig. 2.8), and many have also reduced the migratory distance travelled (Visser et al. 2009). The migratory patterns of large pelagic fishes, such as bluefin tuna, *Thunnus* spp., may also

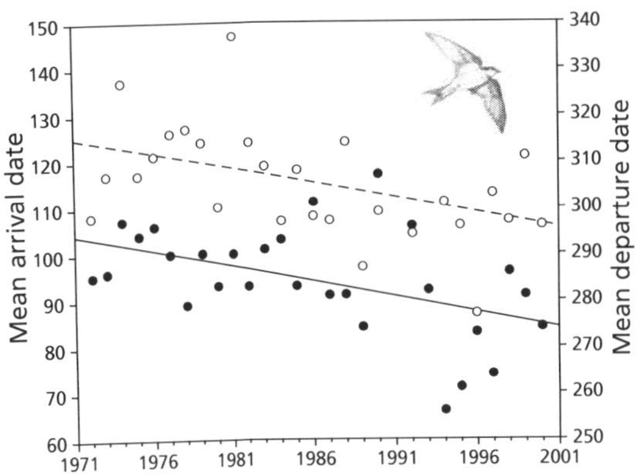

Figure 2.8 Linear regressions on the arrival date (filled circles, solid line) and departure date (open circles, dashed line) of northern house martins, *Delichon urbica*, in Oxfordshire, from 1971 to 2000. Data from Cotton (2003).

be indirectly affected by changing prey distributions as a function of global temperature increases (Polovina 1996). Climate change can also influence the fitness outcomes of migrants and residents in partially migratory populations, and thus alter population patterns of migratory behaviour. This can result in unpredictable outcomes. For example, a 30-year study of blue tit migration showed that, counter to predictions, the proportion of migrants increased as mean temperature increased (Nilsson et al. 2006). Evidence from other partially migratory bird populations shows the opposite trend, however, as many bird populations have become increasingly sedentary, a trend which is correlated with the increasing temperatures associated with global climate change (Berthold 1995).

Rampant post-industrial urbanization has also shifted migratory patterns towards residency in some species, in e.g. blackbirds, *Turdus merula* (Partecke and Gwinner 2007). The emergence of migratory behaviour in sedentary populations is apparently much less common, and has been described in just a few cases, for example in introduced house finches, *Haemorhous mexicanus*, in eastern North America (Able and Belthoff 1998), and in European serin, *Serinus serinus*, which have become more migratory as they have spread further north in Europe (Berthold 1999). In partially anadromous fishes, human exploitation can act as a powerful selective force against migratory behaviour (Thériault et al. 2008), as can the damming of rivers. Human activities can also have indirect effects such as restricting pesticide use. For example, following the recovery of raptor populations across North America which resulted from restricting the use of DDT as a pesticide, western sandpipers, *Calidris mauri*, have reduced the duration of migratory stopovers in the Straits of Georgia, British Columbia (Ydenberg et al. 2004). Increased predation risk has been implicated in these shifts in migratory behaviour.

Other factors besides human behaviour can drive shifts in migratory patterns. From the early 1990s, an increasing number of barnacle geese, *Branta leucopsis*, began to delay the onset of their breeding migration by an incredible four weeks (Eichhorn et al. 2009). These individuals skip spring stopover sites in the Baltic Sea previously used by the whole population, leading to the emergence of an entirely new migration routine. It is possible that this is due to increasing competition for food at the stopover sites, as numbers of geese have increased over the years and then levelled off, which is suggestive that this habitat has reached carrying capacity. Predation risk from white-tailed eagles, *Haliaeetus albicilla*, may also play a role in this shift, as this species has increased in numbers in recent decades.

Alterations to the direction of migratory travel have been recorded several times among bird species in recent decades. Research into a model species of the study of migratory variation in birds, the blackcap, *Sylvia atricapilla*, has revealed the emergence of a novel migratory direction and destination in blackcaps breeding in Western Europe. This new winter migratory direction is perhaps counterintuitively northwestward to the UK, potentially as a consequence of the increasing number of bird feeders at this location (Berthold 1991). Further experiments showed that this new migration route had a genetic basis and was heritable (Berthold 1991), and hence an example of microevolution rather than migratory plasticity (Liedvogel and Lundberg, Chapter 12).

2.7 Future perspectives

Migration is an extraordinarily diverse behaviour and patterns of migration vary within and between species and populations, in their temporal characteristics, and also spatially, in terms of scale and geography. The migratory journeys of animals are fantastically varied, and reflect the natural history, biology, and biogeography of the individuals that make them. Given the taxonomic breadth of migration it is perhaps no surprise that the phenomenon is so wonderfully heterogenous. In a changing world, patterns of migration are not static, but can shift in step with the environment in many cases. However, if migrants are unable to adapt to environmental changes along the migratory route, they are also at risk of population declines and potential extirpation. If we are to successfully conserve the vast biological diversity that encompasses the myriad migratory species, understanding the broader patterns of migration is essential.

References

Aarestrup, K., Okland, F., Hansen, et al. (2009). Oceanic spawning migration of the European eel (*Anguilla anguilla*). *Science*, **325**, 1660.

Able, K. P., and Belthoff, J. R. (1998). Rapid 'evolution'of migratory behaviour in the introduced house finch of eastern North America. *Proceedings of the Royal Society B: Biological Sciences*, **265**, 2063–71.

Alerstam, T. (2000). Bird migration performance on the basis of flight mechanics and trigonometry. In P. Domenici and R. W. Blake (eds), *Biomechanics in Animal Behaviour*, pp. 105–24. Bios Scientific Publishers, London.

Alerstam, T. (2001). Detours in bird migration. *Journal of Theoretical Biology*, **209**, 319–31.

Alerstam, T. (2009). Flight by night or day? Optimal daily timing of bird migration. *Journal of Theoretical Biology*, **258**, 530–6.

Alerstam, T., Hedenström, A., and Åkesson, S. (2003). Long-distance migration: evolution and determinants. *Oikos*, **103**, 247–60.

Alonso J. C., Palacin, C., Alonso, J. A., and Martin, C. A. (2009). Post-breeding migration in male great bustards: low tolerance of the heaviest Palearctic bird to summer heat. *Behavioral Ecology and Sociobiology*, **63**, 1705–15.

Altizer, S., and Davis, A. K. (2010). Populations of monarch butterflies with different migratory behaviours show divergence in wing morphology. *Evolution*, **64**, 1018–28.

Baisez, A., Bach, J. M., Leon, C., et al. (2011). Migration delays and mortality of adult Atlantic salmon *Salmo salar* en route to spawning grounds on the River Allier, France. *Endangered Species Research*, **15**, 265–70.

Bartelt, P. E., Peterson, C. R., and Klaver, R. W. (2004). Sexual differences in the post-breeding movements and habitats selected by western toads (*Bufo boreas*) in southeastern Idaho. *Herpetologica*, **60**, 455–67.

Beguer-Pon, M., Benchetrit, J., Castonguay, M., Aarestrup, K., Campana, S. E., Stokesbury, M. J. W., and Dodson, J. J. (2012). Shark predation on migrating adult American eels (*Anguilla rostrata*) in the Gulf of Lawrence. *PLoS One*, 7(10), e46830. doi:10.1371/journal.pone.0046830.

Bell, P., and Snively, E. (2008). Polar dinosaurs on parade: a review of dinosaur migration. *Alcheringa*, **32**, 271–84.

Bensch, S., Andersson, T., and Åkesson, S. (1999). Morphological and molecular variation across a migratory divide in willow warblers, *Phylloscopus trochilus*. *Evolution*, **53**, 1925–35.

Berge, J., Cottier, F., Last, K. S. et al. (2009). Diel vertical migration of Arctic zooplankton during the polar night. *Biology Letters*, **5**, 69–72.

Berthold, P. (1991). Genetic control of migratory behaviour in birds. *Trends in Ecology and Evolution*, **6**, 254–7.

Berthold, P. (1995). *Control of Bird Migration*. Chapman & Hall, London.

Berthold, P. (1999). A comprehensive theory for the evolution, control and adaptability of avian migration. *Ostrich*, **70**, 1–11.

Bill, R. G., and Herrnkind, W. F. (1976). Drag reduction by formation movement in spiny lobsters. *Science*, **193**, 1146–48.

Bisson, I. A., Safi, K., and Holland, R. A. (2009). Evidence for repeated independent evolution of migration in the largest family of bats. *PLoS One*, **4**, e7504.

Blake, S., Yackulic, C. B., Cabrera, F. et al. (2012). Vegetation dynamics drive segregation by body size in Galapagos tortoises migrating across altitudinal gradients. *Journal of Animal Ecology*, **82**, 310–21.

Bollens, S., Quenette, J. A., and Rollwagen-Bollens, G. (2012). Predator-enhanced diel vertical migration in a planktonic dinoflagellate. *Marine Ecology Progress Series*, **447**, 49–54.

Bonte, D., Maelfait, J. P., and Hoffman, M. (2000). Seasonal and diurnal migration patterns of the spider (Araneae) fauna of coastal grey dunes. *Ekologia (Bratislava)*, **19**, 5–16.

Boyle, W. A. (2008a). Partial migration in birds: tests of three hypotheses in a tropical lekking frugivore. *Journal of Animal Ecology*, **77**, 1122–8.

Boyle, W. A. (2008b). Can variation in risk of nest predation explain altitudinal migration in tropical birds? *Oecologia*, **155**, 397–403.

Boyle, W. A., Norris, D. R., and Guglielmo, C. G. (2010). Storms drive altitudinal migration in a tropical bird. *Proceedings of the Royal Society B: Biological Sciences*, **277**, 2511–19.

Brönmark, C., Chapman, B. B., Hansson, L.-A., Hulthén, K., Skov, C., and Nilsson, P. A. (2014). There and back again: migration in freshwater fishes. *Canadian Journal of Zoology*, **92**, 467–479. doi: 10.1139/cjz-2012-0277.

Brönmark, C., Skov, C., Brodersen, B., Nilsson, P. A., and Hansson, L.-A. (2008). Seasonal migration determined by a trade-off between predator avoidance and growth. *PLoS One*, **3**, e1957.

Burgess, N., and Mlingwa, C. (2000). Evidence for altitudinal migration of forest birds between montane Eastern Arc and lowland forests in East Africa. *Ostrich*, **71**, 184–90.

Burks, R. L., Lodge, D. M., Jeppesen, E., and Lauridsen, T. L. (2002). Diel horizontal migration of zooplankton: costs and benefits of inhabiting the littoral. *Freshwater Biology*, **47**, 343–65.

Cagnacci, F., S. Focardi, M. Heurich, A. et al. (2011). Partial migration in roe deer: migratory and resident tactics are end points of a behavioural gradient determined by ecological factors. *Oikos*, **120**, 1790–1802.

Campbell, H. A., Watts, M. E., Sullivan, S., et al. (2010). Estuarine crocodiles ride surface currents to facilitate long-distance travel. *Journal of Animal Ecology*, **79**, 955–64.

Carpenter, S. J., Erickson, J. M., and Holland, F. D. (2003). Migration of a Late Cretaceous fish. *Nature*, **423**, 70–4.

Chamberlain, C., Bensch, S., Feng, X., Åkesson, S., and Andersson, T. (2000). Stable isotopes examined across a migratory divide in Scandinavian willow warblers (*Phylloscopus trochilus trochilus* and *Phylloscopus trochilus acredula*) reflect their African winter quarters. *Proceedings of the Royal Society B: Biological Sciences*, **267**, 43–8.

Chapman, B. B., Brönmark, C., Nilsson, J-Å., and Hansson, L.-A. (2011). The ecology and evolution of partial migration. *Oikos*, **120**, 1764–75.

Chapman, B. B, Hulthen, K., Brodersen, J., et al. (2012a). Partial migration in fishes: causes and consequences. *Journal of Fish Biology*, **81**, 456–78.

Chapman, B. B., Skov, C., Hulthen, K., et al. (2012b). Partial migration in fishes: definitions, methodologies and taxonomic distribution. *Journal of Fish Biology*, **81**, 479–99.

Chapman, B. B., Erikson, A., Baktoft, H., et al. (2013). A foraging cost of migration in a partially migratory cyprinid fish. *PLoS One*, **8**, e61223.

Chapman, J. W., Nesbit, R. L., Burgin, L.E., et al. (2010). Flight orientation behaviours promote optimal migration trajectories in high-flying insects. *Science*, **327**, 682–5.

Chapman, J. W., Bell, J. R., Burgin, L. E., et al. (2012). Seasonal migration to high latitudes results in major reproductive benefits in an insect. *Proceedings of the National Academy of Sciences USA*, **109**, 14924–9.

Cotton, P. A. (2003). Avian migration phenology and global climate change. *Proceedings of the National Academy of Sciences USA*, **100**, 12219–22.

Cristol, D. A., Baker, M. B., and Carbone, C. (1999). Differential migration revisited: latitudinal segregation by age and sex class. *Current Ornithology*, **15**, 33–88.

Deutsch, C. A., Tewksbury, J. J., Huey, R. B., et al. (2008). Impacts of climate warming on terrestrial ectotherms across latitude. *Proceedings of the National Academy of Sciences USA*, **105**, 6668–72.

Dingle, H., and Drake, V. A. (2007). What is migration? *Bioscience*, **57**, 113–21.

Dodson, S. (1990). Predicting diel vertical migration of zooplankton. *Limnology and Oceanography*, **35**, 1195–1200.

Dodson, J. J. (1997). Fish migration: an evolutionary perspective. In J.-G. Godin (ed.), *Behavioural Ecology of Teleost Fishes*. Oxford University Press, Oxford.

Doney, S. C., Ruckelshaus, M., Emmett Duffy, J., et al. (2012). Climate change impacts on marine ecosystems. *Annual Review of Marine Science*, **4**, 11–37.

Duijns, S., Jukema, J., Spaans, B., Horssen, P. Piersma, T. (2012). Revisiting the proposed leap-frog migration of Bar-tailed Godwits along the East-Atlantic Flyway. *Ardea*, **100**, 37–43.

Egevang, C., Stenhouse, I. J., Phillips, R. A., Petersen, A., Fox., J. W., and Silk, J. R. D. (2010). Tracking of Arctic terns *Sterna paradisaea* reveals longest animal migration. *Proceedings of the National Academy of Sciences USA*, **107**, 2078–81.

Eichhorn, G., Drent, R. H., Stahl, J., Leito, A., and Alerstam, T. (2009). Skipping the Baltic: the emergence of a dichotomy of alternative spring migration strategies in Russian barnacle geese. *Journal of Animal Ecology*, **78**, 63–72.

Emlen, S. T. (1967). Migratory Orientation in the Indigo Bunting, *Passerina cyanea*: Part I: Evidence for Use of Celestial Cues. *Auk*, **84**, 309–42.

Engstedt, O., Stenroth, P., Larsson, P., Ljunggren, L., and Elfman, M. (2010). Assessment of natal origin of pike (*Esox lucius*) in the Baltic Sea using Sr:Ca in otoliths. *Environmental Biology of Fishes*, **89**, 547–55.

Fleming, T. H., Eby, P., Kunz, T. H., and Fenton, M. B. (2003). Ecology of bat migration. In T. H. Kunz and M. B. Fenton (eds), *Bat Ecology*, pp. 156–208. University of Chicago Press, Chicago.

Forman, R. T. T., and Alexander, L. E. (1998). Roads and their major ecological effects. *Annual Review of Ecology and Systematics*, **29**, 207–31.

Fort, J., Pettex, E., Tremblay, Y., et al. (2012). Metapopulation evidence of oriented chain migration in northern gannets (*Morus bassanus*). *Frontiers in Ecology and the Environment*, **10**, 237–42.

Fricke, H. C., Hencecroth, J., and Hoerner, M. E. (2011). Lowland-upland migration of sauropod dinosaurs during the Late Jurassic epoch. *Nature*, **480**, 513–15.

Gill, R. E., Jr., Piersma, T., Hufford, G., Servanckx, R., and Riegen, A. (2005). Crossing the ultimate ecological barrier: evidence for an 11 000-km-long nonstop flight from Alaska to New Zealand and eastern Australia by bar-tailed godwits. *Condor*, **107**, 1–20.

Gill, R. E., Jr., Tibbitts, T. L., Douglas, D. C., et al. (2009). Extreme endurance flights by landbirds crossing the Pacific Ocean: ecological corridor rather than barrier? *Proceedings of the Royal Society B: Biological Sciences*, **276**, 447–57.

Gillis, E. A., Green, D. J., Middleton, H. A., and Morrissey, C. A. (2008). Life history correlates of alternative migratory strategies in American Dippers. *Ecology*, **89**, 1687–95.

Gjelland, K. O., Bohn, T., Horne, J. K., Jensvoll, I., Knudsen, F. R., and Amundsen, P. A. (2009). Planktivore vertical migration and shoaling under a subarctic light regime. *Canadian Journal of Fisheries and Aquatic Sciences*, **66**, 525–39.

Gliwicz, M. (1986). Predation and the evolution of vertical migration in zooplankton. *Nature*, **320**, 746–8.

Grayson, K. L., and Wilbur, H. M. (2009). Sex-and context-dependent migration in a pond-breeding amphibian. *Ecology*, **90**, 306–12.

Groeneveld, J. C., and Branch, G. M. (2002). Long-distance migration of South African deep-water rock lobster *Palinurus gilchristi*. *Marine Ecology Progress Series*, **232**, 225–38.

Gross, M. R., Coleman, R. M., and McDowall, R. M. (1988). Aquatic productivity and the evolution of diadromous fish migration. *Science*, **239**, 1291–3.

Hansson, L.-A., Becares, E., Fernández-Aláez, M., et al. (2007). Relaxed circadian rhythm in zooplankton along a latitudinal gradient. *Oikos*, **116**, 585–91.

Hansson, L.-A., and Hylander, S. (2009). Size-structured risk assessments govern *Daphnia* migration. *Proceedings of the Royal Society B: Biological Sciences*, **276**, 331–6.

Hawkes, L., and Balachandran, S. (2013). The paradox of extreme high-altitude migration in bar-headed geese *Anser indicus*. *Proceedings of the Royal Society B: Biological Sciences*, **280**, 1–8.

Hays, G., Broderick, A. C., Godley, B. J., et al. (2002). Biphasal long-distance migration in green turtles. *Animal Behaviour*, **64**, 895–8.

Hebblewhite, M., and Merrill, E. H. (2007). Multiscale wolf predation risk for elk: does migration reduce risk? *Oecologia*, **152**, 377–87.

Hebblewhite, M., Merrill, E., and McDermid, G. (2008). A multi-scale test of the forage maturation hypothesis in a partially migratory ungulate population. *Ecological Monographs*, **78**, 1–26.

Hedenström, A., and Pettersson, J. (1987). Migration routes and wintering areas of willow warblers *Phylloscopus trochilus* (L.) ringed in Fennoscandia. *Ornis Fennici*, **64**, 137–43.

Helfman, G., Collette, B. B., Facey, D. E., and Bowen, B. W. (2009). *The Diversity of Fishes: Biology, Evolution, and Ecology*. John Wiley and Sons, NJ.

Hendry, A. P., Bohlin, T., Jonsson, B., and Berg, O. K. (2004). To sea or not to sea—anadromy versus non-anadromy in salmonids. In A. P. Henry and S. Stearns (eds), *Evolution Illuminated: Salmon and their Relatives*. Oxford University Press, Oxford.

Herrnkind, W. F. (1985). Evolution and mechanisms of mass single-file migration in spiny lobster: synopsis. *Contributions in Marine Science*, **27**, 197–211.

Holland, R. A., Wikelski, M., and Wilcove, D. S. (2006). How and why do insects migrate? *Science*, **313**, 794–6.

Holt, R. D. (2003). On the evolutionary ecology of species' ranges. *Evolutionary Ecology Research*, **5**, 159–78.

Jephson, T., and Carlsson, P. (2009). Species-and stratification-dependent diel vertical migration behaviour of three dinoflagellate species in a laboratory study. *Journal of Plankton Research*, **31**, 1353–62.

Keller, D., Brodbeck, S., Flöss, I., Vonwil, G., and Holderegger, R. (2010). Ecological and genetic measurements of dispersal in a threatened dragonfly. *Biological Conservation*, **143**, 2658–63.

Ketterson, E. D., and Nolan, V., Jr. (1976). Geographic variation and its climatic correlates in the sex ratio of eastern-wintering dark-eyed Juncos (*Junco hyemalis hyemalis*). *Ecology*, **57**, 679–93.

Kikuchi, K. (1930). Diurnal migration of plankton crustacea. *Quarterly Review of Biology*, **5**, 189–206.

Klaassen, R. H., Alerstam, T., Carlsson, P., Fox, J. W., and Lindström, Å. (2011). Great flights by great snipes: long and fast non-stop migration over benign habitats. *Biology Letters*, **7**, 833–35.

Krause, J., and Ruxton, G. (2002). *Living in Groups*. Oxford University Press, Oxford.

Lammi, A., Siikamäki, P., and Mustajärvi, K. (1999). Genetic diversity, population size, and fitness in central and peripheral populations of a rare plant *Lychnis viscaria*. *Conservation Biology*, **13**, 1069–78.

Lindenmayer, D. B., and Fischer, J. (2006). *Habitat Fragmentation and Landscape Change*. CSIRO, Collingwood, Australia.

Lindsay, D. L., Barr, K. R., Lance, R. F., Tweddale, S. A., Hayden, T. J., and Leberg, P. L. (2008). Habitat fragmentation and genetic diversity of an endangered, migratory songbird, the golden-cheeked warbler (*Dendroica chrysoparia*). *Molecular Ecology*, **17**, 2122–33.

Loxterman, J. L., and Keeley, E. R. (2012). Watershed boundaries and geographic isolation: patterns of diversification in cutthroat trout from western North America. *BMC Evolutionary Biology*, **12**, 38.

Lucas, M. C., and Baras, E. (2001). *Migration of Freshwater Fishes*. Blackwell Science, Oxford.

Madsen, T., and Shine, R. (1996). Seasonal migration of predators and prey—a study of pythons and rats in tropical Australia. *Ecology*, **77**, 149–156.

McDonald, W. R., and Clair, C. C. S. (2004). The effects of artificial and natural barriers on the movement of small mammals in Banff National Park, Canada. *Oikos*, **105**, 397–407.

McDowall, R. (2007). On amphidromy, a distinct form of diadromy in aquatic organisms. *Fish and Fisheries*, **8**, 1–13.

McDowall, R. M. (2008). Why are so many boreal freshwater fishes anadromous? Confronting 'conventional wisdom'. *Fish and Fisheries*, **9**, 208–13.

McKinnon, L., Smith, P. A., Nol, E., et al. (2010). Lower predation risk for migratory birds at high latitudes. *Science*, **327**, 326–7.

McMahon, T. E., and Holanov, S. H. (1995). Foraging success of largemouth bass at different light intensities: implications for time and depth of feeding. *Journal of Fish Biology* **46**, 759–67.

Mehner, T. (2012). Diel vertical migration of freshwater fishes–proximate triggers, ultimate causes and research perspectives. *Freshwater Biology*, **57**, 1342–59.

Moore, A. R. (1912). Concerning negative phototropism in *Daphnia pulex*. *Journal of Experimental Zoology*, **13**, 573–5.

Muska, M. T., Tuser, M., Frouzová, J., et al. (2013). To migrate, or not to migrate: partial horizontal migration of fish in a temperate freshwater reservoir. *Hydrobiologia*, **707**, 17–28.

Mysterud, A., Loe, L. E., Zimmerman, B., et al. (2011). Partial migration in expanding red deer populations at northern latitudes—a role for density dependence? *Oikos*, **120**, 1817–25.

Newton, I. (2007). *The Migration Ecology of Birds*. Academic Press, San Diego, CA.

Newton, I., and Dale, L. (1996a). Relationship between migration and latitude among west European birds. *Journal of Animal Ecology*, **65**, 137–46.

Newton, I., and Dale, L. (1996b). Bird migration at different latitudes in eastern North America. *Auk*, **113**, 626–35.

Nilsson, A. L. K., Alerstam, T., and Nilsson, J.Å. (2008). Diffuse, short and slow migration among Blue Tits. *Journal of Ornithology*, **149**, 365–73.

Nilsson, A. L. K., Alerstam, T., Nilsson, J.Å., and Holberton, R. L. (2006). Do partial and regular migrants differ in their responses to weather? *Auk*, **123**, 537–47.

Olden, J. D., Kennard, M. J., Lawler, J. J., and Poff, N. L. (2011). Challenges and opportunities in implementing managed relocation for conservation of freshwater species. *Conservation Biology*, **25**, 40–7.

Onzo, A. Hanna, R., Zannou, I., Sabelis, M. W., and Yaninek, J. S. (2003). Dynamics of refuge use: diurnal, vertical migration by predatory and herbivorous mites within cassava plants. *Oikos*, **101**, 59–69.

O'Dor, R. (1992). Big squid in big currents. *South African Journal of Marine Science*, **12**, 225–35.

Partecke, J., and Gwinner, E. (2007). Increased sedentariness in European Blackbirds following urbanization: a consequence of local adaptation? *Ecology*, **88**, 882–90.

Pavlova, A., Amos, J. N., Goretskaia, M. I., et al. (2012). Genes and song: genetic and social connections in fragmented habitat in a woodland bird with limited dispersal. *Ecology*, **93**, 1717–27.

Polovina, J. J. (1996). Decadal variation in the trans-Pacific migration of northern bluefin tuna (*Thunnus thynnus*) coherent with climate-induced change in prey abundance. *Fisheries Oceanography*, **5**, 114–19.

Rasmussen, K., Palacios, D. M., Calambokidis, J., et al. (2007). Southern Hemisphere humpback whales wintering off Central America: insights from water temperature into the longest mammalian migration. *Biology Letters*, **3**, 302–5.

Reppert, S. M., Gegear, R. J., and Merlin, C. (2010). Navigational mechanisms of migrating monarch butterflies. *Trends in Neurosciences*, **33**, 399–406.

Richter, H., and Cumming, G. (2005). Food availability and annual migration of the straw-colored fruit bat (*Eidolon helvum*). *Journal of Zoology*, **268**, 35–44.

Richter, H. and Cumming, G. (2008). First application of satellite telemetry to track African straw-coloured fruit bat migration. *Journal of Zoology*, **275**, 172–6.

Robinson, S. J., Samuel, M. D., Lopez, D. L., and Shelton, P. (2012). The walk is never random: subtle landscape effects shape gene flow in a continuous white-tailed deer population in the Midwestern United States. *Molecular Ecology*, **21**, 4190–205.

Roff, D. A., and Fairbairn, D. J. (1991). Wing dimorphisms and the evolution of migratory polymorphisms among the Insecta. *American Zoologist*, **31**, 243–51.

Romare, P., and Hansson, L.-A. (2003). A behavioural cascade: top-predator induced behavioural shifts in planktivorous fish and zooplankton. *Limnology and Oceanography*, **48**, 1956–64.

Rondinini, C., and Doncaster, C. P. (2002). Roads as barriers to movement for hedgehogs. *Functional Ecology*, **16**, 504–9.

Root, J. J., Puskas, R. B., Fischer, J. W., et al. (2009). Landscape genetics of raccoons (*Procyon lotor*) associated with ridges and valleys of Pennsylvania: implications for oral rabies vaccination programs. *Vector-Borne and Zoonotic Diseases*, **9**, 583–8.

Rosenberg, K. V., Rohrbaugh, R. W., Barker, J. S. E., Lowe, J. D., Hames, R. S., and Dhondt, A. A. (1999). A land manager's guide to improving habitat for scarlet tanagers and other forest interior birds. *The Cornell Lab of Ornithology*. Cornell University Press, Ithaca, NY.

Ruegg, K. C., and Smith, T. B. (2002). Not as the crow flies: a historical explanation for circuitous migration in Swainson's thrush (*Catharus ustulatus*). *Proceedings of the Royal Society B: Biological Sciences*, **269**, 1375–81.

Semlitsch, R. D. (2008). Differentiating Migration and Dispersal Processes for Pond-Breeding Amphibians. *Journal of Wildlife Management*, **72**, 260–7.

Semlitsch, R. D., and Bodie, J. R. (2003). Biological criteria for buffer zones around wetlands and riparian habitats for amphibians and reptiles. *Conservation Biology*, **17**, 1219–28.

Semmens, J. M., Pecl, G. T., and Gillanders, B. M. (2007). Approaches to resolving cephalopod movement and migration patterns. *Reviews in Fish Biology and Fisheries*, **17**, 401–23.

Shaw, A. K., and Levin, S. A. (2011). To breed or not to breed: a model of partial migration. *Oikos*, **120**, 1871–9.

Skov, C., Baktoft, H., Brodersen, J., et al. (2011). Sizing up your enemy: individual predation vulnerability predicts migratory probability. *Proceedings of the Royal Society B: Biological Sciences*, **278**, 1414–18.

Skov, C., Brodersen, J., Nilsson, P. A., Hansson, L.-A., and Brönmark, C. (2008). Inter- and size-specific patterns of fish seasonal migration between a shallow lake and its streams. *Ecology of Freshwater Fish*, **17**, 406–15.

Skov, C., Chapman, B. B., Baktoft, H., Brodersen, J., Brönmark, C., Hansson, L.-A., Hulthén, K., and Nilsson, P. A. (2013). Migration confers survival benefits against avian predators for partially migratory freshwater fish. *Biology Letters*, **9**, 20121178.

Smith, H. G., and Nilsson, J. Å. (1987). Intraspecific variation in migratory pattern of a partial migrant, the blue tit (*Parus caeruleus*): an evaluation of different hypotheses. *Auk*, **104**, 109–115.

Southwood, T. R. E., and Henderson, P. A. 2000. *Ecological Methods*. Blackwell Science, Oxford.

Stanley, C. Q., MacPherson, M., Fraser, K. C., McKinnon, E. A., and Stutchbury, B. J. M. (2012). Repeat tracking of individual songbirds reveals consistent migration timing but flexibility in route. *PLoS One*, **7**, e40688.

Stefanescu, C., Paramo, F., Åkesson, S., et al. (2012). Multi-generational long-distance migration of insects: studying the painted lady butterfly in the Western Palaearctic. *Ecography*, **36**, 474–86.

Steinhart, G. B., and Wurtsbaugh, W. A. (1999). Under-ice diel vertical migrations of *Oncorhynchus nerka* and their zooplankton prey. *Canadian Journal of Fisheries and Aquatic Sciences*, **56**, 152–61.

Stervander, M., Lindström, Å., Jonzen, N., and Andersson, A. (2005). Timing of spring migration in birds: long-term trends, North Atlantic oscillation and the significance of different migration routes. *Journal of Avian Biology*, **36**, 210–21.

Stevick, P. T., Neves, M. C., Johansen, F., et al. (2011). A quarter of a world away: female humpback whale moves 10 000 km between breeding areas. *Biology Letters*, **7**, 299–302.

Thériault, V., Dunlop, E. S., Dieckmann, U., Bernatchez, L., and Dodson, J. J. (2008). The impact of fishing-induced mortality on the evolution of alternative life-history tactics in brook charr. *Evolutionary Applications*, **1**, 409–23.

Todd, B. D., and Winne, C. T. (2006). Ontogenetic and interspecific variation in timing of movement and responses to climatic factors during migrations by pond-breeding amphibians. *Canadian Journal of Zoology*, **84**, 715–22.

Trinkel, M., Fleishmann, P. H., Steindorfer, A. F., Kastberger G. (2004). Spotted hyenas (*Crocuta crocuta*) following migratory prey. Seasonal expansion of a clan territory in Etosha, Namibia. *Journal of Zoology*, **264**, 125–33.

van Ginneken, V., Antonissen, E., Muller, U. K., et al. (2005). Eel migration to the Sargasso: remarkably high swimming efficiency and low energy costs. *Journal of Experimental Biology*, **208**, 1329–35.

van Oort, B., Tyler, N., Gerkema, M., Folkow, L., Schytte Blix, A., and Stokkan, K.-A. (2005). Circadian organization in reindeer. *Nature*, **438**, 1095–6.

Visser, M. E., Perdeck, A. C., Van Balen, J. H., and Both, C. 2009. Climate change leads to decreasing bird migration distances. *Global Change Biology*, **15**, 1859–65.

von Huene, F. (1928). Lebensbild des Saurischier-Vorkommens in obersten keuper von Trossingen in Wurtemberg. *Palaeobiologica*, **1**, 103–16.

Weersing, K., and Toonen, R. J. (2009). Population genetics, larval dispersal, and connectivity in marine systems. *Marine Ecology-Progress Series*, **393**, 1–12.

Wikelski, M., Moskowitz, D., Adelman, J. S., Cochran, J., Wilcove, D. S., and May, M. L. (2006). Simple rules guide dragonfly migration. *Biology Letters*, **2**, 325–9.

Wikelski, M., Tarlow, E. M., Raim, A., et al. (2003). Avian metabolism: costs of migration in free-flying songbirds. *Nature*, **423**, 704.

Wilcove, D. S., and Wikelski, M. (2008). Going, going, gone: is animal migration disappearing? *PLoS Biology*, **6**, e188.

Wilcox, B. A., and Murphy, D. D. (1985). Conservation strategy: the effects of fragmentation on extinction. *American Naturalist*, **125**, 879–87.

Williamson, C. E., Olson, O. G., Lott, S. E., Walker, N. D., Engstrom, D. R., and Hargreaves, B. R. (2001). Ultraviolet radiation and zooplankton community structure following deglaciation in Glacier Bay, Alaska. *Ecology*, **82**, 1748–60.

Wurtsbaugh, W. A., and Neverman, D. (1988). Post-feeding thermotaxis and daily vertical migration in a larval fish. *Nature*, **333**, 846–8.

Ydenberg, R. C., Butler, R. W., Lank, D. B., Smith, B. D., and Ireland, J. (2004). Western sandpipers have altered migration tactics as peregrine falcon populations have recovered. *Proceedings of the Royal Society B: Biological Sciences*, **271**, 1263–9.

Zalewski, A., Piertney, S. B., Zalewska, H., and Lambin, X. (2009). Landscape barriers reduce gene flow in an invasive carnivore: geographical and local genetic structure of American mink in Scotland. *Molecular Ecology*, **18**, 1601–15.

CHAPTER 3

Movement and migration in a changing world

Åke Lindström, Ben B. Chapman, Niclas Jonzén, and Marcel Klaassen

Our world is constantly changing and with it the living conditions for life on Earth. Habitats crucial to animal reproduction, migration, and survival are transformed, and at an increasing rate. At the same time, a changing climate alters fundamental ecological settings in both time and space. In addition, and because of habitat and climate change, some species change in numbers and distribution, which in turn affects the biotic environment for other organisms. To survive, animals have to cope or adapt to all these changes. Whereas the *potential* effects a changing world will have on animals on the move have been outlined and forecasted several times (e.g. Lindström and Agrell 1999, Mustin et al. 2007, Altizer et al. 2011, Klaassen et al. 2012), the focus of this chapter is on *documented* effects.

The chapter is divided in two sections. In the first, examples that describe behavioural responses and adaptations in time and space of travelling animals are given. The examples are grouped according to changes in three key ecological variables: habitat quality, climate, and biotic interactions. Although these variables interact, making it difficult to discern the key driving force, they form a useful framework when sorting out in what ways a changing world may act on moving animals. Although most examples concern the actual act of movement or migration, carryover effects from periods of movement and migration on other life stages such as reproduction and wintering are also considered.

We might expect certain species with specific ecological traits to be more prone to global change than others that do not possess these traits and vice versa. In the second section we address how migrants cope (or do not cope) with environmental change, looking for patterns of who has been affected, how, where, and why.

3.1 Habitat change

By dint of their habit to exploit temporal opportunities at geographically separated locations, the world of migrants is large. Just by chance alone, their extensive spatial behaviour makes them more susceptible to habitat change than non-migrants. Among migrants, however, habitat generalists are not very likely to be severely affected by local or specific habitat transformations, whether they take place during migration or at other moments during the annual cycle (Table 3.1). Generalists will most certainly have alternative sites or habitats to exploit. At the other end of the spectrum, however, are species that are highly dependent on very specific habitats or only few geographical sites, resulting in narrow migratory corridors and very little leeway. As we will see, for species with specialized requirements during migration the effects of habitat change can be profound.

3.1.1 Destruction of stopover sites

Arctic shorebirds perform some of the longest and most spectacular migrations in the animal kingdom (Hedenström and Lindström, Chapter 5). Each year individuals of different subspecies migrate back and forth to wintering grounds in the southern hemisphere (Africa, South America, and Australia/New Zealand), making annual round-trips of up

Animal Movement Across Scales. Edited by Lars-Anders Hansson and Susanne Åkesson.
© Oxford University Press 2014. Published 2014 by Oxford University Press.

Table 3.1 Examples of migrating species that have been affected by habitat change, climate change, or changing biotic interactions.

	Animal	Where	Change	Effect
	Red knot *Calidris canutus*	China	Habitat change: destruction of tidal flats critical for fuelling during migration	Decreased survival
	Pacific lamprey *Lampetra tridentata*	Canada	Habitat change: dam construction	Local extinction
	Mongolian gazelle *Procapra gutturosa*	Mongolia	Habitat change: railroad potentially prevents migration	Unknown
	Wildebeest *Connochaetes taurinus*	Botswana	Habitat change: fences prevent drought-induced migration	Dramatic population decline
	Blackbird *Turdus merula*	Germany	Habitat change: urbanization	Reduced migratory disposition
	Blackcap *Sylvia atricapilla*	Sweden	Climate change: warmer spring weather	Earlier spring passage
	Pink salmon *Oncorhyncos gorbuscha*		Climate change: warmer river water in spring	Earlier up-stream migration
	Marbled salamander *Ambystoma opacum*	USA	Climate change: warmer autumn overnight temperatures	Delayed arrival to autumn breeding pond
	Elk *Cervus elaphus*	Canada	Biotic interactions: human settlements reduce predation risk	Increasing proportion of residents in partial migratory population

to 30,000 km (Battley et al. 2012). The common denominator is that they all depend on a few sites containing very special habitats for their migration success—intertidal mudflats. The birds spend a few weeks at these sites, making use of extremely high availabilities of invertebrate prey, fuelling up for continued migration. Being reliant on such specific sites, these species are potentially much more at risk than more generalist migrants, and there are species being driven towards extinction by habitat destruction as you read this.

With respect to specialization to and reliance on specific sites, things can get really extreme among these shorebirds (Table 3.1). On their way from their wintering sites in Australia and New Zealand to their breeding grounds in Asia, individuals of two subspecies of red knots, *Calidris canutus rogersi* and *Calidris canutus piersmai*, make only one long stop during their spring migration. This stop is at the coast of the Yellow Sea (China, South Korea, and North Korea; Fig. 3.1). Rogers et al. (2010) estimated that over 45% of the combined world

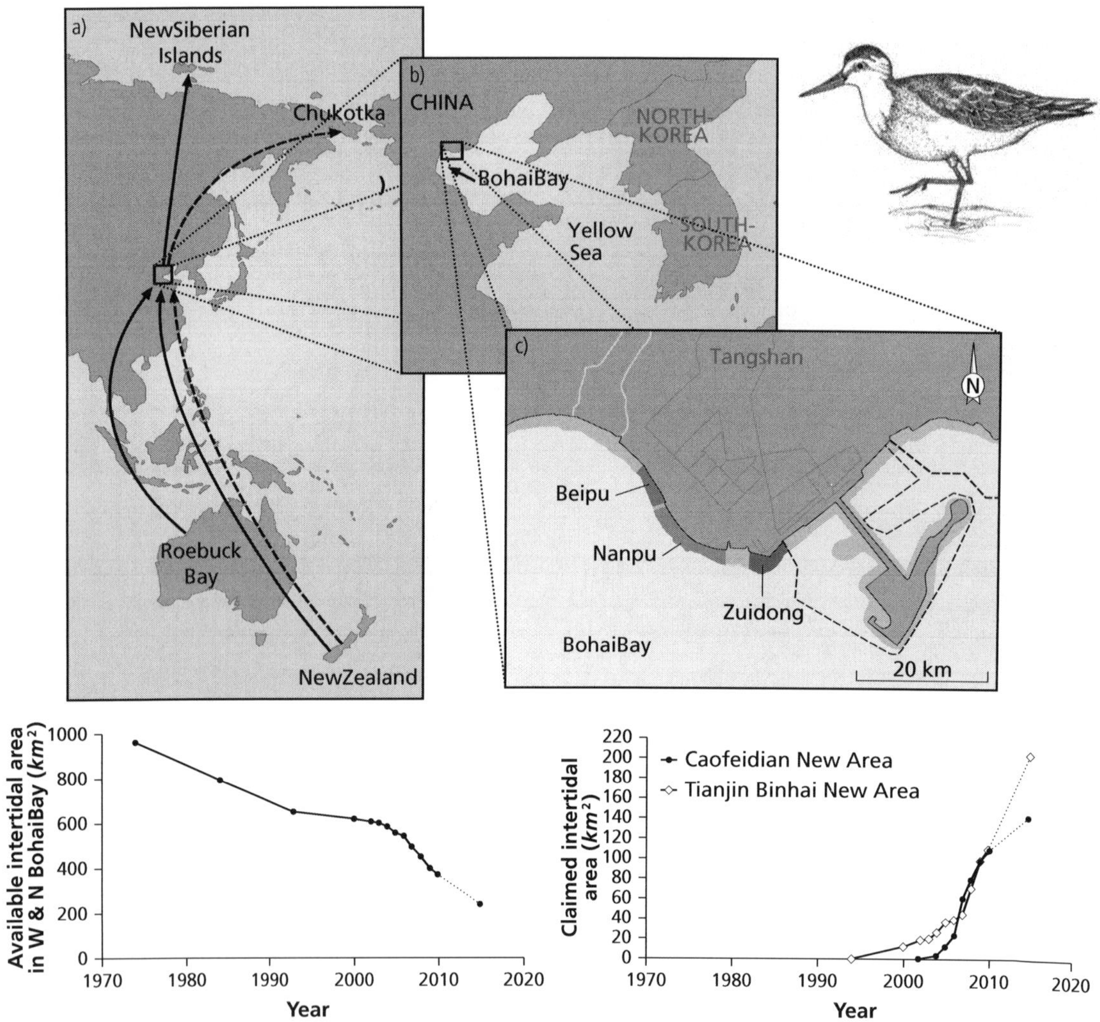

Figure 3.1 Illustrations of the devastating effects of land reclamation on migrating shorebirds in Bohai Bay, Yellow Sea, China. (a) Shorebird migratory paths with the Bohai Bay stopover site indicated (square); (b) and (c) show closeups of the stopover site. (c) Local view of the Bohai Bay with land reclamation sites indicated (dashed line). The graphs show available intertidal areas (km^2) and claimed areas since the 1970s (Rogers et al. 2010, Yang et al. 2011).

population of adult *C. c. rogersi* and *C. c. piersmai* now stopover at only 20 km of coastline in the Bohai Bay. Large parts of this vital staging area are at present being destroyed through land 'reclamation', where tidal flats are walled, pumped, and turned into industrial areas. Many other tidal flats in the Yellow Sea are already destroyed and others are being threatened by reclamation (Yang et al. 2011; Fig. 3.1). Accordingly, these shorebirds are forced into the few remaining suitable areas where densities during peak migration time have risen dramatically in recent years. This most certainly leads to intensified competition and reduced survival rates; in just ten years the world population of the two red knot subspecies has dropped from 220,000 to 105,000 (Yang et al. 2011). Clearly, due to habitat destruction, the red knots migrating along the East Asian coast face an imminent and serious threat to their survival. The ruff, *Philomachus pugnax*, another long-distance migratory shorebird, has recently declined in numbers in Western Europe. Detailed studies of spring stopover in the Netherlands and Ukraine suggest ruffs have shifted their migration routes eastwards in Europe as a response to deteriorating conditions in the Netherlands (Verkuil et al. 2012). Reduced food abundance on the grazed meadows, where the birds feed, has led to decreasing fuelling rates in the Netherlands, but not in Ukraine. The main driving factor is most certainly habitat deterioration due to changing farming practices.

3.1.2 Aquatic habitat change

Not only birds have trouble keeping up with the massive contemporary changes to our environment. For nearlyall migratory taxa examples are easily found. Habitat change in freshwater aquatic environments through the damming of rivers, water withdrawals for irrigation, sedimentation and logging has had powerful and adverse impacts upon many migratory fish species. A combination of habitat modification, in concert with human exploitation, has impacted upon populations of charismatic commercial migrants in the past century (Table 3.1). Before European settlement it is estimated that 175–250 thousand tons of salmon migrated up the rivers of the Pacific Northwest (Wilcove and Wikelski 2008). Currently, the total biomass of spawning salmon in this region is approximately 13–15 thousand tons, a more than tenfold decline. The ecological consequences of the reduction in nutrient transfer accompanying this massive decline in migratory salmon are unclear. However, it has been calculated that the rivers of the Northwest receive only 6–7% of the marine-derived nitrogen and phosphorus compared to levels prior to the huge decline in salmon (Gresh et al. 2000), which may have many different ecological effects.

Of all the human-made changes to freshwater systems, damming has often resulted in some of the most dramatic effects among migrants. Following the construction of the Xinanjiang dam in China the number of fish species fell from 107 to 83 due to the disruptive effect the dam had on the migratory journeys of many fishes (Zhong and Power 1996). Damming can also 'trap' migratory fish populations in upstream habitats. In some cases this leads to a change in the migratory behaviour of the populations, via either microevolution or phenotypic plasticity (Hollander et al., Chapter 7), leading to the emergence of resident behaviour, as has been documented in many salmonids (Morita et al. 2000). However, there are also examples where this has led to local extinction, as in the case of anadromous Pacific lamprey, *Lampeta tridenta*, following the construction of the Elsie Lake dam in Canada (Beamish and Northcote 1989; Table 3.1).

Damming can also have interesting eco-evolutionary consequences for migratory animals and their ecosystems. Landlocked alewives, *Alosa pseudoharengus*, once migratory anadromous fish now trapped in lakes for the whole year, forage voraciously upon zooplankton. The increase in selection pressure (driven by habitat-mediated changes to migratory patterns) from these landlocked fish has led to an evolutionary shift in zooplankton life-history and size (Walsh and Post 2011). This shift in prey phenotype has led to feedbacks to the predator, and contemporary evolutionary effects on the alewives themselves have been documented, with landlocked fish having traits such as reduced gill raker spacing, which allow them to be more effective foragers on small zooplankton.

3.1.3 Terrestrial habitat change

Barriers are also a problem in terrestrial migrations. Roads, fences, pipelines, railways, and reservoirs all act to disrupt migratory patterns and have led to calamitous declines of many migratory ungulates (Bolger et al. 2008). In Mongolia, the Ulaanbaatar–Beijing railway is thought to be the primary driving force behind the loss of the historic east–west migration of Mongolian gazelle (Ito et al. 2005; Table 3.1). In southern Botswana, there have been dramatic declines in wildebeest, *Connochaetes taurinus*, numbers because fences prevented drought-induced migrations (Spinage 1992; Table 3.1). Land use changes in northern Tanzania (particularly agricultural development) have restricted the historical migration routes of many migratory mammals, and as these traditional routes have been disrupted many ungulate populations have declined. For example, in 1988–2001, wildebeest, hartebeest, *Alcelaphus buselaphus*, and oryx, *Oryx spp.*, populations within the Tarangire ecosystem in Tanzania have declined by 88%, 90%, and 95%, respectively (Bolger et al. 2008).

Clearly, many different types of habitats have been modified extensively by mankind, often with devastating effects on animals. Yet another example is the rapid urbanization which followed the industrialization of Europe. But where there are losers due to habitat change, there may also be winners. While urbanization destroyed natural habitats, a completely new habitat was also created. For species that can thrive in close proximity to humans, inhabiting cities can have many advantages. One which pertains especially to migration is that food supply does not fluctuate so radically with the seasons, either intentionally through supplementary feeding by people or unintentionally via waste and food stores. Urbanization also creates 'heat islands' with a warmer temperature than usual for the region (Shochat et al. 2006). Field studies of migratory behaviour have revealed that urban birds such as blackbirds, *Turdus merula*, have become more sedentary in cities (Partecke and Gwinner 2007; Table 3.1). Interestingly, the effect on blackbirds is sex-specific and limited to males.

Another group of migrants that has largely profited from habitat change are herbivorous, migratory geese. In recent decades, shifts in agricultural practices, not least towards more winter-green fields, have increased goose winter survival. This in turn has led to dramatic population increases in many parts of the world (Abraham et al. 2005). Most interestingly, the geese have benefitted so immensely from environmental change that they themselves now constitute a powerful change, with cascading effects further down the system, even to distant places. One of these positively affected geese species is the lesser snow goose, *Chen c. caerulescens*. When returning to their sub-Arctic breeding grounds in coastal Canada, the ever-increasing numbers of geese have now degraded important coastal habitats through overgrazing (Jefferies et al. 2004). This is negatively affecting not only the geese themselves, but also other taxa. For example, migratory savannah sparrows, *Passerculus sandwichensis*, that breed in the same area suffered strong population declines due to loss of breeding habitats (Jefferies et al. 2004).

3.2 Climate change

There are seemingly endless forecasted effects climate change may have on migrants, both during and outside the migratory period (Boyd and Madsen 1997, Lindström and Agrell 1999, Mustin et al. 2007). The most well-known effects of climate change on migrants are the shifts recorded in timing of migration. But animals can also change their routes and the use of specific sites in relation to a changing climate. There is an overwhelming amount of evidence showing that animals have altered their time schedule of migration as a response to a changing climate, not least in birds (Parmesan 2007). At sites where spring has become warmer (earlier), most avian migrants also arrive earlier (Lehikoinen et al. 2004, Jonzén et al. 2006). For example, at Ottenby Bird Observatory in southeast Sweden, migrants such as the thrush nightingale, *Luscinia luscinia*, the lesser whitethroat, *Sylvia curruca*, and the blackcap, *S. atricapilla*, advanced their spring passage with on average 7, 9, and 13 days, respectively, over the period 1952–2002 (Stervander et al. 2005; Table 3.1). Whereas most species did advance their spring migration to some extent, it is important to notice that some species did not. This shows that not all migrants react to climate change in a similar fashion. Do the non-reacting species pay a high price from not reacting, leading to population declines (Møller et al. 2008)? Or is their lack of response actually the best way to react? In fact, we are only starting to understand how different selection pressures work in concert to produce a fitness landscape where alternative phenology strategies can invade (Jonzén et al. 2007, Johansson and Jonzén 2012a).

There seems to be no general consensus on where during migration a changing climate may have the biggest effect on spring arrival to the breeding grounds in long-distance migrant birds. The change in timing may be due to changes taken place already near the tropical wintering areas (Jonzén et al. 2006) or much further north and nearer the breeding grounds (Ahola et al. 2004). It is also not clear to what extent the changes in timing seen among birds are phenotypic, where plastic individuals adjust their migration to match current conditions. Or evolutionary, where the timing of migration is relatively fixed in individuals and genes for early and late migration are rapidly selected for or against in a changing climate (Jonzén et al. 2006, Both 2007, Knudsen et al. 2011).

At least in smaller birds, the lack of firm knowledge depends partly on the difficulty of following individual birds in detail for several migrations. The dramatic long-lasting effects that even relatively local climate perturbations can have on migratory timing throughout migration was shown in a recent study on red-backed shrikes, *Lanius collurio* (Fig. 5.4), and thrush Nightingales, *Luscinia luscinia* (Tøttrup et al. 2012). In spring 2011, a strong drought plagued the Horn of Africa. Individual shrikes and nightingales, tracked by geolocators (Box 5.1), were delayed in this region for on average an additional 9 days compared to the non-drought years of 2010 and 2012, probably because the fuel deposition necessary for the next leg of migration took much longer due to food shortage. The birds did not manage to get back on schedule and the arrival on the breeding grounds about 6,000 km to the north was strongly delayed (Tøttrup et al. 2012).

In a corresponding way to the spring passage, the timing of migrations away from the breeding grounds has in some cases also been affected by climate change (Jenni and Kéry 2003). Interestingly, the shift in autumn migration phenology of birds passing the Swiss Alps differs, depending on the life history of the species. Jenni and Kéry (2003) found that long-distance migrants had advanced their autumn migration in recent decades. Most of these species seem to leave their breeding grounds as soon as breeding is completed. Given an earlier arrival and a fairly constant time spent breeding, departure could be advanced accordingly. In contrast, short-distance migrants had delayed their autumn migration. Most interestingly, species with a variable number of broods generally delayed their passage, possibly because they engage in repeated breeding more frequently.

The effects of climate change on the timing of migration are by no means unique to birds (Parmesan 2007). Over a 34-year period (1972–2005) where temperature increased on average 2°C in an Alaskan river, the upstream migration of adult pink salmon, *Oncorhynchus gorbuscha*, shifted to occur on average 8 days earlier (Taylor 2008). The shift in the downstream migration of newly hatched fry the following spring was even larger, occurring on average 17 days earlier at the end of the study period (Table 3.1). The latter was because of the joint effect of earlier spawning of adults and earlier hatching due to warmer water during incubation. In the same population of salmon, a microevolutionary shift towards earlier migration has been documented. A genetic marker for late migration declined substantially over several decades, suggesting lower fitness in 'late' individuals and a true microevolutionary change in migratory behaviour (Kovach et al. 2012). Such a general pattern of earlier migration, and in addition, a shorter overall migration period, was found in four out of five salmon species inhabiting the same river (Kovach et al. 2013). Despite these rather substantial changes in migration timing, the abundance of adult salmon has not (yet) changed. But there may be other ecological effects, since salmons are a key nutrient source for many other organisms, and the overall time salmon are within the freshwater system has decreased with almost a month over four decades (Kovach et al. 2013). Moreover, there is not always a simple and intuitive correspondence between short-term phenological change and population size (Johansson and Jonzén 2012b).

Also many amphibians making vernal reproductive migrations to wetlands have altered the timing of their journeys in step with temperature changes (in Europe: Chadwick et al. 2006; UK: Beebee 2002; Japan: Kusano and Inoue 2008). Analysis of a 30-year data set of amphibians breeding in the southeast USA showed that four out of ten species studied had significantly shifted the timing of migration (Todd et al. 2011). Two species advanced reproductive migration (as has been reported in many amphibian species), and two species delayed migration and hence reproduction. The estimated rates of phenological change reported in this study of up to 37.2 days per decade represent some of the greatest rates of change reported to date, in any taxa (Parmesan 2007, Todd et al. 2011; Table 3.1).

No doubt, weather and climate are important factors shaping and influencing many spatial aspects of migration (Alerstam 1990). However, there are fewer studies showing how climate change affects spatial aspects of migration, as compared to temporal effects. Many Arctic-breeding birds make a final long flight to their breeding grounds (Alerstam 1990). One reason for these long flights could be that there are no good near-Arctic stopover sites available, mainly because northern springs are short and much of the Arctic becomes available within a very short time period only. It has been predicted that in a climate-warming scenario, new stopover sites should form closer to the Arctic if spring conditions become more favourable (Lindström and Agrell 1999). Such a change may have recently taken place in the southern Baltic Sea region. Over the past few decades where springs have become warmer, dunlins, *Calidris alpina*, wintering around the North Sea and bound for Arctic Russia have started to explore a new stopover site that is 400 km closer to the breeding grounds than the previously used final stopover sites (Stedt and Lindström 2012). In response to a recent rise in spring temperatures, Bauer et al. (2008) found that pink-footed geese, *Anser brachyrhynchus*, in northern Europe have started to depart earlier from their wintering grounds in Denmark and are staying longer at their stopover sites in Norway. Thus, the relative use of already established stopover sites has changed.

A migrating animal must time its migration to fit into the schedule of other important life stages, such as reproduction. If a changing climate affects all stages similarly, and the animal can respond to it, this may not be a big problem. However, if climate changes at different rates in different parts of the life or annual cycle, or if the animals cannot adjust, there may be temporal mismatches between key events, such as between the peak food demand of the young and food abundance. When this occurs the animal's life or annual cycle is no longer (optimally) synchronized to crucial events within its environment. This is what seems to have happened to pied flycatchers, *Ficedula hypoleuca*, in the Netherlands (Both et al. 2006). Here, spring has arrived at gradually earlier dates in recent decades, and with that, the burst of insect larvae that the flycatchers feed their nestlings on. The flycatchers, however, have not advanced their arrival at the same rate. And since it takes roughly the same amount of time from arrival (via egg laying) to the time when the nestlings demand most food, a temporal mismatch has now emerged. For many flycatchers, but not all, the short food peak is already over when the chicks' food demands are at their highest. Both et al. (2006) compared nine populations of flycatchers in The Netherlands and found that the populations breeding in areas with an early caterpillar peak were declining, whereas in populations breeding where the caterpillar peak was late the population declines were much smaller.

However, it may not be as bad for all pied flycatchers in Europe. In Finland, for example, the arrival of flycatchers seems to be highly dependent on the spring weather conditions in central Europe, their last stopover area before making a dash for Finland (Ahola et al. 2004). In recent times the springs have become increasingly warmer there, which has led to an advanced arrival of flycatchers in Finland. At the same time, the spring climate in Finland did not change. Therefore, and in contrast to The Netherlands, the pied flycatchers have been arriving earlier relative to the crucial food peak (Ahola et al. 2004). Whether this is to their advantage or disadvantage is not known, but it is a lucid example on how even strong and clear effects of climate change cannot easily be generalized across geographic scales. One reason is that the expected phenological response is very much dependent on what time of the season there has been a change, and that varies across space (Both and te Marvelde 2007, Both 2010). Using evolutionary game theory to model phenological adaptation in territorial organisms, Johansson and Jonzén (2012a) showed that early warming (affecting breeding success through, e.g., physiological status) has much stronger effect than later warming (affecting the timing of the resource peak needed to feed the young) on the evolutionary stable strategy for spring arrival. Hence, even if the fitness consequences of mismatch with an important phenophase may be severe, changed phenology may not necessarily be selected for (Visser et al. 2012).

Many birds only migrate as far as is strictly necessary to avoid snow-covered ground and ice-covered water (Alerstam 1990). If winters get milder, as they did in recent decades in Europe, these conditions are met at more northerly latitudes. Accordingly, most examples of recent changes in bird migration behaviour are for species that changed their wintering areas northward (Sutherland 1998). As far as the breeding distribution is concerned, there is a large amount of information on how the distribution of birds and other animals have changed in response to recent climate (Parmesan 2006). In Europe the picture is rather clear and consistent (Devictor et al. 2012; Fig. 3.2.). From 1990 to 2008 a given summer temperature moved about 250 km northward in Europe. Over the same period bird communities have only moved about 37 km northward. In Sweden the corresponding shifts from 1975 to 2009 were a 297-km northward shift in temperature and a 105-km shift in bird communities (Lindström et al. 2013). Clearly, birds are lagging behind climate change here. A majority of the birds involved are migrants so, clearly, the capacity to migrate long distances does not preclude a lag in relation to climate. Interestingly, butterflies seem to respond faster, having moved on average 114 km northward over the same period (Devictor et al. 2012; Fig. 3.2), which shows that birds and butterflies do not respond the same way to a given climate change. Devictor et al. (2012) suggest that the more accurate response of butterflies may depend on their relatively short life cycles and that they are ectothermic, enabling (or forcing) them to track changes in temperature regimes very closely. Most birds rely in summer on invertebrates like butterfly caterpillars. If the pattern found for

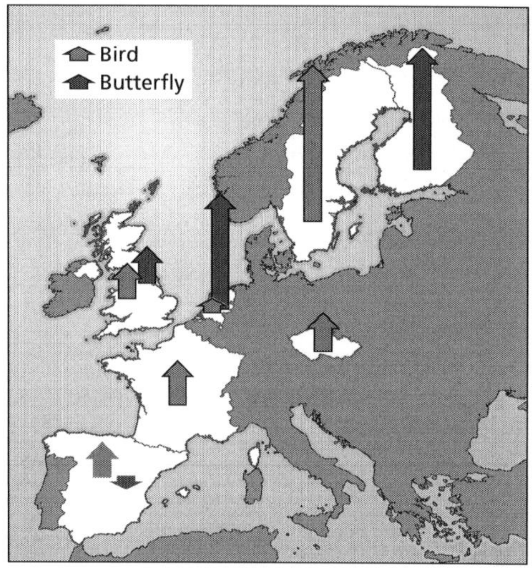

Figure 3.2 European variations in the temporal trend of bird and butterfly CTI (community temperature Index) for each country. The height of an arrow is proportional to the trend and its direction corresponds to the sign of the slope (from south to north for positive trends). Arrows with solid borders indicate significant trends (Devictor et al. 2012).

butterflies is paralleled by similar changes in other invertebrates, there is a risk for a spatial mismatch between the birds and their preferred prey (in addition to the temporal mismatch described earlier). This could have consequences for bird populations.

So why do some animals lag behind climate change? Why is it that birds do not keep up with butterflies? Is it their longer generation time (cf. Devictor et al. 2012)? Clearly, both birds and butterflies are capable flyers, and it would only require one or a few hours of additional flight to keep up with the northward shift in a given temperature regime. First, it remains to be shown whether the lag behind climate is maladapative (Lindström et al. 2013), or if the animals make an adaptive compromise (cf. the temporal arrival mismatch, Jonzén et al. 2007, Visser et al. 2012). However, adaptive responses could also have negative demographic consequences and even result in population declines (Johansson and Jonzén 2012b), especially if there is frequency-dependent selection involved. What we find when analysing migration in relation to climate change is not always expected. Given that migration is often seen as a way of escaping periods of hardship at the breeding grounds, such as cold winters, there are reasons to believe that among partial migrants, where a proportion of individuals stay and the rest migrate (Nilsson et al., Chapter 6; Chapman et al., Chapter 2), warmer winters would lead to fewer migrants. In a study of blue tit, *Parus caeruleus*, a common partial migrant in southern Sweden, Nilsson et al. (2006) found that the proportion of blue tits migrating each year did not change over the study period, or possibly even increased slightly, despite increasing winter temperatures. The most important factors determining the intensity of blue tit migration in a given year was the size of an important winter food source, the beech, *Fagus silvatica*, mast crop (more migrants at lower crops), and the size of the breeding population (more migrants at higher densities), leaving no or little room for the climate to affect the proportion of migrants.

3.3 Changing biotic interactions

Globally, many changes are taking place that directly affect the interactions between animals and their food, and such changes can have profound effects upon migrants. In a recent review of climate change-related causes of local extinctions among animals and plants, the most common identified proximate cause was not the expected abiotic changes, such as higher temperatures, but indeed interactions with other taxa (Cahill et al. 2013). Clearly, changing biotic interactions have their origin in habitat and climate change, but here we will focus on examples where the effects on one species can be directly related to the changing numbers or behaviour of other species.

Imagine you are a small bird. From your very first days in life, to your very last, you will be keeping an eye (or both) open for animals that want to eat you. Now, also imagine that the number of predators around you is constantly increasing. In fact, this is what recently has happened to small birds in large parts of the northern hemisphere. Most bird-eating raptors were decimated to very low numbers in the mid-twentieth century due to pollutants and persecution by man. In the past few decades we have seen a remarkable recovery in predator numbers. The comeback of one prominent predator on birds, the peregrine falcon, *Falco peregrinus*, may have had profound effects on the migration strategies of some North American shorebirds (Ydenberg et al. 2004). At the Pacific coast of British Columbia the number of peregrines seen during spring and late autumn migration increased by on average 2–6 times over a period of 16 years. In parallel, the average body mass of the western sandpipers stopping over in this area declined about 7–10% and the birds stopped for a progressively shorter time. The reduction in body mass, established through shorter periods of fuelling, may be a way of increasing individual agility and the chance to escape a raptor attack (Ydenberg et al. 2004).

Migratory animals are often key players within intricate food webs, and hence disruptions to migratory patterns can have cascading biotic effects with profound impacts upon other species. Take the iconic wildebeest migration through East Africa as an example. Wildebeest provide food for predators along the migratory route, and influence plant community structure by grazing. Some evidence suggests that wildebeest foraging stimulates the new-growth grasses preferred by small-bodied migratory grazers, such as Thomson's gazelles, *Eudorcas thomsonii*, facilitating their migration. Furthermore, many animals perish during the arduous migration. As much as 70% of the wildebeest carcasses are eaten by vultures, providing a key source of food for these scavengers. Reductions in the numbers of migrating wildebeest can thus have adverse effects on vulture populations (Virani et al. 2011), which can have cascading effects further down the ecosystem, as vultures provide essential ecosystem services such as waste removal, disease control, and nutrient cycling.

Hunting by humans has also been shown to influence migratory behaviour. In 1999 and 2000 a spring hunt of greater snow geese, *Anser caerulescens atlanticus*, occurred in North America for the first time since 1916 in an attempt to control the rising population size of this species. In the years without the hunting, migration was strictly unidirectional from south-west to north-east. When hunting was resumed this drove a shift in the direction of goose migration, with birds being four times more likely to migrate westward during hunting years (Bechet et al. 2003). Hence hunting altered the stopover scheduling of snow geese, and triggered a shift in their migratory behaviour as birds changed course to avoid the hunters.

Indirect effects of humans upon migrants can occur in cases where predators are adversely affected by human activities. For example, predators such as wolves avoid human habitation, which in turn creates 'safe zones' for prey such as ungulates. A study of partially migratory elk (partial migration: see Chapman et al., Chapter 2) showed that individuals that adopted the resident strategy (which is usually associated with a higher predation risk) reduced their vulnerability to wolf predation during the migratory period by foraging near human settlements (Hebblewhite and Merrill 2009; Table 3.1). Hence, indirectly, human disturbance acted to increase the fitness of resident elk, which may partially explain the huge declines in the proportion of migratory elk in this population that have been documented since the 1980s (Hebblewhite and Merrill 2011).

Earlier in this chapter we mentioned the Arctic-breeding migratory geese, where increases in goose populations due to agricultural intensification on the wintering grounds led to habitat degradation and dwindling passerine numbers on the breeding grounds (Jefferies et al. 2004). These cascading effects might be more extreme still. Lemming cycles greatly affect the functioning of Arctic ecosystems, and food-web interactions are considered important drivers for these cycles. Nevertheless, climate change effects are currently thought to be the driving force behind fading lemming cycles. Alternatively, however, increased Arctic goose numbers may play a role here too, assisting in dampening predator cycles. Because goose eggs and chicks are present each summer, predators can reproduce better in low lemming years and pose a more constant threat to both goose and other bird populations (Y. Aharon-Rotman et al., unpublished). That the effects of habitat change on one migratory species can influence the fate of other migrants thousands of kilometres away is a good example of the intricate relationship between habitat change, climate change, species interactions, and the fate of migratory animals.

A key biotic (and anthropogenic) interaction which is important to many migratory fishes (particularly partially anadromous species: Chapman et al. 2012; Chapman et al., Chapter 2) is when humans act as extremely effective and abundant predators. For anadromous migrants that migrate from freshwater to sea, human exploitation exerts an immense pressure, and has led to massive declines in fish stocks and even local extirpations. Migratory species, including many salmonids, are particularly threatened as they often show predictable spatial and temporal movements which make them amenable to efficient capture. Moreover, the high harvest rates, with humans selecting against larger-bodied and slow maturing individuals, have led to evolutionary effects promoting reproduction at an increasingly younger age (Kuparinen and Merilä 2007). For partially migratory species evolution towards residency can occur, and some migratory forms of species can dwindle in numbers until at real risk of extinction (e.g. bull charr, *Salvelinus confluentus*, Nelson et al. 2002).

3.3.1 Coping with change

The many regular and seemingly fixed patterns of movement and migration we see today have evolved over thousands of years. Clearly, many animals can adapt to a changing environment, notably if these are slow and gradual, allowing for acclimatization, learning, and evolutionary adaptation. Major historic global changes, such as the quaternary glaciations, occurring 10,000 to 1 million years ago, have had a major impact on the evolution of, for example, bird migration as we know it today (Moreau 1952). But then there were not 7 billion world citizens around, causing very rapid changes that are diverse in their nature and often acting in concert. Are migrants also able to cope with these huge contemporary challenges?

Basically there are two mechanisms by which migrants may cope with change. Firstly, through adaptive behavioural and physiological flexibility. Secondly, through evolutionary adaptation involving natural selection. Earlier we have seen a considerable number of examples of how some migratory species deal with their changing environment and how some do not. But can we explain these discrepancies between species' capacity to adjust, be it through flexibility or evolutionary adaptation? Can we categorize migrants based on some phylogenetic, behavioural, anatomical, physiological, or ecological characteristics into those that are more or less vulnerable to global change? If we could, that would greatly facilitate identifying threatened species and guide conservation efforts.

To start the evaluation of migrants' vulnerability to change on a positive note: phylogenetically, migrants, nomadic species, and non-migrants are often highly related. Many partial migrants, where some individuals of a population migrate and others do not, also exist. This lends credence to the idea that changing the genetic programme for migration may be relatively easily and quickly achieved. And indeed, experiments on the genetics of bird migration also indicate that changes can rapidly evolve (Berthold et al. 1992; Liedvogel and Lundberg, Chapter 12). Still, evolutionary processes take place over evolutionary time and for notably long-lived species, species that are towards the K-selection side of the r/K-selection continuum, the pace of change may be too fast for an evolutionary process to keep up with. Unfortunately, the earlier-mentioned shorebirds along the East-Asian Australasian flyway, where global change takes place at an unprecedented pace (Fig. 3.1), are long-lived and widely considered to be in jeopardy. For long-lived species adaptive flexibility may therefore be a better mechanism to cope with change. And possibly, precisely because of their longevity, these animals may have evolved more flexibility to cope with change than shorter lived species. Indeed, some long-lived migrants have a card up their sleeve to play when conditions change. Among geese and swans, juveniles accompany and learn from adults the art of migration and therefore instantly profit from the experience acquired by the adults during previous migrations. Not surprisingly it is among these species we find most examples of recent rapid changes in migratory behaviour (Sutherland 1998). In contrast, juvenile Arctic shorebirds migrate independently from adults, implying a strong genetic basis for migratory behaviour at least during the first autumn migration. From the first winter onwards they mix with adults and could then potentially benefit from their experience. However, any change in the migratory behaviour of these juveniles would have to arise through natural selection and microevolution.

Now what the future will *exactly* bring for these shorebirds along the East-Asian Australasian flyway, and any migrant confronted with a shifting environment, is hard to say. We have described how animals on the move can be very vulnerable to environmental change, and that some of the responses are highly complex. The case of the landlocked alewives leading to cascading effects through the ecosystem and an evolutionary shift in their zooplankton prey life history and size (Walsh and Post 2011) is a beautiful example of how complicated ecology can be, making it hard to predict firmly the consequences of habitat change. With respect to climate change, the analysis of a 30-year data set of amphibians breeding in the south-eastern USA showed that even highly related species may exhibit opposing responses to change (Todd et al. 2011)—again, a nice example of how difficult it is to make firm predictions on what will happen in individual cases. But by discerning trends we can at least identify the *chances*, the *risks*, with which migrants may go extinct or at least be gravely reduced in numbers. Indeed, it is most likely that the passenger pigeon, *Ectopistes migratorius*, will remain the exception and that the majority of now-abundant migratory species that walk the land, use the skies, and swim in the oceans will 'only' face severe population declines rather than extinction.

Returning now to the issue of who may be best at coping with change. From what we know about the speed of evolutionary processes we thus expect species that are towards the K-selected site of the r/K-selection continuum to be most at risk, and this means species that are often large, long-lived, and endothermic. To make things worse, many long-distance migrants, i.e. those that move over larger distances and are therefore bound to encounter changing conditions, belong to this group. Considering the speed and energy cost of migratory journeys for insects, fish, birds, and mammals of a wide range of sizes (Johansson et al., Chapter 13), travelling by flapping or soaring flight, swimming or running, it appears that really long annual migrations are unlikely to be beneficial for all except for marine mammals and flying birds (McNeill Alexander 2002). Further, at least among birds, the fuelling rate and therefore migration speed (the number of kilometres that can be travelled per day, including the periods of fuel deposition) declines with body mass (Lindström 1991; Hedenström and Lindström, Chapter 5; Johansson et al., Chapter 13). This means that the larger the bird species, the smaller its margins, and the more sensitive it will be to deteriorating conditions. But we should always be wary of exceptions. Sea turtles, for instance, are ectotherms, neither mammal nor bird, but largely migratory, long-lived and all turtle species, without exception, are regarded as threatened and in danger of extinction.

What other generalities can we possibly draw when it comes to identifying species most at risk? When reviewing responses to habitat change we have already suggested that habitat specialists are probably more at risk than habitat generalists. We have also suggested that long-distance migrants might be more at risk, given that during their lifetime they visit more sites than residents. If disturbance at one single site throughout the year is enough to have overall negative effects, they are under increased risk by chance alone. In a study of French birds, Moussus et al. (2011) found that species with the broadest ecological and thermal niches (i.e. habitat generalists) and the shortest migration distances were most able to adjust their breeding phenology to temperature variations. Accordingly, the authors also suggested that long-distance migrants and ecological specialists are more likely to suffer from global change. In line with these hypotheses, long-distance migratory birds were found to have declined on average more in numbers in Europe between 1970 and 2000 than short-distance migrants and residents (Sanderson et al. 2006).

There may also be a relation between vulnerability to change, and notably habitat change, and mode of locomotion. Aquatic and terrestrial migrants, those that swim, run, or walk, largely need areas of continuous suitability to complete their migrations. Any ecological barriers, such as dams in rivers, heavily polluted areas, roads, and build-up areas, may greatly impair their migration and even halt it. For flying animals to cross ecological barriers is less of a problem. Indeed, some ecological barriers, such as deserts and oceans, are naturally crossed by many long-distant migratory birds, some of which with spectacularly long flights (Gill et al. 2009, Klaassen et al. 2011). What is additional bad news for terrestrial migrants is that their cost of transport (i.e. the energy needed to cover a certain distance; Johansson et al., Chapter 13) is much higher than in flyers and even close to an order of magnitude higher than in swimmers (Schmidt-Nielsen 1972). Of all migrants, terrestrial migrants therefore also need much more uniformly distributed food resources along their migration path and are less capable of exploiting foraging hotspots. Habitat fragmentation and deterioration should therefore also be particularly bad news for terrestrial migrants.

Social behaviour may be crucial for the successful completion of a migratory journey. Being in a group may for example allow for better (communal) decisions in orientation and navigation (Simons 2004). If migrants critically rely on conspecifics, this bears the risk that if numbers fall below a critical mass, the population will spiral down in numbers. This is called the Allee effect, after its discoverer, Warder Clyde Allee. A related problem for migratory fish of commercial interest, where high harvest rates result in a reduction of the average age of the fish within the school, was highlighted by Huse et al. (2002). They suggested that the numerical domination of inexperienced individuals may lead to aberrant migration, but equally into the development of new migration routes.

Now we have addressed a number of characters of migrants that may make it more easy, and therefore likely, to cope with global change, we now turn to the extent to which adaptive change is really needed. We have seen that some migrants need to arrive earlier on their breeding grounds in spring to make best use of a short peak in food availability to raise their offspring. Adjustments in the timing of migratory and subsequent breeding schedules in a number of bird species in relation to climate change-driven changes in food availability were apparently imperfect (Visser and Both 2005). However, hitting the food peak is not the only important aspect of a migrant's life. For example, there could be competition for territories, as well as an increased risk of mortality from arriving earlier. Based on a theoretical model, Jonzén et al. (2007) predicted that the optimal shift in arrival date should never be as extreme as the shift in food peak date. One should not take this moderated necessity for change too far though. Arguing that coping with climate change would 'merely' require a northward shift in distribution of populations adapted to warmer environments is too simplistic. Adaptation via immigration of individuals with genotypes adapted to warmer environments will also require microevolution as these genotypes are not adapted to the photoperiod (i.e. the main cue organisms use to forecast future availability of important resources) of their new environment. To what extent such an adaptation to the new photoperiod must happen is, by the way, a complex matter, greatly depending on the exact relationship and reliability of the cue and the optimal timing, and how this relationship will be affected by environmental changes (McNamara et al. 2011).

Acknowledgements

We thank Christiaan Both for commenting upon an earlier draft of the text and Ryan P. Kovach for insightful comments on salmon migration.

References

Abraham, K. F., Jefferies, R. L., and Alisauskas, R.T. (2005). The dynamics of landscape change and snow geese in mid-continent North America. *Global Change Biology*, **11**, 841–55.

Ahola, M., Laaksonen, T., Sippola, K., Eeva, T., Rainio, K., and Lehikoinen, E. (2004). Variation in climate warming along the migration route uncouples arrival and breeding dates. *Global Change Biology*, **10**, 1610–17.

Alerstam, T. (1990). *Bird Migration*. Cambridge University Press, Cambridge.

Altizer, S., Bartel, R., and Han, B. A. (2011). Animal migration and infectious disease risk. *Science*, **331**, 296–302.

Battley, P. F., Warnock, N., Tibbitts, T. L. et al. (2012). Contrasting extreme long-distance migration patterns in bar-tailed godwits *Limosa lapponica*. *Journal of Avian Biology*, **43**, 21–32.

Bauer, S., Van Dinther, M., Høgda, K.-A., Klaassen, M., and Madsen, J. (2008). The consequences of climate-driven stop-over sites changes on migration schedules and fitness of Arctic geese. *Journal of Animal Ecology*, **77**, 654–60.

Beamish, R. J., and Northcote, T. G. (1989). Extinction of a population of anadromous parasitic lamprey, *Lampetra tridentata*, upstream of an impassable dam. *Canadian Journal of Fisheries and Aquatic Sciences*, **46**, 420–5.

Bechet, A., Giroux, J.-F., Gauthier, G., Nichols, J. D., and Hines, J. E. (2003). Spring hunting changes the regional movements of migrating greater snow geese. *Journal of Applied Ecology*, **40**, 553–64.

Beebee, T. J. C. (2002). Amphibian phenology and climate change. *Conservation Biology*, **16**, 1454.

Berthold, P., Helbig, A. J., Mohr, G., and Querner, U. (1992). Rapid microevolution of migratory behaviour in a wild bird species. *Nature*, **360**, 668–70.

Bolger, D. T., Newmark, W. D., Morrison, T. A., and Doak, D. F. (2008). The need for integrative approaches to understand and conserve migratory ungulates. *Ecology Letters*, **11**, 63–77.

Both, C. (2007). Comment on 'Rapid advance of spring arrival dates in long-distance migratory birds'. *Science*, **315**, 598.

Both, C. (2010). Flexibility of timing of avian migration to climate change masked by environmental constraints en route. *Current Biology*, **20**, 243–8.

Both, C., Bouwhuis, S., Lessells, C. M., and Visser, M. E. (2006). Climate change and population declines in a long-distance migratory bird. *Nature*, **441**, 81–3.

Both, C. and te Marvelde, L. (2007). Climate change and timing of avian breeding and migration throughout Europe. *Climate Research*, **35**, 93–105.

Boyd, H., and Madsen, J. (1997). Impacts of global change on Arctic-breeding bird populations and migration. In W. C. Oechel et al. (eds), *Global Change and Arctic Terrestrial Ecosystems*, pp. 201–7. Springer, New York.

Cahill, A. E., Aiello-Lammens, M. E., Fisher-Reid, M. C., et al. (2013). How does climate change cause extinction? *Proceedings of the Royal Society B: Biological Sciences*, **280**, 20121890.

Chadwick, E. A., Slater, F. M., and Ormerod, S. J. (2006). Inter- and intraspecific differences in climatically mediated phenological change in coexisting Triturus species. *Global Change Biology*, **12**, 1069–78.

Chapman, B. B., Hulthén, K., Brodersen, J., et al. (2012). Partial migration in fishes: causes and consequences. *Journal of Fish Biology*, **81**, 456–78.

Devictor, V., van Swaay, C., Brereton, T., et al. (2012). Differences in the climate debts of birds and butterflies at a continental scale. *Nature Climate Change*, **2**, 121–4.

Gill, R. E., Tibbitts, T. L., Douglas, D. C., et al. (2009). Extreme endurance flights by landbirds crossing the Pacific Ocean: ecological corridor rather than barrier? *Proceedings of the Royal Society B: Biological Sciences*, **276**, 447–58.

Gresh, T., Lichatowich, J., and Schoonmaker, P. (2000). An estimation of historic and current levels of salmon production in the Northeast Pacific ecosystem: evidence of a nutrient deficit in the freshwater systems of the Pacific Northwest. *Fisheries*, **25**(1), 15–21.

Hebblewhite, M., and Merrill, E. H. (2009). Trade-offs between predation risk and forage differ between migrant strategies in a migratory ungulate. *Ecology*, **90**, 3445–54.

Hebblewhite, M., and Merrill, E. H. (2011). Demographic balancing of migrant and resident elk in a partially migratory population through forage—predation trade-offs. *Oikos*, **120**, 1860–70.

Huse, G., Railsback, S., and Feronö, A. (2002). Modelling changes in migration pattern of herring: collective behaviour and numerical domination. *Journal of Fish Biology*, **60**, 571–82.

Ito, T. Y., Miura, N., Lhagvasuren, B., et al. (2005). Preliminary evidence of a barrier effect of a railroad on the migration of Mongolian gazelles. *Conservation Biology*, **19**, 945–8.

Jefferies, R. L., Rockwell, R. F., and Abraham, K. F. (2004). Agricultural food subsidies, migratory connectivity and large-scale disturbance in arctic coastal systems: a case study. *Integrative and Comparative Biology*, **44**, 130–9.

Jenni, L., and Kéry, M. (2003). Timing of autumn bird migration under climate change: advances in long-distance migrants, delays in short-distance migrants. *Proceedings of the Royal Society B: Biological Sciences*, **270**, 1467–71.

Johansson, J., and Jonzén, N. (2012a). Effects of territory competition and climate change on timing of arrival to breeding grounds: a game-theory approach. *American Naturalist*, **179**, 463–74.

Johansson, J. and Jonzén, N. (2012b). Game theory sheds new light on ecological responses to current climate change when phenology is historically mismatched. *Ecology Letters*, **15**, 881–8.

Jonzén, N., Hedenström, A., and Lundberg, P. (2007). Climate change and the optimal arrival of migratory birds. *Proceedings of the Royal Society B: Biological Sciences*, **274**, 269–74.

Jonzén, N., Lindén, A., Ergon, T., et al. (2006). Rapid advance of spring arrival dates in long-distance migratory birds. *Science*, **312**, 1959–61.

Klaassen, R. H. G., Alerstam, T., Carlsson, P., Fox, J. W., and Lindström, Å. (2011). Great flights by Great Snipes: long and fast non-stop migration over benign habitats. *Biology Letters*, **7**, 833–5.

Klaassen, M., Hoye, B. J., Nolet, B. A., and Buttemer, W. A. (2012). Ecophysiology of avian migration in the face of current global hazards. *Philosophical Transactions of the Royal Society B: Biological Sciences*, **367**, 1719–32.

Knudsen, E., Lindén, A, Both, C., et al. (2011). Challenging claims in the study of migratory birds and climate change. *Biological Reviews*, **86**, 928–46.

Kovach, R. K., Gharrett, A. J., and Tallmon, D. A. (2012). Genetic change for earlier migration timing in a pink salmon population. *Proceedings of the Royal Society B: Biological Sciences*, **279**, 3870–8.

Kovach, R. K., Joyce, J. E., Echave, J. D., Lindberg, M. S., and Tallmon, D. A. (2013). Earlier migration timing, decreasing phenotypic variation, and biocomplexity in multiple salmonid species. *PLoS One*, **8**(1), e53807.

Kuparinen, A., and Merilä, J. (2007). Detecting and managing fisheries-induced evolution. *Trends in Ecology and Evolution*, **22**, 652–9.

Kusano, T., and Inoue, M. (2008). Long-term trends toward earlier breeding of Japanese amphibians. *Journal of Herpetology*, **42**, 608–14.

Lehikoinen, E., Sparks, T. H., and Zalakevicius, M. (2004). Arrival and departure dates. *Advances in Ecological Research*, **35**, 1–31.

Lindström, Å. (1991). Maximum fat deposition rates in migrating birds. *Ornis Scandinavica*, **22**, 12–19.

Lindström, Å., and Agrell, J. (1999). Global change and possible effects on the migration and reproduction of arctic-breeding waders. *Ecological Bulletins*, **47**, 145–59.

Lindström, Å., Green, M., Paulson, G., Smith, H. G., and Devictor, V. (2013). Rapid changes in bird community composition at multiple spatial scales in response to recent climate change. *Ecography*, **36**, 313–22.

McNamara, J. M., Barta, Z., Klaassen, M., and Bauer, S. (2011). Cues and the optimal timing of activities under environmental changes. *Ecology Letters*, **14**, 1183–90.

McNeill Alexander, R. (2002). The merits and implications of travel by swimming, flight and running for animals of different sizes. *Integrative and Comparative Biology*, **42**, 1060–4.

Moreau, R. E. (1952). The place of Africa in the Palaearctic migration system. *Journal of Animal Ecology*, **21**, 250–71.

Morita, K., Yamamoto, S., and Hoshino, N. (2000). Extreme life history change of white-spotted char (*Salvelinus leucomaenis*) after damming. *Canadian Journal of Fisheries and Aquatic Sciences*, **57**, 1300–6.

Moussus, J.-P., Clavel, J., Jiguet, F., and Julliard, R. (2011). Which are the phenologically flexible species? A case study with common passerine birds. *Oikos*, **120**, 991–8.

Mustin, K., Sutherland, W. J., and Gill, J. A. (2007). The complexity of predicting climate-induced ecological impacts. *Climate Research*, **35**, 165–75.

Møller, A. P., Rubolini, D., and Lehikoinen, E. (2008). Populations of migratory bird species that did not show a phenological response to climate change are declining. *Proceedings of the National Academy of Sciences USA*, **105**, 16195–200.

Nelson, M. L., McMahon, T. E., and Thurow, F. (2002). Decline of the migratory form in Bull Charr, *Salvelinus confluentus*, and implications for conservation. *Environmental Biology of Fishes*, **64**, 321–32.

Nilsson, A. L. K., Lindström, Å., Jonzén, N., Nilsson, S. G., and Karlsson, L. (2006). The effect of climate change on partial migration—the blue tit paradox. *Global Change Biology*, **12**, 2014–22.

Parmesan, C. (2006). Ecological and evolutionary responses to recent climate change. *Annual Review of Ecology, Evolution and Systematics*, **37**, 637–69.

Parmesan, C. (2007). Influences of species, latitudes and methodologies on estimates of phenological response to global warming. *Global Change Biology*, **13**, 1–13.

Partecke, J., and Gwinner, E. (2007). Increased sedentariness in European blackbirds following urbanization: a consequence of local adaptation? *Ecology*, **88**, 882–90.

Rogers, D. I., Yang, H.-Y., Hassell, C. J., et al. (2010). Red Knots (*Calidris canutus piersmai* and *C. c. rogersi*) depend on a small threatened staging area in Bohai Bay, China. *Emu*, **110**, 307–15.

Sanderson, F. J., Donald, P. F., Pain, D. J., Burfield, I. J., and van Bommel, F. P. J. (2006). Long-term population declines in Afro-Palearctic migrant birds. *Biological Conservation*, **131**, 93–105.

Schmidt-Nielsen, K. (1972). Locomotion: energy cost of swimming, flying, and running. *Science*, **177**, 222–8.

Shochat, E., Warren, P. S., Faeth, S., McIntyre, N. E., and Hope, D. (2006). From patterns to emerging processes in mechanistic urban ecology. *Trends in Ecology and Evolution*, **21**, 186–91.

Simons, A. M. (2004). Many wrongs: the advantage of group navigation. *Trends in Ecology and Evolution*, **19**, 453–5.

Spinage, C. A. (1992). The decline of the Kalahari wildebeest. *Oryx*, **26**, 147–50.

Stedt, J., and Lindström, Å. (2012). Spring fuelling by arctic Dunlins *Calidris a. alpina* in the southern Baltic and the potential role of a changing climate. *Ornis Svecica*, **22**, 80–92.

Stervander, M., Lindström, Å., Jonzén, N., and Andersson, A. (2005). Timing of spring migration in birds: long-term trends, North Atlantic Oscillation and the significance of different migration routes. *Journal of Avian Biology*, **36**, 210–21.

Sutherland, W. J. (1998). Evidence for flexibility and constraint in migration systems. *Journal of Avian Biology*, **29**, 441–6.

Taylor, S. G. (2008). Climate warming causes phenological shift in Pink Salmon, *Oncorhynchus gorbuscha*, behavior at Auke Creek, Alaska. *Global Change Biology*, **14**, 229–35.

Todd, B. D., Scott, D. E., Pechmann, J. H. K., and Gibbons, J. W. (2011). Climate change correlates with rapid delays and advancements in reproductive timing in an amphibian community. *Proceedings of the Royal Society B: Biological Sciences*, **278**, 2191–7.

Tøttrup, A. P., Klaassen, R. H. G., Kristensen, M. W., et al. (2012). Drought in Africa caused delayed arrival of European songbirds. *Science*, **338**, 1307.

Verkuil, Y. I., Karlionova, N., Rakhimberdiev, E. N., et al. (2012). Losing a staging area: eastward redistribution of Afro-Eurasian ruffs is associated with deteriorating fuelling conditions along the western flyway. *Biological Conservation*, **149**, 51–9.

Virani, M. Z., Kendall, C., Njoroge, P., and Thomsett, S. (2011). Major declines in the abundance of vultures and other scavenging raptors in and around the Masai Mara ecosystem, Kenya. *Biological Conservation*, **144**, 746–52.

Visser, M. E., and Both, C. (2005). Shifts in phenology due to global climate change: the need for a yardstick. *Proceedings of the Royal Society B: Biological Sciences*, **272**, 2561–9.

Visser, M. E., te Marvelde, L., and Lof, M. E. (2012). Adaptive phenological mismatches of birds and their food in a warming world. *Journal of Ornithology*, 153 (Supplement **1**), S75–84.

Walsh, M. R., and Post, D. M. (2011). Interpopulation variation in a fish predator drives evolutionary divergence in prey in lakes. *Proceedings of the Royal Society B: Biological Sciences*, **278**, 2628–37.

Wilcove, D. S., and Wikelski, M. (2008). Going, going, gone: is animal migration disappearing. *PLoS Biology*, **6**(7), e188.

Yang, H.-Y., Chen, B., Barter, M., et al. (2011). Impacts of tidal land reclamation in Bohai Bay, China: ongoing losses of critical Yellow Sea waterbird staging and wintering sites. *Bird Conservation International*, **21**, 241–59.

Ydenberg, R. C., Butler, R. W., Lank, D. B., Smith, B. D., and Ireland, J. (2004). Western sandpipers have altered migration tactics as peregrine falcon populations have recovered. *Proceedings of the Royal Society B: Biological Sciences*, **271**, 1263–9.

Zhong, Y., and Power, G. (1996). Environmental impacts of hydroelectric projects on fish resources in China. *Regulated Rivers: Research and Management*, **12**, 81–98.

CHAPTER 4

Beyond dispersal: the role of animal movement in modern agricultural landscapes

Henrik G. Smith, Klaus Birkhofer, Yann Clough, Johan Ekroos, Ola Olsson, and Maj Rundlöf

Habitat loss, mostly caused by agricultural expansion, has resulted in local to global declines in biodiversity (Dirzo and Raven 2003, Pereira et al. 2012) and patterns of animal movements (Lindström et al., Chapter 3). As a consequence, conservation of biodiversity will increasingly depend on the management of human-dominated agricultural landscapes (Fischer et al. 2008, but see Phalan et al. 2011). In addition, biodiversity conservation in these landscapes is necessary to maintain ecosystem functioning important for human welfare (Cardinale et al. 2012). However, to successfully manage those ecosystems, it is important to understand how organisms react to the spatial and temporal variation in habitat suitability caused by human management. Many human activities modify habitat suitability for organisms, with negative effects on their populations (Stoate et al. 2009). To mitigate such effects, knowledge of their habitat requirements is needed, which is the focus of much of nature conservation research. However, not only habitat quality, but also its spatial and temporal variation will be affected by, for example, agricultural change. The mobility of organisms will affect their ability to handle these changes in spatiotemporal habitat suitability by shifting between habitats (Fig. 4.1). As a consequence, ecological processes at scales larger than the local patch, i.e. landscape composition and configuration, are crucial to understand the fate of organisms in these landscapes (Schweiger et al. 2005). Hence, by studying the consequences of landscape change for organisms with differing mobility (Fig. 4.1), we may gain understanding of the ecological mechanisms affecting population persistence.

Here we use agricultural changes in Europe and the concomitant loss of animal biodiversity as a framework to show how an understanding of organismal mobility contributes to the efficient conservation of biodiversity and associated ecosystem functions. After providing an overview of the different concepts of landscape suitability for mobile organisms, we review the role of movement in enabling organisms to maintain viable metapopulations in fragmented semi-natural habitat remnants (Hanski 1999, Tscharntke et al. 2002). Thereafter, we address two forms of multiple habitat usage: spillover between different habitat types, on which organisms in highly disturbed environments depend (Bianchi et al. 2006, Tscharntke et al. 2012), and the complementary use of different habitats (Dunning et al. 1992). We highlight the consequences of movement, and its limitations, for species persistence and ecosystem function. For the different sections we primarily give examples from groups where the processes have been particularly well studied, e.g. fragmentation effects on butterfly conservation, spillover effects on natural enemies of pests, and landscape complementation on pollinators, but the theories and their implications are valid across taxa. In particular, we point out the importance of a detailed understanding of how animal movement

Animal Movement Across Scales. Edited by Lars-Anders Hansson and Susanne Åkesson.
© Oxford University Press 2014. Published 2014 by Oxford University Press.

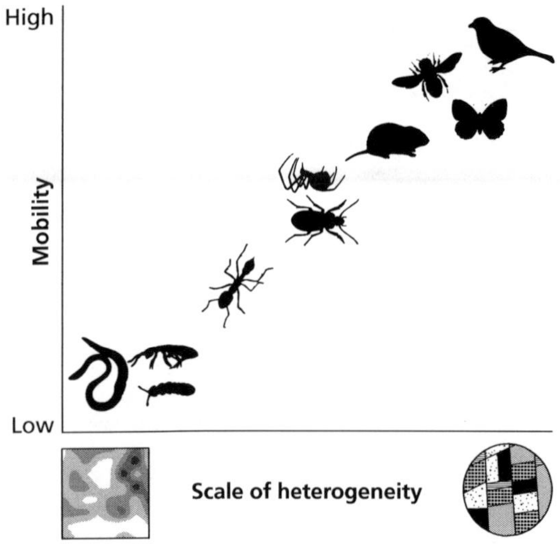

Figure 4.1 Schematic overview of the relationship between the mobility of different taxa and their sensitivity to heterogeneity at different spatial scales ranging from the local field to the landscape scale shown at left and right, respectively, of the *x*-axis.

contributes to their ability to cope with changes in habitat suitability, in order to design suitable measures to preserve functional biodiversity in these landscapes (Flynn et al. 2009).

4.1 Community disassembly and movement

In Europe, with its long tradition of agriculture, much biodiversity is connected to farmland. Not only does agriculture cover close to half of the terrestrial area (e.g. FAO 2003 in Rounsevell et al. 2006), but many species of conservation concern have a long history of cohabitation with agricultural habitats (Poschlod and WallisDeVries 2002). However, agricultural intensification has resulted in dramatic declines of biodiversity across a wide range of taxa, including plants (Kleijn et al. 2009a), pollinating insects (Biesmeijer et al. 2006), and birds (Donald et al.2001), although not all organisms suffer equally. Thus, while many species are declining or even threatened, some species still thrive in simplified, intensively managed farmed landscapes (e.g. some farmland birds, Dänhardt et al. 2010). Various studies have found that habitat specialists (e.g. Ekroos et al. 2010), dietary specialists (e.g. Goulson et al. 2005), large species (e.g. Burel et al. 2004), species at higher trophic levels (e.g. Rand et al. 2012), less mobile species (e.g. Jennings and Pocock 2009), and more mobile species (e.g. Öckinger et al. 2009) have suffered disproportionally from farmland change.

At a first glance these patterns may seem unrelated, but they may be explained by the way organisms react to changes in spatial and temporal variation in resource availability caused by agricultural intensification. The mechanisms by which populations survive in modified agricultural landscapes range from daily and seasonal movements by individuals to utilize scattered foraging resources in habitats varying in quality over time, to population persistence by colonization–extinction dynamics in metapopulation networks suffering from habitat fragmentation. Irrespective of whether we are dealing with individuals using multiple patches or dispersal between habitat patches, the mobility of organisms will affect whether they will be able to sustain changes in the fragmentation and homogenization of agricultural landscapes. The degree of mobility will determine at which spatial scale an organism operates (Fig. 4.1), which in turn affects the distance between habitat patches possible to use by animals (Fig. 4.2). Hence, agricultural landscapes may act as a filter in relation to animal mobility, but due to different processes acting at the level of individuals to populations.

Traditionally, biodiversity conservation in agricultural landscapes has focused on the preservation of semi-natural habitats surrounded by a matrix of less habitable agricultural land. Thus, when habitat loss from agricultural intensification causes fragments of semi-natural habitats to become smaller and more isolated, we expect biodiversity loss in the remaining habitat (Miller et al. 1997, Krauss et al. 2009, Öckinger et al. 2010). In addition, agricultural practices may cause the matrix to be more or less permeable for organisms, thus determining the effective isolation of remaining populations (Vandermeer and Carvajal 2001, Donald and Evans 2006) (Fig. 4.2). Clearly, the ability of organisms to persist in transformed agricultural landscapes may depend on their ability to disperse between

	Concept	Description	Process
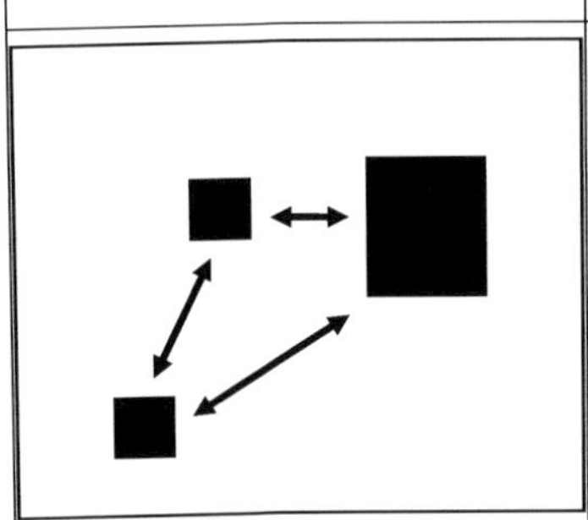	Habitat fragmentation, metapopulations and metacommunities	Binary, fragmented landscape divided between habitat and non-habitable matrix	Extinctions and dispersal-mediated recolonizations of local populations. *Time-scale*: between seasons
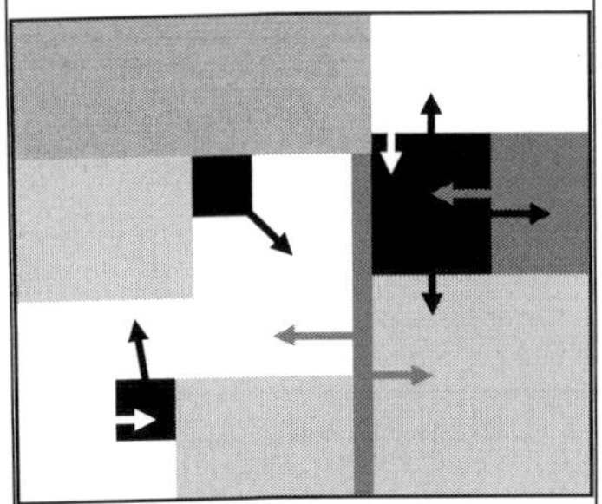	Spill-over	Landscapes with multiple habitats of different quality (mosaic landscapes)	Dispersal between different habitats which fluctuate in quality. *Time-scale*: within seasons
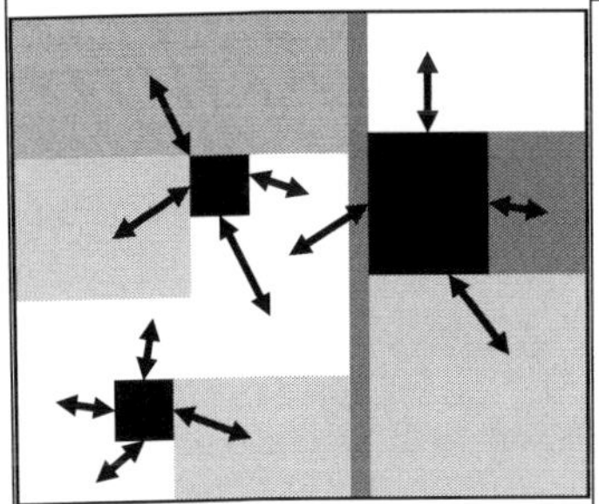	Landscape complementation / supplementation	Mosaic of habitats with complementary resources	Individual organisms using resources in different habitats (e.g. nesting in one habitat and foraging in other habitats). *Time-scale*: within days

Figure 4.2 An overview of concepts linking habitat use and animal movement at landscape scales. For three important concepts (metapopulation/community dynamics, spillover, and landscape complementation) the assumptions about landscapes (description) and the underlying ecological processes (process) are described.

remaining semi-natural habitat fragments, which in turn may be related to, e.g., their degree of specialization with respect to habitat or diet, as well as their size and mobility

However, while this view certainly has merits for many habitat specialists, it has also become increasingly clear that classical metapopulation theory may represent a special case of how organisms utilize mosaic landscapes (Bennett et al. 2006, Fronhofer et al. 2012). Agricultural landscapes consist of a mosaic of habitats of different suitability for organisms, which are linked by dispersal (Fig. 4.2). For example, many populations in the most disturbed habitats, such as regularly tilled agricultural fields, may only be sustained by immigration from more or less permanent semi-natural habitats, which is why the proportion of semi-natural habitats remaining at landscape scales may be instrumental for biodiversity conservation (Tscharntke et al. 2005). Hence, these systems may rather be described as mainland–island or source–sink systems, where populations in the more disturbed habitats are sustained by emigration from the less-disturbed habitats. Furthermore, many organisms may use resources in the so-called matrix (Dennis and Hardy 2006), and even reproduce in regularly disturbed landscapes to the extent that the amount and properties of these disturbed habitats may affect the abundance of organisms in remaining natural habitats (Rand et al. 2006). Hence, meta-community theory, aiming at understanding the composition of interacting communities linked by dispersal (Leibold et al. 2004), is a suitable framework for understanding the preservation of biodiversity in such landscapes (Bengtsson 2009).

However, a conceptual view of mosaic farmland consisting of habitats of different suitability may also be an oversimplification. Because animals need food, shelter, and nesting or hibernation sites, individuals may depend on the presence of different types of habitats within their daily, seasonal, or annual activity area (Fig. 4.2). Agricultural intensification has resulted in loss of habitat heterogeneity at several spatial scales (Benton et al. 2003), and the remaining suitable habitats are often not found in close association to populations in need (Fahrig et al. 2011). Using theories in behavioural ecology, we may predict which animals can persist in contemporary landscapes and how landscapes may be supplemented or complemented to increase the persistency of populations and species (Lima and Zollner 1996, Johst et al. 2001, Olsson et al. 2008).

Regardless of which of the processes mentioned earlier affect populations or individuals, the ability of organisms to move and disperse in the landscape will have profound effects on their ability to persist (Fig. 4.2). The ability to move through a landscape is not a fixed trait, but determined by the interaction between individuals/populations and the composition of the landscape. However, as shown later, it is not as simple as that increased mobility is always an advantage; rather the spatiotemporal variation in resource availability will act as a filter on communities in human-transformed landscapes, resulting in predictable consequences for the composition of animal communities. Incidentally or as being part of a co-adapted suit of traits, other traits may be linked to animal mobility, creating correlated responses in the composition of functional traits in animal communities, with consequences for ecosystem functioning (Lavorel and Garnier 2002, Luck et al. 2012). Below we show how these three perspectives on landscape-animal mobility interactions assist our understanding of contemporary biodiversity loss.

4.2 Effects of landscape fragmentation on movement and dispersal

One of the most common consequences of agricultural intensification is the loss of semi-natural habitats due to agricultural intensification and recently increasingly due to land abandonment (Poschlod and WallisDeVries 2002). Such semi-natural habitats may consist of natural habitat not utilized for agriculture, but often they are remnants of traditional farming such as traditionally managed cut or grazed semi-natural grasslands. Although uncultivated field borders may constitute a natural habitat for many farmland organisms (Lagerlöf and Wallin 1993, Marshall and Moonen 2002, Kuussaari et al. 2007), semi-natural grasslands are by far the most species-rich habitats. Even if the local quality of such grasslands would remain unchanged, habitat loss will normally increase their fragmentation and thus constrain dispersal between remaining habitats. This may result in increased extinction rates of local populations due to a reduced carrying capacity and/or a reduced likelihood of immigration and subsequent recolonization (Hanski and Ovaskainen 2000, Henle 2004). In addition, the likelihood of emigration from a focal patch increases with decreasing habitat size (Kuussaari et al. 1996). Thus, the fragmentation of semi-natural habitat is a concern for conservation in agricultural landscapes, although the consequences of fragmentation per se may be difficult to separate from the effect of habitat loss (Harrison and Bruna 1999, Fahrig 2002).

For species occurring in a network of habitat patches, movement and dispersal between local habitat patches is a crucial component of the long-term population persistence. Accordingly, butterfly species richness in semi-natural grasslands has been shown to depend not only on local habitat size but also on the landscape context, such that an increased availability of habitat at landscape scales correlates positively with local species richness (Bergman et al. 2004, Öckinger and Smith 2006), although there are exceptions (Krauss et al. 2004). While the relative role of fragmentation increases with increasing habitat loss in theory (Hanski 2005), habitat loss and fragmentation will most likely be correlated in practice (Fahrig 2002), which is why any effect of fragmentation on species richness might be difficult to disentangle from the effect of habitat loss. Although species richness may show consistent responses to increased fragmentation, in some cases it will ultimately depend on the species identity, i.e. which species and/or organism groups are subjected to the study. Since animals differ in their ecological attributes, such as degree of habitat specificity and mobility, species can also be expected to respond very differently to land-use change.

4.2.1 Specialization and mobility mediates the impacts of habitat loss and fragmentation

Although habitat loss and fragmentation are expected to have harmful consequences on populations, the exact effects will depend on both the degree of habitat specialization and the mobility of the organism. For strict habitat specialists perceiving the landscapes as consisting of habitat and non-suitable matrix, the loss of particular habitats is always likely to be detrimental. However, other organisms are not strict habitat specialists, and can utilize resources in various farmland habitats. For example, butterfly habitat specialists suffer more from a reduction in the size of calcareous grasslands than habitat generalists and benefit more from increased connectivity to other calcareous grasslands than habitat generalists (Krauss et al. 2003, Brückmann et al. 2010). While the densities of butterflies with specialist (monophagous) diets are negatively affected by fragmentation, the densities of butterflies with broader diets (polyphagous and oligophagous) have somewhat surprisingly shown the opposite response, most likely because they are able to use resources outside the actual grassland fragments (Steffan-Dewenter and Tscharntke 2000) (via 'landscape supplementation', see Fig. 4.2 and discussion in section 4.4.1).

The negative effects of increased habitat fragmentation can be expected to be stronger for species with higher habitat specificity and lower mobility (Warren et al. 2001), or for species with traits correlated with low mobility such as small size (Fischer and Lindenmayer 2007). Thus, the general expectation is that species with poor dispersal abilities show declining population trends driven by land-use changes in contemporary agricultural landscapes. There is evidence to suggest that this indeed is the case. For example, species with poor mobility are over-represented among the declining butterflies in Finland (Fig. 4.3). Moreover, declining butterflies are particularly those preferring semi-natural grasslands (Kuussaari et al. 2007), many of which also are strict habitat specialists. Habitat availability, in terms of semi-natural grasslands located within the dispersing capacity of these butterfly species, therefore plays a key role in maintaining diverse butterfly assemblages (Öckinger and Smith 2007a, Ekroos and Kuussaari 2012). However, patterns may be more complex, and the outcome may depend on the studied focal habitats. In butterfly assemblages within semi-natural grasslands, the relative effects of habitat patch area are strongest for sedentary species, whereas the influence of isolation is strongest for mobile species (Öckinger et al. 2009). This suggests that local processes influence sedentary species to a higher degree.

If land-use change selects stronger against poor dispersal than against habitat specificity, the movement of organisms between local communities

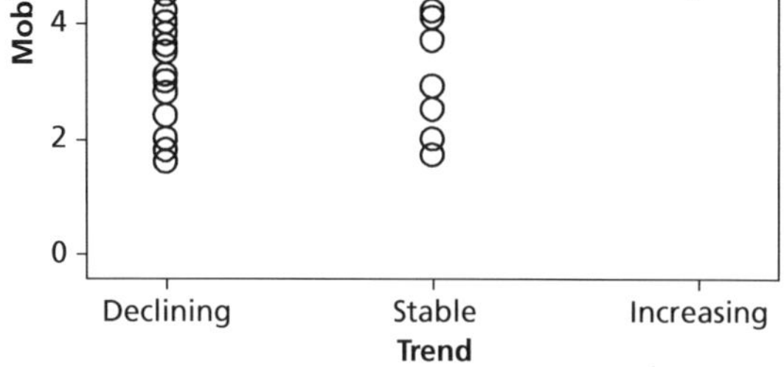

Figure 4.3 Relationship between occupancy trends of butterflies in Finland between 1960 and 2000 and species-specific mobility scores. Declining species tend to be less mobile that increasing species. Data from Kuussaari et al. (2007) and Komonen et al. (2004).

will be impeded but habitat specialists can survive locally (Dormann et al. 2007). Such land-use changes may lead to increasingly dissimilar local communities. If, however, habitat specificity is more strongly selected against, as opposed to poor dispersal, local communities become increasingly similar due to the loss of habitat specialists (biotic homogenization). Concerning butterflies, intensive agricultural land use is known to lead to biotic homogenization due to the loss of habitat specialists and poorly dispersing species (Ekroos et al. 2010).

4.2.2 Differential impact of habitat loss and fragmentation-implications for conservation

From a conservation point of view, it is often important to focus on the most threatened species in farmlands. In terms of specific habitats, semi-natural grasslands are of particular importance for such species. As an example, the vast majority of European butterflies of conservation concern occur in managed semi-natural grasslands (van Swaay et al. 2012). Since these species are often less mobile habitat specialists, suffering from loss and habitat fragmentation (Warren et al. 2001), the conservation of remaining semi-natural grasslands, and increasing the connectivity between such habitats, is a priority. This results in a conservation focus on areas where significant amounts of suitable habitats still exist (Smith et al. 2010), which may be beneficial because of the lower opportunity cost of conservation in such areas (Gabriel et al. 2009). Protecting semi-natural grasslands may, although constituting a static system, enhance the expansion of species benefiting from contemporary climate change (Thomas et al. 2012).

Traditional conservation has focused on habitat preservation, and much less attention has been devoted to increasing the permeability of the matrix to benefit conservation (Vandermeer and Perfecto 2007, Prugh et al. 2008). However, measures such as semi-natural corridors and organic farming may contribute to the conservation of rare and sedentary habitat specialists by allowing increased dispersal and therefore recolonization of habitat patches (Donald and Evans 2006) (Fig. 4.4). In general, corridors have been found to increase the movement of less mobile organisms (Gilbert-Norton et al. 2010). A recent meta-analysis demonstrated that animal movements through the landscape matrix is enhanced when the structure is similar to the species' habitat (Eycott et al. 2012). Although grassland systems has not received as much attention as woodland species (e.g. hedgerows function as corridors for wood-dwelling organisms (Davies and Pullin 2007)), we might expect the presence of semi-natural field borders to increase the permeability

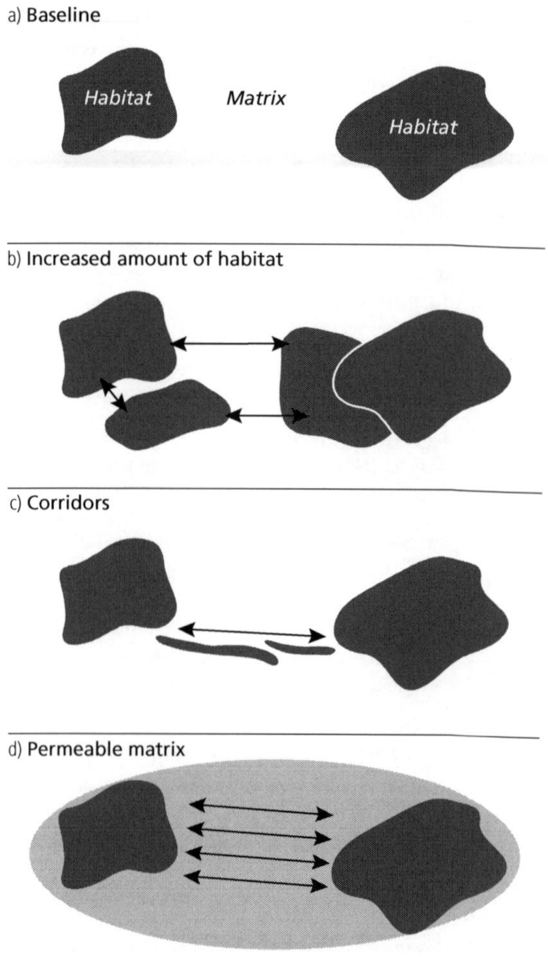

Figure 4.4 (a) Habitat loss results in the fragmentation of remaining habitats, but also the composition of habitats and the hostility of the matrix affect the functional connectivity of habitats. (b) Increased amount of habitats preserved results in reduced fragmentation and benefit recolonization and metapopulation persistence, but (c) a similar effect may be achieved if dispersal is augmented by habitat corridors, or (d) if the matrix is more permeable because of less intensive agriculture.

of agricultural landscapes for grassland specialists. Although empirical evidence for this is scant (Dover et al. 2000, Öckinger and Smith 2008), the colonization of cavity-nesting solitary wasps has been found to increase if patches containing nest sites were connected by grass strips, most likely because this facilitated wasp movement (Holzschuh et al. 2009). A combination of creating relevant habitat and 'smoothing the matrix' may prove to be the most successful strategy for reducing the isolation of fragmented populations in farmland (Baum et al. 2004, Rösch et al. 2013).

4.3 Mobility and spillover in mosaic landscapes

Animal movement in human-dominated landscapes often implies not only movement from one habitat patch to another within the same type of habitat, but also movement through multiple habitats. This is particularly the case in agricultural landscapes, which are characterized by strong temporal landscape dynamics leading to abrupt changes in habitat quality. As a consequence, movement occurs within cropped habitats, e.g. from mown patches to unmown patches of grassland, but also between crop and non-crop habitats, such as from a grassy strip into an arable field, or from a senescing crop into a hedge.

Agricultural intensification has resulted in more intense and/or more widespread disturbance of animal communities in farmland. Intense agricultural management reduces the abundance and species richness of beneficial arthropod groups, such as for example pest antagonists in arable fields (e.g. Geiger et al. 2010). Several farming practices negatively affect populations of ground-living predators (Table 4.1), which are important biological control agents (Symondson et al. 2002). These measures either cause direct mortality or alter habitat conditions to the disadvantage of predaceous species (e.g. Thorbek and Bilde 2004). Insecticide applications, for example, affect predators via both pathways: direct mortality is caused through side effects of pesticides on non-target organisms (Sterk et al. 1999) and sub-lethal effects on beneficials (Desneux et al. 2007). Indirect effects of insecticide applications that result in reduced predator numbers are

Table 4.1 Effects of selected farming practices on ground-living predaceous arthropods in agricultural habitats indicating a general trend on a group's abundance or diversity as positive (+), negative (–), or not consistent (–/+). Sublevels of management practices are moderate, intense, organic, and synthetic. Note that several publications were reviewed to compile this overview, but only key references are cited here.

Management practice		Spiders[1]	Ground beetles[2]	Rove beetles[3]
Cutting/harvest	Moderate	–	+	–
	Intense	–	–	–
Fertilization	Organic	+	+	+
	Synthetic	–	–/+	+
Grazing	Moderate	+	+	+
	Intense	–	–	–
Herbicides		–	–/+	–/+
Insecticides		–	–	–
Soil management		–	–	–

[1]Bell et al. (2001) and Birkhofer et al. (2013).
[2]Kromp (1999) and Rainio and Niemelä (2003).
[3]Krooss and Schaefer (1998) and Bohac (1999).

primarily caused through mortality of potential prey organisms (Birkhofer et al. 2008). Herbicide applications in contrast alter the vegetation structure and the availability of prey that is attracted by floral resources and these effects cascade up to affect consumers at higher trophic levels (Norris and Kogan 2005).

Moving between habitats can offset these negative effects of agricultural management. Thus mass effects (Shmida and Wilson 1985) or 'spillover' (Blitzer et al. 2012; Fig. 4.2) of organisms from more productive habitats sustaining populations to less productive habitats is an important process explaining biodiversity patterns in agricultural landscapes. The ability to offset the negative effects depends on the composition of landscapes. Landscapes, for example, characterized by a diversity of crops among which management intensity and timing differ may provide suitable habitats for spiders and predaceous beetles throughout the growing season. Applied ecologists have long recognized that the biological control of herbivorous pest insects in crops depends on the immigration of natural enemies (predators, parasitoids) from less disturbed habitats (Landis et al. 2000). The complexity of the landscape has been

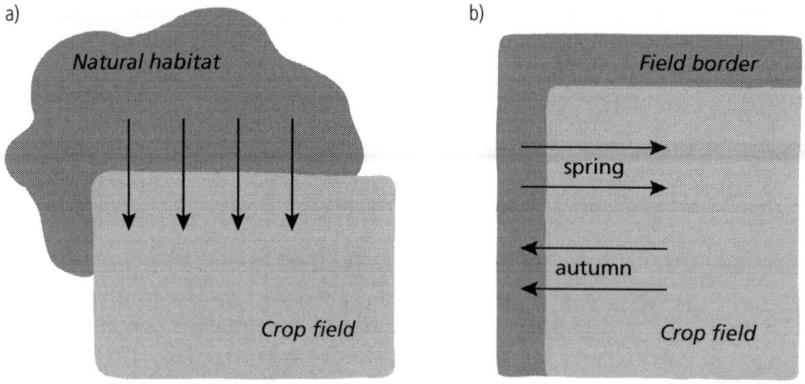

Figure 4.5 Spillover may subsidize frequently disturbed habitats with organisms from less-disturbed habitats, but also the disturbed habitats may experience temporary abundant resources, resulting in reverse spillover of organisms back to semi-natural habitats. (a) Dispersal from semi-natural habitats may sustain populations in regularly disturbed arable fields. (b) Populations in disturbed crop fields may utilize seasonally abundant resources, resulting in a reverse spillover back to semi-natural habitats.

repeatedly shown to increase the predation and parasitism of pests in arable systems. For example, cereal and oil seed rape fields are subsidized in natural enemies by non-crop habitats (Thies 1999, Thies and Tscharntke 2010). In the following section we review the evidence for the extent and effect of movement between habitats as a result of disturbance, the importance of dispersal ability as a filter for predatory arthropod communities, intra- and interspecific variation in dispersal ability and propensity, and impacts of biological control on insect pests (Fig. 4.5).

4.3.1 Spillover between habitats in heterogeneous agricultural landscapes

Heterogeneity at the landscape scale is often characterized by rather sharp borders between adjacent land-use types (Ries et al. 2004), contrasting with the soft borders between patches within habitats (Fig. 4.6). Organisms that move between habitats often pass edge zones, but the change in environmental conditions and resource availability from one habitat to another can be rather drastic, and unfavourable habitat types may even act as a barrier for poor dispersers (Schröter and Irmler 2013). Nevertheless, in many species, arthropods move between resource patches during their lifetime, and in many species individuals will move between habitat patches, i.e. from grassland to arable land and the reverse. This is reflected in that few predaceous arthropod species are true habitat specialists—the majority of the species found in agricultural fields overlap with natural habitat (Duelli and Obrist 2003). Despite this, less intensively or less frequently disturbed habitats act as important sources for the re-colonization of arable fields after local disturbances (Tscharntke et al. 2007). Complex landscapes, including crops with different phenology and semi-natural habitats, allow for the build-up of source populations, whereas simpler landscapes dominated by one crop type (e.g. cereals) do not provide enough source populations (Fig. 4.6) (Kleijn et al. 2009b). The heterogeneity of landscapes, which is usually assessed at distances between 250 and 1,500 m from the sampling area, therefore leads to a higher diversity and abundance of predaceous arthropods than do more homogeneous landscapes (Clough et al. 2005, Purtauf et al. 2005, Drapela et al. 2008). Although farmland heterogeneity has often been defined as the amount of semi-natural habitat in a landscape, it is important to consider heterogeneity in a wider respect, including, for example, the diversity of crops and management practices (Vasseur et al. 2013).

Conversely, crop habitats also subsidize natural habitats, a process that has received little attention even though it may negatively impact the integrity of communities in natural habitats. The same may occur for higher trophic levels. As reviewed by

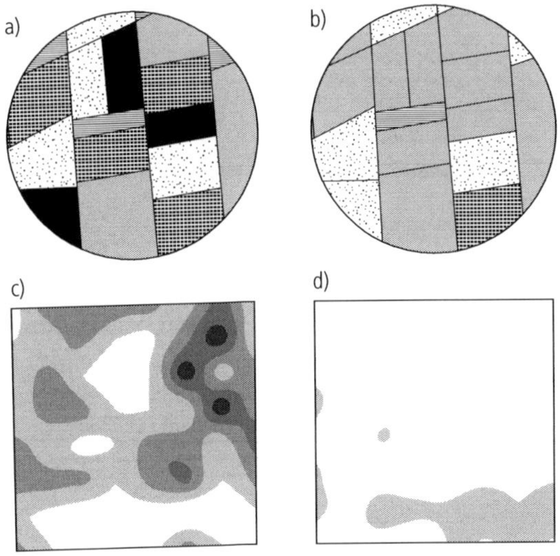

Figure 4.6 Schematic overview of heterogeneity defined at the landscape scale as diversity of land-use types with (a) a diverse landscape and (b) a simple landscape, or heterogeneity defined at the field scale as weed coverage in a cereal field with (c) heterogeneous and (d) homogeneous environmental conditions.

Rand et al. (2006), predators (e.g. coccinellids) may feed on abundant resources in agricultural fields, and then disperse towards natural habitats at field senescence, leading to large increases in population size, and potentially high pressure on herbivores (and other predators) in the natural habitat.

4.3.2 Dispersal ability and the differential response of fragmentation and disturbance frequency

Generally, dispersal allows spreading risk associated with disturbance, especially in generalist species (Kisdi 2002). Heterogeneous landscapes with frequently disturbed habitats can be expected to act as environmental filters which promote species and individuals with higher dispersal ability. Overall, potential dispersal distances vary widely, even for species that can be encountered in disturbed habitats such as arable fields. For instance, the ground-living predators listed in Table 4.1 all have species with rather high (ballooning spiders and long-winged beetles) or low (non-ballooning spiders and short-winged beetles) dispersal abilities with daily ranges of a few metres to several kilometres (Gutiérrez and Menéndez 1997, Weyman et al. 2002). Several studies have estimated dispersal distances by comparing the explanatory power of landscape composition indices based on nested landscape circular sectors of increasing radius, all centred on the point at which the arthropod community was characterized. For spiders in arable fields, the spatial scales with the highest explanatory power for density ranged from 95 m to 3 km radius around the study fields (Schmidt et al. 2008). Despite this variation, there is strong evidence for the importance of dispersal ability as a structuring force for communities of predaceous arthropods in disturbed habitats. For example, more species of spiders, with a higher propensity for aerial dispersal, and a higher proportion of ground beetles with well-developed wings were found in disturbed than in less disturbed habitats (Lambeets et al. 2008, Entling et al. 2011). This is also reflected in patterns of intra-specific variation: individuals of the spider species *Meioneta rurestris* (Linyphiinae) were more prone to engage in aerial dispersal when caught in disturbed habitats (Entling et al. 2011). This is reminiscent of results of an earlier study conducted on planthoppers, a herbivorous group (Denno et al. 1996). Interestingly, in more stable habitats, habitat specialist spiders tend to be less prone to dispersal (Bonte et al. 2003), likely reflecting the higher risk of leaving than staying in the patch.

4.3.3 Differential effects of fragmentation and disturbance frequency

The mobility of species also affects the structure of food webs (Polis et al. 1997). Recent studies across four trophic levels (plant–herbivore–parasitoid–hyperparasitoid systems) showed that diversity in the third and fourth trophic level did not respond to landscape simplification, while secondary parasitism was higher in complex landscapes (Rand et al. 2012). One should be careful while interpreting such studies in terms of dispersal ability. Clearly, predators and parasitoids base their decision on whether to occupy a patch on resource density and quality, thereby leading the observer to underestimate the dispersal distance when patches are reached but not occupied. More problematically, the variables 'resource density' and 'quality' are usually not controlled as is the case with plant resources in studies of herbivores, potentially inducing bias, albeit not necessarily in a systematic manner.

The effects of body size on the ability of natural enemies to disperse are extremely relevant given the small size of parasitoid species, but evidence is ambiguous. Several studies suggest that smaller parasitoid species cope less well with isolation of host patches than larger species (Roland and Taylor 1997, Ellers et al. 1998, van Nouhuys and Hanski 2002, Tscharntke and Brandl 2004). Other studies report that even small parasitoids can disperse surprisingly well (Antolin and Strong 1987), be unsusceptible to increasing isolation (Elzinga et al. 2007), and remain spatially coupled with their host even during long-distance migration, such as egg parasitoids of migratory locusts (Farrow 1981).

4.3.4 The importance of source habitats for landscape-scale biological control

In disturbed agricultural habitats, the biological control of pests depends on the quick recolonization by the natural enemies of these pests at the beginning of the growing season. As seen earlier, both predators and parasitoids are susceptible to isolation from source habitats, resulting in a lower richness and/or population density of these groups in disturbed (monoculture) parts of agricultural landscapes (Thies 1999, Thies and Tscharntke 2010). Specifically, in 74% of the studies reviewed by Bianchi et al. (2006), natural predator populations were higher in complex landscapes than in simple, monoculture, landscapes, while a corresponding reduction in pest pressure was observed in 45% of the studies, suggesting a higher predation pressure in more complex landscapes. Consequently, simpler landscapes, with fewer sources of natural predators, seem to have a higher pest pressure, as suggested by a recent study of insecticide use patterns in the Midwestern USA (Meehan et al. 2011). However, these outcomes may not be generalizable to all systems. Where predators or parasitoids are themselves subject to mortality by natural enemies, such as in food webs including hyperparasitoids or birds feeding preferentially on spiders, isolation from source habitats may result in increases, rather than decreases at the third trophic level, with a dampening of biological control (Hermann et al. 2012, Rand et al. 2012).

4.4 Mobility and multiple patch use

4.4.1 Agricultural change and resource separation

In addition to habitat loss, fragmentation and a general decrease in landscape heterogeneity has resulted in the spatial separation of resources critical for animal breeding and survival. For example, traditional farmland contains many permanent or semi-permanent habitats, such as semi-natural mowed or grazed grasslands, uncultivated field borders, and incidental habitats, which to a large extent have been lost at the expense of annually tilled fields. These habitats often contain complementary resources, such as nesting sites (Öckinger and Smith 2007b) or larval resources (Ouin et al. 2004), for animals that, otherwise, can use resources in the wider landscape. Landscape complementation occurs when habitat patches in a landscape (within the potential activity range of an organism) contain different resources that are required by the organism (Dunning et al. 1992) (Fig. 4.7).

To utilize the non-substitutable resources that fulfil separate needs, such as nesting and foraging, organisms move between habitat patches. For example, butterflies require both larval host plants and adult nectar resources for population persistence (Boggs 2003). Female butterflies move between host plants to oviposit and flowering plants for nectar foraging, suggesting that these two resources needed for different life-stages can be spatially separated (Wiklund 1977, Fred et al. 2006). Similarly, bumblebees need floral resources for nectar and pollen collection, but floral resources need to be within reach of suitable nesting sites (Goulson 2003). Bumblebees may nest in semi-natural grasslands but still utilize surrounding linear habitat elements, such as road verges and uncultivated field borders, for foraging (Öckinger and Smith 2007b), or seek out mass-flowering crops in the wider landscape (Westphal et al. 2003). This is similar to many birds such as wheatears, *Oenanthe oenanthe*, starlings, *Sturnus vulgaris*, white storks, *Ciconia ciconia*, and stone

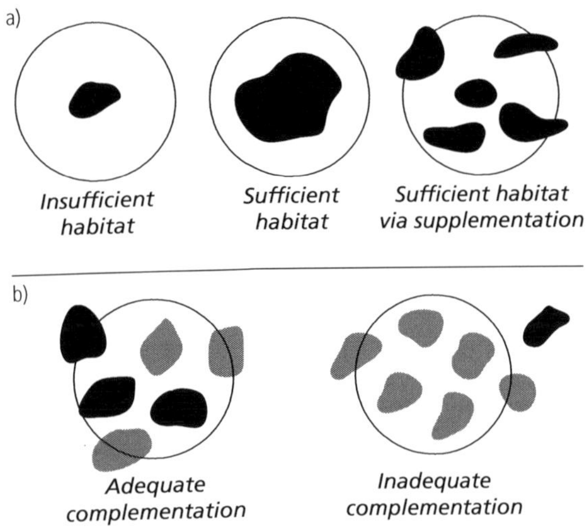

Figure 4.7 (a) Landscape supplementation results from animals unable to find sufficient resources in their prime habitat using additional, for example, human-transformed habitat, where the same resources exist. (b) Landscape complementation stems from the need of animals to utilize complementary resources available in different habitats. When these habitats are distant from each other in relation to the mobility of organisms, populations will go extinct although all necessary resources exist at larger spatial scales.

curlews, *Burhinus oedicnemus*, which are also central place foragers (Olsson et al. 2008) during breeding (Bruun and Smith 2003, Arlt and Pärt 2007, Olsson and Rogers 2009, Dallimer et al. 2012). Such foragers integrate the landscape quality at various scales from tens of metres in some solitary bee species, 3 km in some bumblebees and starlings, and up to over 10 km in, e.g., white storks. Common for these species is that they regularly and repeatedly visit different habitat types for resource extraction, traverse some habitats without using them, while their nests are in yet other habitats. Hence, movement is here considerably different from dispersal.

Agricultural landscapes have also become homogenized at larger spatial scales. At regional scales, the structural rationalization of agriculture has resulted in a polarization of agricultural landscapes, with some areas being dominated by intensive crop production and others by animal husbandry and fodder production (Robinson and Sutherland 2002). Depending on the spatial scale at which animals utilize farmlands, both these processes may result in loss of landscape complementation (cf. Dunning et al. 1992; Fig. 4.7). Hence, this loss of spatial association in critical resources for breeding and feeding within a profitable movement range has therefore contributed to a decline in abundances of farmland animals.

This process of landscape complementation, when animals use spatially distributed non-substitutable resources, should, however, be distinguished from landscape supplementation (Fig. 4.7), where an animal shifts to the same resource in other habitats if the availability in the first is not sufficient. An example of the latter is the switching to alternative pollen resources by solitary bees, which primarily collect pollen from plants in semi-natural habitats, but switch to organically managed farmland when isolated from the semi-natural habitats (Williams and Kremen 2007). Landscape heterogeneity may contribute to both landscape supplementation and complementation (Benton et al. 2003, Fahrig et al. 2011, Vasseur et al. 2013).

Complementary resource use in different habitats may involve resources available at different times, i.e. cross-season complementation of resources. For example, arable fields and field margins are complementary summer and winter habitats for ground beetles (Thomas et al. 2002). The creation of beetle banks, i.e. specially made low-disturbance vegetation strips in agricultural fields, makes use of the ability of natural predators such as ground beetles to overwinter in this habitat and then move to intensively managed arable fields where they become abundant (Holland and Luff 2000). Also parasitoids utilize overwintering sites, alternative hosts and food sources in different habitats. Moreover, herbivores may move between patches of host plants of the same or of different species. For example, host-alternation is common in aphids, of which many species use a primary, woody, winter host and a secondary, grassy summer host. This often implies movement between different habitats, i.e. a colonization of crop fields from perennial vegetation. When landscape complementation involves temporally separated alternative resources in different habitats, the concepts of spillover and landscape complementation becomes partly overlapping (Fig. 4.2). *Sensu stricto*, landscape complementation refers to the use of complementary resources in different habitats by individuals, while spillover refers to population processes acting across habitat boundaries. However, the distinction is less clear when individuals move from, e.g., a hibernation site to utilize resources and reproduce in another habitat, while their offspring may move back to the hibernation site.

4.4.2 Landscape complementation and animal mobility

The scale at which organisms utilize landscapes will affect the scale at which landscape structure affects them (Steffan-Dewenter et al. 2002). Due to these mobility-mediated effects, loss of habitat heterogeneity at different spatial scales has different effects for different organisms. For example, bees can respond by increasing the foraging distance as the distance to high-quality habitat increases (Beekman and Ratnieks 2000) or the resource availability decreases (Steffan-Dewenter and Kuhn 2003). However, this flexibility is only possible to a certain extent, and species that have more restricted foraging ranges will not be able to find sufficient high-quality foraging patches in their foraging range and will go locally extinct. In contrast, bumblebees with larger foraging ranges can utilize scattered temporary resources, such as rich flower patches at organically managed farms or mass flowering crops, and survive also in homogeneous agricultural landscapes (Westphal et al. 2003, Rundlöf et al. 2008). Furthermore, the effects may be more complex due to interacting effects of local and landscape-scale determinants. As an example, both bumblebee species with smaller foraging ranges and sedentary butterfly species are more affected by local habitat factors, independent of landscape context (Rundlöf et al. 2008, Öckinger et al. 2009). Such sedentary species are expected to spend most of their lives within a single patch. More mobile species can freely move between several habitat patches and utilize their resources. Species with intermediate mobility need to move outside of their natal or home patch to find resources, but may not reach other patches within their mobility range if these are too spatially separated. Therefore, species with intermediate movement abilities and foraging ranges may be the most sensitive to habitat loss, fragmentation, and spatial scattering of resources (Thomas 2000).

4.4.3 Differential mobility, structural change, and pollination

The delivery of pollination as an ecosystem service is dependent on source habitats with suitable nesting sites for the (bee) pollinators, as well as temporally complementing flower resources, besides the flowering crop itself. Semi-natural grasslands have been identified as sources for pollinating insects in agricultural landscapes (Öckinger and Smith 2007b). Optimal foraging theory predicts that increasing distance to semi-natural or natural habitats decreases the flower visitation rate by central-place foraging pollinators (cf. Cresswell et al. 2000). Agricultural crops grown in the vicinity of semi-natural or natural habitats often have a lower pollination deficit and higher fruit set due to the pollination provided by wild pollinators living in such habitats (Ricketts et al. 2008). Stability in the provisioning of pollination services can also increase with decreasing distance to natural habitats (Garibaldi et al. 2011). However, pollinators which are not confined to a central place (nest) may show a stronger ability to utilize resources that are spatially disconnected from semi-natural habitats (Jauker et al. 2009, Ekroos et al. 2013).

The ecological traits affecting an organisms' ability to persist in transformed agricultural landscapes, for example mobility, may correlate with traits affecting its impact on ecosytem functions. Therefore, mobility-mediated community changes may result in concomitant changes in ecosystem services. In Sweden, the bumblebee community has changed from a rather evenly distributed community to one dominated by two short-tongued species over the past 70 years (Bommarco et al. 2012). This change is mainly driven by the ability of mobile, and at the same time short-tongued bumblebees to persist in contemporary simplified agricultural landscapes (Rundlöf et al. 2008). Persson (2011) suggested that these trait correlations were not incidental, but that they formed co-adapted trait clusters related to seasonal foraging strategies where high mobility and low foraging specificity co-evolved to meet variable foraging conditions (see also Ekroos et al. 2013). Irrespective of the cause for this functional homogenization of communities, the loss of functional diversity can have consequences for the delivery and stability of pollination services, in particular the pollination of flowers dependent on long-tongued pollinators.

Interactions between organisms that are dependent to a differing degree upon spatiotemporal variation in complementary food resources around the nest may result in unexpected results. Oilseed rape subsidizes mobile short-tongued generalist bumblebees (Westphal et al. 2003), and there are suggestions that this results in reduced visitation rates of long-tongued bumblebees to plants with long-tubed flowers, most likely because of resource depletion by nectar-robbing short-tongued bumblebees (Diekötter et al. 2010). This illustrates the complexity of interacting factors which differentially influence organisms with different ecological traits, including their dispersal capacity.

4.4.4 Landscape complementation and the matrix

For species to benefit from landscape complementation, not only the presence of different habitats within the home range of an animal is vital, but also the ability to move between the patches (Taylor et al. 1993). Thus, increasing the connectivity in landscapes may enable organisms to utilize complementary resources more efficiently and therefore contribute to population persistence. For example, the viability of ground beetle populations may depend not only on the amount of different habitats, but on their distribution in the landscape which affects the ability for organisms to commute between summer and winter habitats (Nabe-Nielsen et al. 2009). However, most studies on, for example, pollination focus on habitat loss; few consider how habitats are functionally connected (Hadley and Betts 2012; Fig. 4.4). This functional connectedness will be affected by habitat loss, habitat configuration, but also the permeability of the matrix, such as in the form of corridors (Hadley and Betts 2012). For example, corridors may facilitate the use of dispersed flowering resources by bumblebees (Cranmer et al. 2012), promote foraging movements by butterflies (Delattre et al. 2013), and facilitate pollen dispersal through pollinator movement (van Geert et al. 2010).

However, it should be noted that our perception of connectivity may sometimes reflect human perception rather than the actual movements of the animal. For example, Krewenka et al. (2011) did not find any positive effect of corridor strips on foraging solitary bees, trap-nesting bees, wasps, or parasitoids. Thus, while we may expect the suitability of corridors to be related to the habitat quality and mobility of focal animals, more research is needed to understand what landscape features may promote movements of animals between complementary habitats.

4.5 Future perspectives

We have here shown that movement is crucial for animals' ability to cope with habitat change, resulting in the loss and fragmentation of high-quality habitats, increased disturbance, and spatial and temporal disassociation of critical resources. Generally mobile species and habitat generalists are better able to endure contemporary agricultural intensification, resulting in a loss of more sedentary farmland species and a simplification of animal communities. This has consequences both for the conservation of species and for the preservation of ecosystem services contributing to crop production and public goods. However, different ecological processes, at levels ranging from individuals to populations to communities, affect different species. Because of this, general recipes such as 'reduce fragmentation', 'increase connectivity', and 'increase ecological heterogeneity' may not provide a universal solution for all organisms. A lack of landscape complementation may not be solved by corridors and extinction resulting from habitat fragmentation may not be resolved by increasing farmland heterogeneity. Instead it is important to understand how species with different movement abilities utilize resources in, and disperse through, agricultural landscapes, and tailor conservation actions to the organism groups that are targeted.

Habitat specialists and generalists may be differently affected by the ecological processes we have discussed (fragmentation, spillover, and landscape complementation/supplementation). Thus, to define efficient conservation measures, it may be important to specify the conservation goals in in terms of biodiversity conservation per se (mostly habitat specialists) and preservation of ecosystem services (often habitat generalists) (Smith et al. 2010, Kleijn et al. 2011). While the conservation of rare species may result in a focus of preserving well-connected natural and semi-natural habitats in areas where much biodiversity still remains (Gabriel et al. 2009, Thomas et al. 2012), the preservation of biodiversity underpinning ecosystem services may result in a focus on intensively farmed areas where marginal gains from mitigation measures increasing ecological heterogeneity or increasing landscape complementation may be largest (Rundlöf and Smith 2006). In both cases it will be important to consider the effect of animal mobility. When conserving rare species, it may be important to not only preserve existing habitat, but to benefit the dispersal of sedentary organisms by, for example, making the matrix more permeable (Vandermeer and Carvajal 2001, Donald and Evans 2006). Given recent evidence that less mobile (Mattila et al. 2011) and/or rare (Pöyry et al. 2009) species may experience difficulties in adjusting ranges to climate, and that speed of range-shift rates may be related to habitat availability (Hill et al. 2001), it may be important to increase the permeability of the matrix at sufficiently large spatial scales (but see Hodgson et al. 2009, Doerr et al. 2011). To instead mitigate general biodiversity loss, increasing ecological heterogeneity and reducing agricultural intensity have proved to be efficient measures in intensively farmed landscapes (Rundlöf and Smith 2006, Batáry et al. 2011). Effects have shown to differ between taxa (Fuller et al. 2005), most likely because of different movement abilities of organisms.

Although animal movement studies have contributed to our understanding of biodiversity conservation, our knowledge about animal movements is limited. Many studies rely on proxies to infer animal mobility, which may, however, be highly uncertain (Sekar 2012). Molecular techniques have improved our knowledge on mobility (Chapman et al. 2003, Carvell et al. 2012; Liedvogel and Lundberg, Chapter 12), and the field of landscape genetics shows promise in increasing our understanding of landscape connectivity (Zeller et al. 2012). However, there is still a need to find techniques for tracking not only larger animals in the landscape (Bruun and Smith 2003; Box 6.1), but also smaller organisms that often provide essential ecosystem services (Cant et al. 2005).

References

Antolin, M. F., and Strong, D. R. (1987). Long distance dispersal by a parasitoid (*Anagrus delicatus*, Mymaridae) and its host. *Oecologia*, **73**, 288–92.

Arlt, D., and Pärt, T. (2007). Nonideal breeding habitat selection: a mismatch between preference and fitness. *Ecology*, **88**, 792–801.

Batáry, P., Báldi, A., Kleijn, D., and Tscharntke, T. (2011). Landscape-moderated biodiversity effects of agri-environmental management: a meta-analysis. *Proceedings of the Royal Society B: Biological Sciences*, **278**, 1894–1902.

Baum, K. A., Haynes, K. J., Dillemuth, F. P., and Cronin, J. T. (2004). The matrix enhances the effectiveness of corridors and stepping stones. *Ecology*, **85**, 2671–5.

Beekman, M., and Ratnieks, F. L. W. (2000). Long-range foraging by the honeybee, *Apis mellifera* L. *Functional Ecology*, **14**, 490–6.

Bell, J. R., Wheater, P., and Rod Cullen, W. (2001). The implications of grassland and heathland management for the conservation of spider communities: a review. *Journal of Zoology, London*, **255**, 377–87.

Bengtsson, J. (2009). Applied (meta)community ecology: diversity and ecosystem services at the intersection of local and regional processes. In P. J. Morin and H. A. Verhoef (eds), *Community Ecology*. Oxford University Press, New York.

Bennett, A. F., Radford, J. Q., and Haslem, A. (2006). Properties of land mosaics: implications for nature conservation in agricultural environments. *Biological Conservation*, **133**, 250–64.

Benton, T. G., Vickery, J. A., and Wilson, J. D. (2003). Farmland biodiversity: is habitat heterogeneity the key? *Trends in Ecology and Evolution*, **18**, 182–8.

Bergman, K. O., Askling, J., Ekberg, O., Ignell, H., Wahlman, H., and Milberg, P. (2004). Landscape effects on butterfly assemblages in an agricultural region. *Ecography*, **27**, 619–28.

Bianchi, F. J. J. A., Booij, C. J. H., and Tscharntke, T. (2006). Sustainable pest regulation in agricultural landscapes: a review on landscape composition, biodiversity and natural pest control. *Proceedings of the Royal Society B: Biological Sciences*, **273**, 1715–27.

Biesmeijer, J. C., Roberts, S. P. M., Reemer, M., et al. (2006). Parallel declines in pollinators and insect-pollinated plants in Britain and the Netherlands. *Science*, **313**, 351–54.

Birkhofer, K., Bezemer, M., Bonkowski, M., et al. (2008). Long-term organic farming fosters below and aboveground biota: implications for soil quality, biological control and productivity. *Soil Biology and Biochemistry*, **40**, 2297–308.

Birkhofer, K., Entling, M., and Lubin, Y. (2013). Agroecology: trait composition, spatial relationships, trophic interactions. In D. Penney (ed.), *Spider Research in the 21st Century: Trends and Perspectives*. Siri Scientific Press, Manchester.

Blitzer, E. J., Dormann, C. F., Holzschuh, A., Klein, A.-M., Rand, T. A., and Tscharntke, T. (2012). Spillover of functionally important organisms between managed and natural habitats. *Agriculture, Ecosystems and Environment*, **146**, 34–43.

Boggs, C. L. (2003). Environmental variation, life histories, and allocation. In C. L. Boggs, W. B. Watt, and P. R. Ehrlich (eds), *Butterflies: Ecology and Evolution Taking Flight*. University of Chicago Press, Chicago.

Bohac, J. (1999). Staphylinid beetles as bioindicators. *Agriculture, Ecosystems and Environment*, **74**, 357–72.

Bommarco, R., Lundin, O., Smith, H. G., and Rundlof, M. (2012). Drastic historic shifts in bumble-bee community composition in Sweden. *Proceedings of the Royal Society B: Biological Sciences*, **279**, 309–15.

Bonte, D., Vandenbroecke, N., Lens, L., and Maelfait, J.-P. (2003). Low propensity for aerial dispersal in specialist spiders from fragmented landscapes. *Proceedings of the Royal Society B: Biological Sciences*, **270**, 1601–7.

Bruun, M., and Smith, H. G. (2003). Landscape composition affects habitat use and foraging flight distances in breeding European starlings. *Biological Conservation*, **114**, 179–87.

Brückmann, S. V., Krauss, J., and Steffan-Dewenter, I. (2010). Butterfly and plant specialists suffer from reduced connectivity in fragmented landscapes. *Journal of Applied Ecology*, **47**, 799–809.

Burel, F., Butet, A., Delettre, Y. R., and Millàn De La Peña, N. (2004). Differential response of selected taxa to landscape context and agricultural intensification. *Landscape and Urban Planning*, **67**, 195–204.

Cant, E. T., Smith, A. D., Reynolds, D. R., and Osborne, J. L. (2005). Tracking butterfly flight paths across the landscape with harmonic radar. *Proceedings of the Royal Society B: Biological Sciences*, **272**, 785–90.

Cardinale, B. J., Duffy, J. E., Gonzalez, A., Gonzalez, A., Hooper, D. U., Perrings, C., Venail, P., Narwani, A., Mare, G. M., Tilman, D., Wardle, D. A., Kinzig, A. P., Daily, G. C., Loreau, M., Grace, J. B., Larigauderie, A., Srivastava, D. S., and Naeem, S., (2012). Biodiversity loss and its impact on humanity. *Nature*, **486**, 59–67.

Chapman, R. E., Wang, J., and Bourke, A. F. G. (2003). Genetic analysis of spatial foraging patterns and resource sharing in bumble bee pollinators. *Molecular Ecology*, **12**, 2801–8.

Clough, Y., Kruess, A., Kleijn, D., and Tscharntke, T. (2005). Spider diversity in cereal fields: comparing factors at local, landscape and regional scales. *Journal of Biogeography*, **32**, 2007–14.

Cranmer, L., Mccollin, D., and Ollerton, J. (2012). Landscape structure influences pollinator movements and directly affects plant reproductive success. *Oikos*, **121**, 562–8.

Cresswell, J. E., Osborne, J. L., and Goulson, D. (2000). An economic model of the limits to foraging in central place foragers with numerical solutions for bumblebees. *Ecological Entomology*, **25**, 249–55.

Dallimer, M., Skinner, A. M. J., Davies, Z. G., Armsworth, P. R., and Gaston, K. J. (2012). Multiple habitat associations: the role of offsite habitat in determining onsite avian density and species richness. *Ecography*, **35**, 134–45.

Dänhardt, J., Green, M., Lindström, Å., Rundlöf, M., and Smith, H. G. (2010). Farmland as stopover habitat for migrating birds—effects of organic farming and landscape structure. *Oikos*, **119**, 1114–25.

Davies, Z. G., and Pullin, A. S. (2007). Are hedgerows effective corridors between fragments of woodland habitat? An evidence-based approach. *Landscape Ecology*, **22**, 333–51.

Delattre, T., Vernon, P., and Burel, F. (2013). An agri-environmental scheme enhances butterfly dispersal in European agricultural landscapes. *Agriculture, Ecosystems and Environment*, **166**, 102–9.

Dennis, R. L. H., and Hardy, P. B. (2006). Support for mending the matrix: resource seeking by butterflies in apparent non-resource zones. *Journal of Insect Conservation*, **11**, 157–68.

Denno, R. F., Roderick, G. K., Peterson, M. A., et al. (1996). Habitat persistence underlies intraspecific variation in the dispersal strategies of planthoppers. *Ecological Monographs*, **66**, 389–408.

Desneux, N., Decourtye, A., and Delpuech, J.-M. (2007). The sublethal effects of pesticides on beneficial arthropods. *Annual Review of Entomology*, **52**, 81–106.

Diekötter, T., Kadoya, T., Peter, F., Wolters, V., and Jauker, F. (2010). Oilseed rape crops distort plant-pollinator interactions. *Journal of Applied Ecology*, **47**, 209–14.

Dirzo, R., and Raven, P. H. (2003). Global state of biodiversity loss. *Annual Review of Environment and Resources*, **28**, 137–67.

Doerr, V. a. A. J., Barrett, T., and Doerr, E. D. (2011). Connectivity, dispersal behaviour and conservation under climate change: a response to Hodgson et al. *Journal of Applied Ecology*, **48**, 143–7.

Donald, P. F., and Evans, A. D. (2006). Habitat connectivity and matrix restoration: the wider implications of agri-environment schemes. *Journal of Applied Ecology*, **43**, 209–18.

Donald, P. F., Green, R. E., and Heath, M. F. (2001). Agricultural intensification and the collapse of Europe's farmland bird populations. *Proceedings of the Royal Society B: Biological Sciences*, **268**, 25–9.

Dormann, C. F., Schweiger, O., Augenstein, I., et al. (2007). Effects of landscape structure and land-use intensity on similarity of plant and animal communities. *Global Ecology and Biogeography*, **16**, 774–87.

Dover, J., Sparks, T., Clarke, S., Gobbett, K., and Glossop, S. (2000). Linear features and butterflies: the importance of green lanes. *Agriculture, Ecosystems and Environment*, **80**, 227–42.

Drapela, T., Moser, D., Zaller, J. G., and Frank, T. (2008). Spider assemblages in winter oilseed rape affected by landscape and site factors. *Ecography*, **31**, 254–62.

Duelli, P., and Obrist, M. K. (2003). Regional biodiversity in an agricultural landscape: the contribution of seminatural habitat islands. *Basic and Applied Ecology*, **4**, 129–38.

Dunning, J. B., Danielson, B. J., and Pulliam, H. R. (1992). Ecological processes that affect populations in complex landscapes. *Oikos*, **65**, 169–75.

Ekroos, J., Heliölä, J., and Kuussaari, M. (2010). Homogenization of lepidopteran communities in intensively cultivated agricultural landscapes. *Journal of Applied Ecology*, **47**, 459–67.

Ekroos, J., and Kuussaari, M. (2012). Landscape context affects the relationship between local and landscape species richness of butterflies in semi-natural habitats. *Ecography*, **35**, 232–8.

Ekroos, J., Rundlöf, M., and Smith, H. G. (2013). Trait-dependent responses of flower-visiting insects to distance to semi-natural grasslands and landscape heterogeneity. *Landscape Ecology*, **28**, 1283–92.

Ellers, J., Van Alphen, J. J., and Sevenster, J. G. (1998). A field study of size–fitness relationships in the parasitoid *Asobara tabida*. *Journal of Animal Ecology*, **67**, 318–24.

Elzinga, J. A., Van Nouhuys, S., Van Leeuwen, D. J., and Biere, A. (2007). Distribution and colonisation ability of three parasitoids and their herbivorous host in a fragmented landscape. *Basic and Applied Ecology*, **8**, 75–88.

Entling, M. H., Stämpfli, K., and Ovaskainen, O. (2011). Increased propensity for aerial dispersal in disturbed habitats due to intraspecific variation and species turnover. *Oikos*, **120**, 1099–109.

Eycott, A. E., Stewart, G. B., Buyung-Ali, L. M., Bowler, D. E., Watts, K., and Pullin, A. S. (2012). A meta-analysis on the impact of different matrix structures on species movement rates. *Landscape Ecology*, **27**, 1263–78.

Fahrig, L. (2002). Effect of habitat fragmentation on the extinction threshold: a synthesis. *Ecological Applications*, **12**, 346–53.

Fahrig, L., Baudry, J., Brotons, L., et al. (2011). Functional landscape heterogeneity and animal biodiversity in agricultural landscapes. *Ecology Letters*, **14**, 101–12.

Farrow, R. A. (1981). Aerial dispersal of *Scelio fulgidus* (hym. Scelionidae), parasite of eggs of locusts and grasshoppers (orth. Acrididae). *Entomophaga*, **26**, 349–55.

Fischer, J., Brosi, B., Daily, G., et al. (2008). Should agricultural policies encourage land sparing or wildlife--friendly farming? *Frontiers in Ecology and the Environment*, **6**, 382–7.

Fischer, J., and Lindenmayer, D. (2007). Landscape modification and habitat fragmentation: a synthesis. *Global Ecology and Biogeography*, **16**, 265–80.

Flynn, D. F. B., Gogol-Prokurat, M., Nogeire, T., et al. (2009). Loss of functional diversity under land use intensification across multiple taxa. *Ecology Letters*, **12**, 22–33.

Fred, M. S., O'Hara, R. B., and Brommer, J. E. (2006). Consequences of the spatial configuration of resources for the distribution and dynamics of the endangered Parnassius apollo butterfly. *Biological Conservation*, **130**, 183–92.

Fronhofer, E. A., Kubisch, A., Hilker, F. M., Hovestadt, T. and Pethke, H. J. (2012). Why are metapopulations so rare? *Ecology*, **93**, 1967–78.

Fuller, R. J., Norton, L. R., Feber, R. E., et al. (2005). Benefits of organic farming to biodiversity vary among taxa. *Biology Letters*, **1**, 431–4.

Gabriel, D., Carver, S. J., Durham, H., et al. (2009). The spatial aggregation of organic farming in England and its underlying environmental correlates. *Journal of Applied Ecology*, **46**, 323–33.

Garibaldi, L. A., Steffan-Dewenter, I., Kremen, C., et al. (2011). Stability of pollination services decreases with isolation from natural areas despite honey bee visits. *Ecology Letters*, **14**, 1062–72.

Geiger, F., Bengtsson, J., Berendse, F., et al. (2010). Persistent negative effects of pesticides on biodiversity and biological control potential on European farmland. *Basic and Applied Ecology*, **11**, 97–105.

Gilbert-Norton, L., Wilson, R., Stevens, J. R., and Beard, K. H. (2010). A meta-analytic review of corridor effectiveness. *Conservation Biology*, **24**, 660–8.

Goulson, D. (2003). *Bumblebees: Behaviour and Ecology*. Oxford University Press, Oxford.

Goulson, D., Hanley, M. E., Darvill, B., Ellis, J. S., and Knight, M. E. (2005). Causes of rarity in bumblebees. *Biological Conservation*, **122**, 1–8.

Gutiérrez, D., and Menéndez, R. (1997). Patterns in the distribution, abundance and body size of carabid beetles (Coleoptera: Caraboidea) in relation to dispersal ability. *Journal of Biogeography*, **24**, 903–14.

Hadley, A. S., and Betts, M. G. (2012). The effects of landscape fragmentation on pollination dynamics: absence of evidence not evidence of absence. *Biological Reviews*, **87**, 526–44.

Hanski, I. (1999). *Metapopulation Ecology*. Oxford University Press, Oxford.

Hanski, I. (2005). *The Shrinking World: Ecological Consequences of Habitat Loss*. International Ecology Institute, Oldendorf/Luhe.

Hanski, I., and Ovaskainen, O. (2000). The meatapopulation capacity of a fragmented landscape. *Nature*, **404**, 755–8.

Harrison, S., and Bruna, E. (1999). Habitat fragmentation and large-scale conservation: what do we know for sure? *Ecography*, **22**, 225–32.

Henle, K. (2004). Species survival in fragmented landscapes: where are we now? *Biodiversity and Conservation*, **13**, 1–8.

Hermann, J. D., Kormann, U., Schüepp, C., Stocker, Y., Herzog, F., and Entling, M. H. (2012). Effects of habitat isolation and predation pressure on an arboreal food-web *Community Ecology*, **13**, 82–7.

Hill, J. K., Collingham, Y. C., Thomas, C. D., et al. (2001). Impacts of landscape structure on butterfly range expansion. *Ecology Letters*, **4**, 313–21.

Hodgson, J. A., Thomas, C. D., Wintle, B. A., and Moilanen, A. (2009). Climate change, connectivity and conservation decision making: back to basics. *Journal of Applied Ecology*, **46**, 964–9.

Holland, J. M., and Luff, M. L. (2000). The effects of agricultural practices on Carabidae in temperate agroecosystems. *Integrated Pest Management Reviews*, **5**, 109–9.

Holzschuh, A., Steffan-Dewenter, I., and Tscharntke, T. (2009). Grass strip corridors in agricultural landscapes enhance nest-site colonization by solitary wasps. *Ecological Applications*, **19**, 123–32.

Jauker, F., Diekötter, T., Schwartzbach, F., and Wolters, V. (2009). Pollinator dispersal in an agricultural matrix: opposing responses of wild bees and hoverflies to landscape structure and distance from main habitat. *Landscape Ecology*, **24**, 547–55.

Jennings, N., and Pocock, M. J. (2009). Relationships between sensitivity to agricultural intensification and ecological traits of insectivorous mammals and arthropods. *Conservation Biology*, **23**, 1195–1203.

Johst, K., Brandl, R., and Pfeifer, R. (2001). Foraging in a patchy and dynamic landscape: human land use and the white stork. *Ecological Applications*, **11**, 60–9.

Kisdi, E. (2002). Dispersal: risk spreading versus local adaptation. *American Naturalist*, **159**, 579–96.

Kleijn, D., Kohler, F., Baldi, A., et al. (2009). On the relationship between farmland biodiversity and land-use intensity in Europe. *Proceedings of the Royal Society B: Biological Sciences*, **276**, 903–9.

Kleijn, D., Rundlöf, M., Scheper, J., Smith, H. G., and Tscharntke, T. (2011). Does conservation on farmland contribute to halting the biodiversity decline? *Trends in Ecology and Evolution*, **26**, 474–81.

Komonen, A., Grapputo, A., Kaitala, V., Kotiaho, J. S., and Päivinen, J. (2004). The role of niche breadth, resource availability and range position on the life history of butterflies. *Oikos*, **105**, 41–54.

Krauss, J., Alfert, T., and Steffan-Dewenter, I. (2009). Habitat area but not habitat age determines wild bee richness in limestone quarries. *Journal of Applied Ecology*, **46**, 194–202.

Krauss, J., Klein, A. M., Steffan-Dewenter, I., and Tscharntke, T. (2004). Effects of habitat area, isolation, and landscape diversity on plant species richness of calcareous grasslands. *Biodiversity and Conservation*, **13**, 1427–39.

Krauss, J., Steffan-Dewenter, I., and Tscharntke, T. (2003). How does landscape context contribute to effects of habitat fragmentation on diversity and population density of butterflies? *Journal of Biogeography*, **30**, 889–900.

Krewenka, K. M., Holzschuh, A., Tscharntke, T., and Dormann, C. F. (2011). Landscape elements as potential barriers and corridors for bees, wasps and parasitoids. *Biological Conservation*, **144**, 1816–25.

Kromp, B. (1999). Carabid beetles in sustainable agriculture: a review on pest control efficacy, cultivation impacts and enhancement. *Agriculture, Ecosystems and Environment*, **74**, 187–228.

Krooss, S., and Schaefer, M. (1998). How predacious are predators? A study on *Ocypus similis*, a rove beetle of cereal fields. *Annals of Applied Biology*, **133**, 1–16.

Kuussaari, M., Heliölä, J., Pöyry, J., and Saarinen, K. (2007). Contrasting trends of butterfly species preferring semi-natural grasslands, field margins and forest edges in northern Europe. *Journal of Insect Conservation*, **11**, 351–66.

Kuussaari, M., Nieminen, M., and Hanski, I. (1996). An experimental study of migration in the Glanville fritillary butterfly *Melitaea cinxia*. *Journal of Animal Ecology*, **65**, 791–801.

Lagerlöf, J., and Wallin, H. (1993). The abundance of arthropods along two field margins with different types of vegetation composition: an experimental study. *Agriculture, Ecosystems and Environment*, **43**, 141–54.

Lambeets, K., Vandegehuchte, M. L., Maelfait, J.-P., and Bonte, D. (2008). Understanding the impact of flooding on trait-displacements and shifts in assemblage structure of predatory arthropods on river banks. *Journal of Animal Ecology*, **77**, 1162–74.

Landis, D. A., Wratten, S. D., and Gurr, G. M. (2000). Habitat management to conserve natural enemies of arthropod pests in agriculture. *Annual Review of Entomology*, **45**, 175–201.

Lavorel, S., and Garnier, S. (2002). Predicting changes in community composition and ecosystem functioning from plant traits: revisiting the Holy Grail. *Functional Ecology*, **16**, 545–56.

Leibold, M. A., Holyoak, M., Amarasekare, P., et al. (2004). The metacommunity concept: a framework for multi-scale community ecology. *Ecology Letters*, **7**, 601–13.

Lima, S. L., and Zollner, P. A. (1996). Towards a behavioral ecology of ecological landscapes. *Trends in Ecology and Evolution*, **11**, 131–5.

Luck, G. W., Lavorel, S., Mcintyre, S., and Lumb, K. (2012). Improving the application of vertebrate trait-based frameworks to the study of ecosystem services. *Journal of Animal Ecology*, **81**, 1065–76.

Marshall, E. J. R., and Moonen, A. C. (2002). Field margins in northern Europe: their functions and interactions with agriculture. *Agriculture Ecosystems and Environment*, **89**, 5–21.

Mattila, N., Kaitala, V., Komonen, A., Päivinen, J., and Kotiaho, J. S. (2011). Ecological correlates of distribution change and range shift in butterflies. *Insect Conservation and Diversity*, **4**, 239–46.

Meehan, T. D., Werling, B. P., Landis, D. A., and Gratton, C. (2011). Agricultural landscape simplification and insecticide use in the Midwestern United States. *Proceedings of the National Academy of Sciences USA*, **108**, 11500–5.

Miller, J. N., Brooks, R. P., and Croonquist, M. J. (1997). Effects of landscape patterns on biotic communities. *Landscape Ecology*, **12**, 137–53.

Nabe-Nielsen, J., Sibly, R. M., Forchhammer, M. C., Forbes, V. E., and Topping, C. J. (2009). The effects of landscape modifications on the long-term persistence of animal populations. *PLoS One*, **5**, e8932.

Norris, R. F., and Kogan, M. (2005). Ecology of interactions between weeds and arthropods. *Annual Review of Entomology*, **50**, 479–503.

Öckinger, E., Franzen, M., Rundlöf, M., and Smith, H. G. (2009). Mobility-dependent effects on species richness in fragmented landscapes. *Basic and Applied Ecology*, **10**, 573–8.

Öckinger, E., Schweiger, O., Crist, T. O., et al. (2010). Life-history traits predict species responses to habitat area and isolation: a cross-continental synthesis. *Ecology Letters*, **13**, 969–79.

Öckinger, E., and Smith, H. G. (2006). Landscape composition and habitat area affects butterfly species richness in semi-natural grasslands. *Oecologia*, **149**, 526–34.

Öckinger, E., and Smith, H. G. (2007a). Asymmetric dispersal and survival indicate population sources for grassland butterflies in agricultural landscapes. *Ecography*, **30**, 288–98.

Öckinger, E., and Smith, H. G. (2007b). Semi-natural grasslands as population sources for pollinating insects in agricultural landscapes. *Journal of Applied Ecology*, **44**, 50–9.

Öckinger, E., and Smith, H. G. (2008). Do corridors promote dispersal in grassland butterflies and other insects? *Landscape Ecology*, **23**, 27–40.

Olsson, O., Brown, J. S., and Helf, K. L. (2008). A guide to central place effects in foraging. *Theoretical Population Biology*, **74**, 22–33.

Olsson, O., and Rogers, D. J. (2009). Predicting the distribution of a suitable habitat for the white stork in southern Sweden: identifying priority areas for reintroduction and habitat restoration. *Animal Conservation*, **12**, 62–70.

Ouin, A., Aviron, S., Dover, J., and Burel, F. (2004). Complementation/supplementation of resources for butterflies in agricultural landscapes. *Agriculture, Ecosystems and Environment*, **103**, 473–9.

Pereira, H. M., Navarro, L. M., and Martins, I. S. (2012). Global biodiversity change: the bad, the good, and the unknown. *Annual Review of Environment and Resources*, **37**, 25–50.

Persson, A. S. (2011). Effects of Landscape Context on Bumblebees. PhD-thesis, Lund university.

Phalan, B., Onial, M., Balmford, A., and Green, R. E. (2011). Reconciling food production and biodiversity conservation: land sharing and land sparing compared. *Science*, **333**, 1289–91.

Polis, G. A., Anderson, W. B., and Holt, R. D. (1997). Toward an integration of landscape and food web ecology: the dynamics of spatially subsidized food webs. *Annual Reviews of Ecology and Systematics*, **28**, 289–316.

Poschlod, P., and WallisDeVries, M. F. (2002). The historical and socioeconomic perspective of calcareous grasslands—lessons learned from the distant and recent past. *Biological Conservation*, **104**, 361–76.

Pöyry, J., Luoto, M., Heikkinen, R. K., Kuussaari, M., and Saarinen, K. (2009). Species traits explain recent range shifts of Finnish butterflies. *Global Change Biology*, **15**, 732–43.

Prugh, L. R., Hodges, K. E., Sinclair, A. R., and Brashares, J. S. (2008). Effect of habitat area and isolation on fragmented animal populations. *Proceedings of the National Academy of Sciences USA*, **105**, 20770–5.

Purtauf, T., Roschewitz, I., Dauber, J., Thies, C., Tscharntke, T., and Wolters, V. (2005). Landscape context of organic and conventional farms: influences on carabid beetle diversity. *Agriculture Ecosystems and Environment*, **108**, 165–74.

Rainio, J., and Niemelä, J. (2003). Ground beetles (Coleoptera: Carabidae) as bioindicators. *Biodiversity and Conservation*, **12**, 487–506.

Rand, T. A., Tylianakis, J. M., and Tscharntke, T. (2006). Spillover edge effects: the dispersal of agriculturally subsidized insect natural enemies into adjacent natural habitats. *Ecology Letters*, **9**, 603–14.

Rand, T. A., Van Veen, F. J. F., and Tscharntke, T. (2012). Landscape complexity differentially benefits generalized fourth, over specialized third, trophic level natural enemies. *Ecography*, **35**, 97–104.

Ricketts, T. H., Regetz, J., Steffan-Dewenter, I., et al. (2008). Landscape effects on crop pollination services: are there general patterns? *Ecology Letters*, **11**, 499–515.

Ries, L., Fletcher, R. J., Battin, J., and Sisk, T. D. (2004). Ecological responses to habitat edges: mechanisms, models, and variability explained. *Annual Review of Ecology, Evolution, and Systematics*, **41**, 491–522.

Robinson, R. A., and Sutherland, W. J. (2002). Post-war changes in arable farming and biodiversity in Great Britain. *Journal of Applied Ecology*, **39**, 157–76.

Roland, J., and Taylor, P. D. (1997). Insect parasitoid species respond to forest structure at different spatial scales. *Nature*, **386**, 710–13.

Rösch, V., Tscharntke, T., Scherber, C., Batáry, P., and Osborne, J. (2013). Landscape composition, connectivity and fragment size drive effects of grassland fragmentation on insect communities. *Journal of Applied Ecology*, **50**, 387–94.

Rounsevell, M., Reginster, I., Araújo, M., et al. (2006). A coherent set of future land use change scenarios for Europe. *Agriculture, Ecosystems and Environment*, **114**, 57–68.

Rundlöf, M., Nilsson, H., and Smith, H. G. (2008). Interacting effects of farming practice and landscape context on bumblebees. *Biological Conservation*, **141**, 417–26.

Rundlöf, M., and Smith, H. G. (2006). The effect of organic farming on butterfly diversity depends on landscape context. *Journal of Applied Ecology*, **43**, 1121–7.

Schmidt, M. H., Thies, C., Nentwig, W., and Tscharntke, T. (2008). Contrasting responses of arable spiders to the landscape matrix at different spatial scales. *Journal of Biogeography*, **35**, 157–66.

Schröter, L., and Irmler, U. (2013). Organic cultivation reduces barrier effect of arable fields on species diversity. *Agriculture, Ecosystems and Environment*, **164**, 176–80.

Schweiger, O., Maelfait, J. P., Van Wingerden, W., et al. (2005). Quantifying the impact of environmental factors on arthropod communities in agricultural landscapes across organizational levels and spatial scales. *Journal of Applied Ecology*, **42**, 1129–39.

Sekar, S. (2012). A meta-analysis of the traits affecting dispersal ability in butterflies: can wingspan be used as a proxy? *Journal of Animal Ecology*, **81**, 174–84.

Shmida, A., and Wilson, M. V. (1985). Biological determinants of species diversity. *Journal of Biogeography*, **12**, 1–20.

Smith, H. G., Öckinger, E., and Rundlöf, M. (2010). Biodiversity and the landscape ecology of agri-environment schemes. *Aspects of Applied Biology*, **100**, 225–32.

Steffan-Dewenter, I., and Kuhn, A. (2003). Honeybee foraging in differentially structured landscapes. *Proceedings of the Royal Society B: Biological Sciences*, **270**, 569–75.

Steffan-Dewenter, I., Munzenberg, U., Burger, C., Thies, C., and Tscharntke, T. (2002). Scale-dependent effects of landscape context on three pollinator guilds. *Ecology*, **83**, 1421–32.

Steffan-Dewenter, I., and Tscharntke, T. (2000). Butterfly community structure in fragmented habitats. *Ecology Letters*, **3**, 449–56.

Sterk, G., Hassan, S. A., Baillod, M., et al. (1999). Results of the seventh joint pesticide testing programme carried out by the IOBC/WPRS-working group 'Pesticides and Beneficial Organisms'. *BioControl*, **44**, 99–117.

Stoate, C., Báldi, A., Beja, P., et al. (2009). Ecological impacts of early 21st century agricultural change in Europe: a review. *Journal of Environmental Management*, **91**, 22–46.

Symondson, W. O. C., Sunderland, K. D., and Greenstone, M. H. (2002). Can generalist predators be effective biocontrol agens? *Annual Review of Entomology*, **47**, 561–94.

Taylor, P. D., Fahrig, L., Henein, K., and Merriam, G. (1993). Connectivity is a vital element of landscape structure. *Oikos*, **68**, 571–3.

Thies, C. (1999). Landscape structure and biological control in agroecosystems. *Science*, **285**, 893–5.

Thies, C., and Tscharntke, T. (2010). Biological rape pest control in spatio-temporally changing landscapes. In I. H. Williams (eds), *Biocontrol-Based Integrated Management of Oilseed Rape Pests*. Springer, New York.

Thomas, C. D. (2000). Dispersal and extinction in fragmented landscapes. *Proceedings of the Royal Society B: Biological Sciences*, **267**, 139–45.

Thomas, C. D., Gillingham, P. K., Bradbury, R. B., et al. (2012). Protected areas facilitate species' range expansions. *Proceedings of the National Academy of Sciences USA*, **109**, 14063–8.

Thomas, C. F. G., Holland, J. M., and Brown, N. J. (2002). The spatial distribution of carabid beetles in agricultural landscapes. In J. M. Holland (ed.), *The Agroecology of Carabid Beetles*. Intercept, Andover.

Thorbek, P., and Bilde, T. (2004). Reduced numbers of generalist arthropod predators after crop management. *Journal of Applied Ecology*, **41**, 526–38.

Tscharntke, T., Bommarco, R., Clough, Y., et al. (2007). Conservation biological control and enemy diversity on a landscape scale. *Biological Control*, **43**, 294–309.

Tscharntke, T., and Brandl, R. (2004). Plant-insect interactions in fragmented landscapes. *Annual Review of Entomology*, **49**, 405–30.

Tscharntke, T., Klein, A. M., Kruess, A., Steffan-Dewenter, I., and Thies, C. (2005). Landscape perspectives on agricultural intensification and biodiversity – ecosystem service management. *Ecology Letters*, **8**, 857–74.

Tscharntke, T., Steffan-Dewenter, I., Kruess, A., and Thies, C. (2002). Characteristics of insect populations on habitat fragments: a mini review. *Ecological Research*, **17**, 229–39.

Tscharntke, T., Tylianakis, J. M., Rand, T. A., et al. (2012). Landscape moderation of biodiversity patterns and processes - eight hypotheses. *Biological Reviews*, **87**, 661–85.

Van Geert, A., Van Rossum, F., and Triest, L. (2010). Do linear landscape elements in farmland act as biological corridors for pollen dispersal? *Journal of Ecology*, **98**, 178–87.

Van Nouhuys, S., and Hanski, I. (2002). Colonization rates and distances of a host butterfly and two specific parasitoids in a fragmented landscape. *Journal of Animal Ecology*, **71**, 639–50.

Van Swaay, C., Collins, S., Dušej, G., et al. (2012). Dos and don'ts for butterflies of the habitats directive of the European union. *Nature Conservation*, **1**, 73–153.

Vandermeer, J., and Carvajal, R. (2001). Metapopulation dynamics and the quality of the matrix. *American Naturalist*, **158**, 211–20.

Vandermeer, J., and Perfecto, I. (2007). The agricultural matrix and a future paradigm for conservation. *Conservation Biology*, **21**, 274–7.

Vasseur, C., Joannon, A., Aviron, S., Burel, F., Meynard, J.-M., and Baudry, J. (2013). The cropping systems mosaic: how does the hidden heterogeneity of agricultural landscapes drive arthropod populations? *Agriculture, Ecosystems and Environment*, **166**, 3–14.

Warren, M. S., Hill, J. K., Thomas, J. A., et al. (2001). Rapid responses of British butterflies to opposing forces of climate and habitat change. *Nature*, **414**, 65–9.

Westphal, C., Steffan-Dewenter, I., and Tscharntke, T. (2003). Mass flowering crops enhance pollinator densities at a landscape scale. *Ecology Letters*, **6**, 961–5.

Weyman, G. S., Sunderland, K. D., and Jepson, P. C. (2002). A review of the evolution and mechanisms of ballooning by spiders inhabiting arable farmland. *Ethology Ecology and Evolution*, **14**, 307–26.

Wiklund, C. (1977). Oviposition, feeding and spatial separation of breeding and foraging habitats in a population of *Leptidea sinapis* (Lepidoptera). *Oikos*, **28**, 56–68.

Williams, N. M., and Kremen, C. (2007). Resource distribution among habitats determine solitary bee offspring reproduction in a mosaic landscape. *Ecological Applications*, **17**, 910–21.

Zeller, K. A., McGarical, K., and Whiteley, A. R. (2012). Estimating landscape resistance to movement: a review. *Landscape Ecology*, **27**, 777–97.

PART II

Movement Strategies and Adaptations

Chapter 5. Migration and flight strategies in animals: new insights from tracking migratory journeys
Chapter 6. Individuality in movement: the role of animal personality
Chapter 7. Dispersal and phenotypic plasticity
Chapter 8. Pathogens and hosts on the move

CHAPTER 5

Migration and flight strategies in animals: new insights from tracking migratory journeys

Anders Hedenström and Åke Lindström

Bird migration has inspired scientific inquiry about how animals manage to navigate over uncharted seas and land, how they can fly non-stop across inhospitable deserts and oceans, as well as more basic questions such as, 'Where do they migrate?' Occasional recoveries of ringed birds from exotic countries have fuelled our curiosity and initiated major or minor expeditions to sites where particular birds may winter, but still their journeys remained a mystery. The study of individual birds was revolutionized by the tracking of birds large enough to lift the weight of the first satellite transmitters (Jouventin and Weimerskirch 1990). With the subsequent reduction in the mass of transmitters, it has been possible to study smaller and smaller species (Gill et al. 2009, Strandberg et al. 2009). However, the smallest songbird and shorebird migrants defied tracking until very recently when lightweight (< 2 g) geolocators entered the scene (Stutchbury et al. 2009; Box 5.1). These devices record and register the ambient light at the position of the bird, and when a bird returns after a full migration, the geolocator contains information about the timing of sunrise/sunset and the length of each day. These data can be used to tell the approximate longitude and latitude of the logger (and hence the bird). Geolocators are less precise than satellite transmitters; a higher precision than ±100 km is difficult to achieve (Phillips et al. 2004), but this technique has nevertheless added another dimension to studies on migration. It has allowed reasonable reconstruction of the migratory routes and wintering areas for birds as small as 20 g. This technique not only allows us to fill the gaps in our knowledge about bird migration routes and wintering areas, but also gives us new opportunities to test predictions about migration strategies as derived from optimal migration theory (Alerstam and Lindström 1990, Alerstam and Hedenström 1998, Alerstam 2011). In this chapter we present some exciting recent findings about migratory journeys in birds, bats, and insects, and ask what they can teach us about migration strategies in general and optimal migration theory in particular.

5.1 Migration strategies

Migration by animals most likely evolved because it conferred improved survivorship to individuals that migrated, in relation to individuals that stayed all year around at one site. The migrants may in this way have reached sites more suitable for reproduction, maintenance (such as moult), or spending the winter. Once migration is the preferred option there are a number of alternative strategies on how to carry out the migration. Animals may migrate according to some alternative rules, for example: (1) minimize the cost of transport (COT), (2) minimize the duration of migration, or (3) minimize the total risk of predation during migration (Alerstam and Lindström 1990; Johansson et al., Chapter 13). Whereas there are also variants of these strategies, and migrants may use more than one strategy

Animal Movement Across Scales. Edited by Lars-Anders Hansson and Susanne Åkesson.
© Oxford University Press 2014. Published 2014 by Oxford University Press.

during a journey, these three are the principle alternatives that are often contrasted against each other in migration studies. For example, when migrating as fast as possible, is it best to make frequent short moves or are a few long flights preferred? Which route is best? In many cases the geography dictates the length of flights, such as across ecological barriers like deserts and oceans, but also the distribution of suitable habitats for feeding plays an important role. Migrants may use many different tactics and behaviours to fulfil their strategies and we are interested in finding tools that help us to understand the adaptive significance of different behaviours. In birds a number of different patterns of migration have evolved, such as partial, leap-frog, and chain migration (Fig. 2.2; Chapman et al., Chapter 2). For general accounts on the natural history of migration patterns see Alerstam (1990) and Dingle (1996).

Migration and flight strategies can be analysed and evaluated on the basis of optimality reasoning, which is a research field that has developed and matured over the past three decades (Alerstam and Lindström 1990, Alerstam and Hedenström 1998, Alerstam 2011). Today 'optimal migration' serves as a natural point of departure when asking questions about migration. A cornerstone of optimal migration theory is the flight mechanical theory developed and refined by Colin J. Pennycuick (e.g. Pennycuick 2008). A fundamental equation emerging from this theory is the relationship between the power required to fly, P_{flight}, and speed through the air, U, which assumes a characteristic U-shape (Fig. 5.1a). From this 'power curve' we can derive the optimal flight speeds associated with minimum power (U_{mp} results in the longest time airborne on a given fuel load) and the speed of minimum cost of transport (U_{mr} results in the longest distance covered on a given fuel load).

Another fundamental concept is the potential flight range, which is described by a diminishing return utility function (Fig. 5.1b). The reason for the negatively accelerated function is that as fuel (mainly fat) is added to the body, the total body weight increases, which increases the power consumption during flight. Another contributing factor is that added fat, which is stored subcutaneously, increases the frontal body area of the bird and therefore the aerodynamic drag, and thereby the power needed for flight increases. In other words, an additional gram of fat will not have the same value measured in flight distance to an already very heavy bird as it will for a bird without any fat reserves. In optimality theory a curve like that in Fig. 5.1b immediately signals an opportunity for making a best possible choice among alternatives.

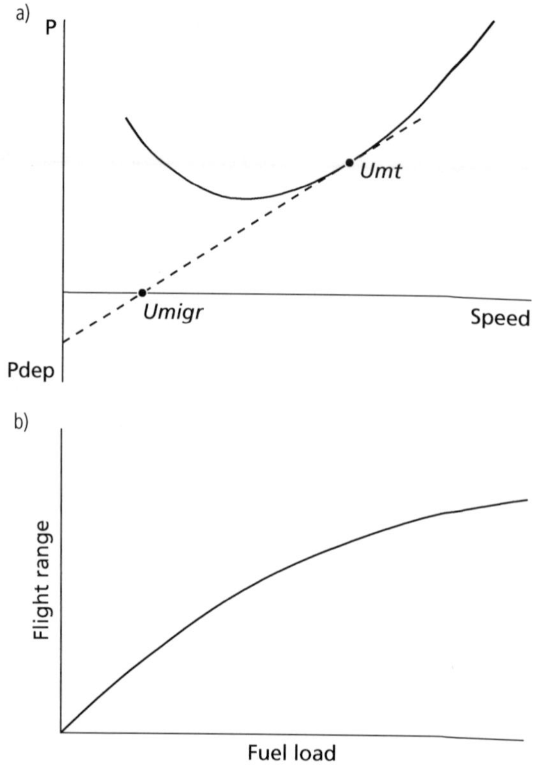

Figure 5.1 (a) Power (P) required for animal flapping flight in relation to airspeed. The negative y-axis shows the rate of energy (fuel) deposition at stopovers in the same unit as power consumption in flight. The dotted line is a tangent from a hypothetical energy deposition rate to the power curve, with the optimal speed of time minimizing migration (U_{mt}) and the overall migration speed (U_{migr}) indicated. (b) Flight range in relation to fuel load for an animal using flapping flight.

The flight range curve can be used to generate testable predictions about migration behaviours. Depending on which strategy is applied and which assumptions exist in regards to the cost to search/settle when arriving at a new stopover, one can predict at what fuel load a bird should depart from a stopover site. By comparing predictions derived from optimal migration theory with the responses of birds exposed to experimental manipulation of food availability and fuel accumulation rate or perceived predation risk, we may conclude whether they migrate according to energy or time selection, or whether survival selection is important (Lindström and Alerstam 1992, Dänhardt and Lindström 2001).

The overall migration speed can be written as

$$U_{migr} = \frac{P_{dep} \cdot U_{flight}}{P_{dep} + P_{flight}}, \quad (1)$$

where U_{flight} is the flight speed through air, P_{dep} is the rate of energy accumulation (at stopovers), and P_{flight} is the power required to fly. This is a very general and yet fundamental formulation of the overall speed of migration, including the time needed for fuel accumulation. In the case of flapping flight, Eq. (1) can be graphically illustrated as shown in Fig. 5.1b, where the curve is the power curve. The power curve is often calculated based on aerodynamic theory (e.g. Pennycuick 2008), while its U-shape has been confirmed by empirical measurements of flight metabolic rate (Engel et al. 2010). In principle it is also possible to tell whether a bird is migrating according to time or energy minimization by measuring the flight (air) speed and comparing it with predicted speeds of U_{mr} or U_{mt} (Fig. 5.1a), but in practice this is difficult because we rarely know the exact weight and wing morphology of a migrant passing an observation point, and there are also some uncertainties regarding the underlying aerodynamic model.

5.2 Examples from tracking studies of migratory journeys

Birds are by far the most well-studied group of migrants when it comes to migration routes, length of migratory flights, and different modes of migration. In this section we give some examples of how avian migrants have solved these issues in various ways, and we discuss which ecological and evolutionary forces may have shaped them.

5.2.1 The giant leaps

It has been long assumed that, in order to cross ecological barriers such as seas and deserts with thousands of kilometres of unsuitable habitat for landing or feeding, birds must be able to make long non-stop flights across the barrier. A few decades ago, the longest non-stop migratory flights known to be carried out by birds were 3,000–4,000 km, normally to be covered in 3–4 days of continuous flight (Alerstam 1990; Lindström et al., Chapter 3). They were in all cases based on the known appearance of birds at each side of an ecological barrier, such as terrestrial birds crossing vast stretches of water or ice, where intermittent fuelling or even landing was impossible. Distances of about 4,000 km were also the maximum distances that the most advanced flight models allowed (Pennycuick 1975), given that the highest fuel loads known to be carried were equivalent to a doubling of body mass. No doubt, from a human perspective, these distances were impressive enough to be accepted as the maximum possible. Nevertheless, Alerstam (1990) speculated, based on the vast ringing recovery data of ruddy turnstones, *Arenaria interpres*, from the northern Pacific (Thompson 1974), that in cases of strong wind assistance and exceptional fuel loads, it could not be ruled out that some individuals fly 5,500 km non-stop, and maybe even a mind-blowing 7,000 km (which was indeed suggested by Thompson 1974).

In the past few years, the tracking of individual birds with satellite transmitters and geolocators has not only repeatedly confirmed the existence of such flights of several thousand kilometres, but shown that the non-stop flight capacity of many migrants is beyond our wildest expectations. For example, great snipes, *Gallinago media*, flew up to 6,800 km non-stop from Sweden to Africa (Klaassen et al. 2011), and ruddy turnstones made 7,600-km non-stop flights in 6–7 days from Australia to Taiwan (Minton et al. 2010). Even more impressive, pacific golden plovers, *Pluvialis fulva*, flew on average 9,800 km non-stop in 5–7 days from Alaska to the American Samoa Islands in the southern Pacific Ocean (Johnson et al. 2012). Most impressive of all, however, are the bar-tailed godwits, *Limosa lapponica baueri*, that in spring fly 10,000 km from New Zealand to China in 6–8 days and in autumn fly 11,000–12,000 km non-stop from Alaska to New Zealand in about 8 days (Gill et al. 2009, Battley et al. 2012). Prior to the giant leap of all these cases, the birds had fuelled extensively and had at least doubled their mass from fuel deposition.

At a first glance, basically everything speaks against natural selection to have favoured such flights, and the general belief has been that long flights evolve when there are simply no other options to reach a certain destination. Flight incurs some of the highest metabolic rates known among animals (5–17 times the basal metabolic rate (BMR) or more; Hedenström 2008), not least for birds with huge fuel loads, and such hard work for a week ought to impose dramatic physiological stress and immunological costs (e.g. Deerenberg et al. 1997). Further, in most of the cases we've just referred to, there are simply very few emergency landing sites along the flight route in case of inclement weather, physiological problems, or if an individual runs out of fuel. The safety margins seem to be much too small for such behaviour to evolve. But actually, this may be a too anthropocentric way of seeing things. It actually seems as if a week of continuous flight for the birds is no big deal.

First, annual survival is high in adult bar-tailed godwits and many other long-distance migrant shorebirds (80–90%; Conklin and Battley 2012, Leyrer et al. 2012), which shows that many individuals do these flights for many years. Second, red knots, *Calidris canutus*, flying the equivalent of 1,500 km in a wind tunnel had no detectable effects of these flights on either their immune response or their stress hormones (Hasselquist et al. 2007). This indicates that long-distance migrants may be well adapted to the workload induced by long flights. Gill et al. (2009) turned the prevailing view upside down and instead suggested that long over-water flights give the migrants several important advantages, such as the complete avoidance of both predators and pathogens along the route. The importance of pathogens is further discussed in Westerdahl et al. (Chapter 8).

5.2.2 Repeated flights and multi-day stops

The most common migration strategy, not least among songbirds, consists of repeated cycles where a few days of foraging are followed by one day or night of flight, until the final destination is reached. Exceptions to this rule may come when the birds face an ecological barrier such as a wide desert or sea. They may then have to temporarily switch to extensive fuelling and a long flight across the barrier. Alternatively, they make a detour and circumvent it in smaller steps (Alerstam 2001). When it comes to the time spent fuelling at stopovers in relation to that spent airborne, Hedenström and Alerstam (1997) estimated a ratio of 7:1. Hence for many birds, most of the time during migration is actually spent on the ground fuelling, which at first may sound somewhat counterintuitive but shows how important fuelling is for migration and that flight is energetically very demanding (Johansson et al., Chapter 13).

The northern wheatear, *Oenanthe oenanthe*, has a nearly circumpolar breeding distribution in the northern hemisphere, with an absence only between 120 and 140°W in central North America. All populations migrate to wintering quarters in tropical Africa, but depending on their breeding area they may approach Africa from opposite directions. The populations breeding in Greenland and eastern North America cross the North Atlantic on their way to Western Europe (Delingat et al. 2008), and may fly directly to West Africa from Greenland (Thorup et al. 2006). At the other extreme of the wheatear's distribution, Eastern Siberia and Alaska, the birds must migrate westwards in the autumn to reach their final winter destinations in East Africa.

When the migration has a large east–west component, especially at high latitudes, it is often beneficial to migrate along orthodromes (great circles; Fig. 5.2) since they are the shortest distances between two locations on the spherical Earth. However, this requires a continuous shift in orientation direction, which in theory can be achieved by using a time-compensated sun compass, without resetting the internal clock as new time zones are entered (Alerstam and Pettersson 1991; Åkesson et al., Chapter 9, and Muheim et al., Chapter 10). Shorebirds breeding in the high Arctic and migrating to the southern hemisphere are likely candidates for using orthodromic routes (Alerstam et al. 2001). The loxodrome (rhumb line; Fig. 5.2) is an alternative route associated with a constant compass course (on a

Mercator map projection loxodromes are straight lines). To migrate along a loxodrome is perhaps easier to achieve by a bird because it only needs to maintain a constant compass direction, but it comes with a cost of being longer than the orthodrome. These two alternative migration routes are often contrasted in relation to the observed migration tracks of birds. Alerstam et al. (2008) suggested that a few passerine species breeding in the high Arctic in the Beringia region (northeastern Russia and Alaska), including the wheatear, actually do follow orthodromes to their topical wintering areas (East Africa in the case of the wheatear). This suggestion was based on initial migration directions obtained by tracking radar onboard an ice-breaker in the Chuckchi Sea, northeastern Russia. Passerines could be distinguished from other birds by using the characteristic radar echo signature produced by the intermittent wing-beat pattern of these birds. From this region the initial flight direction along a great circle is about 311°, i.e. towards NW (Fig. 5.2). This was consistent with the flight directions of passerines over the Chukchi Sea.

In a recent study, Bairlein et al. (2012) used geolocators (Box 5.1) to trace the migratory route taken by three individual wheatears that were breeding in Alaska (Fig. 5.2). Surprisingly the Alaskan wheatears migrated along a route that was intermediate between the orthodrome and the loxodrome (Fig. 5.2). It took these birds on average 91 days to complete the 14,600-km migration, which was 22% longer than the orthodrome but 6% shorter than the loxodrome. Hence, these wheatears do not use any of the simple orientation mechanisms that researchers have thought are the alternatives available. There is actually a third orientation mechanism based on migration according to the apparent inclination angle, which would give curved migration paths like those of the Alaskan wheatears (Kiepenheuer 1984), but this mechanism does not allow for the sharp direction changes as shown by the Nearctic breeding wheatears. Therefore, this alternative is not a likely candidate as a general orientation mechanism. These data certainly raise new questions about the orientation/navigation mechanism(s) used by long-distance migrants. Is there an inherited programme that contains a pre-defined itinerary that takes them from start to goal? Is there an element of a true navigation system where the birds somehow know their GPS position at all times, or is there a vector of clock-and-compass instructions that is executed?

What about the wheatears breeding on Baffin Island in NE America? So far only one bird has been tracked by geolocators (Bairlein et al. 2012), and this bird showed a dog-leg migration via the British Isles to the wintering area in Mauritania, West Africa. The first part of the migration from Baffin Island across the Atlantic to the British Isles is a 3,400-km flight that took the bird 4 days to complete, while the remaining 4,000 km to Africa another 22 days.

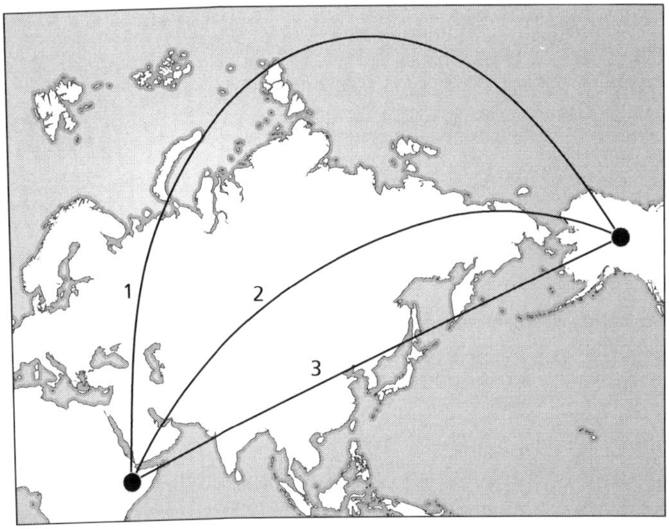

Figure 5.2 Hypothetical migration routes according to the alternative orientation mechanisms orthodrome (1) and loxodrome (3) for wheatears breeding in Alska and migrating to East Africa, with the actual migration by three individuals as recorded by geolocators (2) (Bairlein et al. 2012).

Box 5.1 Geolocation by measuring light

The principle of using ambient light to determine geographical position is well known among navigators. Put simply, the time of sunrise or sunset provides information about longitude, while day (or night) length gives latitude. More sophisticated methods such as analysing the slope of light intensity curves during sunrise and sunset can be used to determine latitude during periods around equinoxes (see end of paragraph). The use of geolocation to study animal movements involves the attachment of an archival tag ('geolocator' or 'lightlogger') to the animal, which is then retrieved for data download and analysis. The geolocator contains a light sensor, a microprocessor and accurate clock, a memory, and a battery (Fig. 5.3). Light intensity levels are typically measured every minute and the maximum values are stored in the memory for each 5-min period. The geolocator is attached to the bird by using a harness to position the geolocator on the bird's back, or in cases of larger birds the geolocator can be attached to the ring. The technique of geolocation was initially developed for marine birds about 20 years ago, but thanks to the miniaturization of electronic components and batteries the reduction of tag weights (currently about 0.5 g) geolocators are now widely used to study migration in small terrestrial birds. A rule of thumb says that the tag weight should not exceed 3% of the bird's own body weight. Longitudinal positions are more accurate than latitudes (typically 50 km versus 150 km), while during periods around the equinoxes latitudinal positions become uncertain due to little geographical variation in day length. Also, close to the poles there are periods of constant daylight that prevent use of geolocation. For general reading about geolocation, please consider Ekstrom (2004) and Bridges et al. (2013).

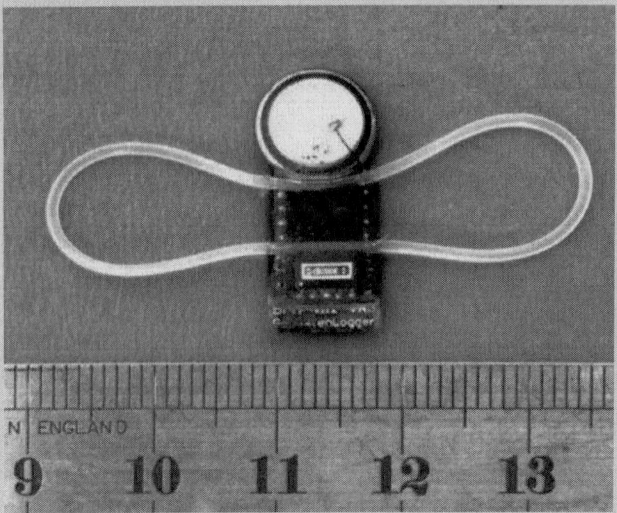

Figure 5.3 A geolocator, or 'lightlogger', is used to study animal movements. The device contains a light sensor, a microprocessor and accurate clock, a memory, and a battery and is attached to the bird by using a harness to position the geolocator.

A direct flight across the Atlantic from Baffin Island to Mauritania is about 6,400 km, which means that the migration route taken is a 16% detour. For comparison, the Alaskan breeding birds made a detour that was 22% longer than the orthodrome (Fig. 5.2).

Assuming a flight speed of 46 km/h the observed migration speed suggests that the birds are flying during 14.5% of the time, corresponding to a flight/stop ratio of 6.9:1 between stopover and flight time (compare with the predicted ratio of 7:1, Hedenström and Alerstam 1997). A single wheatear migrated from Baffin Island to the British Isles in 4 days at 850 km/day (Bairlein et al. 2012), but this exclusively reflects a long non-stop flight and does not account for the preceding fuelling time. During the second migration leg from the British Isles to West Africa the migration speed of this individual was more modest at 182 km/day, which is closer to the expected stopover:flight time ratio. The bird probably arrived quite lean from the transatlantic flight, and thus had to start the second leg of migration from zero. This shows that an accurate estimate of migration speed must include the first fuelling episode—actually migration starts when the bird begins to accumulate fuel for migration (see section 5.2.6)—which means that the first 'stopover' is at the breeding and wintering sites, respectively. However, when analysing tracking data, such as those obtained by geolocators, it is often impossible to know when fuelling, and hence migration, begins.

Thanks to the miniaturization of geolocators and satellite transmitters we have entered a new era of migration natural history studies, and new information about migration systems are reported at a fast rate. Some notable examples are wood thrushes (*Hylocichla mustelina*; Stutchbury et al. 2009) and red-backed shrikes (*Lanius collurio*; Tøttrup et al. 2012a). Wood thrushes appear to cross the Gulf of Mexico from Eastern USA to wintering sites in Central America, while the northbound spring migration may be either across the Gulf again or a circumnavigation of it. Ringers knew the red-backed shrike to be a migrant going southeast in autumn. Using geolocators, a group of researchers confirmed this observation (Tøttrup et al. 2012a). In addition, new information about the detailed travel schedule was revealed. From southern Scandinavia the shrikes travelled to SE Europe, where they stopped for about two weeks, before embarking on the migration across the eastern Mediterranean and the Sahara in one go. Immediately south of the Sahara, red-backed shrikes stayed for on average 53 days before setting out on a new long migration to the final wintering area in southern Africa (Fig. 5.4). That many migrants make such a long stop south of the Sahara had been implied for a long time via the data of migrating birds caught at Ngulia Lodge in Tsavo West National Park, Kenya (Pearson and Backhurst 1976), but never really proven for a small passerine migrant.

In spring the shrikes took a more easterly route than in autumn, leaving Africa already at the Somali coast to fly almost due north across the Arabic Peninsula (Fig. 5.4). After a distinct westward turn through Turkey, the spring migration converged with the autumn route in SE Europe, before the final approach to the breeding sites in Scandinavia (Fig. 5.4). Spring migration was about 22% longer than autumn migration, but the total duration of spring migration was shorter than in autumn, mainly due to the absence of a nearly 2-month stopover in Africa. Actual travel speed, that is, the rate of travel when on the move, was higher in the autumn, but the number of days travelling was higher in the spring (Tøttrup et al. 2012a). Our guess is that seasonal food availability is a major player when it comes to developing such a loop migration, which is supported by the observation of a two-week delay in the 2011 spring migration, when a severe drought in Ethiopia prevented efficient fuel accumulation there (Tøttrup et al. 2012b). Another potentially important factor that could be involved in the formation of a loop migration system is the wind, where new data about migration routes combined with analyses of wind patterns will soon give us answers.

5.2.3 Fly-and-forage

If food suitable for migrants was available consistently along the migration route, the most favourable strategy in terms of both time and energy minimization would be to fly and forage at the same time (Alerstam 2011). Fuel stores that incur extra flight cost and possibly also increased predation risk due to reduced escape flight performance would be unnecessary, and no time would be spent on searching for suitable stopover sites. The most extreme variant of such a strategy could at least theoretically be achieved by aerial foragers such as swallows and swifts. However, such a strategy remains to be described. For example, barn swallows, *Hirundo rustica*, are known to put on substantial fuel loads prior to the crossing of the Mediterranean Sea and the Sahara (Rubolini et al. 2002). During the early spring migration in tropical Africa, swifts, *Apus apus*, concentrate in large numbers in Liberia where they stay for about 10 days before crossing the Sahara, most likely putting on fuel (Gatter 1987, Åkesson et al. 2012). It may be safe to conclude that usually food is neither ubiquitous nor evenly distributed along the migration route of any bird. A strict fly-and-forage strategy therefore seems unlikely.

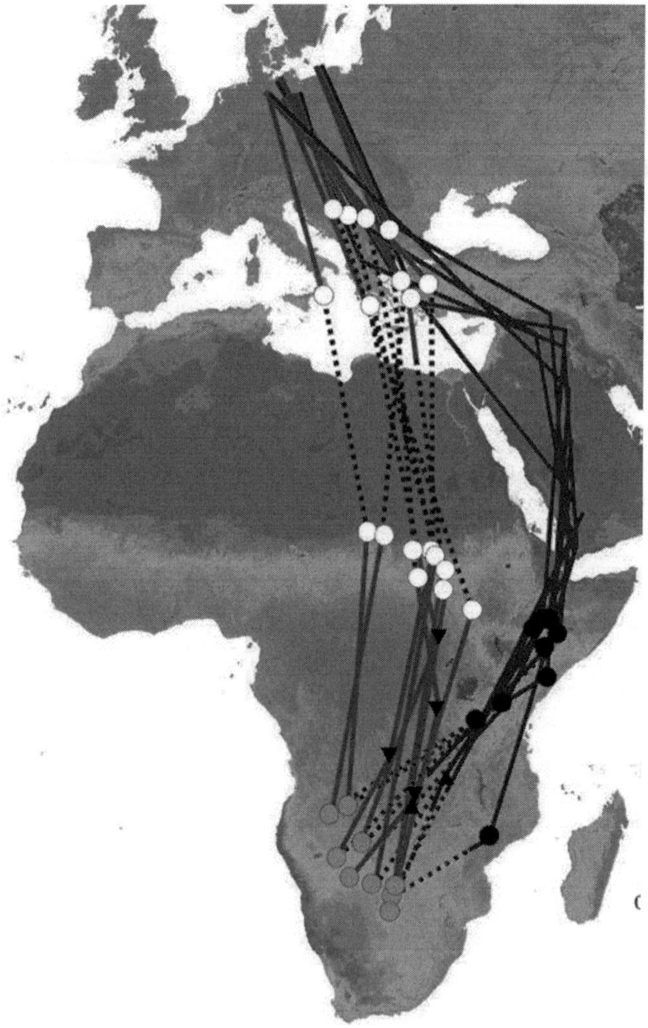

Figure 5.4 Migration by Scandinavian red-backed shrikes, *Lanius collurio*, as revealed by geolocators. Autumn migration is indicated by grey lines and spring migration by black lines. Redrawn from Tøttrup et al. (2012a).

Some birds may nevertheless come close to such a strategy, but in a less extreme form. Satellite tracking of white storks, *Ciconia ciconia*, and ospreys, *Pandion haliaetus*, has revealed that when crossing Europe on their way to Africa in autumn, the birds make migratory flights on almost all days and use a few hours before or after the daily flight for feeding (Berthold et al. 2001, Klaassen et al. 2008). These daytime, mainly soaring migrants can look for good foraging sites while migrating. This minimizes the time and energy loss many other migrants have in connection to searching for foraging sites after landing. Such a variant of a fly-and-forage strategy allows a high speed of travel (Alerstam 2011).

5.2.4 Soaring

So far we have mainly been concerned with migration where flight is powered by an internal engine—the flight muscles. The white storks and ospreys, however, are examples of an alternative flight strategy involving soaring in thermals. The atmosphere is never at rest and both horizontal and vertical winds provide exploitable opportunities for migrants. Any flight behaviour where a flyer extracts energy from the atmosphere, mostly in the form of rising air, is called soaring (Pennycuick 1972a). Hence, in principle a bird flying by flapping flight can soar if it gains energy from flying in an updraught and thereby reduces its own muscle work. There are several types of vertical winds, such as convective thermals, upwinds generated by the deflection upwards of horizontal winds as they hit a mountain ridge, and stationary lee-waves (Bradbury 1989).

Thermals occur when air near the ground surface is warmed by contact with the ground, which is heated by the sun. The warmed air expands and thereby decreases in density, which makes it rise in relation to the surrounding air (Bradbury 1989). The typical migration situation is cross-country soaring by alternate climbs in thermals followed by horizontal glides until the next thermal is encountered. In gliding flight there is no internal work generated to overcome aerodynamic drag, other than the static muscle work needed to keep the wings in an outstretched position. Instead the bird converts potential energy into aerodynamic work during gliding flight; hence the bird loses height when gliding. Gliding flight performance is described by the glide polar (Fig. 5.5), which is a relationship between forward and sinking speed. A bird in gliding flight can adjust the glide polar by flexing the wings to increase the performance at different speeds (Lentink et al. 2007, Pennycuick 2008), where the envelope of all these individual glide polars form a glide 'superpolar'. The cross-country speed in thermal soaring (U_{cc}), the average travel speed when accounting for the time spent in thermals, can be written as

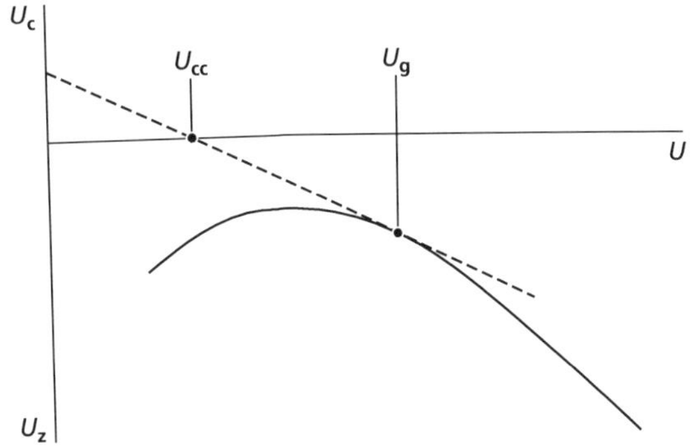

Figure 5.5 A glide polar showing the relationship between sink rate (U_z) and forward speed in gliding flight (U). The positive *y*-axis shows climb rate in thermals (U_c), and the dotted line indicates how the optimal forward glide speed (U_{opt}) is graphically determined, and at the intersection with the *x*-axis the resulting cross-country speed (U_{cc}) is found. U_g is the forward inter-thermal gliding speed.

$$U_{cc} = \frac{U_g U_c}{U_c + U_Z}, \qquad (2)$$

where U_g is the forward inter-thermal gliding speed, U_c is the climb rate in thermals, and U_z is the sink speed during the inter-thermal glides (Fig. 5.5). We may predict that the optimal forward gliding speed between successive thermals depends on the climb rate in thermals (i.e. convective strength), with relatively high glide speed and sink rate at high climb rates Eq. (2) and low glide speed and sink rate when climb rate is low. Migrating raptors have been observed to behave in this predicted pattern (e.g. Spaar 1997, Spaar and Bruderer 1997). To gain altitude in thermals at the maximum possible rate, the bird should reduce the speed to the minimum (U_{ms}, minimum sink), but they also must bank to some degree in order to circle within the core of rising air. Recent studies using GPS loggers to track golden eagles, *Aquila chrysaetos*, during their spring migration confirmed that they do adjust their speed as predicted when circling in thermals (Duerr et al. 2012).

The cross-country performance depends on the rate of climb in thermals, which in turn is a function of sink rate while gliding and the strength of the vertical wind within the thermal. As the flight metabolic rate is generally much lower in gliding flight than in flapping flight, and even if powered flight speed is faster than the cross-country soaring speed, soaring migration will still be the best option from an energy minimization viewpoint (Hedenström 1993). In other words, a soaring migrant can choose a detour of the same energy cost as a direct flight by powered flight. This may explain why many soaring migrants tend to minimize distances across open water, where there are no or very weak thermals (Pennycuick 1972b). Soaring migrants can therefore be seen in large numbers as they concentrate at narrow sea crossings such as at Gibraltar and the Bosporus. In some cases of very high climb rates in thermal soaring, the switching from flapping flight to soaring flight migration can be explained by the time minimization strategy (Hedenström 1993).

The development of animal-attached data loggers, such as GPS loggers combined with accelerometers (devices that record accelerations of an object such as an animal), allow new insights into the three-dimensional flight trajectories and fine-detailed decisions made by soaring birds. For example, this new technology can allow us to determine when a bird will leave a thermal and proceed with gliding flight. The climb rate typically decreases at a certain altitude, and the climb rate can therefore be likened with foraging according to a diminishing return gain function. Climbing in thermals is synonymous with gaining potential energy, and very interestingly, the fact that condors, *Vultur gryphus*, will stop soaring in thermals is broadly in agreement with the marginal value theorem (Shepard et al. 2011).

5.2.5 The unexpected journeys

As explained earlier, the orthodrome (great circle) is the shortest route between two locations on Earth. Distance savings by migrating along orthodromes are particularly pronounced at high latitudes and when the migration has an east–west component. Realized migration routes often deviate from the direct orthodrome, where detours can be quite significant distances. How can detours be reconciled with the notion of optimal migration? Since it is energetically costly to fly with the heavy fuel loads required for crossing a barrier, a bird could instead migrate a longer distance using the same amount of energy if the migration instead is split into many short flights where each one is covered by a very small fuel load. Using this observation, Alerstam (2001) was able to explain the occurrence of detours in several migrating birds, including the wheatears migrating between Greenland and West Africa. However, there are some birds that behave unexpectedly by actually making a detour to a location from where they embark on a *longer* direct flight than what was necessary from the start location. How can such migratory flights be explained from an optimal migration point of view?

The sharp-tailed sandpiper, *Calidris acuminata*, breeds on the tundra in north-east Siberia. The adults migrate an expected route due south via the Yellow Sea to wintering destinations in Australia (Handel and Gill 2010). But the juveniles, unguided by adults, make an initial move of 1,500–3,400 km eastwards to Alaska. From there they make a 7,100- to 9,800-km direct flight across the Pacific to Australia. This is not only an enormous detour but the final flight also requires a huge fuel load with concomitant increased flight costs. This behaviour is compatible with a time minimization strategy, where an advantageous fuelling rate in Alaska offsets the time needed to fuel and fly from the breeding area to Alaska, instead of remaining in the breeding area and fuelling for a shorter migration without the detour (Alerstam 2001, Lindström et al. 2011). Predation pressure could also play a role in shaping this migration pattern. Alaska in September seems to provide a relatively safe haven in terms of predation, adding to its attractiveness for juveniles (Lindström et al. 2011). It is not known why the adults do not make this detour. The key is probably in the seasonal timing in predation pressure and food availability, a compromise also seen in seasonally migrating fish (Chapman et al., Chapter 2). Adults migrate much earlier in the autumn, and, possibly, the predation pressure and fuelling conditions are at this time relatively more favourable along the shorter route, making the detour no longer beneficial (Lindström et al. 2011).

Another intriguing example, a recent discovery made with the help of geolocators (Box 5.1), is the annual migration of short-tailed shearwaters, *Puffinus tenuirostris* (Carey et al. 2014). The post-breeding migration away from the nesting sites on small islands off the Tasmanian coast (40°S) first goes about 2,800 km towards the very productive waters east of the Ross Sea near Antarctica (60–70°S). This is opposite to the direction towards the final 'wintering' grounds in the north Pacific. After an average of 26 days in the south, which most certainly involves substantial fuelling, the shearwaters turn north and race across the Pacific in only 13 days at a speed of 840 km/day. This almost corresponds to the non-stop flights of the bar-tailed godwits (see section 5.2.1). The non-breeding season of almost 5 months is spent in two main areas, either east of northern Hokkaido, Japan, or west of the Pribilof Islands in the Bering Sea. Some populations of this species even enter the Bering Strait as huge flocks (estimated in the tens of thousands; Hedenström, personal observation). Presumably these belong to populations other than those breeding near Tasmania. A migration like this, involving an initial migration in the opposite direction to the main subsequent flight, is best explained as an expression of time-selected migration, where the fuelling conditions in the Antarctic waters compensate for the time it takes to make the detour (cf. the juvenile sharp-tailed sandpipers). Apparently, the vast Pacific Ocean provides little in terms of food en route even to a seabird. Is crossing the Pacific a similar obstacle for shearwaters as it is for more terrestrial species? The southwards return flight in 'spring' is on average somewhat slower with a mean speed of 692 km/day, but still so fast that en route fuelling cannot be a part of this migration either. After finishing the annual round-trip migration, including winter movements, the shearwaters have logged a total flight distance of about 60,000 km, equal to 1.5 laps around the Earth!

Another unexpected migration strategy, again revealed by geolocators (Box 5.1), is the impressive flights of the great snipes, *Gallinago media* (Klaassen et al. 2011). In autumn these secretive waders embark on a 6,000- to 7 000-km non-stop flight between the breeding grounds in central Sweden and stopover sites in sub-Saharan West Africa. These flights are impressively fast, with flight speeds of up to 90 km/h. What is particularly remarkable and unexpected is that the snipes cross large areas of what ought to be suitable stopover sites. Indeed, the Mediterranean Sea and the Sahara are clear ecological barriers, but the other half of the distance over the European continent and parts of Africa cannot be considered classical barriers. This is again an indication of a time minimization strategy, if the northern fuelling sites are simply superior to fuelling sites further south in Europe. It would then pay in time to put on all fuel at the initial site and overfly the others (Gudmundsson et al. 1991). Whether the northern fuelling sites are indeed superior, remains to be shown. Alternatively, again, predation may be a relevant selection force. Great snipes have long been a very popular quarry and shot in such numbers that selection may have favoured a strategy that does not expose the birds to the hunting pressure of continental Europe.

5.2.6 Feeding for flight

As hinted at earlier, the fuelling prior to migratory flights is a crucial part of migration. Indeed, migration starts when fuelling for the first flight starts (Alerstam and Lindström 1990), just as a new car must be fuelled before it can be driven anywhere. In contrast to fuelling a car, however, fuelling in migrants is a much more time-consuming activity. Whereas only a few minutes of petrol-filling can make the car run for hours, the animal migrants actually spend much more time on fuelling than on actual movement. No wonder that a careful choice of fuelling sites is such a crucial aspect of animal migration.

Avian migratory flight is fuelled by fat and protein catabolism (Jenni-Eiermann et al. 2002), and this fuel must be stored prior to flight (Lindström and Piersma 1993). Consider a bird that has a lean body mass (LBM) of 100 g. 'Lean' means that it has no fuel stores. If it adds 25 g fuel, it will have a fuel load of 25% in relation to LBM. If 50 g fuel is added, the fuel load is 50% of LBM. Stored fuel loads can vary dramatically between species and individuals, but is normally between 10% and 100% of LBM. The smallest loads are found in irruptive and partial migrants that only migrate short distances (Chapman et al., Chapter 2), or are forced to migrate due to food deficit. The largest fuel loads are found in birds crossing vast ecological barriers (Alerstam and Lindström 1990).

A fast migration that involves spending a minimum of time on migration may allow migrants to arrive relatively early at the destination and, for example, gain the best breeding or wintering territories. A fast migration may also simply be necessary to allow enough time for breeding or moulting at the destination. Since most days on migration are spent on fuelling (7 days fuelling: 1 day movement; Hedenström and Alerstam 1997) the rate of fuelling is crucial for speed of migration (Alerstam and Lindström 1990).

Lindström (2003) reviewed the causes, consequences, and constraints of fuelling rates in migrating birds. In short, most fuelling rates are in the range of 1–3% of LBM per day (equivalent to a 1- to 3-g mass increase per day in a 100-g bird) and the highest fuelling rates ever recorded in the wild are around 10% of LBM per day. But these peak rates are by no means what is physiologically possible. In experiments where captive migrants were given food *ad libitum* around the clock and artificial light to allow feeding during night, Kvist and Lindström (2000, 2003) found *average* fuelling rates of 10% of LBM per day and maximum values of up to 14% of LBM. This indicates that free-living birds are normally constrained by food abundance and that some periods of the 24-h day are unavailable for feeding, due to darkness or high tides. Accordingly, one way to migrate fast is to migrate when days are long. Long-distance migrant songbirds migrate faster in spring than in autumn, and indeed, spring migration is carried out at dates when days are longer than days of autumn migration (Bauchinger and Klaassen 2005).

Another way of speeding up fuelling, and hence migration, is to find food that can be exploited 24 h per day, independent of day length. The best-documented example so far is the Bewick's swan, *Cygnus bewickii* (Klaassen et al. 2010b; Fig. 5.6). At two out of three studied stopover sites, the birds could feed around the clock on submerged tubers of pondweeds, *Potamogeton pectinatus*. This resulted in extremely high daily energy intake rates and, most certainly, high fuelling rates. At the third stopover site intake rates were lower because periods of high tide precluded around-the-clock feeding.

5.3 Migration in bats and insects

5.3.1 Bats

Some species of bats migrate long distances between summer breeding areas in the north and wintering areas in the south (Hutterer et al. 2005). For example, in northern Europe Nathusius' bats, *Pipistrellus nathusii*, migrate some 2,000 km in the autumn as documented by extensive ringing (Petersons 2004). In North America there are also migrating bats having a similar ecology to that of those migrating in Europe, with a common denominator of belonging to species that mainly have their winter roosts in trees (Fleming and Eby 2003). Altogether, within the family Verspertillionidae, migration occurs in 23 out of 316 (7%) species of known migration status (Bisson et al. 2009). The flight capacity of bats is superficially similar to that of birds, such that the power required to fly follow a U-shaped curve (Fig. 5.1) and flight speeds are comparable among birds and bats (Hedenström 2009). Hence, the optimal migration theory originally developed for birds should be applicable also to bats (Hedenström 2009).

Only in recent times has it been possible to track bats for entire migrations. Richter and Cumming

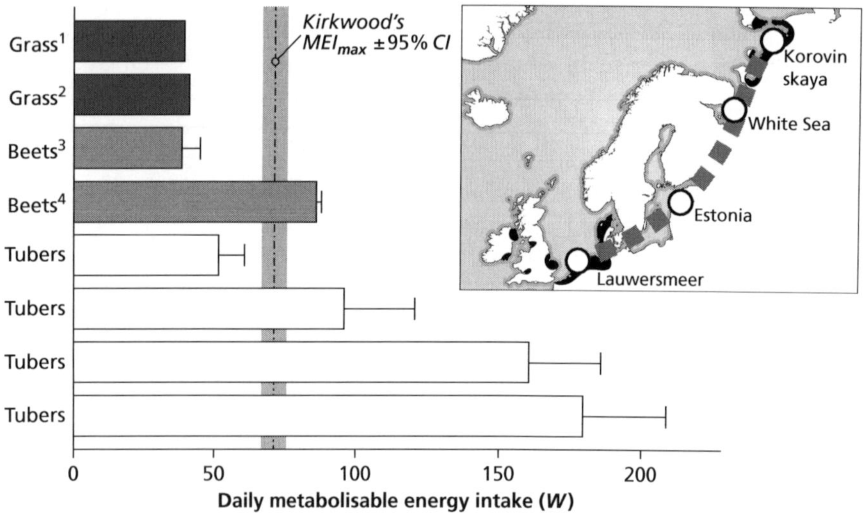

Figure 5.6 Daily metabolizable energy intake rates (± SE estimated using Monte Carlo simulations) for Bewick's swans using three different food sources during migration (grass, harvest remains of sugar beets, and pondweed tubers). Inset shows migratory flyway with wintering and breeding grounds. Data from Klaassen et al. (2010b).

(2008) attached 12-g satellite transmitters on tropical fruit bats and found them to migrate seasonally up to 2,500 km in central Africa. Using a set of radio antennae and 'bar-coded' radio transmitters at Long Point, Lake Erie, Canada, McGuire et al. (2012) found that silver-haired bats, *Lasionycteris noctivagans*, most often stopped over for only one day before continuing migration. Upon arrival at Long Point the bats carried about 19% fat, which is sufficient to cover the remaining 1,500 km of the migration without further refuelling. A few bats stayed for longer if it was raining on the night of arrival, and they were observed feeding on those occasions. It appears that bats become torpid if stopping for 2 days or more (Barclay et al. 1988), which saves a lot of energy that is not needed for thermoregulation. Silver-haired bats departed from Long Point in two main directions, either due south across the lake or westwards along the lakeshore. Departure direction could not be related to temperature, wind speed, or barometric pressure (McGuire et al. 2012), factors that are known to trigger migration in birds.

Migration speed in silver-haired bats was estimated to be about 260 km/day (McGuire et al. 2012), while Nathusius' bats ringed in Latvia showed a mean progress of 46 km/day (Petersons 2004). Migration speeds are generally underestimated by ringing recoveries, while the calculated 260 km/day for silver-haired bats did not include the pre-migratory fuelling period. Hence, the true migration speed for small bats is probably somewhere between these two estimates.

Nathusius' bats migrate along a NE–SW axis from north-eastern to southern Europe (Petersons 2004, Hutterer et al. 2005). Populations from Finland and the Baltic states could fly across the Baltic Sea along their intended migration route, or could follow the coast and thereby make a detour. Hedenström (2009) predicted that it is optimal to cross the Baltic Sea, a calculation that is supported by observations at Ottenby, Öland, in the western Baltic, and some observations at sea (Ahlén et al. 2009). Bats undergo many physiological changes associated with migration, including fat deposition, the reduction of digestive organs, and the enlargement of exercise organs (flight muscles, heart, and lungs) (McGuire et al. 2013). Such physiological changes are very similar to those found in birds, which are associated with the demands of flying with additional fuel reserves and needs to minimize the cost of carrying unnecessary payload. The overall migration strategy of bats is probably closest to a fly-and-forage strategy, with stops related to local weather. Like birds, bats are able to orient by using information provided by the geomagnetic field that is calibrated against the sun direction at dusk (Holland et al. 2010, Åkesson et al. Chapter 9). However, our knowledge about bat migration is still fragmentary and will remain so until new technology makes it possible to track whole migrations in small bats.

5.3.2 Insects

The embodiment of insect migration is that of the monarch butterfly, *Danaus plexippus*, where individuals migrate some 4,000 km from northeast Canada to overwintering sites in Mexico in the autumn (e.g. Brower 1996; Fig. 2.6). However, in contrast to vertebrate migrants, it is not the same individuals that make the round-trip migration, but several generations complete the annual migration cycle (Brower 1996; transgenerational migration: Chapman et al., Chapter 2). There are similar butterfly migrations also between Europe and Africa, such as the recently described migration system in the painted lady, *Vanessa carduii* (Stefanescu et al. 2012). The evolutionary drivers for insect migration are probably similar to those in birds and bats, i.e. increased reproductive output at northerly latitudes more than compensates for the losses during migration compared with a residential strategy at southerly latitudes. For example, the immigrant British summer population of the noctuid moth, the silver Y, *Autographa gamma*, increases four times and some 80% were estimated to successfully return to the wintering area in the Mediterranean region (Chapman et al. 2012).

The migration strategy found in insects shows overall similarities with that of birds. Because the airspeeds are generally relatively low due to their small size (Chapman et al. 2008), insects are potentially more vulnerable to adverse winds than birds. Consequently, migration is more wind sensitive in insects. In a comparative study of high-altitude nocturnal migratory flights between birds and insects, the progress above ground was similar between the two groups (Alerstam et al. 2011), mainly because insects migrated with tail winds to a larger extent than did birds. Diurnal migration in monarch butterflies takes place under light wind conditions, and monarchs are also able to use thermals for cross-country soaring (Gibo 1986). Southbound red admirals, *Vanessa atalanta*, studied in southern Sweden preferred to migrate with light tail winds (Brattström et al. 2008), and clear skies also significantly contributed to the probability to migrate. The orientation in monarchs is by a simple clock-and-compass mechanism, where the compass information is given by the sun position (Reppert et al. 2010, Mouritsen et al. 2013).

Dragonflies constitute another insect group where long-distance migrants are found. The common green darner, *Anax junius*, of North America is such a species, and thanks to the formidable flight capacity of dragonflies, small radio transmitters were used to track individuals for about one week during autumn migration (Wikelski et al. 2006). Even though the transmitters weighed 0.3 g (25% of body mass) the green darners seemed to be able to migrate without any signs of hindrance. Individuals migrated up to 150 km in one day, with an overall migration speed of 58 km/day. Flights were oriented towards the south, but when encountering open water the flight direction reversed, in a way resembling many migrating birds (Åkesson et al. 1996). Migratory flights in green darners were triggered by a drop in night temperature, which predicts favourable northerly winds (Wikelski et al. 2006).

5.4 Future perspectives

In this chapter we have looked at some recent findings, often achieved through technical developments that allow the tracking of migratory journeys in animals. This technical development continues and is likely to accelerate, allowing us to study smaller and smaller animals and to obtain new types of information. For example, future loggers may not only record the position of the animal, but could also collect information about flight modes (accelerometers), altitude, and body temperature. Perhaps some physiological measures could also provide information about the size of fuel stores. What could this wealth of information tell us about migration in general? For example, what are the most likely selective agents in migration? Is time, energy, survival, or other factors most important? Or may a combination of several factors simultaneously mould migration? When does mortality occur in the annual cycle? Indirect findings suggested that mortality may take place primarily during migration (Sillett and Holmes 2002). The first more detailed information is now coming from studies of raptor migration using satellite transmitters (Strandberg et al. 2010). In juveniles, as much as half of the annual mortality may occur during the crossing of the Sahara, indeed a very short episode in the annual cycle of the migrant but perhaps a bottleneck in terms of survival. Of the techniques presently available to track migrants, only satellite transmitters can convey such information. As the number of birds tracked by satellite increases, we will learn much more about this important aspect of migrant life-cycles. If we could measure body mass and/or fuel status of the individuals in parallel to their tracks in time and space, fascinating opportunities will open up to study the dynamics of fuel and migration strategies. In some organisms, such as in bats and most insects, migration research is still in its infancy, and further progress depends on new technological advancements. Great natural history discoveries are still to be made.

References

Ahlén, I., Baagøe, H. J. and Bach, L. (2009). Behavior of Scandinavian bats during migration and foraging at sea. *Journal of Mammalogy*, **90**, 1318–1323.

Åkesson, S., Karlsson, L. Walinder, G., and Alerstam, T. (1996). Bimodal orientation and the occurrence of temporary reverse bird migration during autumn in south Scandinavia. *Behavioural Ecology and Sociobiology*, **38**, 293–302.

Åkesson, S., Klaassen, R., Holmgren, J., Fox, J. W., and Hedenström, A. (2012). Migration routes and strategies in a highly aerial migrant, the common swift *Apus apus*, revealed by light-level geolocators. *PLoS One*, **7**(7), e41195.

Alerstam, T. (1990). *Bird Migration*. Cambridge University Press, Cambridge.

Alerstam, T. (2001). Detours in bird migration. *Journal of Theoretical Biology*, **209**, 319–31.

Alerstam, T. (2011). Optimal bird migration revisited. *Journal of Ornithology*, **152** (Suppl 1), 5–23.

Alerstam, T., Bäckman, J., Strandberg, R., et al. (2008). Great circle migration of arctic passerines. *Auk*, **125**, 831–8.

Alerstam, T., Chapman, J. W., Bäckman, J., et al. (2011). Convergent patterns of long-distance nocturnal migration in noctuid moths and passerine birds. *Proceedings of the Royal Society B: Biological Sciences*, **278**, 3074–80.

Alerstam, T., Gudmundsson, G. A., Green, M., and Hedenström, A. (2001). Migration along orthodromic sun compass routes by Arctic birds. *Science*, **291**, 300–3.

Alerstam, T., and Hedenström, A. (1998). The development of bird migration theory. *Journal of Avian Biology*, **29**, 343–69.

Alerstam, T., and Lindström, Å. (1990). Optimal bird migration: the relative importance of time, energy and safety. In E. Gwinner (ed.), *Bird Migration: Physiology and Ecophysiology*, pp. 331–51. Springer-Verlag, Berlin.

Alerstam, T., and Pettersson, S.-G. (1991). Orientation along great circles by migrating birds using a sun compass. *Journal of Theoretical Biology*, **152**, 191–202.

Bairlein, F., Norris, D. R., Nagel, R., et al. (2012). Cross-hemisphere migration of a 25 g songbird. *Biology Letters*, **8**, 505–7.

Barclay, R. M. R., Faure, P. A., and Farr, D. R. (1988). Roosting behaviour and roost selection by migrating silver-haired bats (*Lasionycteris noctivagans*). *Journal of Mammalogy*, **69**, 821–5.

Battley, P. F., Warnock, N., Tibbitts, T. L., et al. (2012). Contrasting extreme long-distance migration patterns in bar-tailed godwits *Limosa lapponica*. *Journal of Avian Biology*, **43**, 21–32.

Bauchinger, U., and Klaassen, M. (2005). Longer days in spring than in autumn accelerate migration speed of passerine birds. *Journal of Avian Biology*, **36**, 3–5.

Berthold, P., van den Bosche, W., Fiedler, W., et al. (2001). Der Zug des Weißstorchs (*Ciconia ciconia*): eine besondere Zugform auf Grund neuer Ergebnisse. *Journal of Ornithology*, **142**, 73–92.

Bisson, I.-A., Safi, K., and Holland, R. A. (2009). Evidence for repeated independent evolution of migration in the largest family of bats. *PLoS One*, **4**(10), e7504.

Bradbury, T. (1989). *Meteorology and Flight: A Pilot's Guide to Weather*. A & C Black, London.

Brattström, O., Kjellén, N., Alerstam, T., and Åkesson, S. (2008). Effects of wind and weather on red admiral, *Vanessa atalanta*, migration at a coastal site in southern Sweden. *Animal Behaviour*, **76**, 335–44.

Bridge, E. S., Kelly, J. F., Contina, A., Gabrielson, R. M., MacCurdy, R. B., and Winkler, D. W. (2013). Advances in tracking small migratory birds: a technical review of light-level geolocation. *Journal of Field Ornithology*, **84**, 121–37.

Brower, L. P. (1996). Monarch butterfly orientation: missing pieces of a magnificent puzzle. *Journal of Experimental Biology*, **199**, 93–103.

Carey, M. J., Phillips, R.A., Silk, J. R. D., and Shaffer, S. A. (2014). Trans-equatorial migration of short-tailed shearwaters revealed by geolocators. Emu, in press.

Chapman, J. W., Bell, J. R., Burgin, L. E., et al. (2012). Seasonal migration to high latitudes results in major reproductive benefits in an insect. *Proceedings of the National Academy of Sciences USA*, **109**, 14924–9.

Chapman, J. W., Reynolds, D. R., Mouritsen, H., et al. (2008). Wind selection and drift compensation optimize migratory pathways in a high-flying moth. *Current Biology*, **18**, 514–18.

Conklin, J. R., and Battley, P. F. (2012). Carry-over effects and compensation: late arrival on non-breeding grounds affects wing moult but not plumage or schedules of departing bar-tailed godwits *Limosa lapponica* baueri. *Journal of Avian Biology*, **43**, 252–63.

Dänhardt, J., and Lindström, Å. (2001). Optimal departure decisions of songbirds from an experimental stopover site and the significance of weather. *Animal Behaviour*, **62**, 235–43.

Deerenberg, C., Apanius, V., Daan, S., and Bos, N. (1997). Reproductive effort decrease antibody responsiveness. *Proceedings of the Royal Society London B: Biological Sciences*, **264**, 1021–29.

Delingat, J., Bairlein, F., and Hedenström, A. (2008). Obligatory barrier crossing and adaptive fuel management in migratory birds: the case of Atlantic crossing in northern wheatears (*Oenanthe oenanthe*). *Behavioural Ecology and Sociobiology*, **62**, 1069–78.

Dingle, H. (1996). *Migration: The Biology of Life on the Move*. Oxford University Press, New York.

Duerr, A. E., Miller, T. A., Lanzone, M., et al. (2012). Testing an emerging paradigm in migration ecology show surprising differences in efficiency between flight modes. *PLoS One*, **7**, e35538.

Ekstrom, P. A. (2004). An advance in geolocation by light. *Memoirs of the National Institute of Polar Research*, Special Issue, **58**, 210–26.

Engel, S., Bowlin, M. S., and Hedenström, A. (2010). The role of wind-tunnel studies in integrative research on migration biology. *Integrative and Comparative Biology*, **50**, 323–35.

Fleming, T. H., and Eby, P. (2003). Ecology of bat migration. In T. H. Kunz, M. B. Fenton (eds), *Bat Ecology*, pp. 156–208. University of Chicago Press, Chicago.

Gatter, W. (1987). Beobachtungen und Hypothesen zu Zugstrategien und Wanderrouten Vogelzug in Liberia, Teil II. *Vogelwarte*, **34**, 80–92.

Gibo, D. L. (1986). Flight strategies of migrating monarch butterflies (*Danaus plexippus* L.) in southern Ontario.

In W. Danthanarayana (ed.), *Insect Flight: Dispersal and Migration*, pp. 172–84. Springer, Berlin.

Gill, R. E. Jr., Tibbitts, T. L., Douglas, D. C., et al. (2009). Extreme endurance flights by landbirds crossing the Pacific Ocean: ecological corridor rather than barrier? *Proceedings of the Royal Society B: Biological Sciences*, **276**, 447–57.

Gudmundsson, G. A., Lindström, Å., and Alerstam, T. (1991). Optimal fat loads and long-distance flights by migrating knots *Calidris canutus*, sanderlings *C. alba* and turnstones *Arenaria interpres*. *Ibis*, **133**, 140–52.

Handel, C. M., and Gill, R. E. Jr. (2010). Wayward youth: trans-Beringian movement and differential southward migration by juvenile sharp-tailed sandpiper. *Arctic*, **63**, 273–88.

Hasselquist, D., Lindström, Å., Jenni-Eiermann, S., Koolhaas, A., and Piersma, T. (2007). Long flights do not influence the immune responses of a long-distance migrant bird: a wind-tunnel experiment. *Journal of Experimental Biology*, **210**, 1123–31.

Hedenström, A. (1993). Migration by soaring or flapping flight in birds: the relative importance of energy cost and speed. *Philosophical Transactions of the Royal Society B: Biological Sciences*, **342**, 353–61.

Hedenström, A. (2008). Power and metabolic scope of bird flight: a phylogenetic analysis of biomechanical predictions. *Journal of Comparative Physiology A: Neuroethology, Sensory, Neural, and Behavioral Physiology*, **194**, 685–91.

Hedenström, A. (2009). Optimal migration strategies in bats. *Journal of Mammalogy*, **90**, 1298–1309 .

Hedenström, A., and Alerstam, T. (1997). Optimum fuel loads in migratory birds: distinguishing between time and energy minimization. *Journal of Theoretical Biology*, **189**, 227–34.

Holland, R. A., Borissov, I., and Siemers, B. M. (2010). A nocturnal mammal, the greater mouse-eared bat, calibrates a magnetic compass by the sun. *Proceeedings of the National Academy of Sciences*, **107**, 6941–5.

Hutterer, R., Ivanova, T., Meyer-Cords, C., and Rodrigues, L. (2005). *Bat Migrations in Europe: A Review of Banding Data and Literature*. Federal Agency for Nature Conservation, Bonn, Germany.

Jenni-Eiermann, S., Jenni, L., Kvist, A., Lindström, Å., Piersma, T., and Visser, G. H. (2002). Fuel use and metabolic response to endurance exercise: a windtunnel study of a long-distance migrant shorebird. *Journal of Experimental Biology*, **205**, 2453–60.

Johnson, O. W., Fielding, L., Fisher, J. P., et al. (2012). New insight concerning transoceanic migratory pathways of Pacific Golden-Plovers (*Pluvialis fulva*): the Japan stopover and other linkages as revealed by geolocators. *Wader Study Group Bulletin*, **119**, 1–8.

Jouventin, P., and Weimerskirch, H. (1990). Satellite tracking of wandering albatrosses. *Nature*, **343**, 746–8.

Kiepenheuer, J. (1984). The magnetic compass mechanism of birds and its possible association with shifting course directions of migrants. *Behavioural Ecology and Sociobiology*, **14**, 81–99.

Klaassen, R. H. G., Alerstam, T., Carlsson, P., Fox, J. W., and Lindström, Å. (2011). Great flights by great snipes: long and fast non-stop migration over benign habitats. *Biology Letters*, **7**, 833–5.

Klaassen, M., Hangelbroek, H. H., de Boer, T., and Nolet, B. A. (2010b). Insights from the eco-physiological book of records: Bewick's swans outperform the canonical intake-maximizing vertebrate. *Oikos*, **119**, 1156–60.

Klaassen, R. H. G., Strandberg, R., Hake, M., and Alerstam, T. (2008). Flexibility in daily travel routines causes regional variation in bird migration speed. *Behavioural Ecology and Sociobiology*, **62**, 1427–32.

Kvist, A., and Lindström, Å. (2000). Maximum daily energy intake: it takes time to lift the metabolic ceiling. *Physiological and Biochemical Zoology*, **73**, 30–6.

Kvist, A., and Lindström, Å. (2003). Gluttony in migratory waders–unprecedented energy assimilation rates in vertebrates. *Oikos*, **103**, 397–402.

Lentink, D., Müller, U.K., Stamhuis, E. J., et al. (2007). How swifts control their glide performance with morphing wings. *Nature*, **446**, 1082–5. doi:10.1038/nature05733

Leyrer, J., Lok, T., Brugge, M., et al. (2012). Small-scale demographic structure suggests preemptive behavior in a flocking shorebird. *Behavioral Ecology*, **23**, 1226–33.

Lindström, Å. (2003). Fuel deposition rates in migrating birds: causes, constraints and consequences. In P. Berthold, E. Gwinner, and E. Sonnenschein (eds), *Avian Migration*, pp. 307–20. Springer-Verlag, Berlin.

Lindström, Å., and Alerstam, T. (1992). Optimal fuel loads in migrating birds: a test of the time minimization hypothesis. *American Naturalist*, **140**, 477–91.

Lindström, Å., Gill, R. E. Jr., Jamieson, S. E., et al. (2011). A puzzling migratory detour: are fueling conditions in Alaska driving the movement of juvenile sharp-tailed sandpipers? *Condor*, **113**, 129–39.

Lindström, Å., and Piersma, T. (1993). Mass changes in migrating birds: the evidence for fat and protein storage re-examined. *Ibis*, **135**, 70–8.

McGuire, L. P., Fenton, M. B., and Guglielmo, C. G. (2013). Phenotypic flexibility in body composition, organ sizes and fatty acid profiles. *Journal of Experimental Biology*, **216**, 800–8.

McGuire, L. P., Guglielmo, C. G., Mackenzie, S. A., and Taylor, P. D. (2012). Migratory stopover in the long-distance migrant silver-haired bat, *Lasionycteris noctivagans*. *Journal of Animal Ecology*, **81**, 377–85.

Minton, C., Gosbell, K., Johns, P., Christie, M., Fox, J. W., and Afanasyev, V. (2010). Initial results from light level geolocator trials on Ruddy Turnstone *Arenaria interpres* reveal unexpected migration route. *Wader Study Group Bulletin*, **117**, 9–14.

Mouritsen, H., Derbyshire, R., Stalleicken, J., Mouritsen, O. Ø, Frost, B. J., and Norris, D. R. (2013). An experimental displacement and over 50 years of tag-recoveries show that monarch butterflies are not true navigators. *Proceedings of the National Academy of Sciences USA*, **110**, 7348–53.

Pearson, D. J., and. Backhurst, G. C. (1976). The southward migration of Palaearctic birds over Ngulia, Kenya. *Ibis*, **118**, 78–105.

Pennycuick, C. J. (1972a). *Animal Flight*. Studies in Biology no. 33. Edward Arnold, London.

Pennycuick, C. J. (1972b). Soaring behaviour and performance of some East African birds observed from a motor-glider. *Ibis*, **114**, 178–218.

Pennycuick, C. J. (1975). Mechanics of flight. In D. S. Farner and J. R. King (eds), *Avian Biology*, pp. 1–75. Academic Press, New York.

Pennycuick, C. J. (2008). *Modelling the Flying Bird*. Academic Press, London.

Petersons, G. (2004). Seasonal migrations of north-eastern populations of Nathusius' bat *Pipistrellus nathusii* (Chiroptera). *Myotis*, **41**/42, 29–56.

Phillips, R. A., Silk, J. R. D., Croxall, J. P., Afanasyev, V., and Briggs, D. R. (2004). Accuracy of geolocation estimates for flying seabirds. *Marine Ecology Progress Series*, **266**, 265–72.

Reppert, S. M., Gebear, R. J., and Merlin, C. (2010). Navigational mechanisms of migrating monarch butterflies. *Trends in Neuroscience*, **33**, 399–406.

Richter, H. V., and Cumming, G. S. (2008). First application of satellite telemetry to track African straw-coloured fruit bat migration. *Journal of Zoology*, **275**, 172–6.

Rubolini, D., Pastor, A. G., Pilastro, A., and Spina, F. (2002). Ecological barriers shaping fuel stores in barn swallows *Hirundo rustica* following the central and western Mediterranean flyways. *Journal of Avian Biology*, **33**, 15–22.

Shepard, E. L. C., Lambertucci, S. A., Vallmitjana, D., and Wilson, R. P. (2011). Energy beyond food: foraging theory informs time spent in thermals by a large soaring bird. *PLoS One*, **6**(11), e27375.

Sillett, T. S., and Holmes, R. T. (2002). Variation in survivorship of a migratory songbird throughout its annual cycle. *Journal of Animal Ecology*, **71**, 296–308.

Spaar, R. (1997). Flight strategies of migrating raptors; a comparataive study of intraspecific variation in flight characteristics. *Ibis*, **139**, 523–35.

Spaar, R., and Bruderer, B. (1997). Optimal flight behavior of soaring migrants: a case study of migrating steppe buzzards, *Buteo buteo vulpinus*. *Behavioural Ecology*, **8**, 288–97.

Stefanescu, C., Páramo, F., Åkesson, S., et al. (2012). Multi-generational long-distance migration of insects: studying the painted lady butterfly in the Western Palaearctic. *Ecography*, **35**, 1–14.

Strandberg, R., Klaassen, R. H. G., Hake, M., Olofsson, P., and Alerstam, T. (2009). Converging migration routes of Eurasian hobbies *Falco subbuteo* crossing the African equatorial rain forest. *Proceedings of the Royal Society B: Biological Sciences*, **276**, 727–33.

Strandberg, R., Klaassen, R. H. G., Hake, M., and Alerstam, T. (2010). How hazardous is the Sahara desert crossing for migratory birds? Indications from satellite tracking of raptors. *Biology Letters*, **6**, 297–300.

Stutchbury, B. J. M., Tarof, S. A., Done, T., et al. (2009). Tracking long-distance songbird migration by using geolocators. *Science*, **232**, 896.

Thompson, M. C. (1974). Migratory patterns of Ruddy Turnstones in the Central Pacific region. *Living Bird*, **12**, 5–23.

Thorup, K., Ortvad, T. E., and Rabøl, J. (2006). Do Nearctic northern wheatears (*Oenanthe Oenanthe leucorhoa*) migrate nonstop to Africa? *Condor*, **108**, 446–51.

Tøttrup, A. P., Klaassen R. H. G., Kristensen, M. W., et al. (2012b). Drought in Africa caused delayed arrival of European songbirds. *Science*, **338**, 1307.

Tøttrup, A. P., Klaassen R. H. G., Strandberg, R., et al. (2012a). The annual cycle of a trans-equatorial Eurasian-African passerine migrant: different spatio-temporal strategies for autumn and spring migration. *Proceedings of the Royal Society B: Biological Sciences*, **279**, 1008–16.

Wikelski, M., Moskowitz, D., Adelman, J. S., Cochran, J., Wilcove, D. S., and May, M. L. (2006). Simple rules guide dragonfly migration. *Biology Letters*, **2**, 325–9.

CHAPTER 6

Individuality in movement: the role of animal personality

Jan-Åke Nilsson, Christer Brönmark, Lars-Anders Hansson, and Ben B. Chapman

That individuals vary in their behaviour is universally recognized among biologists, and has been since the dawn of scientific enquiry (Aristotle 350 BC). Individuals differ in their expression of a variety of behaviours, such as aggression, risk-taking, activity, and exploration and this may be a function of many variables, including sex, age, size, and state. However, in addition to differences in behaviour attributable to such factors, many animals also exhibit surprisingly consistent individual differences in behaviour for a wide range of traits, across time and contexts (Sih et al. 2004). Many studies have now reported such behavioural consistency (also termed 'personality') between individuals (Bell et al. 2009) and this area of research continues to attract a great deal of interest from scientists from various disciplines. Besides accumulating more and more examples of the existence of animal personalities from a larger and larger range of animals it is becoming increasingly clear that variation in personality has the potential to have major implications for evolutionary and ecological processes (Sih et al. 2012, Wolf and Weissing 2012). Arguably some of the best evidence for the relevance of personality for ecological processes comes from the links between animal personality and animal movement. In this chapter we aim to elucidate some of these links, and reflect upon how personality variation can relate to animal movement at different scales, and what the ecological and evolutionary consequences of this variation may be. However, before we go into the links between personality and movement patterns we need to know in more detail what animal personality is.

6.1 Animal personality

The phenomenon of individual differences in behaviour has become known as 'animal personality', and has many synonyms in the literature, such as temperament, coping style, and behavioural type (Sih et al. 2004, Réale et al. 2007). The many recent observations of surprising behavioural consistency, along with discussions about the ecological and evolutionary importance of such variation, has led to a surge of interest in the causes and consequences of animal personality. Behavioural consistency has been documented in a bewilderingly broad array of animals, from birds to ants to fishes to crustaceans, and across a diverse suite of behavioural traits, many of which can be directly linked to movement ecology (e.g. boldness, activity, exploratory behaviour; see section 6.4). For example, in an elegant laboratory study, three-spined sticklebacks, *Gasterosteus aculeatus*, tested for boldness (i.e. an individual's propensity to take risks) showed individual consistency over time and across contexts. Individual fish were tested to measure how quickly they resumed feeding following a predator stimulus, with bold individuals being faster to feed. This behaviour was consistent over time and also correlated with willingness to join a group, such that bold individuals in the feeding trial also displayed lower shoaling tendencies (Ward et al. 2004).

Animal Movement Across Scales. Edited by Lars-Anders Hansson and Susanne Åkesson.
© Oxford University Press 2014. Published 2014 by Oxford University Press.

Hence, fish were consistent in their propensity to take risks in both a simulated predator encounter and a social situation (where leaving the safety of a group can have significant risks associated with it; Krause and Ruxton 2002). Field studies have also shown that behavioural differences can be stable in the wild (Dingemanse et al. 2012), although this is less well studied, at least in part due to the logistical challenges involved. In cooperatively breeding meerkats, *Suricata suricatta*, a number of helping behaviours were repeatable, particularly the frequency of provisioning the young (English et al. 2010). Finally, studies also show (in some cases) that behaviours tested in the lab are consistent in wild situations (Herborn et al. 2010). Thus, evidence continues to mount that personality variation both is extremely widespread and can have significant importance to ecological and evolutionary processes (Sih et al. 2012, Wolf and Weissing 2012).

6.2 Behavioural syndromes and behavioural types

Another important aspect of animal personality, besides behavioural consistency over time and context, is that different behavioural traits may be correlated at the population level. Hence, we often see specific combinations of behaviours within individuals; for example, highly active individuals may also be highly aggressive, while inactive individuals exhibit low aggressive tendencies. This between-individual correlation is known as a 'behavioural syndrome' (Dingemanse and Dochtermann 2013; Fig. 6.1). A common and widespread behavioural syndrome is the boldness–activity–aggression syndrome, which has been reported in a variety of taxa (e.g. Dingemanse et al. 2007). Behavioural syndromes are also evident in 'simple' organisms such as invertebrates: in a study of red ants, *Myrmica spp.*, Chapman et al. (2011a) found evidence of correlations between aggression, activity, and boldness. Hence, individual ants that exhibited high levels of aggression in response to an encounter with a heterospecific *Myrmica* ant were also highly active when placed in a novel environment, and were less risk averse in a boldness trial when exposed to ant alarm cues (see also Garamszegi et al. 2012 for a review on the topic). At an individual level,

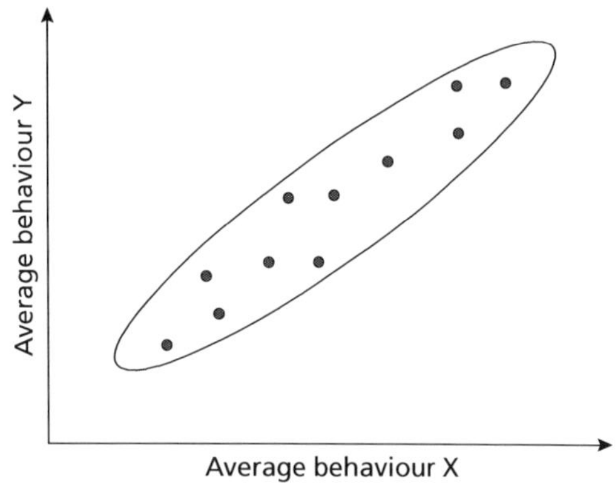

Figure 6.1 Between-individual correlation between an individual's average for behaviour Y and X, forming a behavioural syndrome. Each point represents individuals with a particular combination of behaviours Y and X, forming a behavioural type. Modified from Bell (2007).

a combination of traits is often referred to as an individual's 'behavioural type' (Fig. 6.1). Hence individuals may have a bold, aggressive behavioural type, or a shy, passive behavioural type. Essentially, animal personality is a form of behavioural specialization in which different individuals from a single population may potentially occupy varying behavioural niches (Bergmüller and Taborsky 2010).

6.2.1 Are behavioural syndromes adaptive?

On one hand it can be argued that syndromes restrict independent combinations of the traits correlated in the syndrome and thereby reduce the scope for shaping unique phenotypes in response to selection for each environment. Such constraints to plasticity are potentially costly if a suite of behaviours differs in relation to an optimal combination of behaviours for given environments or contexts. However, syndromes might still be adaptive if the information that the choice of optimal behaviours should be based on is poor (Sih et al. 2004), when optimal behaviours depend on the state of an individual, and/or when a behavioural polymorphism is based on frequency-dependent selection, resulting in, for example, exploiting a larger part of the behavioural niche, potentially reducing competition (Dingemanse and Wolf 2010). This latter scenario seems to select for different syndromes in different phases of metapopulation dynamics, range expansions, and invasions (see section 6.7). In this case the variation in syndromes is most probably maintained by spatiotemporal variation in selection.

If the advantages of behavioural syndromes are specific to certain circumstances, we would expect to find syndromes of only short-term consistency and contexts without a firm correlation between personality traits. Both of these examples occur. For example, the dispersing propensity of root voles, *Microtus oeconomus*, can be predicted by their higher scores of exploration and activity before the dispersing event (Hoset et al. 2011; Hollander et al., Chapter 7). However, after dispersal, these behaviours did not differ between dispersers and residents; thus, the personality traits were of short duration. This suggests that it is important for a dispersing individual to have certain personality traits, but during the sedentary phase of life other personality traits are advantageous. Furthermore, several studies have reported variation in the existence of behavioural syndromes in relation to predation risk. In different populations of the three-spined stickleback populations, those inhabiting ponds with predators present demonstrated tight correlations between aggression, activity, and exploration, i.e. behavioural syndromes, whereas populations from predator-free ponds did not exhibit these or had only weak correlations (Bell 2005, Dingemanse et al. 2007). One potential reason for this difference may be that the presence of predators made the environment more complex, selecting for frequency-dependent behavioural polymorphism (Dingemanse et al. 2007). At any rate, these examples suggest that the presence of syndromes is not universal and that population-specific selection pressures select for syndromes in some but not in all populations.

6.2.2 Heritability of personality traits and behavioural syndromes

Several personality traits linked to movement have been found to be heritable (Réale et al. 2007). For example, variation in exploration behaviour has a significant heritable component in great tits, *Parus major*, in three different populations (Drent et al. 2003, Quinn et al. 2009, Nicolaus et al. 2012); thus, this additive genetic variance makes the trait liable for evolutionary change (e.g. van Oers et al. 2004a). In line with this, results from an artificial selection experiment conducted on great tits are a particularly illuminating example of the genetic basis for exploratory behaviour (Drent et al. 2003). Lines selected for increased and decreased exploration, respectively, diverged significantly already after four generations of selection. However, although additive genetic variance was significant in the great tit populations just mentioned, non-genetic effects, i.e. permanent environmental effects, have also been shown to be important in influencing exploration behaviour (Quinn et al. 2009, Nicolaus et al. 2012), highlighting that personality traits can also exhibit plasticity. In line with this, it was possible to manipulate great tits to become faster explorers by reducing their life expectancy. The adaptive rationale behind this change was suggested to depend on the willingness of individual great tits to take larger risks when their survival prospects had decreased (Nicolaus et al. 2012).

To formally estimate the ecological and evolutionary implications of behavioural syndromes, genetic correlations between personality traits or between personality traits and movement traits are essential. Although, phenotypic correlations between behaviours have been shown to reliably reflect underlying genetic correlations (Dochtermann 2011), estimates of genetic correlations are scarce. In a study on dispersal and explorative behaviour, Dingemanse et al. (2003) found longer natal dispersal distances in great tit descendants from fast-exploring parents. Furthermore, in the selection lines referred to in the previous paragraph, fast explorers also tended to take larger risks (measured as the latency to resume feeding after being startled), indicative of correlated selection on these two traits. In a second selection experiment, risk-taking behaviour was shown to be heritable and to diverge significantly after two generations of selection (van Oers et al. 2004b). High direct estimates of genetic correlations have been found between, for example, risk-taking and explorative behaviour in great tits (van Oers et al. 2004c) and between aggression and dispersal in western bluebirds (Duckworth and Kruuk 2009). Such between-individual genetic correlations between behaviours (syndromes) are important for evaluating to what extent syndromes constrain evolutionary trajectories (Dochtermann and Dingemanse 2013).

6.3 Plasticity as a trait

A final component of individual behavioural variation in animals is variation in behavioural plasticity, or responsiveness. While the notion of personality suggests limited behavioural flexibility, this does not preclude that animals are responsive to environmental shifts, and that individuals can vary in their plasticity (Hollander et al., Chapter 7). A number of studies suggest that flexibility to stimuli or environmental changes is a trait in itself, which can be integrated into a broader behavioural syndrome (see Dingemanse and Wolf 2013 for a review). Evidence from behavioural neuroscience suggests that certain personality types (bold, aggressive types, known as 'proactive' individuals) are less responsive to environmental changes than individuals that are shyer and less aggressive (i.e. 'reactive' individuals: Coppens et al. 2010). For example, proactive and reactive laboratory mice experienced a shift in their usual light–dark schedule. Proactive individuals took 8 days to adapt to the new situation, continuing with their original schedule despite the radically changed conditions. Conversely, reactive mice responded in half this time and their schedules were soon in step with the new light regime (Benus et al. 1988). Recent reviews have suggested that such variation in individual behavioural plasticity is widespread (Dingemanse et al. 2010), and can be linked to personality type (Mathot et al. 2012). For example, bold fish showed little response to the experience of being caught by an angler, whereas shy fish showed a strong response and changed their behaviour significantly (Stålhammer et al. personal communication). One way to explain these links is that bold animals lack behavioural inhibition and have more engrained responses to situations as they trade-off speed and accuracy differently to shy individuals. Bold animals use old information to make decisions and are thus able to act quickly, whereas shy individuals gather information about their current environment and update regularly. This means that they may exhibit responses that are a better fit to current conditions, but at a cost to speed. Whatever the underlying reason for links between personality and plasticity, it is clear that in many animals, individual flexibility is an important component of a behavioural phenotype.

As we have already discussed, animal personality is an incredibly widespread phenomenon, encompassing within-behaviour consistency, between-behaviour correlations, and behavioural responsiveness. Personality is strongly linked to animal movement patterns at different scales, and the ecological and evolutionary consequences of movement. In the next section, we review studies dealing with the links between animal movement and personality, and we start with local movements and then scale up, to dispersal and finally migratory movements. We conclude by discussing the consequences of the links between personality and movement processes, and also reflect a little on how viewing animals as individuals can shed light on many questions in movement ecology.

6.4 Animal personality and local movements

6.4.1 Movement variables as components of animal personality

Most animals are mobile at some stage in their lives, and move around their local environments to procure food, hide from predators, and seek mating opportunities. These local patterns of movements can vary within species and also within populations, and there is a growing recognition that, in many cases, individual differences in movement are consistent. A plethora of laboratory studies investigating animal personality have incorporated movement variables as measures of personality traits, such as activity and exploration. Activity is considered the baseline level of general movement that an animal makes in a non-novel and non-risky environment (Réale et al. 2007), and exploration can be quantified by measuring the speed of exploration of a novel environment, or alternatively the proportion of a new environment explored (Minderman et al. 2009). Both of these traits have been shown to be repeatable, e.g. activity level in barnacle geese, *Branta leucopsis* (Kurvers et al. 2009), and exploration in starlings, *Sturnus vulgaris* (Minderman et al. 2009).

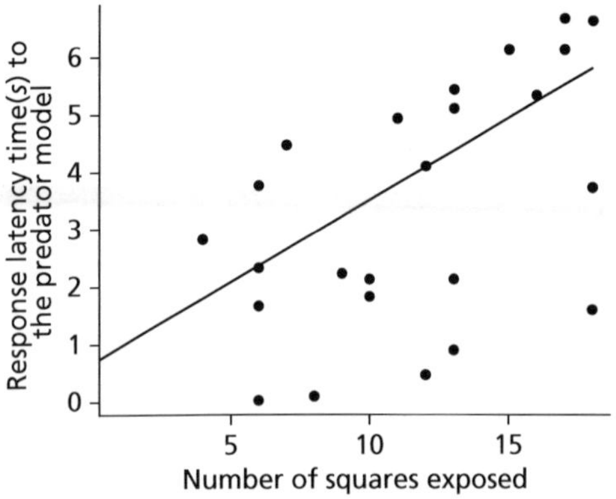

Figure 6.2 Relationship between the exploratory behaviour of individual test fish in the novel environment test and their response latency time to an approaching fish predator model while foraging. Based on Jones and Godin (2010).

These traits have been shown to be positively correlated as part of a behavioural syndrome in some cases, for example in high predation populations of sticklebacks (Dingemanse et al. 2007). Measurements of activity and exploration in experimental setups also have shown that mobility indices can be components of a broader behavioural syndrome. For example, a study of juvenile convict cichlids, *Amatitlania nigrofasciata*, revealed that exploration speed and anti-predator response were correlated, with fast explorers being slower to respond to a robotic predator (Jones and Godin 2010; Fig. 6.2).

Recent discussions have proposed that more diverse and ecologically relevant movement traits could also be explicitly integrated into a personality framework (Liedvogel et al. 2013). One example of a study that has attempted this in the lab is an experiment involving the large white butterfly, *Pieris brassicae* (Ducatez et al. 2012). In this study, three mobility-related traits were quantified for 125 adult butterflies. Flight willingness and flight endurance under stressful conditions, and flight distance were measured, and shown to be individually consistent over time. Furthermore, these traits were correlated in a 'mobility syndrome'.

6.4.2 Personality, consistency, and local movements in the wild

A key question in personality studies and movement ecology is how personality traits co-vary with movement strategies in the wild. This area of research is in its infancy, and only a handful of studies have made links between personality and local movements in free-ranging animals. A key opportunity to study consistency in movement lies with individual differences in foraging behaviour. Individuals from many species potentially vary in their space use in a foraging context, in the locations travelled to, and in the search strategies used to find food. Furthermore, foraging bouts or trips are frequent and repeated, allowing for the collection of data amenable to analyse in a personality framework over relatively short time-spans. A powerful example of personality-dependent local movement comes from a detailed telemetry study of foraging movements in black-browed albatrosses, *Thalassarche melanophrys*. By tracking individual birds that had been assayed for boldness, researchers showed that bolder, risk-prone birds foraged closer to the colony, where competition probably is intense, and shy, risk-averse birds made longer foraging trips further away from the colony (Patrick et al. under review). Furthermore, fast-exploring great tits reacted to an experimental reduction in food availability by more rapidly shifting to other foraging areas than did their slow-exploring conspecifics (van Overveld and Matthysen 2010).

Individuality in local movements has also been documented in other species. Piscivorous pike, *Esox lucius*, that were individually tracked using radio telemetry methods showed highly conserved individual movement strategies. Analysis of the data showed that individuals could be classified into three behavioural types based on activity patterns and habitat use. Two of the three behavioural types in pike showed strong habitat fidelity to either reed-bed or submerged macrophyte habitats, while the third type was an opportunistic mover, using all habitats including the pelagic zone (Kobler et al. 2009). Growth rates of the different strategies did not differ, although activity did, suggesting that fish that adopted the active, opportunistic strategy were potentially able to compensate for the higher energy demands of increased movement with an increase in prey consumption in the pelagic habitat.

Migratory animals spend different seasons in often geographically distinct areas, and so the question of whether personality traits are consistent at these kinds of geographical and temporal scales is of interest to students of both animal personality and migration biology. Do migratory animals show consistency in behaviour or do they adopt location-specific behaviours? A study following the migratory behaviour of individually marked common cranes, *Grus grus*, showed that birds that hatched in undisturbed habitats in Finland choose undisturbed stopover sites in Hungary (Vegvari et al. 2011). Four of five disturbance tolerance variables (which included proximity to road and human population density) were highly repeatable within and between years for individual cranes. Hence, for this species behavioural sensitivity to disturbance can be considered a personality trait that is consistent over large temporal and geographical scales. This fascinating analysis also highlights important questions—why did birds from undisturbed natal habitats choose undisturbed roosting sites? Was this due to early experience and imprinting, or learning from parents? Or alternatively, does habitat preference have a strong genetic component? It is in this way that migratory systems can be used for asking critical questions about the development and evolution of personality variation in animals.

In social species, decisions about movement between food patches (when to leave, and if to join a patch) may also be influenced by the social context. Sociability is a key personality trait (Cote and Clobert 2007), and the social tolerance of an individual may contribute to shaping movement patterns across a landscape of varying resources and social conditions (Michelena et al. 2010). We will discuss this in relation to dispersal later in the chapter, but it is also potentially relevant in the context of more local movements, although few studies have addressed this.

6.4.3 Fitness consequences of personality-dependent local movement

Personality differences in space use can have wider consequences that may be associated with individual fitness. In a fascinating study of personality and parasite load in Siberian chipmunks, *Tamias sibiricus*, Boyer et al. (2010) showed that more active and exploratory individuals had more severe hard tick infestations than less active conspecifics. Activity and exploration in an experimental arena was related to space use at a landscape scale, with more active and exploratory animals (in the assay) also using more space in the local habitat. This correlated with parasite load, highlighting a potential fitness cost for highly active individuals. Making a precise link between an animal's behavioural phenotype and its fitness is challenging, however. To highlight this, while increased activity may be costly and increase parasite load in chipmunks, in another sciurid, the red squirrel, *Sciurus vulgaris*, activity has potential fitness benefits, with the offspring of active females exhibiting higher growth rates (Boon et al. 2007). Furthermore, the costs and benefits of different personality types may not be temporally stable (Dingemanse et al. 2004), and so detailed and painstaking longitudinal studies are needed to investigate relationships between personality (i.e. between-individual variation in behaviour), local movement strategies, and fitness.

6.5 Personality and dispersal

Dispersal is the active or passive movement from the natal or breeding site of an individual to a new site where the individual breeds or could potentially breed would it be alive during the next breeding season (Clobert et al. 2009; Hollander et al., Chapter 7). From a functional point of view, the process of dispersal is commonly divided into three different stages: (i) leaving the current site (emigration), (ii) moving between sites, and (iii) establishing in a new site (immigration), each of which potentially depending on different characteristics of the individual (Bowler and Benton 2005).

Causes of dispersal, in relation to a non-dispersive, sedentary strategy or in relation to variation in dispersal distances, include intra-specific competition, kin interactions, inbreeding avoidance, and improvement of habitat quality (Bowler and Benton 2005, Clobert et al. 2009). However, within this evolutionary framework, a variety of factors have been found to explain individual differences in dispersal propensity. Besides sex and age differences in dispersal tendencies (Greenwood 1983), these factors can be grouped in morphological, parental life-history decisions and behaviour. The most obvious of these may be those related to morphology. Examples of such characteristics are winged morphs in insects, where individuals with wings also attain larger flight muscles and a fat metabolism geared at fuel flight (Fairbairn and Roff 1990, Zhao and Zera 2002). Another spectacular example is the density-mediated transition from a solitary and sedentary phase to a gregarious and dispersive phase in plague locusts (e.g. *Chortoicetes terminifera*; Cullen et al. 2012). In both of these examples, morphological and biochemical traits interact to form a dispersal syndrome. Parental life-history decisions may affect conditions for the young that promote individual variation in dispersal. Such life-history decisions include clutch size (Pärt 1990), date of birth (Nilsson 1989a), and fledging size (Nilsson 1989a, van de Jeugd 2001); the relation between dispersal and both date of birth and fledging size are probably due to their effect on dominance (Nilsson 1990, Altwegg et al. 2000). In recent years, growing interest in behavioural types or personality traits has been connected to dispersal propensity.

6.5.1 Dispersal and behavioural traits

Dispersal has been found to be affected primarily by three behavioural traits: exploration/boldness, aggression, and sociality (e.g. Cote et al. 2010a). In general, these traits, as usually measured, contain both within-individual variation and between-individual variation, only in the latter case representing a personality trait. Commonly, fast explorers and bold individuals disperse further than slow and/or shy individuals in a wide range of organisms. In two species of fish, individuals of the Trindidad killifish, *Rivulus hartii*, and the southern leatherside chub, *Lepidomeda aliciae*, which scored as bold and fast explorers, respectively, in a behavioural assay before dispersal, dispersed longer distances than did their shy and slow conspecifics (Fraser et al. 2001, Rasmussen and Belk 2012). Dispersing root voles were more active and were faster explorers (scored before dispersal) than resident ones (Hoset et al. 2011). In two separate studies on great tits, one in a Dutch (Dingemanse et al. 2003) and one in an English (Quinn et al. 2011) population, birds immigrating to the study site were faster explorers than the resident ones. However, within-population dispersal in great tits from the English population could not be explained by variation in exploratory behaviour, whereas this was the case for females in the Dutch population. Therefore, the relation between exploration and dispersal seems to be both state- and scale-dependent.

Several mechanisms might explain the relation between exploration/boldness and dispersal propensity. For example, in the Dutch great tit population, fast-exploring juveniles without a territory (i.e. the category that makes up the highest proportion of dispersing birds) were more sub-dominant than slow explorers (Dingemanse and de Goede 2004). As juveniles dispersing long distances (immigrants) were fast explorers, they may be constrained in their establishment success and thereby will end up further from their natal site. Furthermore, slow explorers are better at dealing with social defeat and stress than are fast explorers (Carere et al. 2001). Thus, also at the beginning of the dispersal process, when individuals decide to emigrate, fast explorers are more apt to leave a natal area with high densities of already established dominant adult birds as they need to reduce the stress of social defeat more than slow explorers (Dingemanse and de Goede 2004, Quinn et al. 2011). Also the intermediate phase during the dispersal process, moving between sites, is influenced by exploration/boldness. Bold juvenile bluegill sunfishes, *Lepomis macrochirus*, moved in longer and faster bursts of swimming than did shy individuals (Wilson and Godin 2010). Thus, in all three of the phases composing the dispersal process, fast-exploring and/or bold individuals seem to behave in a way promoting long dispersal distances.

6.5.2 Aggression

A common observation during the emigration phase of dispersal is that aggressive individuals force less aggressive conspecifics away from the natal area. However, these interactions are probably due to context-dependent aggression as already established individuals are dominant to newcomers through the effect of prior occupancy (Nilsson 1989b) and, thus, at least not primarily due to an inherent behavioural type. However, a recent study on natural range expansion in western bluebirds, *Sialia mexicana*, forms a nice example of how consistently aggressive personalities bias the propensity to disperse. In this species, aggressive males dispersed to new areas, which had been made suitable by the erection of nest boxes, whereas less aggressive males remained resident in the natal area (Fig. 6.3d; Duckworth and Badyaev 2007). The mechanism behind this variation can be found in the variable success of the two personalities during both the emigration and immigration phase. Aggressive birds are at a disadvantage in resident areas because they are less likely to get established in the natal population. Establishment success in such areas is dependent on facultative cooperative behaviour towards kin, increasing the possibilities for less aggressive juveniles to remain but resulting in the emigration of aggressive birds. On the other hand, aggressive juveniles are at an advantage in newly colonized areas as they are superior in competition for territories towards both less aggressive conspecifics and less aggressive mountain bluebirds, *Sialia currucoides* (Fig 6.3a), resulting in an increased success during immigration (Duckworth 2006a).

6.5.3 Sociality

An individual's tendency to be attracted to conspecifics (social individuals) or to be sensitive to crowding (asocial individuals) has been found to be related to propensity to disperse, thereby linking effects of density to dispersal behaviours (Cote et al. 2010a). Mosquitofish, *Gambusia affinis*, that tended to spend short periods of time near a shoal (asocial), dispersed further in ensuing dispersal assays than did fish spending longer periods of time near shoals (Cote et al. 2010b). Another aspect of sociality, the number of individuals a focal animal had

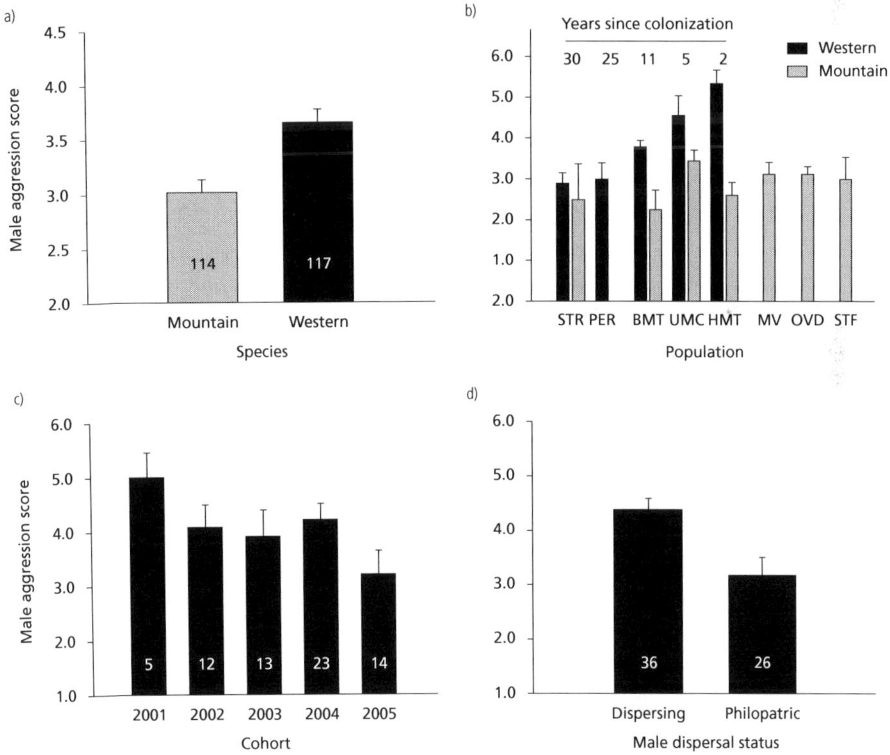

Figure 6.3 Variation in aggression across species, populations, generations, and dispersal classes. (a) Western bluebirds, *Sialia mexicana*, were more aggressive than mountain bluebirds. (b) Aggression differed significantly among western bluebird populations and was related to the number of years since the colonization of a population (numbers above bars) (c) Aggression significantly decreased across cohorts in the Blue Mountain population. (d) Males that dispersed away from their natal population to breed (Left) were more aggressive than males that remained in their natal population (Right). Shown are means ± SE. Numbers on bars indicate the number of males sampled (Duckworth and Badyaev 2007).

affiliative interactions with predicted dispersal in female yellow-bellied marmots, *Marmota flaviventris*; females with few such interactions dispersed away from the group (Blumstein et al. 2009). The relationship between sociality, density, and dispersal was also nicely illustrated in an experimental study on common lizards, *Lacerta vivipara*. Juvenile lizards were scored according to their attraction towards the odour of conspecifics and these scores were used as an estimate of social tolerance (Cote and Clobert 2007). By releasing these juveniles into experimental populations of different densities, Cote and Clobert (2007) found that juveniles dispersing away from high density populations had low social tolerance scores and those dispersing away from low density populations instead had high scores. Thus, dispersal behaviour in this case depended on an interaction between density and social personality. In a similar, although non-experimental study, Jokela et al. (2008) found that sociality also predicted dispersal in humans. Finnish women and men with high sociality scores were found to disperse to urban areas, whereas those with low scores moved out of urban areas. However, in this case individuals with high social scores had longer dispersal distances.

6.6 Behavioural syndromes and dispersal

Besides the correlations between single personality traits and dispersal exemplified in section 6.5, certain traits, for example boldness, aggressiveness, and activity, have been found to correlate positively (Cote et al. 2010b). As outlined already, such suites of correlated behavioural characters form behavioural syndromes (Fig 6.1) and when correlated to dispersal, they have been called dispersal behavioural syndromes (Cote et al. 2010a). As the consistency of these personality traits over time and context will potentially have powerful consequences for spatial population dynamics, this issue has attracted a lot of recent interest and most of the published reports actually provide evidence for such consistencies (Réale et al. 2007). Mosquitofish exhibited consistent personality traits over a 3-week period that spanned the dispersal stage (Cote et al. 2010b). In western bluebirds, aggression was highly repeatable within individuals irrespective of age or breeding status, indicating the existence of distinct behavioural morphs (Duckworth and Badyaev 2007). Both sociality and exploration was found to be consistent over a one-year period, including the dispersal event in common lizards and great tits, respectively (Cote and Clobert 2007, Quinn et al. 2011). Further, personality traits may also be consistent over major life stage transitions. Activity, exploration, and boldness were consistent between the larvae and adult stage in the hemimetabolous firebugs, *Pyrrhocoris apterus* (Gyuris et al. 2012), and across metamorphosis in the lake frog, *Rana ridibunda* (Wilson and Krause 2012). In contrast, although root voles that dispersed were more explorative and active than their non-dispersing conspecifics before dispersal, this difference had vanished after the dispersal event (Hoset et al. 2011; Hollander et al., Chapter 7). Thus, the consistency of a dispersal behaviour syndrome seems to be context-dependent, but more studies must be performed before we can explain this variation.

6.7 Consequences of dispersal behavioural syndromes at the population level

Dispersal behavioural syndromes will potentially be important for explaining processes at the population level, including variation in metapopulation dynamics, range expansions, and the success of invasive species (Sih et al. 2012). In the next sections, we will provide some examples of the influence of personality traits on these processes.

6.7.1 Metapopulation dynamics

In a heterogeneous environment, populations often inhabit patches of sustainable habitats surrounded by relatively inhospitable areas, creating a metapopulation structure (Hanski 1999; Smith et al., Chapter 4). The dynamic of such systems involve local extinctions and re-colonizations from other sub-populations. The existence of consistent personalities of dispersers will then predict a non-random distribution of personalities among sub-populations, depending on the time since the last re-colonization.

As already discussed, sociality is a personality trait that might be especially important in this respect; asocial mosquitofish dispersed further than more social ones (Cote et al. 2010b) and common lizards colonizing empty habitats were of an asocial behavioural type (Cote and Clobert 2007). Thus, patches in which the local population has gone extinct will be re-colonized by an asocial and/or aggressive personality type. In the case of the lizards, this will be a transient phase as social personalities disperse towards high-density patches. As density starts to build up in the new patch, an increasing number of social lizards will establish, further increasing the density and the asocial personality types start to emigrate (Cote and Clobert 2007). Thus, the asocial, dispersive individuals will facilitate the local buildup of density in newly established patches. In the case that the sorting of personalities according to density is connected to fitness advantages for each personality type in its preferred patch, as has been shown for western bluebirds (see next section; Duckworth 2008), selection and differential dispersal propensity may maintain variation in personalities within the population.

6.7.2 Range expansion

As human-induced changes in the environment (e.g. habitat alterations and global climate change) become increasingly common, certain species will be favoured with range expansions as a result (Smith et al., Chapter 4; Hollander et al., Chapter 7). In parallel with metapopulation dynamics discussed in the previous section, dominating personality syndromes may differ between core distribution areas and the expansion front. To illustrate this we return to the example of the range expansion by the western bluebird. This species, favoured by human-provided nest boxes, is expanding its range into previously non-occupied areas. Male bluebirds settling in these new areas are more aggressive than the remaining, philopatric part of the population (Fig. 6.3d; Duckworth and Badyaev 2007). Thus, at the expansion front, male bluebirds have a more aggressive personality than in core areas. However, with time the average male aggression score decreases (Fig. 6.3b), indicating either flexibility in individual aggression or a slow replacement of aggressive males by less aggressive ones. Duckworth and Badyaev (2007) provided evidence for the latter explanation as consecutive cohorts established in an area had progressively lower aggression scores (Fig. 6.3c), rather than individual males altering their aggression. Thus, selection favours aggressive personalities at the expansion front, probably because this behaviour facilitates establishment in competition with already established mountain bluebirds. However, as the population increases, negative selection on high aggression scores intensifies, partly because cooperative behaviour favours less aggressive males and partly because aggressive males do not provide very much parental care, resulting in low reproductive success (Duckworth 2006a, b). Thus, dispersal-driven personality sorting may facilitate the range expansions of populations through initial successful competition with already resident species (Hollander et al., Chapter 7).

Also explorative behaviour has been found to differ between core and expansion edge populations. House sparrows, *Passer domesticus*, were introduced to Mombasa in Kenya around 1950. By using distance from Mombasa as a cue for the age of the house sparrow population (the further away, the more recently the populations were established), Liebl and Martin (2012) found increasingly higher scores of exploratory behaviour towards the expansion edge. Explorative personalities might have an advantage in new areas by being better at discovering and using novel resources in an unfamiliar environment.

6.7.3 Invasive species and reintroductions

During recent decades, concomitant with an increase in the transportation of humans and goods there has also been an increasing risk of unintentionally transporting species between regions (Hollander et al., Chapter 7). Some species have been introduced several times to a region without being able to establish stable populations, whereas others have been very successful, outcompeting native species and becoming an invasive species. General morphological and life-history traits that characterize successful and unsuccessful invaders have been hard to find, but it is only recently that personality traits have been evaluated in this respect (Chapple et al. 2012). The success of an introduced species depends on a chain of processes that must be successfully completed including transport, initial establishment, population growth, and secondary spread. Bold, aggressive, exploratory, and asocial individuals have proved to be more dispersive and able to establish populations, at least in low density environments. Hence, these behavioural traits may potentially form an invasion syndrome that characterize successful invaders (Sih et al. 2004). However, the general conclusion from studies on metapopulation dynamics and range expansions is that different personalities are at an advantage at different times, i.e. in different ecological contexts. Thus, the prediction from these studies would be that successful invaders should be composed of a mix of personalities, some with an advantage in the beginning of the invasion chain of processes and some with an advantage in later stages (Chapple et al. 2012, Sih et al. 2012). This is also what was found in a model simulating the invasion process in a mixture of two different sociality types differing in dispersal and fitness responses to density (Fogarty et al. 2011). Populations that were polymorphic with regard to these sociality types both were faster to invade and had a larger impact in the new environment (Fig. 6.4). This is in line with that the number of invasion events and the number of individuals invading at each invasion are the best predictors of a successful invasion (Kolar and Lodge 2001, Hayes and Barry 2008). The more tries and the higher number of individuals at each try, the higher the probability that the right mix of personalities is included in the founder population.

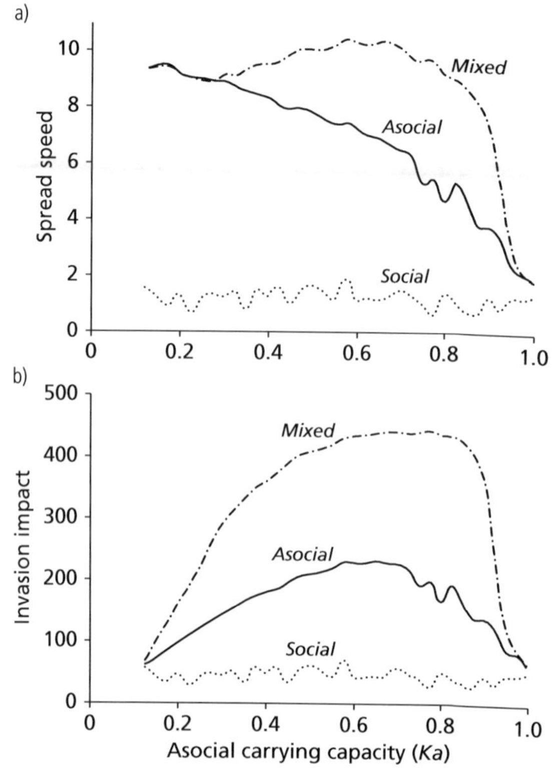

Figure 6.4 Spread speed (a) and invasion impact (b) during invasion for the three behavioural mixes as a function of asocial carrying capacity (K_a). When K_a is smaller, types are more dissimilar. Spread speed is measured as the rate of patch establishment after 10% of patches are established. Invasion impact is measured as the total population size in the network 50 time steps after 10% of patches are established. This measurement takes both local patch density and spread speed into account (Fogarty et al. 2011).

In conservation and restoration projects we can find a similar concern, namely how to increase the success of intentional reintroductions of species. Drawing on the lesson from successful invasions, a mix of personalities among the group going to be reintroduced is probably very important, although often hard to achieve. If individuals are captured in the wild and then translocated to new areas, there is a potential risk that only bold and aggressive personalities are reintroduced, as such individuals are easier to capture (Biro et al. 2004). Individuals originating from captive breeding schemes have potentially been under intentional or unintentional artificial selection, which may result in, for example, less aggressive personalities, as these are better adapted to captive conditions (McDougall et al. 2006). Furthermore, individuals bred in captivity may lose some of the anti-predator behaviours of wild individuals, thus being bolder. Extremely bold individuals do not cope with life in the wild, and this effect has been detrimental to reintroduced individuals of swift foxes, *Vulpes velox* (Bremner-Harrison et al. 2004) and to hatchery-reared fishes (Conrad et al. 2011).

6.8 Individuality, personality, and migration

Migration is perhaps the most visible and spectacular form of animal movement. Each year billions of organisms make migratory journeys in step with temporal changes to their habitats, across a vast range of spatial and temporal scales. Understanding the causes and consequences of such migratory movements is an important goal of modern biology. Yet studying migratory behaviour in the wild is problematic as individuals can travel for thousands of miles across rugged and inhospitable terrain. Similarly, tracking small organisms at the millimetre-scale is generally impossible since any tracking device will be too large for the animal to carry. However, new tracking techniques have enabled researchers to move to new frontiers in our understanding of movement and patterns as individual animals can now be followed both when making arduous trips and by e.g. using recent advances in nanotechnology to track crustacean zooplankton (Box 6.1). Focusing in on this individual level has uncovered a wealth of individual variation in migratory behaviour. Within a single population, individuals can differ in many migratory traits, such as the timing of migration or the destination an individual travels to, a phenomenon known as differential migration (Ketterson and Nolan 1976, Newton 2008; Liedvogel and Lundberg, Chapter 12). In many populations, not all individuals migrate, leading to a common form of migration known as partial migration (Chapman et al. 2011b; Chapman et al., Chapter 2). This individual variation in migratory behaviour so often reported in animal populations allows scientists to investigate phenotypic differences between individuals and relate these to variation in migratory behaviour. Such approaches have led to the discovery that, in many cases, individual variation can be consistent over time, and correlated with other behaviours as part of a behavioural syndrome, i.e. that animal personality variation can be important in explaining patterns of migration in the wild. In this way we can take a personality view of animal migration that can give insights into the genesis and evolution of migratory behaviour. Few empirical studies have yet addressed this potential link, for the most part due to the logistical challenges in quantifying personality traits in addition to garnering individual level data on migratory movements. In this section we review what is known about the relationship between animal personality and migratory behaviour, detailing evidence from studies that assess within-population and between-species variation in migratory behaviours.

6.9 Partial migration and animal personality

Partial migration is an extremely widespread form of animal migration, and occurs when populations are composed of both migratory and resident individuals (Lack 1943, Lundberg 1987, Chapman et al. 2011b). Partially migratory populations have been recorded from all migratory taxa, including fishes (Chapman et al. 2012), birds (Lundberg 1988), mammals (Mysterud et al. 2011), amphibians (Grayson and Wilbur 2009), crustaceans (Hansson and Hylander 2009), and insects (Attisano et al. 2013). This phenomenon provides a powerful opportunity to study the processes that shape the evolution of migration more generally, as partial migration is thought to be an incipient stage in the evolution of migration (Chapman et al. 2011b). Hence, researchers can correlate the probability of individual migration with phenotypic traits to shed light upon the ecological forces at play in the evolution of animal migration more broadly. This approach has begun to be used with respect to consistent individual differences in behaviour in animals (i.e. animal personality). As we have detailed previously in this chapter, a great deal of evidence links personality variation to variation in animal movement. Do animals with different behavioural types also differ in their migratory behaviour? Here we review the evidence for this question with two case studies.

6.9.1 To boldly go: partial migration and personality in roach

A recent study made some of the first links between animal personality and migration in a two-year field study of roach, *Rutilus rutilus*, a partially migratory cyprinid fish. During the spring and summer months, the entire study population inhabits a

Box 6.1 Using nanotechnology to track small organisms

Various tracking techniques have provided important information on the movement of large animals, such as mammals and birds, whereas our understanding of the individual behaviour of small (millimetre-scale) organisms is extremely limited due to the large sizes of existing tracking devices; i.e. small animals simply cannot carry them. However, recent advances in nanotechnology have opened up possibilities for three-dimensional (3D) tracking of individual millimetre-sized aquatic organisms. The method is based on *in vivo* labelling of the organisms with fluorescent nanoparticles, so-called quantum dots (Q-dots; Lard et al. 2010), and tracking of the organisms in 3D via the quantum-dot fluorescence using a synchronized multiple-camera system. At the individual level, it is now possible to quantify tracks (e.g. speed and distance moved) when the organism, for example a *Daphnia magna* (Fig. 6.5; see also colour plate section), is exposed to a threat, for example ultraviolet radiation (UV; Ekvall et al. 2013). At normal (white) light, the *Daphnia* utilizes the upper part of the aquarium for feeding (Fig. 6.5a), whereas when it senses the threat (UV radiation) it rapidly, and actively, escapes to the bottom (Fig. 6.5b). Hence, by using marking by nanotechnology, in combination with advanced camera systems, it is now also possible to simultaneously study the individual behaviour of multiple individuals and thereby address personality differences even among small organisms (Ekvall et al. 2013). In a broader perspective this new technique will advance plankton ecology and be applicable for tracking other small organisms in water, but also small terrestrial organisms, such as insects. Hence, by combining biology and nanotechnology, the tracking of small animal movements and migration can now be raised to the same scientific level as for larger animals, such as birds and fish.

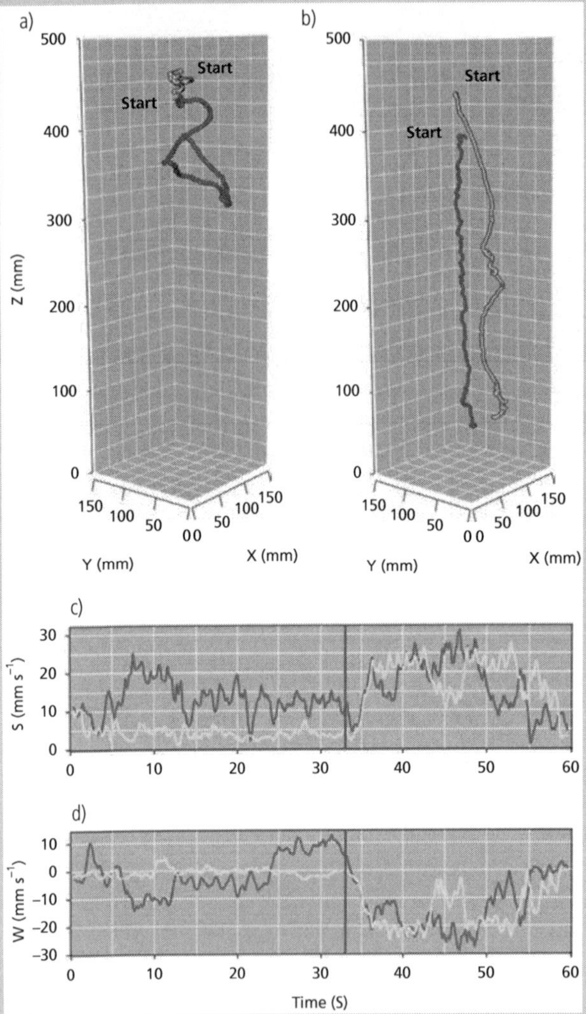

Figure 6.5 Tracking of the position, speed, and vertical displacement of two *Daphnia magna* individuals in the absence (a) and presence (b) of UV radiation. The animals were marked with yellow and red nanoparticles (quantum dots), and monitored simultaneously to obtain 3D trajectories. Swimming speed (*S*) for both animals is shown in (c) and vertical speed (*w*) in (d). From Ekvall et al. (2013). (See Plate 1)

lake in southern Sweden, Lake Krankesjön. It has one outlet and two inlet streams into which cyprinids such as roach migrate during the autumn and over the winter (Hansson et al. 2007, Brodersen et al. 2008, Skov et al. 2008). Pike, the dominant predator of roach in the lake, do not follow their prey into the streams and hence roach migration is thought to function as a predator avoidance behaviour during the winter when food availability is low in the lake (Brönmark et al. 2008). However, not all individuals migrate into the streams during winter. By assaying fish for boldness that were tagged with uniquely coded PIT tags prior to migration, researchers showed that bold roach have a higher probability of migration than shy ones (Fig. 6.6), hence linking individual personality to migratory behaviour in a wild population (Chapman et al. 2011c). The experiment was repeated in the following year with the same results, giving strong support for the role of personality in shaping these migratory movements. One explanation for why bold fish migrate is that they have higher predation vulnerability due to their risk-prone behaviour, and that migration has evolved as a strategy to mediate some of the costs of being bold. Alternatively, boldness could be part of a wider behavioural syndrome that includes exploratory behaviour, which might also explain why bold fish are more likely to migrate.

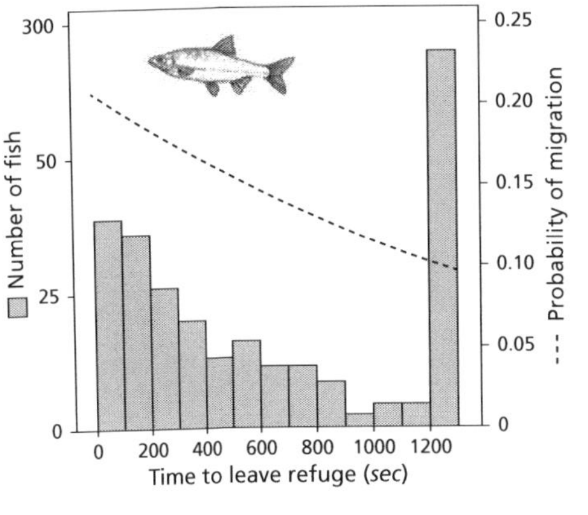

Figure 6.6 Distribution of boldness scores for fish assayed and tagged in 2009 and 2010, overlaid with the migratory probability of individuals with a given boldness score. Note that there is a high frequency of fish that had not left the refuge at 1200 s. These shy fish were given a ceiling value for analysis (Chapman et al. 2011c). Reproduced with permission from John Wiley & Sons.

6.9.2 Boldness and blue tit migration

Similar to roach, blue tits, *Cyanistes caeruleus*, are also partial migrants, with only some individuals migrating each autumn and others remaining resident. However, in this species migrants are subdominant and seem to adopt a 'best-of-a-bad-job' strategy when migrating (Smith and Nilsson 1987). A comparison of resident and migrant birds in their reaction to a novel object showed that migratory blue tits had shorter approach latencies than resident birds (Nilsson et al. 2010). Hence, in this example an index of boldness also predicted which birds were migrants or residents. One explanation for this result is that migratory birds are likely to encounter many novel situations and environments during the migratory journey, and so the adaptive value of neophilia is higher for migrants than residents. However, work is needed to confirm that neophilia, i.e. being attracted to novel objects or situations, is consistent over time and part of a behavioural syndrome in this species.

6.9.3 Comparing species-level behaviours

Another approach to studying potential links between personality and migration is to compare the behaviour of closely related migratory and non-migratory species. A comparison of the behaviour of two closely related warbler species (the resident Sardinian warbler, *Sylvia melanocephala*, and the migratory garden warbler, *Sylvia borin*) showed differences in both mean behaviour and behavioural consistency (Mettke-Hofmann et al. 2005). Resident Sardinian warblers were less neophobic, i.e less afraid of novel objects, and more exploratory than individuals of the migratory species. This is perhaps a surprising result, as intraspecific comparisons showed that resident blue tits were more neophobic than migratory conspecifics (Nilsson et al. 2010). In addition, resident Sardinian warblers showed consistent behaviours over a ten-month time period, whereas migratory garden warblers did not. While the data are based on a small sample size and just includes two species, the results are interesting and may suggest that residency promotes the evolution of behavioural consistency. One might speculate that as obligate migratory animals encounter a diverse range of habitats and situations throughout their lives, consistency in behaviour would be selected against, and a high degree of plasticity selected for in the place of personality. This notion is supported by a species-level comparative analysis of palaearctic passerines, which links residency to a high degree of innovative feeding behaviour (Sol et al. 2005). Bird species with relatively larger brains and more behavioural flexibility were more likely to be residents.

6.10 Individuality and migration

Migration biology has entered an exciting new era, as technological developments now allow scientists to study migration at an individual level for many species. This has revealed a surprising degree of intraspecific variation in migratory traits. For example, in the Eleonora's falcon (*Falco eleonorae*), tracking data revealed that, unlike previously thought, this species does not migrate along a historical route but shows a high degree of individual variation (Gshweng et al. 2008). Even in marine fishes, whose movements are notoriously difficult to track, electronic storage tags revealed details of the migrations of individual plaice, *Pleuronectes platessa* (Hunter et al. 2003). These advances mean that we can now not only begin to link personality variation to migratory traits, but also consider migratory traits themselves from a personality perspective. How widespread are individual differences in migratory behaviour? Is migratory behaviour consistent over time within individuals? Do migratory behaviours correlate to form part of a wider behavioural syndrome?

Evidence from a range of taxa now reveals that individuals can be surprisingly consistent in migratory behaviours over time and context. A groundbreaking study of individuality in the migratory journeys of marsh harriers, *Circus aeruginosus*, showed that this long-distance migrant showed a high degree of consistency in timing of its movements. The annual timing of migration varied less for repeated journeys of the same individual than for that of conspecific migrants (Vardanis et al. 2011). Repeatability in space was less pronounced, and route fidelity was low, with considerable within-individual variation. Similar patterns were shown in black-browed albatross (*Thalassarche melanophrys*): a study following 24 individuals over two consecutive years, highlighted consistent timing of movements, but plasticity in space use and routes on the return journey, with variable use of intermediate staging sites depending on local conditions (Phillips et al. 2005). This difference in temporal versus spatial patterns of consistency perhaps suggests a stronger endogenous control of timing, compared with a more flexible approach to spatial movements, with routes being determined more by environmental conditions during the migratory journey. This also means that route recapitulation by landmark recognition is an unlikely navigational mechanism for these birds. Timing was also shown to be consistent in migratory roaches, in their migratory journeys from lakes into streams during the winter months (Brodersen et al. 2012). In this study the authors were also able to investigate individual plasticity, and found intriguing evidence that this may be a more general property that spans several migratory traits. In an analysis of 6 years of migration data of close to 3,000 fish it was shown that, at a population level, roach migrants were consistent in the timing of migration and also the destination (i.e. which stream to migrate into). Interestingly, fish that were consistent in one migratory trait were also consistent in the other; thus, fish with highly conserved timing also showed strong site fidelity (Brodersen et al. 2012). This fits with the concept of 'coping styles', where proactive individuals are routine forming and reactive individuals are more responsive to current conditions. Hence, in this analysis migrants can potentially be placed along the proactive–reactive personality axis.

6.11 Future perspectives

Partial migration offers a fantastic opportunity to study how migratory behaviour relates to personality type. Data are difficult to gather but technological advances make individual level analyses more possible with each year that passes. A number of interesting questions are unresolved. For example, how do migratory traits fit into a broader behavioural syndrome? Both of the intraspecific, partial migration case studies presented here looked at just one behavioural axis. Gathering behavioural syndrome data is time-consuming and combined with following individual patterns of migration is certainly a logistical challenge. However, we hope that in the future ambitious studies attempt to tackle these questions. Also studies which combine behavioural traits such as boldness with more detailed migratory traits such as timing, distance travelled, migratory speed, and migratory destination would shed further light on the links between personality and migration. In addition, migratory species offer an opportunity to assess the stability of animal personality at different stages of the migratory cycle.

The importance of the social environment in determining which individuals migrate is a fascinating question. It is intriguing to investigate whether the behavioural phenotypic landscape responsible for dispersal decisions in the mosquitofish (Cote et al. 2011) might be similarly true for migratory systems. Can the mix of personality types at the winter grounds influence which individuals migrate? The competitive release hypothesis (Chapman et al. 2011b, Gauthreax 1982) suggests that in partially migratory species, competitively inferior individuals would be predicted to migrate as they are unable to access reduced resources available during winter. However, much of the evidence for this uses proxies of competitive ability such as body size or age. Recent evidence indicates that animals of different personality types can be more or less competitive than other conspecifics, and so potentially personality may help explain patterns of partial migration. However, boldness (which was a trait associated with migrants in blue tits and roach) is often correlated with aggressiveness as part of a behavioural syndrome. Hence, the competitive release hypothesis might generate a prediction that bold and aggressive animals remain resident as they can outcompete more shy and passive conspecifics. However, we can just as well predict that bold and aggressive individuals will migrate as they may be worse dealing with the stress of losing interactions than shy individuals (Carere et al. 2001). Furthermore, at least with the roach example, it is thought that predation, rather than competition, is a key ecological factor driving migratory dynamics. This shows the importance of understanding the ecological forces that shape migratory behaviour before making predictions about personality-dependent migration. As data are currently so sparse, it is too early to uncover any general predictive patterns governing which behavioural types should migrate under which conditions. We hope that the future will yield exciting new insights that will allow these kinds of analyses in the coming years.

References

Altwegg, R., Harald, T., Ringsby, H., and Sæther, B. E. (2000). Phenotypic correlates and consequences of dispersal in a metapopulation of house sparrows *Passer domesticus*. *Journal of Animal Ecology*, **69**, 762–70.

Aristotle, *c*.350 BC. A history of animals. Translated by D'Arcy Wentworth Thompson. Available at http://classics.mit.edu/Aristotle/history_anim.8.viii.html.

Attisano, A., Tregenza, T., Moore, A. J., and Moore, P. J. (2013). Oosorption and migratoty strategy of the milkweed bug, *Oncopeltus fasciatus*. *Animal Behaviour*, **86**, 651–7.

Bell, A. M. (2005). Behavioral differences between individuals and two populations of stickleback (*Gasterosteus aculeatus*). *Journal of Evolutionary Biology*, **18**, 464–73.

Bell, A. M. (2007). Future directions in behavioural syndromes research. *Proceedings of the Royal Society B: Biological Sciences*, **274**, 755–61.

Bell, A. M., Hankison, S. J., and Laskowski, K. L. (2009). The repeatability of behaviour: a meta-analysis. *Animal Behaviour*, **77**, 771–83.

Benus, R. F, Koolhaas, J. M., and Vanoortmerssen, G. A. (1988). Aggression and adaptation to the light-dark cycle: role of intrinsic and extrinsic control. *Physiology and Behavior*, **43**, 131–7.

Bergmüller, R., and Taborsky, M. (2010). Animal personality due to social niche specialisation. *Trends in Ecology and Evolution*, **25**, 504–11.

Biro, P. A., Abrahams, M. V., Post, J. R., and Parkinson, E. A. (2004). Predators select against high growth rates and risk-taking behaviour in domestic trout populations. *Proceedings of the Royal Society B: Biological Sciences*, **271**, 2233–7.

Blumstein, D. T., Wey, T. W., and Tang, K. (2009). A test of the social cohesion hypothesis: interactive female marmots remain at home. *Proceedings of the Royal Society B: Biological Sciences*, **276**, 3007–12.

Boon, A. K., Réale, D., and Boutin, S. (2007). The interaction between personality, offspring fitness and food abundance in North American red squirrels. *Ecology Letters*, **10**, 1094–104.

Bowler, D. E., and Benton, T. G. (2005). Causes and consequences of animal dispersal strategies: relating individual behaviour to spatial dynamics. *Biological Reviews*, **80**, 205–25.

Boyer, N., Réale, D., Marmet, J., Pisanu, B., and Chapuis, J.-L. (2010). Personality, space use and tick load in an introduced population of Siberian chipmunks *Tamias sibiricus*. *Journal of Animal Ecology*, **79**, 538–47.

Bremner-Harrison, S., Prodohl, P. A., and Elwood, R. W. (2004). Behavioural trait assessment as a release criterion: boldness predicts early death in a reintroduction programme of captive-bred swift fox (*Vulpes velox*). *Animal Conservation*, **7**, 313–20.

Brodersen, J., Nilsson, P. A., Chapman, B. B., Skov, C., Hansson, L.-A., and Brönmark, C. (2012). Variable individual consistency in timing and destination of winter migrating fish. *Biology Letters*, **8**, 21–3.

Brodersen, J., Nilsson, P. A., Hansson, L.-A., Skov, C., and Brönmark, C. (2008). Condition-dependent individual decision-making determines cyprinid partial migration. *Ecology*, **89**, 1195–200.

Brönmark, C., Skov, C., Brodersen, J., Nilsson, P. A., and Hansson, L.-A. 2008. Seasonal migration determined by a trade-off between predator avoidance and growth. *PLoS One*, **3**, 1953–7.

Carere, C., Welink, D., Drent, P. J., Koolhaas, J. M., and Groothuis, T. G. G. (2001). Effect of social defeat in a territorial bird (*Parus major*) selected for different coping styles. *Physiology and Behavior*, **73**, 427–33.

Chapman, B. B., Brönmark, C., Nilsson, J.-Å., and Hansson, L.-A. (2011b). The ecology and evolution of partial migration. *Oikos*, **120**, 1764–75.

Chapman, B. B., Hulthen, K., Blomqvist, D. R., et al. (2011c). To boldly go: individual differences in boldness influence migratory tendency. *Ecology Letters*, **14**, 871–6.

Chapman, B.B., Skov, C., Hulthén, K., et al. (2012). Partial migration in fishes I: definitions, methodologies and taxonomic distribution. *Journal of Fish Biology*, **81**, 479–99.

Chapman, B. B., Thain, H., Coughlin, J., and Hughes, W. O. H. (2011a). Behavioural syndromes at multiple scales in ants. *Animal Behaviour*, **82**, 391–7.

Chapple, D. G., Simmonds, S. M., and Wong, B. B. M. (2012). Can behavioral and personality traits influence the success of unintentional species introductions? *Trends in Ecology and Evolution*, **27**, 57–64.

Clobert, J., Le Galliard, J.-F., Cote, J., Meylan, S., and Massot, M. (2009). Informed dispersal, heterogenity in animal dispersal syndromes and the dynamics of spatially structured populations. *Ecology Letters*, **12**, 197–209.

Conrad, J. L., Weinersmith, K. L., Brodin, T., Saltz, J. B., and Sih, A. (2011). Behavioural syndromes in fishes: a review with implications for ecology and fisheries management. *Journal of Fish Biology*, **78**, 395–435.

Coppens, C. M., de Boer, S. F., and Koolhaas, J. M. (2010). Coping styles and behavioural flexibility: towards underlying mechanisms. *Philosophical Transactions of the Royal Society B: Biological Sciences*, **365**, 4021–8.

Cote, J., and Clobert, J. (2007). Social personalities influence natal dispersal in a lizard. *Proceedings of the Royal Society B: Biological Sciences*, **274**, 383–90.

Cote, J., Clobert, J., Brodin, T., Fogarty, S., and Sih, A. (2010a). Personality-dependent dispersal: characterization, ontogeny and consequences for spatially structured populations. *Philosophical Transactions of the Royal Society B: Biological Sciences*, **365**, 4065–76.

Cote, J., Fogarty, S., Brodin, T., Weinersmith, K., and Sih, A. (2011). Personality dependent dispersal in the invasive mosquitofish: group composition matters. *Proceedings of the Royal Society B: Biological Sciences*, **278**, 1670–8.

Cote, J., Fogarty, S., Weinersmith, K., Brodin, T., and Sih, A. (2010b). Personality traits and dispersal tendency in the invasive mosquitofish (*Gambusia affinis*). *Proceedings of the Royal Society B: Biological Sciences*, **277**, 1571–9.

Cullen, D. A., Sword, G. A., and Simpson, S. J. (2012). Optimizing multivariate behavioural syndrome models in locusts using automated video tracking. *Animal Behaviour*, **84**, 771–84.

Dingemanse, N. J., Both, C., Drent, P. J., and Tinbergen, J. M. (2004). Fitness consequences of avian personalities in a fluctuating environment. *Proceedings of the Royal Society B: Biological Sciences*, **271**, 847–52.

Dingemanse, N. J., Both, C., van Noordwijk, A. J., Rutten, A. L., and Drent, P. J. (2003). Natal dispersal and personalities in great tits (*Parus major*). *Proceedings of the Royal Society B: Biological Sciences*, **270**, 741–7.

Dingemanse, N. J., Bouwman, K. M., van de Pol, M. et al. (2012). Variation in personality and behavioural plasticity across four populations of the great tit *Parus major*. *Journal of Animal Ecology*, **81**, 116–26.

Dingemanse, N. J., and de Goede, P. (2004). The relation between dominance and exploratory behavior is context-dependent in wild great tits. *Behavioral Ecology*, **15**, 1023–30.

Dingemanse, N. J., and Dochtermann, N. A. (2013). Quantifying individual variation in behaviour: mixed-effect modelling approaches. *Journal of Animal Ecology*, **82**, 39–54.

Dingemanse, N. J., Kazem, A. J. N., Réale, D., and Wright, J. (2010). Behavioural reaction norms: animal personality meets individual plasticity. *Trends in Ecology and Evolution*, **25**, 81–89.

Dingemanse, N. J., and Wolf, M. (2010). Recent models for adaptive personality differences: a review. *Philosophical Transactions of the Royal Society B: Biological Sciences*, **365**, 3947–58.

Dingemanse, N. J., and Wolf, M. (2013). Between-individual differences in behavioural plasticity within populations: causes and consequences. *Animal Behaviour*, **85**, 1031–9.

Dingemanse, N. J., Wright, J., Kazem, A. J. N., Thomas, D. K., Hickling, R., and Dawnay, N. (2007). Behavioural syndromes differ predictably between 12 populations of three-spined stickleback. *Journal of Animal Ecology*, **76**, 1128–38.

Dochtermann, N. A. (2011). Testing Cheverud's conjecture for behavioral correlations and behavioral syndromes. *Evolution*, **65**, 1814–20.

Dochtermann, N. A., and Dingemanse, N. J. (2013). Behavioral syndromes as evolutionary constraints. *Behavioral Ecology*, **24**, 806–11.

Drent, P. J., van Oers, K., and van Noordwijk, A. J. (2003). Realized heritability of personalities in the great tit (*Parus major*). *Proceedings of the Royal Society B: Biological Sciences*, **270**, 45–51.

Ducatez, S., Legrand, D., Chaput-Bardy, A., Stevens, V. M., Freville, H., and Baguette, M. (2012). Inter-individual variation in movement: is there a mobility syndrome in the large white butterfly *Pieris brassicae*? *Ecological Entomology*, **37**, 377–85.

Duckworth, R. A. (2006a). Aggressive behaviour affects selection on morphology by influencing settlement patterns in a passerine bird. *Proceedings of the Royal Society B: Biological Sciences*, **273**, 1789–95.

Duckworth, R. A. (2006b). Behavioral correlations across reproductive contexts provide a mechanism for a cost of aggression. *Behavioral Ecology*, **17**, 1011–19.

Duckworth, R. A. (2008). Adaptive dispersal strategies and the dynamics of a range expansion. *American Naturalist*, **172**, S4–S17.

Duckworth, R. A., and Badyaev, A. V. (2007). Coupling of dispersal and aggression facilitates the rapid range expansion of a passerine bird. *Proceedings of the National Academy of Sciences USA*, **104**, 15017–22.

Duckworth, R. A., and Kruuk, L. E. B. (2009). Evolution of genetic integration between dispersal and colonization ability in a bird. *Evolution*, **63**, 968–77.

Ekvall, M. T., Bianco, G., Linse, S., Linke, H., Bäckman, J., and Hansson, L.-A. (2013). Three-dimensional tracking of small aquatic organisms using fluorescent nanoparticles. *PloS One*, **8**, e78498.

English, S., Nakagawa, S., and Clutton-Brock, T. H. (2010). Consistent individual differences in cooperative behaviour in meerkats (*Suricata suricatta*). *Journal of Evolutionary Biology*, **23**, 1597–604.

Fairbairn, D. J., and Roff, D. A. (1990). Genetic correlations among traits determining migratory tendency in the sand cricket, *Gryllus firmus*. *Evolution*, **44**, 1787–95.

Fogarty, S., Cote, J., and Sih, A. (2011). Social personality polymorphism and the spread of invasive species: a model. *American Naturalist*, **177**, 273–87.

Fraser, D. F., Gilliam, J. F., Daley, M. J., Le, A. N., and Skalski, G. T. (2001). Explaining leptokurtic movement distributions: intrapopulation variation in boldness and exploration. *American Naturalist*, **158**, 124–35.

Garamszegi, L. Z., Markó, G., and Herczeg, G. (2012). A meta-analysis of correlated behaviours with implications for behavioural syndromes: mean effect size, publication bias, phylogenetic effects and the role of mediator variables. *Evolutionary Ecology*, **26**, 1213–35.

Gauthreaux, S. A. Jr. (1982). The ecology and evolution of avian migration systems. In D. S. Farner et al., (eds), *Avian Biology*, pp. 93–168. Academic Press, London.

Grayson, K. L., and Wilbur, H. M. (2009). Sex- and context-dependent migration in a pond-breeding amphibian. *Ecology*, **90**, 306–11.

Greenwood, P. J. (1983). Mating systems and evolutionary consequences of dispersal. In I. R. Swingland and P. J. Greenwood, (eds), *The Ecology of Animal Movement*, pp. 116–31. Clarendon Press, Oxford.

Gschweng, M., Kalko, E. K. V., Querner, U., Fiedler, W., and Berthold, P. (2008). All across Africa: highly individual migration routes of Eleonara's falcon. *Proceedings of the Royal Society B: Biological Sciences*, **275**, 2887–96.

Gyuris, E., Feró, O., and Barta, Z. (2012). Personality traits across ontogeny in firebugs, *Pyrrhocoris apterus*. *Animal Behaviour*, **84**, 103–9.

Hanski, I. (1999). *Metapopulation Ecology*. Oxford University Press, Oxford.

Hansson, L.-A., and Hylander, S. (2009). Size-structured risk assessments govern *Daphnia* migration. *Proceedings of the Royal Society B: Biological Sciences*, **276**, 331–6.

Hansson, L.-A., Nicolle, A., Brodersen, J., Romare, P., Nilsson, P. A., and Brönmark, C. (2007). Consequences of fish predation, migration, and juvenile ontogeny on zooplankton spring dynamics. *Limnology and Oceanography*, **52**, 696–706.

Hayes, K. R., and Barry, S. C. (2008). Are there any consistent predictors of invasion success? *Biological Invasions*, **10**, 483–506.

Herborn, K. A., McLeod, R., Miles, W. T. S., Schofield, A. N. B., Alexander, L., and Arnold, K. E. (2010). Personality in captivity reflects personality in the wild. *Animal Behaviour*, **79**, 835–43.

Hoset, K. S., Ferchaud, A.-L., Dufour, F., Mersch, D., Cote, J., and Galliard, J.-F. L. (2011). Natal dispersal correlates with behavioral traits that are not consistent across early life stages. *Behavioral Ecology*, **22**, 176–83.

Hunter, E., Metcalfe, J. D., and Reynolds, J. D. (2003). Migration route and spawning area fidelity by North Sea plaice. *Proceedings of the Royal Society B: Biological Sciences*, **270**, 2097–103.

Jokela, M., Elovainio, M., Kivimäki, M., and Keltikangas-Järvinen, L. (2008). Temperament and migration patterns in Finland. *Psychological Science*, **19**, 831–7.

Jones, K. A., and Godin, J. G. J. (2010). Are fast explorers slow reactors? Linking personality type and anti-predator behaviour. *Proceedings of the Royal Society B: Biological Sciences*, **277**, 625–32.

Ketterson, E. D., and Nolan, V., Jr. (1976). Geographic variation and its climatic correlates in the sex ratio of eastern-wintering dark-eyed juncos (*Junco hyemalis hyemalis*). *Ecology*, **57**, 679–93.

Kobler, A., Klefoth, T., Mehner, T., and Arlinghaus, R. (2009). Coexistence of behavioural types in an aquatic top predator: a response to resource limitation? *Oecologia*, **161**, 837–47.

Kolar, C. S., and Lodge, D. M. (2001). Progress in invasion biology: predicting invaders. *Trends in Ecology and Evolution*, **16**, 199–204.

Krause, J., and Ruxton, G. D. (2002). *Living in Groups*. Oxford University Press, Oxford.

Kurvers, R., Eijkelenkamp, B., van Oers, K., et al. (2009). Personality differences explain leadership in barnacle geese. *Animal Behaviour*, **78**, 447–53.

Lack, D. (1943). The problem of partial migration. *British Birds*, **37**, 122–30.

Lard, M., Bäckman, J., Yakovleva, M., Danielsson, B., and Hansson, L.-A. (2010). Tracking the small with the smallest—using nanotechnology in tracking zooplankton. *PLoS One*, **5** (10), e13516. doi:10.1371/journal.pone.0013516.

Liebl, A. L., and Martin, L. B. (2012). Exploratory behaviour and stressor hyper-responsiveness facilitate range expansion of an introduced songbird. *Proceedings of the Royal Society B: Biological Sciences*, **279**, 4375–81.

Liedvogel, M., Chapman, B. B., Muheim, R., and Åkesson, S. (2013). The behavioural ecology of animal migration: reflections on potential synergies. *Animal Migration*, **1**, 39–46.

Lundberg, P. (1987). Partial bird migration and evolutionarily stable strategies. *Journal of Theoretical Biology*, **125**, 351–60.

Lundberg, P. (1988). The evolution of partial migration in birds. *Trends in Ecology and Evolution*, **3**, 172–5.

McDougall, P. T., Réale, D., Sol, D., and Reader, S. M. (2006). Wildlife conservation and animal temperament: causes and consequences of evolutionary change for captive, reintroduced, and wild populations. *Animal Conservation*, **9**, 39–48.

Mathot, K. J., Wright, J., Kempenaers, B., and Dingemanse, N. J. (2012). Adaptive strategies for managing uncertainty may explain personality-related differences in behavioural plasticity. *Oikos*, **121**, 1009–20.

Mettke-Hofmann, C., Ebert, C., Schmidt, T., Steiger, S., and Stieb, S. (2005). Personality traits in resident and migratory warbler species. *Behaviour*, **142**, 1357–75.

Michelena, P., Jeanson, R., Deneubourg, J. L., and Sibbald, A. M. (2010). Personality and collective decision-making in foraging herbivores. *Proceedings of the Royal Society B: Biological Sciences*, **277**, 1093–9.

Minderman, J., Reid, J. M., Evans, P. G. H., and Whittingham, M. J. (2009). Personality traits in wild starlings: exploration behavior and environmental sensitivity. *Behavioral Ecology*, **20**, 830–7.

Mysterud, A., Loe, L. E., Zimmerman, B., Bischof, R., Veiberg, V., and Meisingset, E. (2011). Partial migration in expanding red deer populations at northern latitudes–a role for density dependence? *Oikos*, **120**, 1817–25.

Newton, I. (2008). *Migration Ecology of Birds*. Academic Press, London.

Nicolaus, M., Tinbergen, J. M., Bouwman, K. M., et al. (2012). Experimental evidence for adaptive personalities in a wild passerine bird. *Proceedings of the Royal Society B: Biological Sciences*, **279**, 4885–92.

Nilsson, J.-Å. (1989a). Causes and consequences of natal dispersal in the marsh tit, *Parus palustris*. *Journal of Animal Ecology*, **58**, 619–36.

Nilsson, J.-Å. (1989b). Establishment of juvenile marsh tits in winter flocks: an experimental study. *Animal Behaviour*, **38**, 586–95.

Nilsson, J.-Å. (1990). Establishment success of experimentally delayed juvenile marsh tits *Parus palustris*. *Ethology*, **85**, 73–9.

Nilsson, A. L. K., Nilsson, J-Å., and Alerstam, T. (2010). Migratory and resident blue tits *Cyanistes caeruleus* differ in their reaction to a novel object. *Naturwissenschaften*, **97**, 981–5.

Pärt, T. (1990). Natal dispersal in the collared flycatcher: possible causes and reproductive consequences. *Ornis Scandinavica*, **21**, 83–8.

Patrick, S. C., and Weimerskirch, H. (2014). Personality, foraging and fitness consequences in a long lived seabird. *PLoS ONE*, **9(2)**, e87269. doi: 10.1371/journal.pone.0087269.

Phillips, R. A., Silk, J. R. D., Croxall, J. P., Afanasyev, V., and Bennett, V. J. (2005). Summer distribution and migration of nonbreeding albatrosses: individual consistencies and implications for conservation. *Ecology*, **86**, 2386–96.

Quinn, J. L., Cole, E. F., Patrick, S. C., and Sheldon, B. C. (2011). Scale and state dependence of the relationship between personality and dispersal in a great tit population. *Journal of Animal Ecology*, **80**, 918–28.

Quinn, J. L., Patrick, S. C., Bouwhuis, S., Wilkin, T. D., and Sheldon, B. C. (2009). Heterogeneous selection on a heritable temperament trait in a variable environment. *Journal of Animal Ecology*, **78**, 1203–15.

Rasmussen, J. E., and Belk, M. C. (2012). Dispersal behavior correlates with personality of a North American fish. *Current Zoology*, **58**, 260–270.

Réale, D., Reader, S. M., Sol, D., McDougall, P. T., and Dingemanse, N. J. (2007). Integrating animal temperament within ecology and evolution. *Biological Reviews*, **82**, 291–318.

Sih, A., Bell, A. M., Johnson, J. C., and Ziemba, R. E. (2004). Behavioral syndromes: an integrative overview. *Quarterly Review of Biology*, **79**, 241–77.

Sih, A., Cote, J., Evans, M., Fogarty, S., and Pruitt, J. (2012). Ecological implications of behavioural syndromes. *Ecology Letters*, **15**, 278–89.

Skov, C., Brodersen, J., Nilsson, P. A., Hansson, L-A., and Brönmark, C. (2008). Inter- and size-specific patterns of fish seasonal migration between a shallow lake and its streams. *Ecology of Freshwater Fish*, **17**, 406–15.

Smith, H. G., and Nilsson, J.-Å. (1987). Intraspecific variation in migratory pattern of a partial migrant, the blue tit (*Parus caeruleus*): an evaluation of different hypotheses. *Auk*, **104**, 109–15.

Sol, D., Duncan, R. P., Blackburn, T. M., Cassey, P., and Lefebvre, L. (2005). Big brains, enhanced cognition, and response of birds to novel environments. *Proceedings of the National Academy of Sciences USA*, **102**, 5460–5.

Vegvari, Z., Barta, Z., Mustakallio, P., and Szekely, T. (2011). Consistent avoidance of human disturbance over large geographical distances by a migratory bird. *Biology Letters*, **7**, 814–17.

van der Jeugd, H. P. (2001). Large barnacle goose males can overcome the social costs of natal dispersal. *Behavioral Ecology*, **12**, 275–82.

van Oers, K., Drent, P. J., de Jong, G., and van Noordwijk, A. J. (2004a). Additive and nonadditive genetic variation in avian personality traits. *Heredity*, **93**, 496–503.

van Oers, K., Drent, P. J., de Goede, P., and van Noordwijk, A. J. (2004b). Realized heritability and repeatability of risk-taking behaviour in relation to avian personalities. *Proceedings of the Royal Society B: Biological Sciences*, **271**, 65–73.

van Oers, K., de Jong, G., Drent, P. J., and van Noordwijk, A. J. (2004c). A genetic analysis of avian personality traits: correlated response to artificial selection. *Behavior Genetics*, **34**, 611–19.

van Overveld, T., and Matthysen, E. (2010). Personality predicts spatial responses to food manipulations in free-ranging great tits (*Parus major*). *Biology Letters*, **6**, 187–90.

Vardanis, Y., Klaasen, R. H. G., Strandberg, R., and Alerstam, T. (2011). Individuality in bird migration: routes and timing. *Biology Letters*, **7**, 502–5.

Ward, A. J. W., Thomas, P., and Krause, J. (2004). Correlates of boldness in three-spined sticklebacks. *Behavioural Ecology and Sociobiology*, **55**, 561–8.

Wilson, A. D. M., and Godin, J.-G. J. (2010). Boldness and intermittent locomotion in the bluegill sunfish, *Lepomis macrochirus*. *Behavioral Ecology*, **21**, 57–62.

Wilson, A. D. M., and Krause, J. (2012). Personality and metamorphosis: is behavioural variation consistent across ontogenetic niche shifts? *Behavioral Ecology*, **23**, 1316–24.

Wolf, M., and Weissing, F. J. (2012). Animal personalities: consequences for ecology and evolution. *Trends in Ecology and Evolution*, **27**, 452–61.

Zhao, Z., and Zera, A. J. (2002). Differential lipid biosynthesis underlies a trade-off between reproduction and flight capability in a wing-polymorphic cricket. *Proceedings of the National Academy of Sciences USA*, **99**, 16829–34.

CHAPTER 7

Dispersal and phenotypic plasticity

Johan Hollander, Machteld Verzijden, Erik Svensson,
and Christer Brönmark

Dispersal is the movement of individuals from their native habitat to another habitat (Box 1.1), and both theoretical models and empirical investigations suggest that high dispersal rate can favour the evolution of phenotypic plasticity (Via and Lande 1985, Hollander 2008). All animals move, at least to some extent, during their life cycle, but whereas some only move within their native habitat, others leave their native habitat for part of their life and return at a later stage (migration). Still others leave their native habitat for good and reproduce elsewhere (dispersal). Thus, an important consequence of dispersal is that, in addition to the movement of the individuals, there will also be gene flow from one population to another. The effects of dispersal on population dynamics is a major research field of its own and treated in two excellent book volumes published in recent years (Clobert et al. 2001, 2012). In this chapter, we will discuss how dispersal influences gene flow and the evolution of phenotypic plasticity versus local adaptation and, conversely, how phenotypic plasticity and local genetic adaptations can affect dispersal and thereby patterns of gene flow. In order to describe individuals with inflexible constitutive traits, we will hereafter use the term 'local adaptation', in contrast to the strategy of a more flexible plastic genotype. Our intention is to examine different spatial scales in the context of phenotypic plasticity, dispersal rate, and gene flow. The rate of dispersal and the distance that individuals disperse may cause a dynamic continuum between the evolution of local adaptation and phenotypic plasticity. Accordingly, both the evolution of phenotypic plasticity and sexual selection can ultimately affect patterns of gene flow, and combined together could even cause population divergence and ultimately speciation.

7.1 Phenotypic plasticity

Much of the variability in adaptive traits within and among populations was previously considered as noise, but recently the understanding of plasticity has considerably improved, and today we know that the expression of a flexible phenotype can be beneficial for many individuals and clearly adaptive. Phenotypic plasticity is the ability of a single genotype to produce various phenotypes within an array of environments (Agrawal 2001). If plasticity is to be considered adaptive, there should be a close match between the acquired phenotype and the environment (Aubret and Shine 2009, Hollander and Butlin 2010). Phenotypic plasticity is studied in the framework of reaction norms, which is a function of the trait mean in different environments (Pigliucci 2001). It is essential to point out that different individuals, even within species, may show different reaction norms—variation in the slope and in the trait mean would accordingly denote genetic variation between individuals for phenotypic plasticity and represent a genotype by environment interaction. Thus, natural selection may act on reaction norms (Schlichtling and Pigliucci 1998), resulting in differences in plasticity among individuals and potentially populations. The reaction norm framework is a valuable tool for graphically representing the expression of the phenotype. Moreover, any trait can demonstrate phenotypic plasticity;

Animal Movement Across Scales. Edited by Lars-Anders Hansson and Susanne Åkesson.
© Oxford University Press 2014. Published 2014 by Oxford University Press.

e.g. morphological, behavioural, and physiological traits have the potential to show environmental sensitivity. Among animals, plasticity has frequently been studied in the context of predator–prey interactions (e.g. Tollrian and Harvell 1999), and the plastic response can be either reversible or irreversible. Often, plastic traits related to morphological developments are irreversible after a certain point in time; once the organism has chosen a developmental path their ontogeny is locked into that course. However, behavioural and physiological traits are usually more flexible and thus reversible in time. Moreover, the plastic trait mean could have a continuous response over a gradient between two extremes, but alternatively show polyphenism, discrete trait means with no intermediates.

7.2 Phenotypic plasticity and environmental heterogeneity

The discussion of scale is well integrated in the concept of phenotypic plasticity since plasticity requires some sort of environmental heterogeneity to evolve. Such heterogeneity can be derived from environmental fluctuations in both time and space. Heterogeneity is often described using grain-size theory, which is an allegory for how an organism experiences spatial or temporal variability in its environment (Levins 1968). Levins modelled the optimal strategy for the evolution of a phenotype under fine-grained and coarse-grained conditions. More recently, contemporary theoreticians have developed models to include dispersal rate, showing that the rate of dispersal is a crucial component for the evolution of phenotypic plasticity (de Jong 2005), since it sets the environmental grain and determines the rate of gene flow (Via and Lande 1985, Scheiner 1998, De Jong 1999). The grain-size theory states that species with broad dispersal and a large geographical range experience a fine-grained environment, since most likely such species will be exposed to significant spatial heterogeneity. In contrast, species with a limited geographical range may be subjected to less environmental variation and therefore experience a coarse-grained environment. In a fine-grained environment the theory predicts the evolution of a generalist or a plastic phenotype in order to increase fitness of the mean phenotype in contrasting environments, while in a coarse-grained environment, when the organism experiences a narrow range of environmental conditions, selection favours a constitutive, canalized development to provide local adaptation (Scheiner 1998, De Jong 1999, Berrigand and Scheiner 2004). With the added perspective of dispersal, between-generation temporal heterogeneity and spatial heterogeneity are two sides of the same coin, since dispersal influences heterogeneity in space and produces variation in selection regimes over successive generations via changes in habitats.

The intertidal rough periwinkle, *Littorina saxatilis*, is an ovoviviparous gastropod, widespread on North Atlantic shores, and is a suitable example of an organism that experiences a coarse-grained environment and has, over succeeding generations, evolved local adaptation (Johannesson and Johannesson 1996). Due to its low vagility and lack of larval dispersal, succeeding generations experience a single habitat. Specific ecotypes of *L. saxatilis* have evolved, showing high morphological and genetic variation over a steep environmental gradient that comprises crab predation and powerful hydrodynamic forces (Panova et al. 2006). The ecotypes are divided only by a few metres, but still demonstrate local adaptation in several traits, such as shell shape, aperture size, and thickness of the shell, which is related to strong divergent selection (Hollander et al. 2005, Panova et al. 2006). Behavioural preferences in mate choice and trail-following further diminish the gene flow between ecotypes and strengthen the genetic structure associated with traits under selection (Hollander et al. 2005, Johannesson et al. 2008). Yet *L. saxatilis* still exhibits plasticity (albeit a rather small amount), which may still be adaptive and potentially evolved to fine-tune morphology on a microhabitat scale (Hollander and Butlin 2010).

The importance of dispersal for creating coarse- or fine-grained environments, and thereby affecting the evolution of different traits, has also been studied in freshwater environments. Here, the expression of anti-predator traits may vary on a spatial scale between discrete habitat units, such as ponds or lakes, due to differences in predator assemblages among ponds. Within a pond system, temporal variability in predation pressure, e.g.

extinction/colonization cycles of predatory fish, may result in a fine-grained selection environment favouring the development of phenotypically plastic defence adaptations in prey organisms. However, permanent ponds are typically consistent over time with regards to the presence or absence of fish predators (Brönmark, personal observation). Hence, they constitute a homogeneous, coarse-grained environment regarding predation regime, thereby favouring the evolution of genetically fixed, local adaptations in prey defences. In a study of ponds with and without predatory fish, it was found that amphipod populations (*Gammarus pulex*) from ponds with fish increased their refuge use in an experimental arena in response to chemical cues from fish, whereas *Gammarus* from populations inhabiting ponds without fish instead showed a decreased refuge use (Åbjornsson et al. 2004). A breeding experiment suggested that this behavioural response was a heritable trait, i.e. an example of a local adaptation to the prevailing predator regime. A key feature is of course the low dispersal rate among ponds in *Gammarus*, a species that depends on passive dispersal, e.g. by birds for dispersal. A contrasting result was found in the freshwater snail *Radix balthica*, which commonly coexist with *Gammarus* in ponds with and without fish predators. Freshwater snails have evolved both behavioural (Rundle and Brönmark 2001, Turner and Montgomery 2003) and morphological (DeWitt et al. 2000, Hoverman and Relyea 2007) anti-predator adaptations. Brönmark et al. (2011) studied whether *R. balthica* has evolved locally constitutive morphological defence traits or, alternatively, showed phenotypic plasticity in morphological defences across habitats with different predator regimes. The authors sampled snails from natural ponds with and without predatory fishes, as well as conducted common garden experiments, and found that all snails showed significant reaction norms in a morphological defence (shell shape) against predatory fish cues, independent of their population of origin. Thus, no genetic local adaptation has evolved, and phenotypic plasticity governs any phenotypic differences between local populations. This was somewhat paradoxical in comparison to the *Gammarus* example discussed earlier, as *R. balthica* also depends on passive dispersal. However, *R. balthica* is known to have a remarkably strong dispersal ability and, further, molecular analyses of genetic diversity among populations, as a proxy for dispersal rate (F_{ST} estimates with neutral markers), have not detected evidence of local adaptation in *R. balthica* despite high environmental heterogeneity (Evanno et al. 2006). Conceivably, this could be the result of high gene introgression among populations. Thus, a high dispersal rate among ponds with different predator regimes may create a fine-grained environment favouring the evolution of phenotypic plasticity in *R. balthica*.

7.2.1 Agents of disturbance

The three examples from the previous section demonstrate that natural communities are dynamic systems, which means that organisms are exposed to disturbance and stressors in various forms, causing environmental heterogeneity at different degrees and at different scales. Temporal dynamics, for example, will affect natural communities from the level of continents down to small spatial scales (Sousa 1984). Agents of disturbance include both physical and biological processes and their effects on ecological communities seem fairly similar: reducing the number of individuals and hence facilitating new recruitments. Disturbance can arise periodically or in a stochastic fashion and may lead to stressful and heterogeneous conditions. In case of large unpredictable shifts in the environment, a population must rely upon standing genetic variation to achieve adaptive evolution and avoid extinction. Although catastrophic events take place with dramatic implications for ecosystem communities, the usual state experienced by populations is a more or less heterogeneous periodic fluctuation around a trend or an average mean (Marquet et al. 1993, Scheffer and Carpenter 2003). Disturbance is an agent of natural selection but, in the context of the evolution of phenotypic plasticity, adaptive plasticity does not evolve in unpredictable environments. Under stochastic environmental conditions, plasticity is not considered likely to evolve since available cues are uninformative to the organism (Moran 1992, Tufto 2000). The evolution of adaptive phenotypic plasticity requires predictable environmental cues, such as seasonal or systematic spatial variation, or variation in the presence or absence of a co-evolved predator.

Furthermore, the period of the temporal fluctuation determines whether plasticity will evolve, because if short temporal fluctuations exceed the generation time of the organism, an irreversible trait would lag behind the environment and result in a maladaptive response (Lande and Shannon 1996). In such a case, evolution should select against plasticity and favour a generalist strategy (intermediate phenotype) (Levins 1968).

The concept of grain size and the evolution of phenotypic plasticity can obviously be implemented to spatial heterogeneity and particular to the dynamics of dispersal rate. In populations that disperse there will be gene flow over a wider geographic area, compared to populations that do not, and consequently, dispersing populations are exposed to a larger variation under abiotic and/or biotic conditions; they live in a more heterogeneous, fine-grained environment. As a result, individuals in populations that disperse experience a larger range of selection pressures, resulting in the evolution of different adaptations and possibly a difference in the degree of phenotypic plasticity.

7.3 Plasticity and costs related to habitat selection

Phenotypic plasticity allows individuals to cope with a wider variety of environmental conditions, because they are able to adjust to many different habitats. This increased ability to cope with environmental variation may enable plastic individuals both to survive and to reproduce in one single habitat, negating the necessity for dispersing. On the other hand, plasticity enables individuals to cope with new and perhaps less predictable habitats after dispersal. In other words, plasticity can be a strategy to cope with heterogeneity in time or space, and phenotypic plasticity may thus favour both dispersal and philopatry. The level of phenotypic plasticity that evolves in a population may depend on the particular mode of dispersal, i.e. directed dispersal or random dispersal, and the ability to predict future environmental conditions. Theory predicts that in species where future habitat conditions are predictable at some intermediate level, phenotypic plasticity should evolve. Possibly, both philopatry and dispersal strategies may have similar optimal levels of phenotypic plasticity (Murren et al. 2001), and dispersal may not always invoke increased selection of increased phenotypic plasticity.

An important argument for dispersers to not evolve phenotypic plasticity is that both plasticity and dispersal are likely to carry costs, and when individuals that disperse rely on phenotypic plasticity to establish a new population, they are paying a double cost. To avoid this, species could evolve a habitat selection strategy (Davis and Stamps 2004, Karpestam et al. 2012) and a subsequent specialist adaptation to a specific habitat (van Tienderen 1997). Such a habitat selection strategy may be mediated through natal habitat preference induction, where (positive) experience in the natal habitat increases the preference for settlement in a similar habitat after dispersal, a phenomenon observed in a wide range of taxa (Davis and Stamps 2004, Stamps et al. 2009). Genetic variation for habitat selection also appears to be common, particularly in certain taxa (Jaenike and Holt 1991), and has similar effects as natal habitat preference inductions. Whatever the proximate mechanism is, habitat selection potentially directs gene flow between similar habitats, and could promote the evolution of specialist strategies for particular habitat types (Fig. 7.1; Edelaar et al. 2008, Edelaar and Bolnick 2012). Habitat

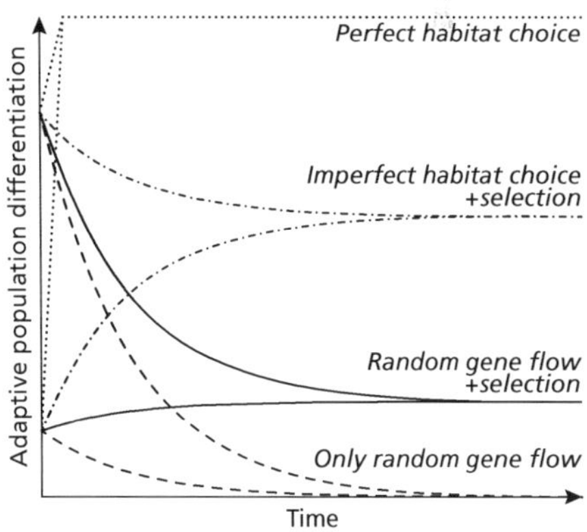

Figure 7.1 An example of how random gene flow reduces population differentiation, while non-random gene flow promotes adaptive population divergence. This illustrates how dispersal can drive adaptive evolution without natural selection (Edelaar and Bolnick 2012).

selection would thus decrease selection for phenotypic plasticity in dispersing species.

7.3.1 Timing of dispersal

Habitat selection strategies may not only influence the costs of plasticity but also the timing of dispersal to a particular habitat. Scheiner et al. (2012) explored the evolution of phenotypic plasticity and dispersal in an individual-based simulation model and discovered that the probability for phenotypic plasticity to evolve depends to a large extent on life-history patterns, i.e. if dispersal occurs either before or after selection. The timing of dispersal to a novel habitat is central because animals risk to be exposed to selection in an environment different from their development. For example, when individuals are exposed to selection prior to dispersal, which means that development and selection can occur in the same environment, dispersal favours the evolution of phenotypic plasticity. However, if costs for plasticity are high, local adaptation is favoured—unless at the highest initial dispersal rate, where plasticity evolves regardless of high costs. When selection occurs after occupying the new habitat the conflict between plasticity and local adaptation is determined exclusively by the cost of plasticity. Thibert-Plante and Hendry (2011) used the same type of model, but were interested in examining the effects of plasticity on reproductive isolation and ecological speciation. Again in this model the timing between the expressed plastic phenotype and dispersal is vital. However, if plasticity occurs after dispersal, reproductive barriers will be degraded because migrants will have a higher adaptive value in the new environment and contribute to a stronger gene flow among local populations. If plasticity is, however, expressed prior to dispersal, barriers to gene flow will remain unaffected or enhanced, and plasticity may, accordingly, promote the evolution of reproductive isolation and influence the process of ecological speciation. These models confirm the discussion we outline in this chapter, that high dispersal may select for phenotypic plasticity as long as the benefits outweigh the costs. However, if dispersal is associated with costs, for example if the phenotype is maladaptive in the new habitat, which would be the result if plasticity is expressed before dispersal, these individuals would have a lower adaptive value and hence gene flow among local populations would be reduced. As a consequence divergent selection could operate on different local populations with the potential for lineage splitting.

7.4 The dynamics and evolution of phenotypic plasticity and genetic adaptation

Whether an organism will evolve traits that are locally adapted or phenotypically plastic depends not only on the rate of dispersal and spatial variation, but also on the extent of gene flow across populations, which can be reduced when certain individuals experience natural selection or assortative mating. In the absence of gene flow, divergent natural selection across environments can potentially influence each local population to evolve focal traits that correspond to its local environmental conditions (Schluter 2000, Kawecki and Ebert 2004, Garant et al. 2007). As a consequence, each local population is expected to, on average, show a higher relative fitness in its home environment than in alternative environments. This would thus select against exogenous colonizers, a phenomenon defined as 'phenotype-environment mismatch' (DeWitt 1998, Marshall et al. 2010). Consequently, if the local environment does not express heterogeneity in time and space, i.e. it is constant, there is no need for plasticity and local adaptation should evolve (Kawecki and Ebert 2004). However, if there is demographic connectivity among local populations, dispersal will facilitate the exchange of genes and this gene flow will have a homogenizing effect, eroding the structure of local adaptations, and natural selection will favour the evolution of phenotypic plasticity. Alpert and Simms (2002) listed the following three criteria in order for plasticity to evolve: The environmental variation should be predictable; the plastic response should not lag behind the environmental change; and finally, the environment should be heterogeneous in either time or space. Sultan and Spencer (2002) recognized that this process would be further catalysed in a meta-population structure with multi-connectivity among local populations.

However, in a situation where gene flow is restricted in one direction between a donor and a recipient population, the evolution of local adaptation is still conceivable in the donor population, since there is no introgression of genes from the adjacent environment. On the other hand, if the donor population shows variability in plasticity and the new environment differs from the native habitat in the selection regime, then increased plasticity may evolve among the recipients (Crispo 2008). This has been shown empirically by Aubret and Shine (2009), who examined whether adaptive plasticity had evolved in response to a novel environment in the tiger snake, *Notechis scutatus*. In this snake, the head and jaw lengths are directly related to prey size and thus reflect a local adaptation to available prey. Snakes from the mainland feed on small prey while introduced island populations mostly feed on larger prey. Common garden experiments revealed that snakes from the mainland population showed a canalized development regardless of experimental treatment (Fig. 7.2.). An island population, however, revealed plasticity in head size and jaw length, indicating the integration of a genotype-by-environment effect. This example is important since it confirms that phenotypic plasticity is governed genetically and can evolve like any other trait (Scheiner 1993; Pigliucci 2001, 2005).

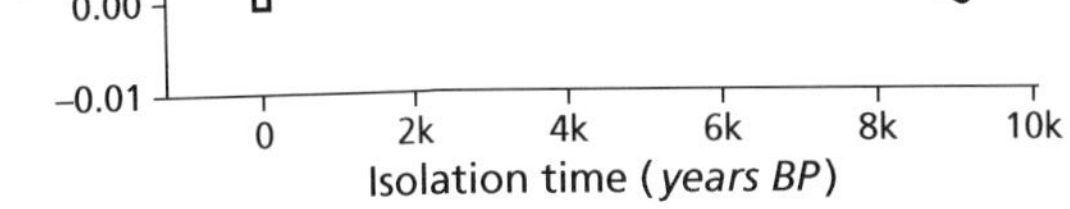

Figure 7.2 An empirical example of the evolution of phenotypic plasticity. Tiger snakes, *Notechis scutatus*, from islands (circles) with different colonization time, compared with reference populations (squares) show strong genetic assimilation—going from phenotypic plasticity expression to a more canalized constitutive development (least-squares regression $n = 5$; $R = 0.96$; $F_{1,3} = 37.54$; $p < 0.009$; Aubret and Shine 2009). Reproduced with permission from Elsevier.

A second example that shows that phenotypic plasticity is genetic and under selection is from the northern swordtail fish, *Xiphophorus birchmanni* and *Xiphophorus malinche*, which inhabit mountain streams in Mexico, and where multiple species often share river drainages, with one species living upstream from the other (Verzijden et al. 2012). Males have species-specific mating cues, including several visual and olfactory cues. In a reciprocal experiment involving two such parapatric species, exposure of adult females to males of the parapatric species mimicked dispersal. Females were plastic in their mating preferences for males, but the two species showed different reaction norms in their plasticity. Females of one of the two species changed their preference towards the males of their new population, while females of the other species also shifted their mate preference, but in the opposite direction, a decreased preference of new males. The direction of the change of preferences may in this case relate to behavioural differences among the two species (Verzijden et al. 2012). A highly similar pattern was found in these two same species in preferences formed through social experience before or during sexual development (Verzijden and Rosenthal 2011), which mimicked a situation that may arise when a pregnant female of these live bearing species releases her young after dispersal. This example clearly shows that the plastic response is highly similar in both species, but that the plasticity reaction norms have evolved in different directions. A similar asymmetry in plasticity of mate preferences has been found in field crickets (Izzo and Gray 2011). Hence, selection will favour the evolution of plasticity when the organisms experience a heterogeneous environment, but as we have emphasized here, gene flow is also a strong driver as to whether phenotypic plasticity will have the potential to evolve. In the following sections we will discuss the importance of this heterogeneity in the framework of dispersal and gene flow.

7.5 Phenotypic plasticity, dispersal rate, and the complexity of gene flow

High dispersal resulting in connections among local populations may have a homogenizing effect on populations. In other words, exclusive gene complexes break up if dispersal happens across several generations, and the different environments that successive generations experience would select for increased plasticity. However, this homogenizing effect relies on the assumption that all individuals in the dispersing population have the same prospects and fitness values, which in most instances is not the case. When individuals select a new habitat the dispersal of the individual may be completed, but if the individual does not produce offspring in the new habitat, there will not be any lasting evolutionary effect of the movement of the individual. Thus, just because individuals disperse this does not mean by default that they will always reproduce in the new habitat. For that reason, gene flow between populations may not be random, and there are a number of mechanisms that influence this, both at the new site and before the dispersal phase. First, this may be because the dispersers themselves are not a random sample from the native population (Fig. 7.1; Edelaar and Bolnick 2012). Second, of the individuals that disperse, a non-random set may end up reproducing. Third, the individuals reproducing after dispersal may not be reproducing with a random sample from the adopted population. In other words, it is not just the question of whether they are able to mate, but also with whom. Assortative mating by immigrants can profoundly influence evolutionary dynamics in populations that receive significant numbers of immigrants (M'Gonigle and FitzJohn 2010, Shine et al. 2011). Such a pattern of assortative mating by immigrants has, for example, been found in populations of great tits, *Parus major* (Verhulst and van Eck 1996). Since the tendency to disperse is likely to be heritable (Doligez et al. 2009), an interesting hypothetical effect of such assortative mating might be that dispersal rates increase among a subpopulation of dispersers, which in turn could lead to a range expansion of the species (Shine et al. 2011).

Additionally, post-dispersal mortality may also produce genetic structures in populations if certain genes show greater fitness in association with specific habitat types. Accordingly, strong purifying selection can cause population divergence, resulting in a genetic structure similar to that of poorly dispersed species that achieve local adaptation (reviewed by Marshall et al. 2010). Thus, it is not the dispersal rate per se that determines whether selection—causing adaptation to local environmental conditions—drives evolution in the direction of local adaptation or phenotypic plasticity, but this process depends essentially on the relative magnitudes of gene flow, costs, and the strength of selection. For that reason, we will in the next section illustrate why the link between dispersal and the evolution of phenotypic plasticity is complicated.

7.6 The complex relation between dispersal and gene flow

Studies on migratory birds and recent work in marine systems suggest that species with a potentially large dispersal range of propagules can actually show fine-scale genetic structure and thus do not conform to the predictions of the standard model discussed earlier, where theory would predict evolution of phenotypic plasticity (Bensch 1999, Grosberg and Cunningham 2001). Such long distance dispersal thus clearly does not lead to random gene flow, but rather directed gene flow, governed by either natural or sexual selection processes, preventing dispersing individuals to reproduce in certain areas. In this scenario selection would as an alternative favour the evolution of local adaptation. Reproduction is the end product of a multifaceted process, which could involve mate attraction, nest site finding, gamete production, and offspring rearing, each of which could be affected by ecological and social differences between habitats—and thus have the potential to produce fine-scale genetic structures within species. The question is, then, which set of dispersers is likely to reproduce? For mating opportunities, individuals need to be recognized as potential mates. They may face difficulties in attracting a mate due to different abiotic factors, which can affect mate recognition signals and other courtship communication. For instance, ambient noise can obscure mating signals, which can significantly limit mate attraction. A well-studied example of this is song in birds, which can be obscured by ambient noise from natural abiotic sources (Ryan and Brenowitz 1985, Kirschel et al. 2009), as well as human-induced traffic noise (Halfwerk et al. 2011). Similar processes can operate in other communication modalities, such as vision and olfaction (Seehausen et al. 1997, Fisher et al. 2006).

7.6.1 Case study: dispersal in the terrestrial environment

While abiotic and general ecological differences between habitats can deter mating in dispersed individuals, social interactions may also disfavour immigrants. In particular, the mating preferences of both the disperser and the native individuals may differ between populations. Sexual traits, and the mate preferences for them, often show geographical variation even at relatively small distances, such as in the case of rufous-collared sparrow, *Zonotrichia capensis*, populations (Danner et al. 2011), Amazonian frogs (Funk et al. 2009), and damselflies (Wellenreuther et al. 2010). The ultimate causes for such local differences vary from Fisherian sexual selection dynamics, to cultural evolution (Uyeda et al. 2009, Cardoso and Atwell 2011), and may not even be due to ecological differences. The net result is, however, that non-local individuals face sexual selection against their phenotype, and thus a certain degree of reproductive isolation between populations is in place. Also in species without discrete geographic variation of mating traits or preferences, immigrants can face lower mating success (Bensch et al. 1998), although this may be sex specific (Verhulst and van Eck 1996, Hansson et al. 2004), and the exact mechanisms for this are unclear.

Furthermore, dispersers themselves may have mate preferences that do not match the available pool of potential mates at the arrival location, preventing them from effectively courting or accepting mates. However, in many cases individuals show lower choosiness for mates (i.e. the degree to which preferences are expressed) when their preferred phenotype is absent. An elegant empirical example of this is shown in a species of treehoppers, *Enchenopa binotata*, in which males advertise with an acoustic signal, of which the host plant influences the properties of the exact frequency of the signal. Between species, females show strong preferences for certain frequency ranges, but also within a species frequencies can vary according to the particular host plant. Females adjusted their choosiness, but not their actual preferences, after experience with non-preferred mating signals. This indicates that females do not necessarily change their preferences, but are instead flexible in how much they express them, assuring that mating takes place under a wide range of social conditions (Fowler-Finn and Rodriguez 2012).

Newly arrived individuals may also be less successful than the native population in establishing breeding sites or territories. Signals used to attract mates are often also essential in territorial interactions, and thus worse signal-to-noise ratios in new habitats will also affect territory establishment. Furthermore, local competition for breeding opportunities may disfavour immigrants if they are late in acquiring nest sites, or if nest sites are rare and subject to inheritance.

In some cases, dispersers may be more likely to mate with native individuals than expected at random. This has been observed in a number of instances such as in jumping spiders (Hebets and Maddison 2005), field crickets (Izzo and Gray 2011), and guppies (Hughes et al. 1999). This pattern has been explained mainly by inbreeding avoidance in small populations, and is likely to differ strongly between populations and closely related species (Izzo and Gray 2011, Verzijden et al. 2012).

The match between mating traits, such as courtship signals, and preferences for such traits, i.e. mate preferences, between potential mating partners determines the chances of individuals mating. In many species, either (or both) male mating traits or female preferences are plastic. This means that preferences or traits of dispersers can re-shape to match the variation in the population they disperse to, which potentially increases their chances of finding a mate and subsequently reproducing. There are several ways in which traits and preferences can be plastic. They could be condition-dependent, context-dependent, or learned. Different physiological and social mechanisms underlie such behavioural plasticity, but two factors in particular may determine the degree to which preferences or traits can converge to the local population after dispersal. The first factor is the time window for such plasticity. If traits or preferences are plastic in their development, but thereafter fixed, then the timing of the dispersal event relative to the development of the individual will determine the match. In contrast, if the mating behaviour is plastic throughout the lifetime, the chances are that traits or preferences will converge to the population the individual dispersed to. In addition, once an individual has mated, plasticity will allow its offspring to converge to the variation in traits in the native population, even if the plasticity in this species is restricted to ontogeny, and thus did not influence the parents' chances of mating (Verzijden et al. 2012).

7.6.2 Case study: dispersal in the marine environment

An additional example that does not follow the conventional model (i.e. where large dispersal propagules still demonstrate fine-scaled genetic structure) is taken from the marine environment where new findings call for the reconsideration of previous paradigms. The prevalent view of the marine systems is that they consist of large, demographically 'open' populations, which show panmixia without any genetic structure (Grosberg and Cunningham 2001, Marshall et al. 2010). In contrast to terrestrial organisms with more limited dispersal, marine species have the potential to disperse as larvae in the plankton without obvious boundaries. However, there are numerous regulating factors acting as barriers to larval transport which, together with post-dispersal mortality, restrict the connectivity among local populations and thus diminish gene flow. Such post-dispersal mortality is probably high in nature, since leaving the native habitat or niche is always associated with severe risks (Nosil et al. 2005). For example, in south-western England along 200 km of the coast of Devon and Cornwall, there is a hybrid zone of the blue mussels, *Mytilus edulis* and *Mytilus galloprovincialis*. The hybrid zone is maintained by environmentally dependent selection favouring each parental population in the contrasting habitats (Gardner et al. 1993, Wilhelm and Hilbish 1998). A number of diagnostic loci show a strong relationship to *M. edulis* while showing consistently low frequencies in *M. galloprovincialis*, and vice versa. This pattern is present despite high rates of immigration from the hybrid zone, for example more than 10% of the larvae that settled in the *M. galloprovincialis* population carried *M. edulis*-specific alleles. Gilg et al. (2009) investigated the genetic structure across the hybrid zone and discovered a negative correlation between *M. edulis*-specific alleles in the *M. galloprovincialis* population with age. Larvae that originated from the hybrid zone settled within the *M. galloprovincialis* population, but over nine months' time selection had removed *M. edulis*-specific alleles, thus post-settlement selection favouring *M. galloprovincialis* alleles in the region it occupies and keeping the population distinct even in the face of gene flow.

This has also been noted among bird species where one would intuitively assume that species with long distance dispersal would also have a larger colonization range. Numerous species of migratory birds travel annually thousands of kilometres (Hedenström and Lindström, Chapter 5 and Hansson et al., Chapter 14) but disperse and colonize new breeding areas less often than resident birds. This pattern held true in the boreal forest of the Palaearctic and among introduced birds in New Zealand (Veltman et al. 1996, Bensch 1999). This low colonization success in new areas among migratory birds is probably related to strict migratory programmes: the birds have no flexibility to change migratory distance, direction, or the timing to fit the new environment.

7.7 Plasticity and species invasion

In recent years there have been a growing number of alien introductions of animals and plants, owing to the rapid globalization of the world that has increased species' mobility. The phenomenon is occurring in both terrestrial and aquatic environments (Richardson and Pysek 2006, Rahel and Olden 2008, Kenis et al. 2009). Non-indigenous species, and especially harmful alien species, cause significant problems in pristine environments and have become a major concern for biodiversity, economics, and human health and welfare (Smith et al., Chapter 4). In the majority of cases, successful invasions of exotic species are the result of human-mediated vectors and such transport is often unintentional. Human-mediated vectors move alien species through, e.g., agricultural products, horticultural export, and plankton-rich ballast water in ships. Thousands of species have been introduced to new areas in these ways. Only a small fraction of the alien species introduced to new regions become harmful to the native biota (Nilsson et al., Chapter 6). However, the actual numbers of species causing problems is nevertheless large, because of the great number of species that are transported between continents.

An example is the Pacific oyster, *Crassostrea gigas*, a native of the subtropical northwestern Pacific Ocean but introduced to Europe in the 1960s for commercial farming. *Crassostrea gigas* became popular for farming because of its fast growth and its resistance to the diseases that had almost eliminated the native European oyster, *Ostrea edulis*, from Europe except Scandinavia. The Pacific oyster is notoriously plastic, particularly in shell shape morphology, but also in life-history traits (Ernande et al. 2004), growth in response to food abundance (Strathmann et al. 1993), and robustness against temperature variation among both adults and dispersing larvae (Hamdoun et al. 2003). Because of its subtropical origin, the larvae of *C. gigas* were not thought to survive in northern European waters outside the commercial nurseries. However, wild populations are now found from France to Norway, and may have negative effects since the Pacific oyster outcompetes the native individuals of *O. edulis* and the blue mussel *Mytilis edulis* (Wang et al. 2007, Wrange 2008). Moreover, climate change may influence the survival of subtropical larvae and increased water temperatures resulting from global warming should facilitate the future reproductive success of *C. gigas* in higher latitudes. Further, non-indigenous species dispersing to new localities due to environmental change may experience a non-optimal match between the environment and the phenotype, but may still survive until reproduction if they are phenotypically plastic. Cryptic genetic variation and classical natural selection across generations can then subsequently be selected upon and evolve favourable traits. In other words, plasticity may buffer and sustain survival from novel selection in the new environment. However, selection could still occur if certain cryptic genetic variation is expressed in the particular environment and provides exclusive raw material for phenotypic novelties. Over time phenotypic plasticity can either increase towards a phenotypic optimum (directional selection on the plastic genotype to increase the match between the phenotype and the environment) or decay by genetic assimilation, leading to the evolution of local adaptation with a less flexible phenotype (see discussion in section 7.8). Warmer oceans may also influence native species negatively, adding to the stress an introduced species may cause. Global warming may additionally have the power to change the distribution of native species, which may become less abundant due to biotic and abiotic stress (Lindström et al., Chapter 3; and Smith et al., Chapter 4). The composition of a native community may also play an important role in its susceptibility to invasions. Stachowicz et al. (1999) studied communities of sessile marine invertebrates and found that while species-rich communities were buffered against successful invasions, species-poor ones were susceptible to invasions, with potential loss of biodiversity (Smith et al. 2009, Chapter 4). Abiotic stressors like temperature fluctuations, changes in salinity, and alterations in solar radiation impose physiological challenges on organisms dispersing along a latitudinal transect, while biotic factors potentially expose them to novel heterospecific competition and novel predation pressure. The Pacific oyster, given as an example earlier, demonstrates high levels of plasticity (e.g. Strathmann et al. 1993) that may have facilitated its documented global spread.

Non-indigenous, successfully invasive species have been observed to show high rates of survival and reproduction and, for that reason, become established efficiently in new habitats (Crawley et al. 1987, Mack et al. 2000). This is likely due in part to more favourable conditions, including release from natural enemies and relaxed competition (Maron and Vila 2001, Keane and Crawley 2002). However, beyond that, we still do not completely understand the ultimate cause of the invaders' success. Non-indigenous species are by definition relocated from their native habitat and the new environment most likely differs from the indigenous one, providing a dissimilar selection regime. Non-indigenous species are therefore experiencing a fine-grained environment and, as discussed earlier, the grain-size theory predicts that selection should favour the evolution of phenotypic plasticity among invasive species (Smith 2009, Edgell and Hollander 2011). Moderate levels of phenotypic plasticity may facilitate a population's expansion into novel environments, because the trait may place the population on the slope of an adaptive peak from which natural selection can advance (Fig. 7.3; Price et al. 2003). Given that phenotypic plasticity allows an organism to

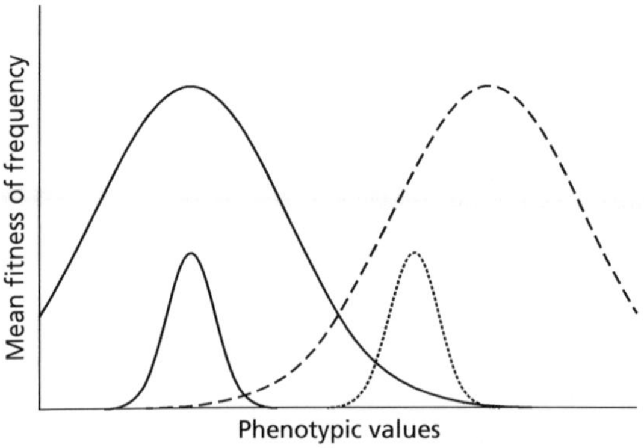

Figure 7.3 The graph illustrates how plasticity contributes to a peak shift in a changing environment (after Fear and Price 1998). The bold, solid line illustrates mean fitness, whereas the dashed bold line shows mean fitness in the new environment. The thin solid line shows the population distribution in the old environment and the thin dotted line illustrates the distribution after a plastic response. Note that the population will go extinct without a plastic reponse.

invade and survive in numerous novel environments, phenotypic plasticity per se may thus be an important facilitator for species to invade novel locations. Plasticity may act as a buffer against the environment even if its adaptive value differs from an adaptive optimum. By adopting the strategy of 'Jack-of-all-trades and master of none', species may be able to sustain themselves in the environment long enough to assimilate local adaptation by natural selection (Waddington 1942, Rollo 1994, Braendle and Flatt 2006). However, if environmental cues are less predictable in the novel environment there will be a lower probability of a correct phenotype–environmental match (DeWitt and Scheiner 2004). In addition, if the plastic genotype originally evolved in response to highly specialized cues, plasticity may provide little or no benefit in the novel habitat (Smith 2009).

Parker et al. (2003) questioned whether invasions are promoted by local adaptation or by plasticity. Successful invaders can achieve local adaptation from a high outcrossing rate, or from a large founder population with high levels of genetic variation and the subsequent fixation of novel genotypes. These authors studied the expansion of a species in the Sierra Nevada of California and found no evidence of locally specialized adaptation to specific environmental conditions. Instead, they showed that phenotypic plasticity played a significant role in the invader's success. However, it is important to point out that other traits besides plasticity may play a crucial role for successful invaders. These include the ability to reproduce by both asexual and sexual means, fast growth rate, and rapid reproduction; almost certainly, several of these traits in concert would increase fitness in a novel environment.

7.8 Plasticity and the potential of speciation

Earlier we showed how a high dispersal rate can allow individuals to cope with a heterogenous environment and thus facilitate the evolution of plasticity. Once plasticity has evolved, the flexibility to change between different adaptive phenotypes may accordingly increase fitness among species with a high dispersal rate and, especially, allow non-indigenous species to survive in novel habitats. However, in the past few years, we have recognized that plasticity is not only an important factor allowing organisms to increase their fitness in heterogeneous environments, but also plays a major role in evolution, by introducing and promoting cryptic genetic variation. If plasticity can drive diversification among populations and potentially cause incipient speciation, this would be a major change in how we view phenotypic plasticity.

In essence, phenotypic plasticity allows an organism to explore a larger array of environments as the organism facultatively switches between adaptive phenotypes. These dissimilar environments will represent a range of selection regimes that simultaneously act on the complete phenotype, not only on focal plastic traits. Traditionally, theoretical models of population divergence and speciation are often based on divergent natural selection causing adaptive divergence, with subsequent reproductive isolation (Schluter 2000, Rundle and Nosil 2005, Nosil 2007). Plasticity, on the other hand, has been assumed to relax selection as it can mask genetic variation and thus slow down the evolution of adaptive genetic divergence (de Jong 2005; but see Thibert-Plante and Hendry 2011). However, in recent years several authors have argued that phenotypic plasticity may be a route for local adaptation (Price et al. 2003, West-Eberhard 2005, Crispo 2007, Pfennig et al.

2010, Thibert-Plante and Hendry 2011, Moczek et al. 2011). West-Eberhard has been one of the strongest advocates of this model, a theory initiated by the long-standing idea of phenotypic accommodation first suggested by Baldwin (1902). Phenotypic accommodation is a novel plastic adjustment that an organism can develop within a generation (Hirasaki et al. 2004). The new innovation will have a greater likelihood of being subject to natural selection since often the new trait is strongly related to adaptive morphologies that have already experienced purifying selection in the past (West-Eberhard 2005). For that reason the new trait is the first step in neo-Darwinian evolution, because plasticity places the phenotype into a new adaptive landscape in which traditional natural selection can proceed on standing genetic variation (Price et al. 2003). Accordingly, cryptic genetic variation can be subject to natural selection and accumulated by phenotypically plastic novelties (Gibson and Dworkin 2004, Le Rouzic and Carlborg 2008), a process described as 'genetic accommodation' and illustrated by Suzuki and Nijhout (2006). Standing genetic variation can thus be selected so that evolution can proceed in the direction of the induced plastic response. This should by no means be considered as inheritance of acquired characters, as it is the result of traditional neo-Mendelian and neo-Darwinian processes and relies on pre-existing genetic variation and conventional natural selection. Genetic accommodation may increase plasticity or have no effect on the level of plasticity, while in contrast, genetic assimilation decreases plasticity and development becomes ontogenetically canalized against environmental perturbation (Waddington 1961, Pigliucci and Murren 2003, Pigliucci et al. 2006, Crispo 2007). Genetic assimilation is generally defined as the evolution of a phenotype that is invariable even in variable environments (Rollo 1994, Price et al. 2003) and is a mechanism for the evolution of biological diversity, whereby specialist genotypes evolve in specific environments from ancestral plastic genotypes (Schlichting 2004, Braendle and Flatt 2006, Pigliucci et al. 2006). A possible outcome is the evolution of genetically distinct ecotypes and over generations the potential for lineage splitting (Strathmann et al. 1993).

With this chapter our goal has been to illustrate the evolutionary causes and consequences of dispersal. The developmental trajectory and the evolution of an organism's phenotype is to a large extent driven by the scale of movement as well as the magnitude of genetic introgression among populations. The process of plasticity is important both prior to and following dispersal, as on the one hand plasticity can be selected for in dispersing species, as long as the dispersal rate sustains gene flow between local populations, while on the other hand plasticity per se enables individuals to explore and colonize novel environments. Additionally, plasticity is no longer seen as the conflicting counterpart of local adaptation, but the concepts are well integrated and the ultimate product for such evolution is the formation of new species.

References

Åbjornsson, K., Hansson, L.-A., and Brönmark, C. (2004). Responses of prey from habitats with different predator regimes: local adaptation and heritability. *Ecology*, **85**, 1859–66.

Agrawal, A. A. (2001). Phenotypic plasticity in the interactions and evolution of species. *Science*, **294**, 321–5.

Alpert, P., and Simms, E. L. (2002). The relative advantages of plasticity and fixity in different environments: when is it good for a plant to adjust? *Evolutionary Ecology*, **16**, 285–97.

Aubret, F., and Shine, R. (2009). Genetic assimilation and the postcolonization erosion of phenotypic plasticity in island tiger snakes. *Current Biology*, **19**, 1932–6.

Baldwin, J. M. (1902). *Development and Evolution*. MacMillan, London.

Bensch, S. (1999). Is the range size of migratory birds constrained by their migratory program? *Journal of Biogeography*, **26**, 1225–1235.

Bensch, S., Hasselquist, D., Nielsen, B., and Hansson, B. (1998). Higher fitness for philopatric than for immigrant males in a semi-isolated population of great reed warblers. *Evolution*, **52**, 877–83.

Berrigand, D., and Scheiner, S. M. (2004). Modeling the evolution of phenotypic plasticity. In T. J. Dewitt and S. M. Scheiner (eds). *Phenotypic Plasticity, Functional and Conceptual Approaches*, pp. 82–97. Oxford University Press, New York.

Braendle, C., and Flatt, T. (2006). A role for genetic accommodation in evolution? *Bioessays*, **28**, 868–73.

Brönmark, C., Lakowitz, T., and Hollander, J. (2011). Predator-induced morphological plasticity across local populations of a freshwater snail. *PLoS One*, 6.

Cardoso, G. C., and Atwell, J. W. (2011). Directional cultural change by modification and replacement of memems. *Evolution*, **65**, 295–300.

Clobert, J., Baguette, M., Benton, T. G., Bullock, J. M., and Ducatez, S. (2012). *Dispersal Ecology and Evolution*. Oxford University Press, Oxford.

Clobert, J., Danchin, E., Dhondt, A. A., and Nichols, J. D. (2001). *Dispersal*. Oxford University Press, New York.

Crawley, M. J., Gray, A. J., and Edwards, P. J. (1987). *Colonization, Succession and Stability*. Blackwell Science, Oxford.

Crispo, E. (2007). The Baldwin effect and genetic assimilation, revisiting two mechanisms of evolutionary change mediated by phenotypic plasticity. *Evolution*, **61**, 2469–79.

Crispo, E. (2008). Modifying effects of phenotypic plasticity on interactions among natural selection, adaptation and gene flow. *Journal of Evolutionary Biology*, **21**, 1460–9.

Danner, J. E., Danner, R. M., Bonier, F., Martin, P. R., Small, T. W., and Moore, I. T. (2011). Female, but not male, tropical sparrows respond more strongly to the local song dialect, implications for population divergence. *American Naturalist*, **178**, 53–63.

Davis, J. M., and Stamps, J. A. (2004). The effect of natal experience on habitat preferences. *Trends in Ecology and Evolution*, **19**, 411–16.

De Jong, G. (1999). Unpredictable selection in a structured population leads to local genetic differentiation in evolved reaction norms. *Journal of Evolutionary Biology*, **12**, 839–51.

De Jong, G. (2005). Evolution of phenotypic plasticity: patterns of plasticity and the emergence of ecotypes. *New Phytologist*, **166**, 101–17.

DeWitt, T. J. (1998). Costs and limits of phenotypic plasticity: tests with predator-induced morphology and life history in a freshwater snail. *Journal of Evolutionary Biology*, **11**, 465–80.

DeWitt, T. J., Robinson, B. W., and Wilson, D. S. (2000). Functional diversity among predators of a freshwater snail imposes an adaptive trade-off for shell morphology. *Evolutionary Ecology Research*, **2**, 129–48.

DeWitt, T. J., and Scheiner, S. M. (2004). *Phenotypic Plasticity, Functional and Conceptual Approaches*. Oxford University Press, Oxford.

Doligez, B., Gustafsson, L., and Part, T. (2009). 'Heritability' of dispersal propensity in a patchy population. *Proceedings of the Royal Society B: Biological Sciences*, **276**, 2829–36.

Edelaar, P., and Bolnick, D. I. (2012). Non-random gene flow: an underappreciated force in evolution and ecology. *Trends in Ecology and Evolution*, **27**, 659–65.

Edelaar, P., Siepielski, A. M., and Clobert, J. (2008). Matching habitat choice causes directed gene flow: a neglected dimension in evolution and ecology. *Evolution*, **62**, 2462–72.

Edgell, T. C., and Hollander, J. (2011). The evolutionary ecology of European green crab (*Carcinus maenas*) in North America. In B. Galil and P. Clark (eds), *In the Wrong Place, Alien Marine Crustaceans—Distribution, Biology and Impacts*. Springer Series, New York.

Ernande, B., Boudry, P., Clobert, J., and Haure, J. (2004). Plasticity in resource allocation based life history traits in the Pacific oyster, *Crassostrea gigas*. I. Spatial variation in food abundance. *Journal of Evolutionary Biology*, **17**, 342–56.

Evanno, G., Castella, E., and Goudet, J. (2006). Evolutionary aspects of population structure for molecular and quantitative traits in the freshwater snail *Radix balthica*. *Journal of Evolutionary Biology*, **19**, 1071–82.

Fear, K. K. and Price, T. (1998). The adaptive surface in ecology. *Oikos*, **82**, 440–448.

Fisher, H. S., Wong, B. B. M., and Rosenthal, G. G. (2006). Alteration of the chemical environment disrupts communication in a freshwater fish. *Proceedings of the Royal Society B: Biological Sciences*, **273**, 1187–93.

Fowler-Finn, K. D., and Rodriguez, R. L. (2012). Experience-mediated plasticity in mate preferences, mating assurance in a variable environment. *Evolution*, **66**, 459–68.

Funk, W. C., Cannatella, D. C., and Ryan, M. J. (2009). Genetic divergence is more tightly related to call variation than landscape features in the Amazonian frogs *Physalaemus petersi* and *P. freibergi*. *Journal of Evolutionary Biology*, **22**, 1839–53.

Garant, D., Forde, S. E., and Hendry, A. P. (2007). The multifarious effects of dispersal and gene flow on contemporary adaptation. *Functional Ecology*, **21**, 434–43.

Gardner, J. P. A., Skibinski, D. O. F., and Bajdik, C. D. (1993). Shell growth and viability differences between the marine mussels *Mytilus edulis* (l), *Mytilus galloprovincialis* (lmk), and their hybrids from 2 sympatric populations in SW England. *Biological Bulletin*, **185**, 405–16.

Gibson, G., and Dworkin, I. (2004). Uncovering cryptic genetic variation. *Nature Reviews Genetics*, **5**, 681–90.

Gilg, M. R., O'Connor, M., Norris, R., and Hilbish, T. J. (2009). Maintenance of parental populations bordering a blue mussel hybrid zone by post-settlement selection. *Journal of Molluscan Studies*, **75**, 207–14.

Grosberg, R. K., and Cunningham, C. W. (2001). Genetic structure in the sea, from populations to communities. In M. D. Bertness, S. D. Gaines, and M. E. Hay (eds), *Marine Community Ecology*, pp. 61–84. Sinauer Associates, Sunderland, MA.

Halfwerk, W., Bot, S., Buikx, J., et al. (2011). Low-frequency songs lose their potency in noisy urban conditions. *Proceedings of the National Academy of Sciences USA*, **108**, 14549–54.

Hamdoun, A. M., Cheney, D. P., and Cherr, G. N. (2003). Phenotypic plasticity of HSP70 and HSP70 gene expression in the Pacific oyster (*Crassostrea gigas*): implications for thermal limits and induction of thermal tolerance. *Biological Bulletin*, **205**, 160–9.

Hansson, B., Bensch, S., and Hasselquist, D. (2004). Lifetime fitness of short- and long-distance dispersing great reed warblers. *Evolution*, **58**, 2546–57.

Hebets, E. A., and Maddison, W. P. (2005). Xenophilic mating preferences among populations of the jumping spider *Habronattus pugillis* Griswold. *Behavioral Ecology*, **16**, 981–8.

Hirasaki, E., Ogihara, N., Hamada, Y., Kumakura, H., and Nakatsukasa, M. (2004). Do highly trained monkeys walk like humans? A kinematic study of bipedal locomotion in bipedally trained Japanese macaques. *Journal of Human Evolution*, **46**, 739–50.

Hollander, J. (2008). Testing the grain-size model for the evolution of phenotypic plasticity. *Evolution*, **62**, 1381–9.

Hollander, J., and Butlin, R.K. (2010). The adaptive value of phenotypic plasticity in two ecotypes of a marine gastropod. *BMC Evolutionary Biology*, 107, **10**: 333.

Hollander, J., Lindegarth, M., and Johannesson, K. (2005). Local adaptation but not geographic separation promotes assortative mating in a snail. *Animal Behaviour*, **70**, 1209–19.

Hoverman, J. T., and Relyea, R. A. (2007). The rules of engagement: how to defend against combinations of predators. *Oecologia*, **154**, 551–60.

Hughes, K. A., Du, L., Rodd, F. H., and Reznick, D. N. (1999). Familiarity leads to female mate preference for novel males in the guppy, *Poecilia reticulata*. *Animal Behaviour*, **58**, 907–16.

Izzo, A. S., and Gray, D. A. (2011). Heterospecific courtship and sequential mate choice in sister species of field crickets. *Animal Behaviour*, **81**, 259–64.

Jaenike, J., and Holt, R. D. (1991). Genetic variation for habitat preference—evidence and explanations. *American Naturalist*, **137**, S67–90.

Johannesson, K., Havenhand, J. N., Jonsson, P. R., Lindegarth, M., Sundin, A., and Hollander, J. (2008). Male discrimination of female mucous trails permits assortative mating in a marine snail species. *Evolution*, **62**, 3178–84.

Johannesson, B., and Johannesson, K. (1996). Population differences in behaviour and morphology in the snail *Littorina saxatilis*: phenotypic plasticity or genetic differentiation? *Journal of Zoology*, **240**, 475–93.

Karpestam, E., Wennersten, L., and Forsman, A. (2012). Matching habitat choice by experimentally mismatched phenotypes. *Evolutionary Ecology*, **26**, 893–907.

Kawecki, T. J., and Ebert, D. (2004). Conceptual issues in local adaptation. *Ecology Letters*, **7**, 1225–41.

Keane, R. M., and Crawley, M. J. (2002). Exotic plant invasions and the enemy release hypothesis. *Trends in Ecology and Evolution*, **17**, 164–70.

Kenis, M., Auger-Rozenberg, M. A., Roques, A., et al. (2009). Ecological effects of invasive alien insects. *Biological Invasions*, **11**, 21–45.

Kirschel, A. N. G., Blumstein, D. T., Cohen, R. E., Buermann, W., Smith, T. B., and Slabbekoorn, H. (2009). Birdsong tuned to the environment, green hylia song varies with elevation, tree cover, and noise. *Behavioral Ecology*, **20**, 1089–95.

Lande, R., and Shannon, S. (1996). The role of genetic variation in adaptation and population persistence in a changing environment. *Evolution*, **50**, 434–37.

Le Rouzic, A., and Carlborg, O. (2008). Evolutionary potential of hidden genetic variation. *Trends in Ecology and Evolution*, **23**, 33–7.

Levins, R. (1968). *Evolution in Changing Environments*. Princeton University Press, Princeton, NJ.

Mack, R. N., Simberloff, D., Lonsdale, W. M., Evans, H., Clout, M., and Bazzaz, F. A. (2000). Biotic invasions: causes, epidemiology, global consequences, and control. *Ecological Applications*, **10**, 689–710.

Maron, J. L., and Vila, M. (2001). When do herbivores affect plant invasion? Evidence for the natural enemies and biotic resistance hypotheses. *Oikos*, **95**, 361–73.

Marquet, P. A., Fortin, M.-J., Pineda, J., et al. (1993). Ecological and evolutionary consequences of patchiness, a marine-terrestrial perspective In S. Levin, T. Powell, and J. Steeele (eds), *Patch Dynamics* (Lecture Notes in Biomathematics). Springer-Verlag, New York.

Marshall, D. J., Monro, K., Bode, M., Keough, M. J., and Swearer, S. (2010). Phenotype-environment mismatches reduce connectivity in the sea. *Ecology Letters*, **13**, 128–40.

M'Gonigle, L. K., and FitzJohn, R. G. (2010). Assortative mating and spatial structure in hybrid zones. *Evolution*, **64**, 444–55.

Moczek, A. P., Sultan, S., Foster, S., et al. (2011). The role of developmental plasticity in evolutionary innovation. *Proceedings of the Royal Society of London Series B: Biological Sciences*, **278**, 2705–13.

Moran, N. A. (1992). The evolutionary maintenance of alternative phenotypes. *American Naturalist*, **139**, 971–89.

Murren, C. M., Julliard, R., Schlichting, C. D., and Clobert, J. (2001). Dispersal, individual phenotype, and phenotypic plasticity In J. Clobert, E. Danchin, A. A. Dhondt, and J. D. Nichols (eds), *Dispersal*. Oxford University Press, New York.

Nosil, P. (2007). Divergent host plant adaptation and reproductive isolation between ecotypes of *Timema cristinae* walking sticks. *American Naturalist*, **169**, 151–62.

Nosil, P., Vines, T. H., and Funk, D. J. (2005). Perspective: reproductive isolation caused by natural selection against immigrants from divergent habitats. *Evolution*, **59**, 705–19.

Panova, M., Hollander, J., and Johannesson, K. (2006). Site-specific genetic divergence in parallel hybrid zones suggests nonallopatric evolution of reproductive barriers. *Molecular Ecology*, **15**, 4021–31.

Parker, I. M., Rodriguez, J., and Loik, M. E. (2003). An evolutionary approach to understanding the biology of invasions, local adaptation and general-purpose genotypes in the weed *Verbascum thapsus*. *Conservation Biology*, **17**, 59–72.

Pfennig, D. W., Wund, M. A., Snell-Rood, E. C., Cruickshank, T., Schlichting, C. D., and Moczek, A. P. (2010). Phenotypic plasticity's impacts on diversification and speciation. *Trends in Ecology and Evolution*, **25**, 459–67.

Pigliucci, M. (2001). *Phenotypic Plasticity, beyond Nature and Nurture*. Johns Hopkins University Press, Baltimore, MD.

Pigliucci, M. (2005). Evolution of phenotypic plasticity: where are we going now? *Trends in Ecology and Evolution*, **20**, 481–6.

Pigliucci, M., and Murren, C. J. (2003). Perspective: genetic assimilation and a possible evolutionary paradox: can macroevolution sometimes be so fast as to pass us by? *Evolution*, **57**, 1455–64.

Pigliucci, M., Murren, C. J., and Schlichting, C. D. (2006). Phenotypic plasticity and evolution by genetic assimilation. *Journal of Experimental Biology*, **209**, 2362–7.

Price, T. D., Qvarnström, A., and Irwin, D. E. (2003). The role of phenotypic plasticity in driving genetic evolution. *Proceedings of the Royal Society B: Biological Sciences*, **270**, 1433–40.

Rahel, F. J., and Olden, J. D. (2008). Assessing the effects of climate change on aquatic invasive species. *Conservation Biology*, **22**, 521–33.

Richardson, D. M., and Pysek, P. (2006). Plant invasions, merging the concepts of species invasiveness and community invasibility. *Progress in Physical Geography*, **30**, 409–31.

Rollo, C. D. (1994). *Phenotypes: Their Epigenetics, Ecology and Evolution*. Chapman & Hall, London.

Rundle, S. D., and Brönmark, C. (2001). Inter- and intraspecific trait compensation of defence mechanisms in freshwater snails. *Proceedings of the Royal Society B: Biological Sciences*, **268**, 1463–8.

Rundle, H. D., and Nosil, P. (2005). Ecological speciation. *Ecology Letters*, **8**, 336–52.

Ryan, M. J., and Brenowitz, E. A. (1985). The role of body size, phylogeny, and ambient noise in the evolution of bird song. *American Naturalist*, **126**, 87–100.

Scheffer, M., and Carpenter, S. R. (2003). Catastrophic regime shifts in ecosystems: linking theory to observation. *Trends in Ecology and Evolution*, **18**, 648–56.

Scheiner, S. M. (1993). Genetics and evolution of phenotypic plasticity. *Annual Review of Ecology and Systematics*, **24**, 35–68.

Scheiner, S. M. (1998). The genetics of phenotypic plasticity. VII. Evolution in a spatially-structured environment. *Journal of Evolutionary Biology*, **11**, 303–20.

Scheiner, S. M., Barfield, M., and Holt, R. D. (2012). The genetics of phenotypic plasticity. Xi. Joint evolution of plasticity and dispersal rate. *Ecology and Evolution*, **2**, 2027–39.

Schlichting, C. D. (2004). The role of phenotypic plasticity in diversification. In T. J. DeWitt and S. M. Scheiner (eds), *Phenotypic Plasticity, Functional and Conceptual Approaches*, pp. 191–200. Oxford University Press, New York.

Schlichting, C. D., and Pigliucci, M. (1998). *Phenotypic Evolution: A Reaction Norm Perspective*. Sinauer, Massachusetts.

Schluter, D. (2000). *The Ecology of Adaptive Radiation*. Oxford University Press, Oxford.

Seehausen, O., van Alphen, J. J. M., and Witte, F. (1997). Cichlid fish diversity threatened by eutrophication that curbs sexual selection. *Science*, **277**, 1808–11.

Shine, R., Brown, G. P., and Phillips, B. L. (2011). An evolutionary process that assembles phenotypes through space rather than through time. *Proceedings of the National Academy of Sciences USA*, **108**, 5708–11.

Smith, L. D. (2009). The role of phenotypic plasticity in marine biological invasions In G. Rilov and J. A. Crooks (eds), *Biological Invasions in Marine Ecosystems, Ecological, Management, and Geographic Perspectives*. Springer, New York.

Sousa, W. P. (1984). The role of disturbance in natural communities. *Annual Review of Ecology and Systematics*, **15**, 353–91.

Stachowicz, J. J., Whitlatch, R. B., and Osman, R. W. (1999). Species diversity and invasion resistance in a marine ecosystem. *Science*, **286**, 1577–9.

Stamps, J. A., Krishnan, V. V., and Willits, N. H. (2009). How different types of natal experience affect habitat preference. *American Naturalist*, **174**, 623–30.

Strathmann, R. R., Fenaux, L., Sewell, A. T., and Strathmann, M. F. (1993). Abundance of food affects relative size of larval and postlarval structures of a molluscan veliger. *Biological Bulletin*, **185**, 232–9.

Sultan, S. E., and Spencer, H. G. (2002). Metapopulation structure favors plasticity over local adaptation. *American Naturalist*, **160**, 271–83.

Suzuki, Y., and Nijhout, H. F. (2006). Evolution of a polyphenism by genetic accommodation. *Science*, **311**, 650–2.

Thibert-Plante, X., and Hendry, A. P. (2011). The consequences of phenotypic plasticity for ecological speciation. *Journal of Evolutionary Biology*, **24**, 326–342.

Tollrian, R., and Harvell, D. (1999). *The Ecology and Evolution of Inducible Defences*. Princeton University Press, Princeton, NJ.

Tufto, J. (2000). The evolution of plasticity and nonplastic spatial and temporal adaptations in the presence of imperfect environmental cues. *American Naturalist*, **156**, 121–30.

Turner, A. M., and Montgomery, S. L. (2003). Spatial and temporal scales of predator avoidance, experiments with fish and snails. *Ecology*, **84**, 616–22.

Uyeda, J. C., Arnold, S. J., Hohenlohe, P. A., and Mead, L. S. (2009). Drift promotes speciation by sexual selection. *Evolution*, **63**, 583–94.

van Tienderen, P. H. (1997). Generalists, specialists, and the evolution of phenotypic plasticity in sympatric populations of distinct species. *Evolution*, **51**, 1372–80.

Veltman, C. J., Nee, S., and Crawley, M. J. (1996). Correlates of introduction success in exotic New Zealand birds. *American Naturalist*, **147**, 542–57.

Verhulst, S., and van Eck, H. M. (1996). Gene flow and immigration rate in an island population of great tits. *Journal of Evolutionary Biology*, **9**, 771–82.

Verzijden, M. N., Culumber, Z. W., and Rosenthal, G. G. (2012). Opposite effects of learning cause asymmetric mate preferences in hybridizing species. *Behavioral Ecology*, **23**, 1133–9.

Verzijden, M. N., and Rosenthal, G. G. (2011). Effects of sensory modality on learned mate preferences in female swordtails. *Animal Behaviour*, **82**, 557–62.

Via, S., and Lande, R. (1985). Genotype-environment interaction and the evolution of phenotypic plasticity. *Evolution*, **39**, 505–22.

Waddington, C. H. (1942). Canalization of development and the inheritance of acquired characters. *Nature*, **150**, 563–5.

Waddington, C. H. (1961). Genetic assimilation. *Advanced Genetics*, **10**, 257–90.

Wang, J., Christoffersen, K., Buck, S., and Tao, Y. (2007). The pacific oyster (*Crassostrea gigas*) in the Isefjord Denmark. Report, Roskilde University.

Wellenreuther, M., Tynkkynen, K., and Svensson, E. I. (2010). Simulating range expansioin, male species recognition and loss of premating isolation in damselflies. *Evolution*, **64**, 242–52.

West-Eberhard, M. J. (2005). Developmental plasticity and the origin of species differences. *Proceedings of the National Academy of Sciences USA*, **102**, 6543–9.

Wilhelm, R., and Hilbish, T. J. (1998). Assessment of natural selection in a hybrid population of mussels, evaluation of exogenous vs endogenous selection models. *Marine Biology*, **131**, 505–14.

Wrange, A. L. (2008). Undersökning av det japanska jätteostronet (*Crassostrea gigas*) längs den svenska västkusten 2007–2008. Report for Aqualiens, Göteborg University.

CHAPTER 8

Pathogens and hosts on the move

Helena Westerdahl, Staffan Bensch, Jan-Åke Nilsson, Emily O´Connor, Ravinder Sehgal, Sylvie Tesson, and Dennis Hasselquist

Biodiversity is in a constant state of flux, with all life forms involved in some type of movement. The distances covered by moving organisms can range from microscopic to incredible pole-to-pole journeys by some migrating birds and marine mammals (Chapman et al., Chapter 2). Most, if not all organisms are host to one or more pathogen species that can result in a host-pathogen evolutionary arms race; pathogens adapt to invade their host species and the hosts adapt to avoid infections. The study of why and when hosts move in relation to pathogen exposure and how pathogens move with their hosts is an emerging field of research with potential implications for the distribution and abundance of host species, as well as the spread and emergence of diseases. In this chapter, our aim is to consider both hosts and pathogens and how their movements may influence genetic, ecological and evolutionary patterns in natural systems. We use the terms 'pathogen' and 'parasite' interchangeably, and we define them in a broad sense as infectious agents (e.g. viruses, bacteria, fungi, blood parasites and gut parasites).

On a global scale the diversity and abundance of parasites is higher in the tropics compared with temperate regions, and higher in temperate than in arctic regions (Fig. 8.1). This pattern is surprisingly consistent when studying parasites in a wide range of different hosts, such as plants, fish, birds, primates and humans (Guegan et al. 2008). Thus, sedentary hosts in the tropics are exposed to a wider range of parasites than sedentary hosts in temperate regions. Hosts that move between regions (e.g. biomes) are naturally exposed to the parasite regimes of all the regions they visit, and it is therefore likely that moving hosts have developed unique adaptations to withstand the selection pressure of encountering a wider range of pathogens than sedentary hosts (Møller et al. 2003, Hasselquist 2007). This increased risk of contracting infections in long-distance moving animals has been recognized as a potential 'cost of migration' (Hasselquist 2007), which may affect decisions over if, when, where and how to move/migrate. Pathogen distribution is known to vary across landscapes and water systems (Bensch and Åkesson 2003, Yohannes et al. 2008, Krkosek et al. 2013), but it can also vary on a more local scale, even within a forest (Wood et al. 2007). Hence, short-distance movements on a local scale can also increase or decrease a host's exposure to pathogens, with important implications for decisions about what distances to move and in which type of habitat to settle.

With host movement comes the inevitable co-movement of their pathogens: a type of 'assisted movement'. There are well-studied cases of pathogens moving with migratory butterflies, fish, birds and mammals (Olsen et al. 2006, Altizer et al. 2011, Fuller et al. 2012), but there is little consensus about the general implications that can be drawn from these examples. Moreover, not only pathogens but also other non-pathogenic micro-organisms can 'hitch-hike' with larger animals (here the host is a vector), for example in the digestive apparatus, on the body or even in soil that may cover the body (Table 8.1). The ecological and community consequences of these micro-organismal transfers can be considerable. In this chapter, we outline prevailing hypotheses that are relevant for movements

Animal Movement Across Scales. Edited by Lars-Anders Hansson and Susanne Åkesson.
© Oxford University Press 2014. Published 2014 by Oxford University Press.

Figure 8.1 On a global scale the diversity and abundance of parasites is higher in the tropics than in the temperate regions (Guegan et al. 2008), here shown as species richness, which is the total number of human diseases known within the boundary limits of each country (here averaged for a certain region; Guernier et al. 2004), in relation to latitude across the two hemispheres for human parasites.

Table 8.1 Examples of parasite/propagule–host/vector associations that have resulted in short- or long-distance dispersals. The table includes information on a selection of pathogenic (parasites) and non-pathogenic (propagules) micro-organisms, potential hosts or 'hosts as vectors' that are involved in passive or active dispersal/migration. Moreover, we give examples of where the parasite can be located on (or within) the host ('target') and the type of interaction between the parasite/propagule–host/vector (e.g. pathogenic, non-pathogenic, and/or 'mastermind', i.e. affect the movement of the host).

Parasite/Propagules	Host/Vector	Target	Interaction	Papers
Virus	Protist	Cytoplasm	Pathogen	Brussaard 2004, Bratbak et al. 1993
	Bird	Digestive apparatus	Both	Brochet et al. 2009a, Lam et al. 2012
Bacteria	Protist	Exoskeleton and cytoplasm	Both	Park et al. 2004, Bates et al. 2004
		Cytoplasm	Master mind	Fellous et al. 2011
	Invertebrate	Oesophage wall	Both	Grossart et al. 2010
Fungi	Bird	Toes	Non-pathogen	Garmyn et al. 2012
	Protist	Cytoplasm	Pathogen	Park et al. 2004
Protist	Invertebrate	Gut and Oesophage	Pathogen	Skovgaard and Saiz 2006, Skovgaard 2005, Duneau et al. 2011
	Protist	Cell	Pathogen	Coats 1999, Kim et al. 2004, Fraga et al. 1989, Chambouvet et al. 2008
	Bird	Blood	Both	Bensch et al. 2000, Palinauskas et al. 2011
	Worm	Cytoplasm	Pathogen	Schärer et al. 2007
	Human	Equipment	Non-pathogen	Broady and Smith 1994, Gollash 2007
Plant propagule	Mammal	Fur	Non-pathogen	Sorensen, 1986
	Bird	Gut	Non-pathogen	Brochet et al. 2009b, c, 2010; Gollash 2007; Viana et al. 2012
	Human	Equipment	Non-pathogen	Waterkeyn et al. 2010b
	Fish	Gut	Non-pathogen	Pollux 2011
Invertebrate	Bird	Gut	Non-pathogen	Figuerola and Green 2002
	Salamander	Gut	Non-pathogen	Bohonak and Whiteman 1999
	Insect		Non-pathogen	van de Meutter et al. 2008
	Invertebrate	Cytoplasm (1st host)	Pathogen	Prugnolle et al. 2005b
	Mammal	Fur	Non-pathogen	Waterkeyn et al. 2010a
		Skin (2nd host)	Non-pathogen	Prugnolle et al. 2005b
	Human	Equipment	Non-pathogen	Waterkeyn et al. 2010b

of pathogens and animal hosts, provide examples testing these hypotheses and suggest potential avenues for future research. Questions that we have identified as central in the context of pathogens and moving hosts are: 1. How do host movements affect the risk of infections? 2. What adaptations have moving hosts developed to reduce the impact of increased pathogen exposure? 3. To what extent are movement strategies a means to avoid infections? 4. To what extent do moving hosts contribute to the spread of pathogens? 5. Which strategies have pathogens and non-pathogenic micro-organisms developed to take advantage of moving hosts?

8.1 Host movements and the risk of infections

One of the most important factors that induce movements is the need for tracking resources in a world where resources vary extensively in abundance and distribution over the annual cycle (Alerstam et al. 2003). Such resource-driven movements could cover very long distances, forcing animals into new environments and climate zones, exposing them to different communities of pathogens and, as is the case for temperate animals travelling to tropical regions, expose them to more parasite-rich environments (Møller and Erritzoe 1998, Møller et al. 2003). Thus, temperate breeders moving to tropical wintering areas run the risk of being exposed to both a generally higher abundance of vectors and pathogens, and a higher diversity of pathogen species (Fig. 8.1). However, small-scale movements may also alter the risk of contracting parasites, and a 'risk of contracting pathogen cost' of movement may therefore be relevant for all types of moving animals.

8.1.1 Expansion of range limits

One situation when animals move into novel areas is during range expansions, which might, for example, be initiated by climatic changes (Lindström et al., Chapter 3). At range edges, organisms may face a number of novel conditions (abiotic and biotic) that require different adaptations compared to in the centre of the range. Dispersal from the range centre outwards may prevent the populations at the range edge from adapting, which ultimately will set the range limit of a species (Kirkpatrick and Barton 1997). One such biotic condition at the species range limit is that individuals might be more likely to be exposed to novel parasites. This could slow down the speed of expansion, an effect that we will call the 'mismatch to novel parasites' hypothesis. However, low host population size and movement into previously unoccupied areas could also represent less favourable conditions for parasites adapted to the conditions in the core distribution area of its host species. This potential restriction on parasites could lead to a lowered cost of infection for hosts settling in the range expansion zone, an effect that we will refer to as the 'leading-edge enemy release' hypothesis (see also Clay 2003). Of these two hypotheses, the one which will be most important during a range expansion event is largely context-dependent, where factors such as speed of range expansion, type of parasite (e.g. mode of transmission) and occurrence of related host species in the newly established areas will affect the outcome of a host–parasite interaction.

Another interesting aspect of range expansions is the fact that individuals at the expansion edge may show consistent differences compared to individuals that constitute the core population (Nilsson et al., Chapter 6). For example, individuals at the 'leading edge' could belong to a sub-group of behavioural types (personality) compared with individuals from the core population. This could lead to formation of a 'dispersal behavioural syndrome' where for example aggression and dispersal are coupled behaviours in individuals that are particularly successful colonizers (Duckworth and Badyaev 2007, Cote et al. 2010; Nilsson et al., Chapter 6; Chapman et al., Chapter 2). If this is the case, it is tempting to speculate that individuals exhibiting different behavioural types may differ in their defence against pathogens. Thus individuals with the dispersal characters may have an immune system that is adapted to a broader range of parasites, an idea that we will call the 'dispersal phenotype immunocompetence' hypothesis. Increased activity is usually part of the dispersal syndrome, which might increase the risk of contracting a pathogen. However, virtually no data exist on behaviour-dependent immune defences or vulnerability to infection.

An alternative possibility is that the range expansion is conducted mainly by low quality or young individuals, which in turn could result in 'leading-edge' individuals having lower condition-dependent immunocompetence, which would subsequently slow down the speed of the range expansion.

A particularly intriguing type of 'range expansion' can come about through human transportation of organisms. From a niche-theory perspective, it is surprising that organisms transported by humans, often to very different environments, can establish themselves and even thrive to the extent of becoming pests. One hypothesis proposed to explain this phenomenon is the 'enemy release' hypothesis (Clay 2003). This hypothesis relies upon the idea that a human-transported species may become a successful invader if it is released from the control of common parasites in its native range. In these cases, the resources that were previously used for defence against these common natural parasites can now be redirected to other activities which may give the invasive species a competitive advantage against the native species. This has been termed the 'evolution of increased competitive ability' (EICA) hypothesis (Blossey and Notzold 1995, Lee and Klasing 2004). Support for the 'enemy release' hypothesis comes from a comparative study where the prevalence of helminth parasites was compared between the introduced and native ranges of 26 host species, ranging from molluscs to birds and mammals. On average 16 helminth species per host species were found in native ranges as compared with an average of three helminth species in introduced ranges (Torchin et al. 2003). Another hypothesis proposed to explain the success of invasive species has been called the 'novel weapon' hypothesis and it builds on the idea that invasive species bring their own parasites to the new environment, which may have severe consequences for related native host species (Callaway and Ridenour 2004). One such example is the grey squirrel, *Sciurus carolinensis*, which was introduced to Europe from North America. Grey squirrels carry a parapoxvirus that is thought to be responsible for the subsequent dramatic decline of the red squirrel, *S. vulgaris*, the species native to Europe (Tompkins et al. 2003).

However, successful invasions are not the only possible outcome of host–parasite interactions of human-transported species. There are numerous cases when human-transported species have failed to become established in a novel area (Lee and Klasing 2004). Human-transported species that encounter a novel parasite community could suffer heavily from infections because their immune system is not adapted to these parasites. This is in line with the 'mismatch to novel parasites' hypothesis mentioned earlier. So far, there is no general framework to help us predict under which circumstances we may expect the 'enemy release' and 'novel weapons' hypotheses to be stronger factors facilitating invasion than the 'mismatch to novel parasites' hypothesis hindering invasion. Such theory and modelling work would be extremely valuable given the potentially high negative impact of invasive species on the distribution and extinction of native species, as well as on human economic resources (Wilcove et al. 1998, Palumbi 2001).

8.1.2 Increased vulnerability to parasites as a result of host movement

In general, the strength of an immune response is negatively affected when an individual simultaneously needs to perform stressful and energetically costly behaviours (Hasselquist and Nilsson 2012). The proximate basis for this trade-off between hard work and mounting an immune response could be either *long-* or *short-term* fitness costs. *Long-term fitness costs* could include lower work performance and/or reduced probability of survival at a later stage in life due to the consequences of excess oxidative stress and an increased risk of autoimmunity (delayed costs; Råberg et al. 1998, Hasselquist and Nilsson 2012). *Short-term fitness costs* may result from energy or nutrient limitations directly constraining performance (immediate costs; Hasselquist and Nilsson 2012). The trade-off between hard work and mounting immune responses has been noted for parental effort (reviewed by Knowles et al. 2009), mating effort (Kilpimaa et al. 2004), general activity (Ewenson et al. 2003), and thermoregulation (Svensson et al. 1998).

As all modes of movements entail high energetic costs (Alexander 1998), with flying being especially costly (Norberg 1996; Johansson et al., Chapter 13), we predict that organisms which move by means of flying for extensive time periods have reduced immunocompetence and thus be more likely to contract pathogens when engaged in these activities. Some evidence in favour of this prediction has been found by manipulating parental effort in birds. Increased brood size, which in birds results in increased energy expenditure (Nilsson 2002) due to an increased number of feeding flight journeys, results in an overall compromised immune response (Knowles et al. 2009). Furthermore, by experimentally increasing the cost of flying, Hasselquist et al. (2001) found a reduced immune response in tree swallows, *Tachycineta bicolor*. However, in a direct simulation of migration in a wind tunnel, where red knots, *Calidris canutus* flew 1500 km in 6 days, the migration effort did not affect the strength of their immune response (Hasselquist et al. 2007). It should be noted that this study was conducted in a laboratory setting, with food available *ad libitum* and a >36 h long period of recovery after each flying bout. Hence, these birds may have been better able to recover from their exertions than wild birds. This suggestion is supported by results from a study on starlings, *Sturnus vulgaris*, flying in a wind tunnel. Immediately after a flying bout of 1 to 4 h, immune function was significantly reduced, but this reduction was less pronounced after a 48-h resting period (Nebel et al. 2012). Thus, although not universal, hard work in the form of moving seems to adversely affect the immune response and thereby increase the risk of contracting pathogens during periods of intense movement. Bar-tailed godwits, *Limosa lapponica*, make a non-stop migration flight of 9 days from breeding areas in Alaska to wintering areas in New Zealand (Gill et al. 2009). This outstanding work performance might be possible due to very low exposure to parasites at both breeding and overwintering sites, allowing these shorebirds to make relatively small investments in the immune-machinery without having to pay the cost of an increased risk of contracting pathogens (Piersma 2011).

8.2 Moving host adaptations to reduce the impact of increased pathogen exposure

8.2.1 'Genetic footprints' in host immunity

Selection pressure from pathogens can shape the genes of their host and vice versa. This process generates 'genetic footprints' that show evidence of past and present selection events. In the host, this may be especially true for genes of the immune system and there is compelling evidence that exposure to pathogens affects the polymorphism of immune genes (Hill et al. 1991, Prugnolle et al. 2005a, Obbard et al. 2009). Immune genes have several functions; they can be effector or receptor molecules, or involved in signalling (Murphy et al. 2008). Selection from parasites is primarily seen in host genes that encode receptor molecules that bind parasite epitopes. Adaptation at the genetic level can be seen as positive selection, and positively selected sites have been reported in genes of both the innate and the adaptive immune system across a wide range of hosts (Hughes and Nei 1988, Obbard et al. 2009, Hellgren and Sheldon 2011, Tschirren et al. 2011). It could be expected that movements of animals also generate genetic footprints, and that such footprints often are found in immunity genes, because different geographical areas hold different communities of parasites (discussed earlier). In humans, malaria has been, and still is, a major selective agent in our evolution. Malaria is common primarily in the tropics and several independent gene mutants have been shown to protect against severe malaria (i.e. the mutated haemoglobin (HbS) and a mutation resulting in loss of the Duffy blood group (Allison 1954, Miller et al. 1976)). These two mutations have negative effects on human physiology and are therefore selected against in parts of the world where malaria is not present, while being overrepresented (positively selected) in humans originating from tropical areas where malaria is common (Hamblin et al. 2000, Guegan et al. 2008). Thus, these two mutations are examples of host adaptations to overcome severe infections and have generated genetic footprints.

8.2.2 Highly polymorphic host immunity genes

The Major Histocompatibility Complex (MHC) genes encode cell surface proteins of vital importance in adaptive immunity. Classical MHC molecules are receptor molecules with an antigen binding site (or peptide binding region) that is extremely polymorphic. The genes encoding these MHC molecules are the most variable genes

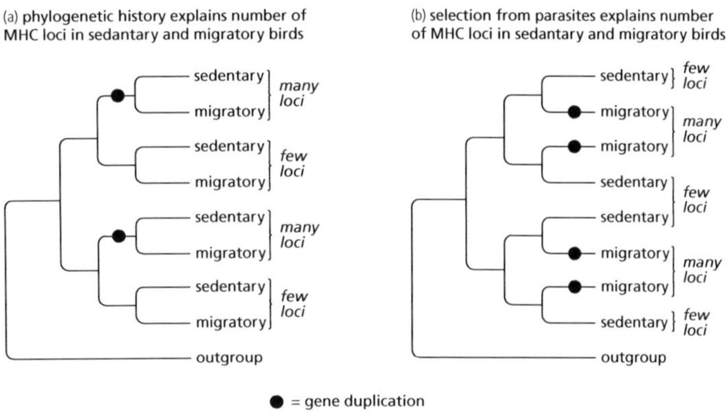

Figure 8.2 Migratory birds may have evolved a larger number of MHC alleles/loci (number of immunity genes) than sedentary birds since parasite selection pressure comes from a broader range of parasites in migratory birds. A larger number of MHC alleles/loci could hold a greater MHC repertoire per individual and equip these birds with adaptive immunity that is better able to combat a diverse set of parasites. This hypothesis could be tested by taking a phylogenetic approach whereby the number of MHC loci is compared within and between migratory and sedentary bird species. (a) If MHC genes became duplicated in an ancestral species then all species that are younger will have the duplication, independent of whether the species are migratory. In these circumstances phylogenetic history would explain the number of MHC loci found in a species. (b) However, if the number of MHC loci varies within bird families and is low or high, depending on whether the species is sedentary or migratory, respectively, then the number of MHC alleles/loci may be under selection from parasites. In this scenario selection from parasites explains the number of MHC alleles/loci and not the phylogenetic history of the birds.

known in vertebrates; in humans there are more than 4,000 alleles reported at the MHC loci (Buhler and Sanchez 2011). The peptide binding region is subject to selection from pathogens which can be seen on the DNA sequence level as positive selection (Apanius et al. 1997, Yeager and Hughes 1999). Each MHC molecule can bind a limited repertoire of pathogen peptides and it is most likely advantageous for an individual to be heterozygous at a MHC locus. Exposure to a wider range of pathogens affects the evolution of host MHC diversity. For example, human populations in Africa that have been exposed to a wider range of pathogens than human populations in Europe have higher MHC diversity (Prugnolle et al. 2005a). Furthermore, specific MHC haplotypes are associated with resistance to severe malaria and these haplotypes are overrepresented in tropical Africa (Hill et al. 1991, Gilbert et al. 1998).

Although there is convincing evidence that the exposure of a wide range of pathogens affects the evolution of immune genes in humans, there is still limited information from other hosts. This is particularly true for comparative studies of host groups that contain both sedentary and migratory species. So far a handful of studies have reported associations between specific MHC alleles/MHC diversity and resistance/susceptibility to avian malaria infections (*Plasmodium, Haemoproteus and Leucocytozoon*) in songbirds. These associations are reported for long distance migratory birds (great reed warbler; Westerdahl et al. 2005, 2012), for partial migrants (blue tits; Westerdahl et al. 2013) and also for sedentary birds (great tits and house sparrows; Bonneaud et al. 2006, Loiseau et al. 2010, Sepil et al. 2013). Recent studies have further investigated to what extent migratory songbirds are exposed to a wider range of pathogens than residents. Migratory species harbour a greater diversity of *Leucocytozoon* (a genus of blood parasites infecting birds) than resident species (Jenkins et al. 2012). Furthermore, in a migratory songbird, the garden warbler, *Sylvia borin*, individuals were infected by a higher diversity of avian malaria parasites at their wintering sites in Africa compared to at their breeding sites in Europe and during migration (Hellgren et al. 2013). Moreover, in some well-studied long-distance migratory songbirds, the vast majority of the lineages/species of blood parasites were transmitted at their wintering sites in tropical Africa (Bensch et al. 2007, Hellgren et al. 2013).

Taken together, these data imply that songbirds are exposed to more avian malaria parasites when wintering in tropical Africa, confirming the general pattern of higher parasite diversity and abundance in the tropics discussed earlier. This pattern implies that migratory birds need an immune system that can cope with a wider range of pathogens than sedentary birds. One such adaptation could be a higher MHC gene diversity, which would allow for recognition of a wider range of parasites. MHC diversity could be maintained by different mechanisms: negative frequency dependent selection, heterozygote advantage and fluctuating selection (Bodmer 1972, Doherty and Zinkernagel 1975, Hedrick 2002). An additional mechanism is gene duplication, a common phenomenon in gene families like the MHC. If the selection pressure comes from a broader range of parasites in migratory as compared with sedentary birds, we can predict that migratory birds will have a higher MHC diversity, i.e. larger number of MHC loci, than sedentary birds. This could be tested taking a phylogenetic approach where the number of MHC loci is compared within and between families of migratory and sedentary birds (Fig. 8.2).

An important take home message from this discussion on genetic footprints in host immunity is to remember that there are multiple ways that hosts can cope with their parasites and screening genetic variation in a single host gene, like the MHC, will give a very limited picture of the total potential of the immune system in a particular host. Thus genome-wide association studies based on the next-generation sequencing (Liedvogel and Lundberg, Chapter 12) of non-model organisms offers promise for this field of research.

8.2.3 Tolerance instead of resistance to infection

An alternative strategy for a host to cope with pathogens could be to tolerate the parasites instead of evolving resistance against them (Råberg et al. 2007). Genetic variation for tolerance has been found in an experimental setup where different strains of mice suffered to different degrees (measured as weight loss and anaemia) from shared malaria parasites. Interestingly, resistance to infection and tolerance were negatively correlated in this study, suggesting that the different host genotypes had evolved two completely different mechanisms to combat malaria infections (Råberg et al. 2007). Given that the pathogen selection pressure is likely to differ for migratory and sedentary animals, it is possible that sedentary and migratory animals have evolved different strategies to combat infections. Resistance may for example be more common in migratory species since they often encounter a wider diversity of parasites and can spend shorter time with their parasites compared to sedentary species, factors which probably make migratory species less able to co-evolve with each of their parasites. Furthermore, only healthy individuals may have the ability to migrate (Altizer et al. 2011). Tolerance could potentially be more common in sedentary species because they encounter their parasites for longer time intervals and very intricate associations could evolve between the host and the parasite. However, these scenarios are just two outcomes of how resistance and tolerance may evolve in migratory and sedentary species, respectively. Another possibility is, for example, that tolerance is more common in migrants since mounting strong immune responses may be too costly. Investigating different mechanisms to overcome infections with respect to sedentary and migratory organisms could indeed be an interesting avenue for future research.

8.2.4 Adjustments of immune function to meet infection risk

Given that immune system activation is costly (e.g. Hasselquist and Nilsson 2012), and the costs and benefits of immune responses vary considerably in space and time, it is likely that a flexible immune system could be advantageous. It has been observed that activation of the immune system reduces work rates in both insects and birds (e.g. Ilmonen et al. 2000, Moret and Schmid-Hempel 2000, Råberg et al. 2000; reviewed in Hasselquist and Nilsson 2012), a finding with likely implications for animal movement. Based on these general findings, we predict that animals should modulate their investment in immune responses depending on their need to perform intense work versus risk of contracting parasites. One such situation is migration, where there is a potential trade-off between work rate and immune function (discussed earlier). We also expect immune system modulation when the risk of contracting infections varies over the annual cycle, and between regions (Hasselquist 2007). Seasonal modulation of immune function has been found in many animals (Nelson et al. 2002), and can be related to both movement patterns and risk of contracting an infection. For example, when short-distance migratory red-winged blackbirds, *Agelaius phoeniceus*, were kept in cages indoors for one year (under the normal light regime that they naturally experience over an annual cycle) they showed a day-length regulated pattern in their immune response. This was characterized by a low humoral immune response during winter, an increased response during early spring and then a peak in response in late spring and summer (Hasselquist et al. 1999, Hasselquist unpublished data). In sedentary house sparrows, different aspects of the immune system peaked in different seasons with most components of the immune system being highest during the breeding season in spring/summer (Pap et al. 2010). In the red knot, *Calidris canutus*, a long-distance migratory shorebird, different aspects of immune function co-varied and showed some consistent patterns over time (Buehler et al. 2010). For example, aspects of immunity assumed to be costly were only used during breeding and when accumulating fat for migration, and not during moult (Buehler et al. 2008). These studies on seasonal modulation of immune function show that immune function can be internally modulated over the annual cycle in both sedentary and migrating birds. These modulations may be related to a reduced pathogen and/or vector activity during colder parts of the year. In contrast, in some species of mice, immune functions are up-regulated in winter, possibly mediated by melatonin levels in the circulation, the so called 'winter immunoenhancement' or 'melatonin-induced immunoenhancement' hypothesis (Nelson and Demas 1996, Nelson et al. 2002, Hasselquist 2007). During the winter these mice are non-territorial and they move more and interact with more conspecifics than they do in the summer. Hence, they have up-regulated immune responses during a period when they are at an increased risk of contracting pathogens (Nelson and Demas 1996, Nelson et al. 2002). In red knots melatonin levels also changed over the annual cycle although it did not co-vary with immune function as in the mice (Buehler et al. 2009).

Investment in immune function in relation to the risk of pathogen exposure can also be predicted to apply on a spatial scale for long-distance migrating organisms. The 'tropical wintering–parasite exposure' hypothesis (Hasselquist 2007) predicts that long-distance migrants to tropical areas have adapted to the higher parasite pressure of these regions by retaining an enhanced immune function in the winter (n.b., not migrants wintering in relatively parasite-free marine habitats and deserts, Piersma 2003). This is in contrast to species that spend the winter in northern temperate regions that should down-regulate immune function as a consequence of lower parasite pressure at this time of the year. Another scenario is related to marine environments that are considered to be parasite-poor habitats. It has been predicted that birds wintering in such areas should invest less in immune function than birds wintering in freshwater environments (Piersma 1997, 2003; see section 8.3.2). A recent example is dunlins, *Calidris alpina*, wintering in saltwater vs. freshwater habitats, where there was evidence for stronger immune responses in the birds occurring in the freshwater habitat (Gutiérrez et al. 2013).

8.2.5 Spatial adaptations to local variation in pathogen distribution

Dispersal is an important mechanism to avoid inbreeding and competition with relatives and conspecifics. However, dispersal can be costly in terms of energy consumption, predation risk, time loss and a mismatch between locally-adapted genes and the new environment. The latter type of cost can be particularly relevant when it comes to the adaptation of the immune system of hosts to their local parasite community at the site of birth (see Møller and Szep 2011). A host immune system that has co-evolved with local pathogens can recognize epitopes of local pathogens (see genetic footprints discussion in section 8.2.1 and Loiseau et al. 2010). However, this recognition could also be achieved at the phenotypic level, for example as a maternal effect where the transfer of maternal antibodies specific to

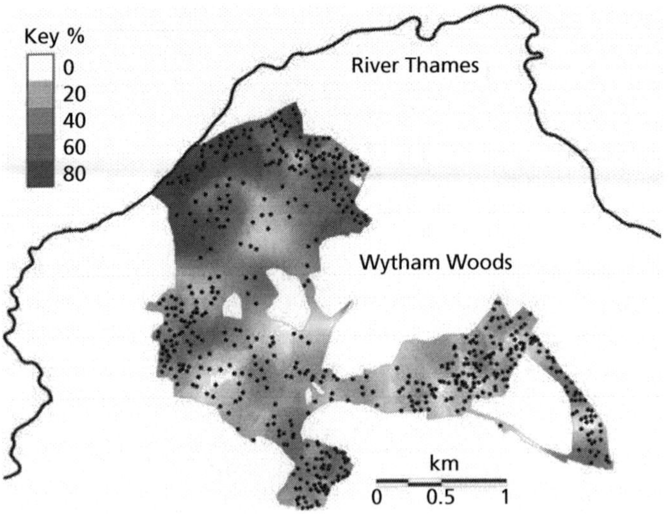

Figure 8.3 Large differences in the prevalence of avian malaria have been observed within a small spatial scale. Wood et al. (2007) examined spatial variation in the prevalence of malaria infection within a population of blue tits, *Cyanistes caeruleus*, inhabiting a woodland area in England (Wytham Woods). The figure depicts an interpolated prevalence map showing variation in malaria infection across the woodland by differential shading. Dots indicate nest box locations at which sampling was conducted. From this map it can be seen that malaria prevalence varied substantially across the woodland (more shaded areas indicates higher prevalence). Some specific associations were observed between landscape factors and malaria infection. For example, the prevalence of malaria infection decreased with increasing distance from the River Thames. Reproduced with permission of John Wiley & Sons.

the home range of the mother can potentially prime the immune system of the offspring to be better at controlling these local pathogens (Hasselquist and Nilsson 2009, Hasselquist et al. 2012). As pathogen faunas may differ at a local scale (Fig. 8.3), we hypothesize that dispersing individuals should preferably establish themselves in areas with a similar pathogen and vector fauna as in their natal area (see Wood et al. 2007). The knowledge of the spatial scale of parasite communities is scarce in most systems. However, recent work on blue tits, *Cyanistes caeruleus*, in Wytham Wood, UK has shown large differences in parasite composition within a very limited area (Wood et al. 2007, Knowles et al. 2011). Work on a closely related species, the great tit, *Parus major*, at the same location, has revealed patterns of genetic differentiation within this population at a surprisingly small spatial scale, which is thought to be a result of the short dispersal distances of these birds (Garant et al. 2005). Thus, in certain species and populations, even very small movements may be influenced by the interaction between hosts and parasites.

8.3 Movement strategies as a mean to avoid infections

Although animal movements that are driven by a need to track resources or avoid predation can increase an individual's risk of contracting infections, we may also expect that animal movements are sometimes initiated as a means of avoiding infections. Whether or not movement is an effective strategy to avoid infection will depend on a number of complex and often inter-related factors, such as the transmission mode of given pathogens and the existence of locally adapted genes in either the host or the pathogen. However, the decision to move in order to avoid pathogens will ultimately be determined by the trade-off between the benefits of movements to escape infection weighed against the costs of moving. For example, when a pathogen-rich area offers sufficient benefits, such as high food abundance, there may be selection on hosts to resist or tolerate infections instead of moving. A significant challenge in this field of research is determining whether movements that result in reduced infection risk are pathogen avoidance strategies or a response to other causal factors.

Moving to escape pathogens may entail relatively subtle movements, as in the example of soay sheep, *Ovis aries*, that choose to graze on patches of vegetation with lower parasite abundances, despite these patches have a lower nutritional value than nearby patches with a greater number of parasites (Hutchings et al. 2002). At the other end of the scale, extremely long-range movements have also been associated with pathogen avoidance; for example, migratory birds that breed and/or over-winter in sites with low parasite abundance (Piersma 1997,

Ramey et al. 2012). The role of pathogen avoidance in migratory behaviour has recently been highlighted as a neglected research area (Altizer et al. 2011, Møller and Szép 2011). Indeed, the hypothesis that 'parasitism begets migration' (Poulin et al. 2012) is yet to be robustly tested.

8.3.1 Mechanisms by which movement may reduce pressure from pathogens

The transmission mode of pathogens plays a pivotal role in whether or not moving on from a location can reduce the risk of infection. In the case of pathogens with infective stages persisting in the local environment, such as many helminths, ectoparasites and microbes with fecal-oral transmission routes, it would benefit a host to leave an infested habitat to avoid infection (Altizer et al. 2011). This is observed in reindeer, *Rangifer tarandus*, where a negative relationship exists between the distances they have migrated from their calving grounds and infection by warble fly, *Hypoderma tarandi*. This is thought to be a consequence of escaping the high density of warble fly larvae that accumulates in the soil during the calving period (Folstad et al. 1991). However, for pathogens which survive only short periods outside of their hosts, such as many bacteria and viruses, host movements between locations may not directly reduce the risk of infection because, in most cases, the parasites will be transported with their hosts (Guegan et al. 2008). However, if the new environment has a negative impact on transmission rates, infection risk from these types of pathogens may be reduced by moving. For example, if hosts move from an area of high to low population density there may be less contact between hosts, reducing the average risk of infection. Additionally, if the new environment alters the physiology of hosts in a manner that better equips them to fight infections, for example better available nutrition, then infection intensity could be reduced.

An indirect reduction in the risk of infection may result from moving to a new location if highly infected hosts, or those infected with particularly virulent pathogen strains, are unable to withstand the physiological costs of moving. By this pruning mechanism, the number of infected individuals will be lower at the new location. This is particularly relevant for long distance movements which pose a significant physiological challenge, such as migration (Altizer et al. 2011). An elegant example of such 'migratory culling' comes from monarch butterflies, *Danaus plexippus*, in which individuals infected with a common protozoan parasite, *Ophryocystis elektroscirrha*, have significantly reduced long-distance flight performance compared to uninfected butterflies (Fig. 8.4; Bradley and Altizer 2005). This has been proposed as one possible explanation for the lower prevalence of *O. elektroscirrha* observed in migratory populations as compared to non-migratory populations of monarch butterflies (Bradley and Altizer 2005).

Another way in which animals may escape infections through moving is if the new location provides less favourable environmental conditions for a given parasite. For instance, the pre-spawning migration of the anadromous fish Japanese grenadier anchovy, *Coilia nasus*, from the ocean to rivers is associated with a reduction in the prevalence of certain marine ectoparasitic helminths (Li et al. 2012). However, this is not the case for intestinal helminths of these fish, whose internal location appears to confer protection from the change in salinity (Li et al. 2012). It has also been suggested that reindeer move to higher altitudes, in part, to avoid harassment from insects which cannot withstand the cooler temperatures of high altitudes (Vistnes et al. 2008). The effectiveness of this type of escape from parasitism will be determined by the host having the upper-hand in the new environment, either by being pre-adapted to the new conditions or by possessing a superior ability to adapt as compared to the parasite.

8.3.2 The role of pathogen avoidance in the habitat-selection of migratory birds

Parasite avoidance may influence the habitat-selection patterns of migratory birds. For example, in a recent study of a large number of songbirds in Alaska it was reported that resident and juvenile birds from high latitudes were not infected with *Plasmodium* (Loiseau et al. 2012). Thus birds may benefit from breeding in the far north where their offspring are less likely to be infected. Similarly, tundra swans, *Cygnus columbianus*, that breed in

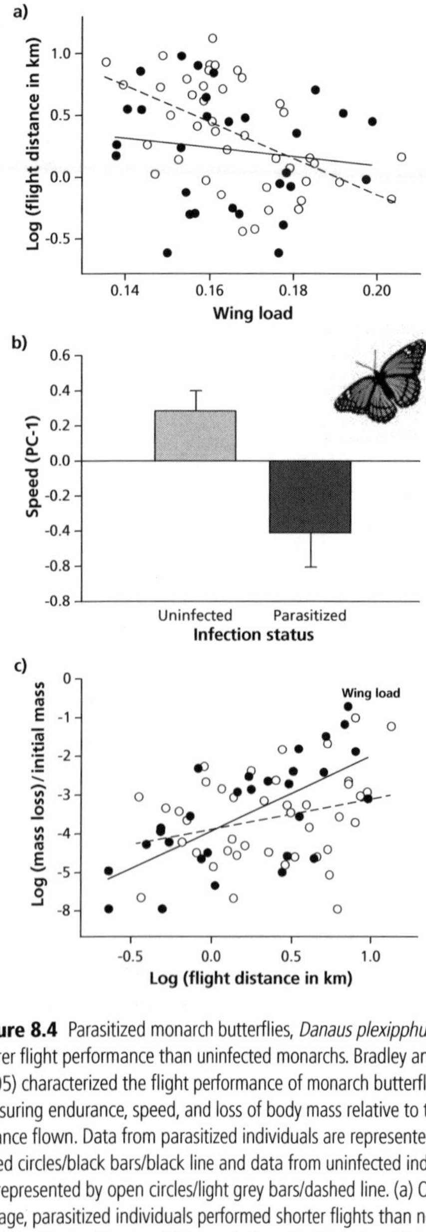

Figure 8.4 Parasitized monarch butterflies, *Danaus plexipphus*, have poorer flight performance than uninfected monarchs. Bradley and Altizer (2005) characterized the flight performance of monarch butterflies by measuring endurance, speed, and loss of body mass relative to total distance flown. Data from parasitized individuals are represented by closed circles/black bars/black line and data from uninfected individuals are represented by open circles/light grey bars/dashed line. (a) On average, parasitized individuals performed shorter flights than non-parasitized butterflies. In addition, uninfected individuals exhibited an inverse relationship between wing load and distance flown, whereas no such relationship was observed in parasitized butterflies. Thus, in parasitized monarchs a low wing load did not improve their ability to endure long flights. (b) Parasitized monarchs flew significantly slower than uninfected individuals. (c) Parasitized butterflies lost more weight per kilometre flown than unparasitized individuals. Taken together these results indicate that parasitized monarch butterflies have reduced flight performance. This study provided the first evidence for a mechanistic link between parasitism and reduced migratory ability, which may explain 'migratory culling' in this species (Bradley and Altizer 2005). Reproduced with permission of John Wiley & Sons.

the arctic coastal plain of Alaska had a low prevalence of blood parasites when compared to birds breeding at more southern sites (Ramey et al. 2012). Thus, birds may breed in the North not only because of the abundance of resources and longer days, but also because of the relative paucity of disease vectors and parasites (Piersma 1997). In addition, habitat selection could be evident in the hosts' choice of overwintering grounds, as available sites are likely to differ in the abundance of parasites. The prevalence of malaria parasites in great reed warblers, *Acrocephalus arundinaceus*, has been shown to differ depending on which region the birds underwent their winter moult in sub-Saharan Africa, as revealed by feather isotope analyses (Yohannes et al. 2008). This raises the possibility that birds avoid overwintering in particularly parasite-rich areas. Piersma (1997) observed that shorebirds breeding in the high Arctic also tend to overwinter in salt-water marshes along the coastlines of temperate and tropical areas, which are considered to harbour fewer parasites than freshwater environments. Piersma (1997) argued that the selection of both breeding and overwintering sites with low parasite pressure could enable birds to maintain a low investment in immunocompetence. This idea was tested in a comparative study of five wader species differing in migration habits as well as in risk of parasite exposure during breeding and overwintering (Mendes et al. 2006). However, there was no indication that the species inhabiting parasite-sparse habitats in summer and winter had lower immune responses than the species inhabiting presumably parasite-rich freshwater habitats in summer and winter (Mendes et al. 2006). Furthermore, large aggregations of wader species at a stopover site in seemingly parasite-poor marine environment such as Delaware Bay, USA, have been shown to be ecological 'hotspots' for influenza virus in waders (Krauss et al. 2010).

Another intriguing possibility for the role of pathogen avoidance in the habitat-selection of migratory birds was proposed by Møller and Szép (2011). They argued that migratory connectivity, whereby populations perform seasonal movements between particular breeding and non-breeding grounds (Webster et al. 2002), may be driven in part by host–parasite interactions. This concept centres on the idea that hosts become locally adapted to the parasite faunas at their breeding and wintering grounds, putting individuals that stray from these locations at a selective disadvantage. This would maintain site fidelity at both wintering and breeding sites, resulting in migratory connectivity. Overall, it is clear that there are cases in which animal movement can lead to a reduction in exposure to parasites. Further work is required to determine when parasite avoidance is a causal factor. This is particularly the case for migration which may be influenced by multiple factors and tends to be a 'hardwired' behaviour often observed in all individuals of a given population. However, partial migrant species (Chapman et al., Chapter 2) can offer a good opportunity to investigate the role of parasites in selecting for migration. A great deal more research is required to better understand the scenarios in which parasite avoidance and adaptation to specific parasites influence different types of animal movement.

8.4 Spreading of pathogens through movement

Moving hosts are potential vehicles for the pathogens they harbour. So far the most appreciated effect is the possibility for pathogens to be transported between distant geographical areas with moving hosts, although this effect can also be relevant at the scale of local movements. The interest in pathogen transport is partly due to the large effect that novel pathogens - if established in new, unprotected host species or populations - can have on food webs and even on entire ecosystems. A drastic example is the spread of smallpox by the early European colonizers into the Americas, which is believed by some researchers to have reduced the indigenous populations by as much as 90% in just a few hundred years (Mann 2005). Other examples include the introduction of avian malaria parasites to Hawaii and the resulting severe decline and even extinction of resident bird species (van Riper et al. 1986, Beadell et al. 2006), and the recent introduction of West Nile virus into North America with severe consequences for Corvidae populations in particular (LaDeau et al. 2007). Although these examples refer to the human-induced spread of pathogens, they highlight the effects that might emerge if wild migrating animals transport and introduce new pathogens into areas with naive hosts.

Rapid changes of species' ranges in animals are likely to result in plenty of opportunities for their pathogens to switch into naive hosts without any adaptive change, a process known as 'ecological fitting' (Brooks and Hoberg 2007). Ecological fitting of a pathogen requires that (1) it can survive the biotic conditions in the new area, (2) that there are suitable vectors for pathogen transmission, and (3) that the pathogen can be efficiently transmitted among the new potential hosts. If novel pathogens are transported into new areas providing a suitable ecological fitting, it is likely that the pathogens could have large impacts (high virulence) on its novel hosts. However, every host species is a unique environment for a pathogen, and will therefore often require precise adaptations in the pathogen for successful multiplication and transmission. The efficiency with which a pathogen can utilize a host depends primarily on how well it can invade the host cells and its ability to evade the host's immune system. It is therefore not surprising that the vast majority of parasites are host specialists, such as among avian malaria parasites (Beadell et al. 2004, Perez-Tris et al. 2007, Bensch et al. 2009, Hellgren et al. 2009). More surprising is the existence of some remarkable host generalists, such as the two mitochondrial cytochrome *b* lineages of the morphospecies *Plasmodium relictum*, GRW4 and SGS1 (Bensch et al. 2009). The mitochondrial lineage GRW4 is found in 53 hosts across 17 bird families and SGS1 in 58 hosts across 19 bird families (Fig. 8.5). How some parasites can manage to invade a large range of taxonomically distantly related host species, given that it requires some kind of a universal key into various immune systems, remains to be understood. Perhaps unsurprisingly, it is the most generalist parasites (GRW4) that successfully colonized remote islands like New Zealand, Hawaii, and the Seychelles islands (Ewen et al. 2012).

8.4.1 Emerging diseases in wild animals

The rapid alterations of species' ranges caused by global change and human assisted introduction of

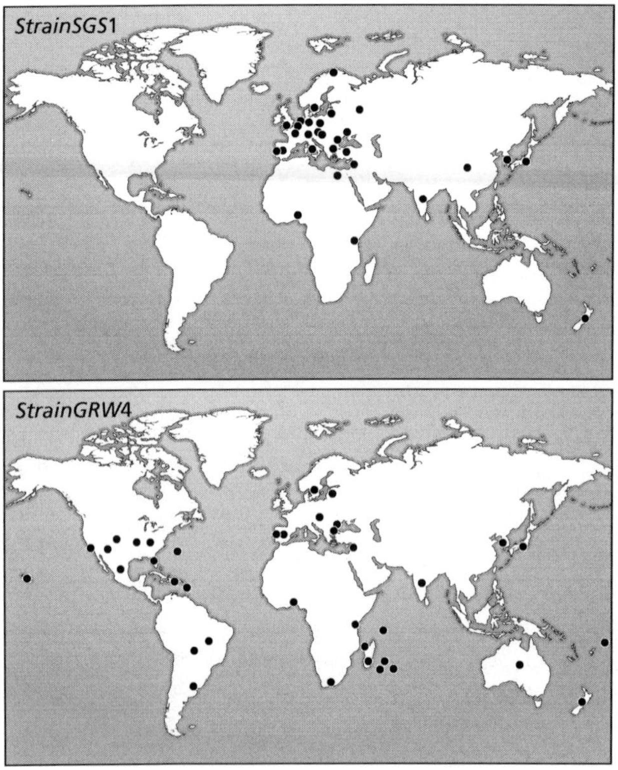

Figure 8.5 Most avian malaria parasites are host specialists, although there are some extraordinary host generalists like the two mitochondrial cytochrome *b* lineages of the morphospecies *Plasmodium relictum*, GRW4 and SGS1 (Bensch et al. 2009). The mitochondrial lineage GRW4 is found in 53 hosts across 17 bird families and SGS1 in 58 hosts across 19 bird families. The worldwide distribution of GRW4 and SGS1, respectively, are indicated with dots.

species to new continents and remote islands are likely to result in plenty of opportunities for pathogens to switch into naive hosts. These anthropogenic increases of novel host–parasite interactions might be further amplified by the natural long-distance migrations of birds or other animals. A severe constraint when predicting potential disease outbreaks and host consequences is our restricted knowledge of the natural occurrence of most pathogens. In fact, the vast majority of pathogens of wild animals have either not been scientifically described or are only known from a fraction of the hosts which they infect. Because birds and their malaria parasites stand out as one of the better known host-parasite systems, we will take this system as an example to outline general principles on how long-distance migration in combination with global change may affect naïve communities of hosts with novel pathogens.

Each year, migratory birds carry numerous species of tropical avian malaria parasites to breeding grounds at northern latitudes (Valkiūnas 2005). Although many of these parasites are only actively transmitted in the tropics, they are present in the blood of the migrant birds during the summer months and can thus potentially be spread to resident species. However, such shifts of transmission area appear to be rare (Hellgren et al. 2007) and may depend on a lack of competent vectors or, if competent vectors are present, that low ambient summer temperatures prevent parasite development somewhere on the way from the ingested gametocytes in the mosquito's gut to infective sporozoites in their salivary glands (LaPointe et al. 2010). With increasing summer temperatures, as a result of global warming, many new tropical parasites might overcome these transmission barriers. Indeed, there is evidence of that the prevalence of bird malaria is increasing in Europe as a consequence of climate warming during the past 20 years (Garamszegi 2011).

It has recently been demonstrated that tropical malaria parasites have the potential to severely affect resident bird species in northern Europe. Siskins *Carduelis spinus*, and crossbills, *Loxia curvirostra* that were inoculated with the parasite, *Plasmodium ashfordi* (transmitted within Africa), developed lethal infection intensities (Palinauskas et al. 2011). Whether this will also happen in the wild depends on the presence of competent vector species and amenable temperature conditions for *P. ashfordi* to develop at northern latitudes. Many tropical parasites are already knocking on the door, of which malaria is probably just a small subset, and it is likely that resident bird populations at northern latitudes are at risk of becoming negatively affected by parasites brought in by migratory birds from the tropics. It has been difficult to confirm that migratory birds contribute to the spread of malaria, but one recent study verified that migratory birds in Europe have indeed contributed to the spread of another protozoal pathogen, *Trichomonas gallinae*, which causes Finch trichomonosis (Lawson et al. 2012). This pathogen, which may have spread from pigeons to songbirds, moves with migratory birds and has resulted in a significant decline in greenfinches, *Carduelis chloris*, in Great Britain (Lawson et al. 2012).

Migrating animals may also contribute to the spread of pathogenic fungi. In the past 20 years it has been recognized that pathogenic fungi have resulted in major losses of biodiversity (Fisher et al. 2012). One of the most devastating of these fungi is *Batrachochytrium dendrobatidis* (chytrid fungus) that globally has caused the decline or extinction in 200 species of frogs (Skerratt et al. 2007). As this pathogen is only known to infect amphibians, which have restricted dispersal, the rapid global spread of this pathogen is puzzling. However, since waterfowl of the family Anatidae may transport the fungi over larger distances (Garmyn et al. 2012) it is possible that such passive pathogen transportation has played a role in the spread of *B. dendrobatidis*. Although this scenario has not been firmly established, it emphasizes that long distance dispersal of pathogens, or other micro-organisms, can have dramatic effects on food webs and even entire ecosystems even when the dispersing species, in this example the waterfowl, is not affected (see further discussion in section 8.5.2). Other threatening fungi that appear to be spreading worldwide are *Geomyces destructans*, which causes white-nosed syndrome of bats, *Fusarium solani*, which infects sea turtles and *Nosema*, which may contribute to the colony collapse of bees, *Bombus* (Fisher et al. 2012). These fungal diseases appear to have expanded their distributions due to human and/or host movements. We can be certain that the movement of animals facilitates the spread of diseases and that this has occurred for millions of years. However, now with human-induced changes in land use and the introduction of invasive species, disease will likely become a significant factor in the extinction of species (Wilcove et al. 1998). Conservation policy makers must recognize the importance of studying animal movements in the context of spreading parasites and particularly pathogens that pose a potential hazard to human health (so called zoonotic diseases, see Box 8.1).

8.5 Micro-organism strategies to take advantage of moving hosts

8.5.1 Parasites that influence and 'master-mind' the movement of their hosts

Some parasites have evolved the ability to exploit and even manipulate the movements of their hosts to favour their own survival. These parasites are sometimes capable of master-minding their host's movement (Table 8.1). For instance, baculoviruses master-mind their caterpillar hosts by inducing a climbing behaviour in the caterpillars to ensure that infected individuals die and liquefy at the top of vegetation (Goulson 1997). This movement of the caterpillars maximizes the spread of infective virus particles when the cadavers are exposed to rainfall. Malaria parasites are known to manipulate the feeding rates of mosquitoes, as mosquitoes act as vectors to transport malaria to a suitable host, and this behavioural manipulation is believed to increase transmission rates (Koella et al. 1998). Rats infected by toxoplasma parasites lose their normal fear of moving into open habitats which improves the chances of toxoplasma reaching its final host—the cat (Berdoy et al. 2000). Killifish infected by a certain trematode (a fluke) alter their swimming behaviour in a manner that makes them more vulnerable to predation by birds, which are the final hosts of this parasite (Lafferty and Morris 1996). Yet another example is when parasites manipulate the dispersal

Box 8.1 Zoonotic diseases

Zoonotic diseases with potentially severe consequences for human health include for example avian influenza (Olsen et al. 2006, Hill et al. 2012), West Nile virus (LaDeau et al. 2007), and Borrelia (Gylfe et al. 2000, Mathers et al. 2011). Although there are many examples of zoonotic diseases, these three cases involve pathogens that can be spread by birds, which is noteworthy as the spread of zoonotic disease may be facilitated by the high potential for rapid, long-range movements of wild bird hosts (Georgopoulou and Tsiouris 2008, Møller and Szep 2011, Fuller et al. 2012). Similarly, it is recognized that bats, which also fly long distances (Hedenström and Lindström, Chapter 5), can be reservoirs of human pathogens, such as rabies and related Nipah, Hendra, and Lyssa viruses, as well as SARS-CoV-like viruses (Calisher et al. 2006).

When considering the spread of zoonotic diseases with migratory animals, the spread of the pathogen will strongly depend on how the pathogen is transmitted. For example, many viruses are spread directly, such as in influenza, while other zoonotic pathogens require an arthropod vector for transmission. Arboviruses, such as West Nile virus, Eastern Equine Encephalitis virus, and Yellow Fever, require mosquitoes for transmission, and thus, if compatible vectors are not found in habitats where the pathogen is introduced, there is little risk of transmission. For this reason, some pathogens are not found in certain habitats; sleeping sickness, spread by the parasite *Trypanosoma brucei*, is only found in Africa, most likely because its vector, the tsetse fly, is only found on this continent.

Although it is highly possible that birds and other migrating animals spread pathogens, there has been little empirical evidence that migratory patterns contribute to major outbreaks of zoonotic disease. One such potential case, however, is the rapid spread of the recently introduced West Nile virus across the USA, where wild birds seem to have been the main hosts and also transporters of the virus (LaDeau et al. 2007). Another example is the highly pathogenic avian influenza, H5N1, where there is evidence that both migratory birds, and the human movement of domestic chickens, *Gallus gallus*, contributed to the spread of the virus (Kilpatrick et al. 2006, Hill et al. 2012).

capacity of their hosts, like the bacterium *Holospora undulata* that affects the phenotypic plasticity of its host, a ciliate, *Paramecium caudatum*, and thereby reduces the capacity of host dispersal (Fellous et al. 2011; Hollander et al., Chapter 7). Moreover, parasites on dinoflagellates (chain-forming flagellate unicellular organisms) can break the swimming dynamics of the dinoflagellate chain and thereby reduce the ability of their host to move, which in turn favours reinfection (Fraga et al. 1989, Kim et al. 2004).

8.5.2 Passive dispersal of micro-organisms

Micro-organisms can either disperse actively or passively. Recent studies have shown that passive dispersal is often underestimated as micro-organisms (including viruses, bacteria, fungi, micro eukaryotes and some larval stages of macro-organisms; Table 8.1) have adopted many different strategies to disperse (Box 8.2). Birds, amphibians (Bohonak and Whiteman 1999), zooplankton (van de Meutter et al. 2008), fishes (Pollux 2011), and mammals (Waterkeyn et al. 2010a), including humans (e.g. Gollash et al. 2007, Waterkeyn et al. 2010b, Perrigo et al. 2012), are all known vectors for the passive dispersal of micro-organisms. Micro-organisms can either be carried in the digestive apparatus of the vector after ingestion, or directly or indirectly (e.g. soil) attached to e.g., fur, skin, nostrils, human boots and vehicles (Table 8.1), as parasites or propagules (i.e. living material such as spores, seeds, resting stages, and vegetative cells that are used for the purpose of dispersal).

Birds in particular can act as vectors during both local movement and long-distance migration. Only a minute fraction of resilient propagules (see Table 8.1 and mini-review in van Leeuwen et al. 2012) have been recorded being dispersed via migratory flyways since most propagules are released just a few hours after ingestion (Brochet et al. 2010). However, the maximum retention time monitored in water birds, such as teals *Anas crecca*, is 60 h (Pollux et al. 2005 in Brochet et al. 2010). It has been estimated that, during these retention times, a waterbird can release about five viable propagules over a 100- to 300-km flying distance (van Leeuwen et al. 2012). Certain studies therefore suggested that waterfowl influence the patchiness of micro-organism distribution when dispersing them from pond-to-pond (Coleman et al. 1996, Taylor et al. 1998, Figuerola and Green 2002, Brochet et al. 2009b, Viana et al. 2012).

Box 8.2 Dispersal of micro-organisms

Dispersal is a natural phenomenon that involves active or passive movement of an organism from an initial habitat to another. For micro-organisms, dispersal can be triggered or induced, and if triggered, the distance travelled will depend on the organism size and its capacity for dispersal. Active movement can be produced using one or more flagella (e.g. dinoflagellates, cryptophytes), cilia (e.g. ciliates), polysaccharide exudations (e.g. pennate diatoms), or appendices (e.g. copepods). The movement of these small organisms is active at local scales, i.e. a few micrometres to metres (e.g. diurnal migration) and passive over large distances. Long-distance dispersal is usually related to abiotic factors. Water and wind currents can separate and disperse plankton communities (Martiny et al. 2006), pollen (Howe and Smallwood 1982), aquatic/terrestrial micro-organisms (Schlichting 1969), and even some macro-organisms (e.g. Szymkowiak et al. 2007). For example, bubbling can facilitate the dispersal of microbes (Schlichting 1974) and dry sand and dust can facilitate dispersal of protists attached to the particles when the wind blows (Broady and Smith 1994).

Micro-organisms have developed different strategies for being passively dispersed and one of them is the use of another organism as a vehicle for dispersal (e.g. an animal or a protist (Nathan 2006)). This micro-organism–vehicle interaction can be commensal or parasitic. A commensal relationship is a 'peaceful' relationship between two organisms with only one beneficial organism, whereas a parasitic relationship is a relationship with unidirectional benefits. In commensal relationships, the micro-organism can be found attached to the external body of the host: in the fur (Sorensen 1986), on the feet and the exoskeletons of, e.g., ducks or insects (Garmyn et al. 2012, Skovgaard and Saiz 2006), or carried intracellularly, such in the nucleus or the cellular cytoplasm (Kim et al. 2004, Chambouvet et al. 2008, Schärer et al. 2007), or in the gut or oesophagus wall (Duneau et al. 2011, Brochet et al. 2009b, c, Figuerola and Green 2002, Grossart et al. 2010, Lam et al. 2012). In parasitic relationships, the micro-organisms are most often found inside the host, such as in the blood circulation (Bensch et al. 2000, Palinauskas et al. 2011) (Table 8.1).

The metapopulation genetics of micro-organisms is known to be influenced by the migratory strategy of their vector (e.g. a bird; Viana et al. 2012), the landscape, and the kind of propagule carried. The morphology of the propagule (e.g. shape, size, cell wall composition) affects the capacity for viability after transportation/ingestion (Hamm et al. 2003). Van Leeuwen et al. (2012) estimated that a '1-mm particle has twice as much chance of being dispersed after ingestion than a particle 10 times larger'. Viable propagules may influence the population structure and community diversity of local and large-scale ecosystems located up to several hundred kilometres from the initial site (see Brochet et al. 2009a, c; van Leeuwen et al. 2012).

8.6 Future perspectives

Host movements can play a fundamental role in the risk of contracting an infection from a given pathogen, and once infected, whether or not the pathogen will be transported to a new area. In this chapter, we have considered both the host and parasite perspectives on infections, and reviewed the consequences of animal movement for the spread of both pathogenic and non-pathogenic micro-organisms. The general latitudinal gradient of increasing parasite diversity towards the tropics means that many animal species moving long distances along this gradient towards the equator are exposed to a greater variety of pathogens than sedentary individuals. On a smaller scale, animals that move into new areas may also encounter novel pathogens. This could either put the host at a disadvantage, if they are not equipped to cope with new pathogens, or at an advantage if novel pathogens are unable to infect the host. When animal movement increases their exposure to parasites, adaptations may develop in the host in response to a more diverse parasite pressure. Such adaptations could leave genetic footprints in the immunity genes of the host as well as in the genes that are important for being able to infect the host in the parasite.

The risk of contracting infections varies not only between geographical regions but also over the annual cycle. For example, long-distance migration is physiologically demanding, which potentially results in a trade-off between work rate and immune function. Therefore, one aspect of the 'cost of migration' could be reduced immunocompetence. Animals may even move to escape pathogens and there is the intriguing possibility that pathogen-avoidance is one of the driving forces behind animal migrations, as well as the phenomenon of migratory connectivity (i.e. the high connectivity between breeding and wintering sites in many birds).

Rapid changes of species' ranges caused by global change are likely to result in plenty of opportunities for pathogens to invade naive hosts. However, each year migratory birds carry numerous species of parasites, for example, tropical avian malaria, to their breeding grounds at northern latitudes yet these parasites are rarely spread to resident species. This could be attributable to a lack of competent vectors to spread the infection or low ambient summer temperatures preventing parasite development. Both of these restrictions have the potential to alter with increasing global temperature leading to the emergence of disease.

There are extreme examples in nature of parasites that manipulate the movement of their host to maximize their own transmission. In this way these parasites effectively master-mind their host's behaviour to their advantage. At the other end of the scale, there are non-pathogenic micro-organisms which simply use animals as vectors for their dispersal without causing any observable harm to them.

There are numerous future avenues for research on hosts and parasites in relation to movement. For example, there is an urgent need to investigate the diversity of immune genes and seasonal adjustments of immune function in migrant and non-migrant hosts and to investigate if there are relationships with the risk of contracting infections. In addition, more experiments measuring the effect of movements and migration on host immunity are required to elucidate the potential costs of migration. When it comes to host range expansions and invasiveness, it is important to understand more about interactions between hosts and the novel pathogens they encounter in new regions. For example, what influences the likelihood of the host avoiding or becoming infected by these novel parasites? In terms of the spread of parasites, it is central to understand more about what determines a parasite's ability to become established in novel areas. For example, what effects will an increasing global temperature have? It is also important to study moving animals as passive dispersers of microrganism propagules, which may influence the dynamics of the ecosystem in which the propagule is deposited. Finally, we do not know which micro-organisms will become severe pathogens in the future, or which pathogens may become new zoonoses affecting humans. Therefore, it may be particularly crucial to monitor pathogens of moving animals that can act as efficient long-distance transporters of potential pathogens, to learn more about this risk and possibly to forecast the emergence of new zoonotic diseases. The study of animal movements in relation to hosts and parasites is to a large extent in its infancy, and we believe that this field of research will grow and mature over the next decade.

References

Alerstam, T., Hedenström, A., and Åkesson, S. (2003). Long-distance migration: evolution and determinants. *Oikos*, **103**, 247–60.

Alexander, R.McN. (1998). When is migration worthwhile for animals that walk, swim or fly? *Journal of Avian Biology*, **29**, 387–94.

Allison, A. C. (1954). Protection afforded by sicklecell trait against subtertian malarial infection. *British Medical Journal*, **1**, 290–4.

Altizer, S., Bartel, R., and Han, B.A. (2011). Animal migration and infectious disease risk. *Science*, **331**, 296–302.

Apanius, V., Penn, D., Slev, P.R., Ruff, L.R. and Potts, W.K. (1997). The nature of selection on the major histocompatibility complex. *Critical Reviews in Immunology*, **17**, 179–224.

Bates, S., Gaudet, J., Kaczmarska, I., and Ehrman, J. M. (2004). Interaction between bacteria and the domoic-acid-producing diatom *Pseudo-nitzschia multiseries* (Hasle) Hasle; can bacteria produce domoic acid autonomously? *Harmful Algae*, **3**, 11–20.

Beadell, J. S. E., Gering, J., Austin, J. P., et al. (2004). Prevalence and differential host-specificity of two avian blood parasite genera in the Australo-Papuan region. *Molecular Ecology*, **13**, 3829–44.

Beadell, J. S., Ishtiaq, F., Covas, R., et al. (2006). Global phylogeographic limits of Hawaii's avian malaria. *Proceedings of the Royal Society B: Biological Sciences*, **273**, 2935–44.

Bensch, S., and Åkesson, S. (2003). Temporal and spatial variation of hematozoans in Scandinavian willow warblers. *Journal of Parasitology*, **89**, 388–91.

Bensch, S., Hellgren, O., and Pérez-Tris, J. (2009). MalAvi: a public database of malaria parasites and related haemosporidians in avian hosts based on mitochondrial cytochrome *b* lineages. *Molecular Ecology Resources*, **9**, 1353–8.

Bensch, S., Stjernman, M., Hasselquist, D., et al. (2000). Host specificity in avian blood parasites: a study of *Plasmodium* and *Haemoproteus* mitochondrial DNA amplified from birds. *Proceedings of the Royal Society B: Biological Sciences*, **267**, 1583–9.

Bensch, S., Waldenstrom, J., Jonzen, N. et al. (2007). Temporal dynamics and diversity of avian malaria parasites in a single host species. *Journal of Animal Ecology*, **76**, 112–122.

Berdoy, M., Webster, J. P., and MacDonald, D. W. (2000). Fatal attraction in *Toxoplasma*-infected rats: a case of parasite manipulation of its mammalian host. *Proceedings of the Royal Society B: Biological Sciences*, **267**, 1591–4.

Blossey, B., and Notzold, R. (1995). Evolution of increased competitive ability in invasive nonindigenous plants: a hypothesis. *Journal of Ecology*, **83**, 887–9.

Bodmer, W. F. (1972). Evolutionary significance of the HLA system. *Nature*, **237**, 139–45.

Bohonak, A. J., and Whiteman, H. H. (1999). Dispersal of the fairy shrimp *Branchinecta coloradensis* (Anostraca): effects of hydroperiod and salamanders. *Limnology and Oceanography*, **44**, 487–93.

Bonneaud, C., Perez-Tris, J., Federici, P., Chastel, O., and Sorci, G. (2006). Major histocompatibility alleles associated with local resistance to malaria in a passerine. *Evolution*, **60**, 383–9.

Bradley, C. A., and Altizer, S. (2005). Parasites hinder monarch butterfly flight: implications for disease spread in migratory hosts. *Ecology Letters*, **8**, 290–300.

Bratbak, G., Egge, J. K., and Heldal, M. (1993). Viral mortality of the marine alga *Emiliania huxleyi* (Haptophyceae) and termination of algal blooms. *Marine Ecology Progress Series*, **93**, 39–48.

Broady, P. A., and Smith, R. A. (1994). A preliminary investigation of the diversity, survivability and dispersal of algae introduced into Antarctica by Human activity. *Proceedings of the NIPR Symposium on Polar Biology*, **7**, 185–97.

Brochet, A. L., Gauthier-Clerc, M., Guillemain, M., et al. (2009b). Field evidence of dispersal of branchiopods, ostracods and bryozoans by teal (*Anas crecca*) in the Camargue (southern France). *Hydrobiologia*, **637**, 255–61.

Brochet, A. L., Guillemain, M., Fritz, H., Gauthier-Clerc, M., and Green, A. J. (2009c). The role of migratory ducks in the long-distance dispersal of native plants and the spread of exotic plants in Europe. *Ecography*, **32**, 919–28.

Brochet, A. L., Guillemain, M., Gauthier-Clerc, M., Fritz, H., and Green, A. J. (2010). Endozoochory of Mediterranean aquatic plant seeds by teal after a period of desiccation: determinants of seed survival and influence of retention time on germinability and viability. *Aquatic Botany*, **93**, 99–106.

Brochet, A. L., Guillemain, M., Lebarbenchon, C., et al. (2009a). The potential distance of highly pathogenic avian influenza virus dispersal by mallard, common teal and Eurasian pochard. *EcoHealth*, **6**, 449–57.

Brooks, D. R., and Hoberg, E. P. (2007). How will global climate change affect parasite-host assemblages? *Trends in Parasitology*, **23**, 571–4.

Brussaard, C. P. D. (2004). Viral control of phytoplankton population—a review. *Journal of Eukaryotic Microbiology*, **51**, 125–38.

Buehler, D. M., Koolhaas, A., Van't Hof, T. J., et al. (2009). No evidence for melatonin-linked immunoenhancement over the annual cycle of an avian species. *Journal of Comparative Physiology A: Neuroethology, Sensory, Neural, and Behavioral Physiology*, **195**, 445–51.

Buehler, D. M., Piersma, T., Matson, K., and Tieleman, B. I. (2008). Seasonal redistribution of immune function in a shorebird: annual cycle effects override adjustments to thermal regimes. *American Naturalist*, **172**, 783–96.

Buehler, D. M., Tieleman, B. I., and Piersma, T. (2010). How do migratory species stay healthy over the annual cycle? A conceptual model for immune function and for resistance to disease. *Integrative and Comparative Biology*, **50**, 346–57.

Buhler, S., and Sanchez-Mazas, A. (2011). HLA DNA sequence variation among human populations: molecular signatures of demographic and selective events. *PLoS One*, **6**, e14643.

Calisher, C. H., Childs, J. E., Field, H. E., Holmes, K. V., and Schountz, T. (2006). Bats: important reservoir hosts of emerging viruses. *Clinical Microbiology Reviews*, **19**, 531–45.

Callaway, R. M., and Ridenour, W. M. (2004). Novel weapons: invasive success and the evolution of increased competitive ability. *Frontiers in Ecology and the Environment*, **2**, 436–43.

Chambouvet, A., Morin, P., Marie, D., and Guillou, L. (2008). Control of toxic marine dinoflagellate blooms by serial parasitic killers. *Science*, **322**, 1254–57.

Clay, K. (2003). Conservation biology: parasites lost. *Nature*, **421**, 585–6.

Coats, D. W. (1999). Parasitic life styles of marine dinoflagellates. *Journal of Eukaryotic Microbiology*, **46**, 402–9.

Coleman, A. W. (1996). Are the impacts of events in the earth's history discernable in the current distributions of freshwater algae? *Hydrobiologia*, **336**, 137–42.

Cote, J., Clobert, J., Brodin, T., Fogarty, S., and Sih, A. (2010). Personality-dependent dispersal: characterization, ontogeny and consequences for spatially structured populations. *Philosophical Transactions of the Royal Society B: Biological Sciences*, **365**, 4065–76.

Doherty, P. C., and Zinkernagel, R. M. (1975). Enhanced immunological surveillance in mice heterozygous at the H-2 gene complex. *Nature*, **256**, 50–2.

Duckworth, R. A., and Badyaev, A. V. (2007). Coupling of dispersal and aggression facilitates the rapid range expansion of a passerine bird. *Proceedings of the National Academy of Sciences USA*, **104**, 15017–22.

Duneau, D., Luijckx, P., Ben-Ami, F., Laforsch, C., and Ebert, D. (2011). Resolving the infection process reveals striking differences in the contribution of environment, genetics and phylogeny to host-parasite interactions. *BMC Biology*, **9**, 11.

Ewen, J. G., Bensch, S., Blackburn, T. M., et al. (2012). Establishment of exotic parasites: the origins and characteristics of an avian malaria community in an isolated island avifauna. *Ecology Letters*, **15**, 1112–19.

Ewenson, E., Zann, R., and Flannery, G. (2003). PHA immune response assay in captive zebra finches is modulated by activity prior to testing. *Animal Behaviour*, **66**, 797–800.

Fellous, S., Quillery, E., Duncan, A. B., and Kaltz, O. (2011). Parasitic infection reduces dispersal of ciliate host. *Biology Letters*, **7**, 327–9.

Figuerola, J., and Green, A. J. (2002). Dispersal of aquatic organisms by waterbirds: a review of past research and priorities for future studies. *Freshwater Biology*, **47**,483–94.

Finlay, B. J. (2002). Global dispersal of free-living microbial eukaryote species. *Science*, **296**, 1061–3.

Fisher, M. C., Henk, D. A., Briggs, C. J., et al. (2012). Emerging fungal threats to animal, plant and ecosystem health. *Nature*, **484**, 186–94.

Folstad, I., Nilssen, A. C., Halvorsen, O., and Andersen, J. (1991). Parasite avoidance: the cause of post-calving migrations in *Rangifer*? *Canadian Journal of Zoology*, **69**, 2423–9.

Fraga, S., Gallagher, S. M., and Anderson, D. M. (1989). Chain-forming dinoflagellates: an adaptation to red tides. In T. Okaichi, D. M. Anderson, and T. Nemoto, (eds), *Red Tides: Biology, Environmental Science and Toxicology*, pp. 281–4. Elsevier, New York.

Fuller, T., Bensch, S., Muller, I., et al. (2012). The ecology of emerging infectious diseases in migratory birds: an assessment of the role of climate change and priorities for future research. *Ecohealth*, **9**, 80–8.

Garamszegi, L. Z. (2011). Climate change increases the risk of malaria in birds. *Global Change Biology*, **17**, 1751–9.

Garant, D., Kruuk, L. E. B., Wilkin, T. A., McCleery, R. H., and Sheldon, B. (2005). Evolution driven by differential dispersal within wild bird population. *Nature*, **433**, 60–5.

Garmyn, A., van Rooij, P., Pasmans, F., et al. (2012). Waterfowl: potential environmental reservoirs of the chytrid fungus *Batrachochytrium dendrobatidis*. *PLoS One*, **7**, e35038.

Georgopoulou, I., and Tsiouris, V. (2008). The potential role of migratory birds in the transmission of zoonoses. *Veterinaria Italiana*, **44**, 671–7.

Gilbert, S. C., Plebanski, M., Gupta, S., et al. (1998). Association of malaria parasite population structure, HLA, and immunological antagonism. *Science*, **279**, 1173–7.

Gill, R. E. Jr., Tibbitts, T. L., Douglas, D. C., et al. (2009). Extreme indurance flights by landbirds crossing the Pacific Ocean: ecological corridor rather than barrier? *Proceedings of the Royal Society B: Biological Sciences*, **276**, 447–57.

Gollash, S. (2007). Is ballast water a major dispersal mechanism for marine organisms? In W. Nentwig (ed.), *Biological Invasions*. Springer-Verlag, Berlin.

Goulson, D. (1997). Wipfelkrankheit: modification of host behaviour during baculoviral infection. Oecologia, **109** (2), 219–28.

Grossart, H. P., Dziallas, C., Leunert, F., and Tang, K. W. (2010). Bacteria dispersal by hitchhiking on zooplankton. *Proceedings of the National Academy of Sciences USA*, **107**, 11959–64.

Guegan, J.-F., Prugnolle, F., and Thomas, F. (2008). Global spatial patterns of infectious diseases and human evolution. In S. C. Stearns and J. C. Koella (eds), *Evolution in Health and Disease* (2nd edn), pp. 19–29. Oxford University Press, Oxford.

Guernier, V., Hochberg, M. E., and Guégan, J.-F. (2004). Ecology drives the worldwide distribution of human diseases. *PLoS Biology*, **2**, 740–6.

Gutiérrez, J. S., Abad-Gómez, J. M., Villegas, A., Sánchez-Guzmán, J. M., and Masero, J. A. (2013). Effects of salinity on the immune response of an 'osmotic generalist' bird. *Oecologia*, **171**, 61–9.

Gylfe, A., Bergstrom, S., Lundstrom, J., and Olsen, B. (2000). Reactivation of *Borrelia* infection in birds. *Nature*, **403**, 724–5.

Hamblin, M. T., and Di Rienzo, A. (2000). Detection of the signature of natural selection in humans: evidence from the Duffy blood group locus. *American Journal of Human Genetics*, **66**, 1669–79.

Hamm, C. E., Merkel, R., Springer, O., et al. (2003). Architecture and material proprieties of diatom shells provide effective mechanical protection. *Nature*, **421**, 841–3.

Hasselquist, D. (2007). Comparative immunoecology in birds: hypotheses and tests. *Journal of Ornithology*, **148**, S571–82.

Hasselquist, D., Lindström, Å., Jenni-Eiermann, S., Koolhaas, A., and Piersma, T. (2007). Long flights do not influence the immune responses of a long-distance migrant bird: a wind tunnel experiment. *Journal of Experimental Biology*, **210**, 1123–1131.

Hasselquist, D., Marsh, J. A., Sherman, P. W., and Wingfield, J. C. (1999). Is avian humoral immunocompetence suppressed by testosterone? *Behavioral Ecology and Sociobiology*, **45**,167–175.

Hasselquist, D., and Nilsson, J.-Å. (2009). Maternal transfer of antibodies in vertebrates: trans-generational effects on offspring immunity. *Philosophical Transactions of the Royal Society B: Biological Sciences*, **364**, 51–60.

Hasselquist, D., and Nilsson, J.-Å. (2012). Physiological mechanisms mediating costs of immune responses:

what can we learn from studies of birds? *Animal Behaviour*, **83**, 1303–12.

Hasselquist, D., Tobler, M., and Nilsson, J.-Å. (2012). Maternal modulation of offspring immune function in vertebrates. In R. M. Nelson and G. Demas (ed.), *Eco-Immunology*, pp. 165–224. Oxford University Press, Oxford.

Hasselquist, D., Wasson, M. F., and Winkler, D. W. (2001). Humoral immunocompetence correlates with date of egg-laying and reflects work load in female tree swallows. *Behavioral Ecology*, **12**, 93–7.

Hedrick, P. W. (2002). Pathogen resistance and genetic variation at MHC loci. *Evolution*, **56**, 1902–8.

Hellgren, O., Pérez-Tris, J., and Bensch, S. (2009). A jack-of-all-trades and still a master of some: prevalence and host range in avian malaria and related blood parasites. *Ecology*, **90**, 2840–9.

Hellgren, O., and Sheldon, B. C. (2011). Locus-specific protocol for nine different innate immune genes (antimicrobial peptides: beta-defensins) across passerine bird species reveals within-species coding variation and a millercase of trans-species polymorphisms. *Molecular Ecology Resources*, **11**, 686–92.

Hellgren, O., Waldenström, J., Peréz-Tris, J., et al. (2007). Detecting shifts of transmission areas in avian blood parasites—a phylogenetic approach. *Molecular Ecology*, **16**, 1281–90.

Hellgren, O., Wood, M. J., Waldenstrom, J., et al. (2013). Circannual variation in blood parasitism in a sub-Saharan migrant passerine bird, the garden warbler. *Journal of Evolutionary Biology*, **26**, 1047–59.

Hill, A. V. S., Allsopp, C. E. M., Kwiatkowski, D., et al. (1991). Common West African Hla antigens are associated with protection from severe malaria. *Nature*, **352**, 595–600.

Hill, N. J., Takekawa, J. Y., Cardona, C. J., et al. (2012). Cross-seasonal patterns of avian influenza virus in breeding and wintering migratory birds: a flyway perspective. *Vector-Borne and Zoonotic Diseases*, **12**, 243–53.

Howe, H. F., and Smallwood, J. (1982). Ecology of seed dispersal. *Annual Review of Ecology, Evolution, and Systematics*, **13**, 201–28.

Hughes, A. L., and Nei, M. (1988). Patterns of nucleotide substitution at major histocompatibility complex class I loci reveals overdominant selection. *Nature*, **335**, 167–70.

Hutchings, M. R., Milner, J. M., Gordon, I. J., Kyriazakis, I., and Jackson, F. (2002). Grazing decisions of Soay sheep, *Ovis aries*, on St Kilda: a consequence of parasite distribution? *Oikos*, **96**, 235–44.

Ilmonen, P., Taarna, T., and Hasselquist, D. (2000). Experimentally activated immune defence in female pied flycatchers results in reduced breeding success. *Proceedings of the Royal Society B: Biological Sciences*, **267**, 665–70.

Jenkins, T., Thomas, G. H., Hellgren, O., and Owens, I. P. F. (2012). Migratory behavior of birds affects their coevolutionary relationship with blood parasites. *Evolution*, **66**, 740–51.

Kilpatrick, A. M., Chmura, A. A., Gibbons, D. W., Fleischer, R. C., Marra, P. P., and Daszak, P. (2006). Predicting the global spread of H5N1 avian influenza. *Proceedings of the National Academy of Sciences USA*, **103**, 19368–73.

Kilpimaa, J., Alatalo, V. A., and Siitari, H. (2004). Trade-offs between sexual advertisement and immune function in the pied flycatcher (*Ficedula hypoleuca*). *Proceedings of the Royal Society B: Biological Sciences*, **271**, 245–50.

Kim, S., Gil Park, M., Yih, W., and Coats, D. W. (2004). Infection of the bloom-forming thecate dinoflagellates *Alexandrium affine* and *Gonyaulax spinifera* by two strains of *Amoebophrya* (Dinophyta). *Journal of Phycology*, **40**, 815–22.

Kirkpatrick, M., and Barton, N. H. (1997). Evolution of a species' range. *American Naturalist*, **150**, 1–23.

Knowles, S. C. L., Nakagawa, S., and Sheldon, B. C. (2009). Elevated reproductive effort increases blood parasitaemia and decreases immune function in birds: a meta-regression approach. *Fuctional Ecology*, **23**, 405–15.

Knowles, S. C. L., Wood, M. J., Alves, R., Wilkin, T. A., Bensch, S., and Shen, C.-K. J. (2011). Molecular epidemiology of malaria prevalence and parasitemia in a wild bird population. *Molecular Ecology*, **20**, 1062–76.

Koella, J. C., Sorensen, F. L., and Anderson, R. A. (1998). The malaria parasite, *Plasmodium falciparum*, increase the frequency of multiple feeding of its mosquito vector, *Anopheles gambiae*. *Proceedings of the Royal Society B: Biological Sciences*, **265**, 763–8.

Krauss, S., Stallknecht, D. E., Negovetich, N. J., Niles, L. J., Webby, R. J., and Webster, R. G. (2010). Coincident ruddy turnstone migration and horseshoe crab spawning creates an ecological 'hot spot' for influenza viruses. *Proceedings of the Royal Society B: Biological Sciences*, **277**, 3373–9.

Krkosek, M., Revie, C. W., Gargan, P. G., Skilbrei, O. T., Finstad, B., and Todd, C. D. (2013). Impact of parasites on salmon recruitment in the Northeast Atlantic Ocean. *Proceedings of the Royal Society B: Biological Sciences*, **280**, 20122359.

LaDeau, S. L., Kilpatrick, A. M., and Marra, P. P. (2007). West Nile virus emergence and large-scale declines of North American bird populations. *Nature*, **447**, 710–13.

Lafferty, K. D., and Morris, A. K. (1996). Altered behavior of parasitized killifish increases susceptibility to predation by bird final hosts. *Ecology*, **77**, 1390–7.

Lam, T. T., Ip, H. S., Ghedin, E., et al. (2012). Migratory flyway and geographical distance are barriers to the gene flow of influenza virus among North American birds. *Ecology Letters*, **15**, 24–33.

LaPointe, D. A., Goff, M. L., and Atkinson, C. T. (2010). Thermal constraints to the sporogonic development and altitudinal distribution of avian malaria *Plasmodium relictum* in Hawai'i. *Journal of Parasitology*, **96**, 318–24.

Lawson, B., Robinson, R. A., Colvile, K. M., et al. (2012). The emergence and spread of finch trichomonosis in the British Isles. *Philosophical Transactions of the Royal Society B: Biological Sciences*, **367**, 2852–63.

Lee, K. A., and Klasing, K. C. (2004). A role for immunology in invasionbiology. *Trends in Ecology and Evolution*, **19**, 523–9.

Li, W. X., Zou, H., Wu, S. G., Song, R., and Wang, G. T. (2012). Richness and diversity of helminth communities in the Japanese grenadier anchovy, *Coilia nasus*, during its anadromous in the Yangtze River, China. *Journal of Parasitology*, **98**, 449–52.

Loiseau, C., Harrigan, R. J., Cornel, A. J., et al. (2012). First evidence and predictions of *Plasmodium* transmission in Alaskan bird populations. *PLoS One*, **7**, e44729.

Loiseau, C., Zoorob, R., Robert, A., Chastel, O., Julliard, R., and Sorci, G. (2011). *Plasmodium relictum* infection and MHC diversity in the house sparrow (*Passer domesticus*). *Proceedings of the Royal Society B: Biological Sciences*, **278**, 1264–72.

Mann, C. C. (2005). *1491: New Revelations of the Americas before Columbus*. Vintage Books, New York.

Martiny, J. B., Bohannan, B. J., Brown, J. H., et al. (2006). Microbial biogeography: putting microorganisms on the map. *Nature Reviews Microbiology*, **4**, 102–12.

Mathers, A., Smith, R. P., Cahill, B., et al. (2011). Strain diversity of *Borrelia burgdorferi* in ticks dispersed in North America by migratory birds. *Journal of Vector Ecology*, **36**, 24–9.

Mendes, L., Piersma, T., Hasselquist, D., Matson, K. D., and Ricklefs, R. E. (2006). Variation in the innate and acquired arms of the immune system among five shorebird species. *Journal of Experimental Biology*, **209**, 284–91.

Miller, L. H., Mason, S. J., Clyde, D. F., and McGinniss, M. H. (1976). The resistance factor to *Plasmodium vivax* in blacks. The Duffy-blood-group genotype, FyFy. *New England Journal of Medicine*, **295**, 302–4.

Møller, A. P., and Erritzoe, J. (1998). Host immune defence and migration in birds. *Evolutionary Ecology*, **12**, 945–53.

Møller, A. P., Erritzoe, J., and Saino, N. (2003). Seasonal changes in immune response and parasite impact on hosts. *American Naturalist*, **161**, 657–71.

Møller, A. P., and Szep, T. (2011). The role of parasites in ecology and evolution of migration and migratory connectivity. *Journal of Ornithology*, **152**, 141–50.

Moret, Y., and Schmid-Hempel, P. (2000). Survival for immunity: the price of immune system activation of bumblebee workers. *Science*, **290**, 1166–8.

Murphy, K., Travers, P., and Walport, M. (2008). *Janeway's Immunobiology*, 7th edn. Garland Science, US.

Nathan, R. (2006). Long-distance dispersal of plants. *Science*, **313**, 786–8.

Nebel, S., Bauchinger, U., Buehler, D. M., et al. (2012). Constitutive immune function in European starlings, *Sturnus vulgaris*, is decreased immediately after an endurance flight in a wind tunnel. *Journal of Experimental Biology*, **215**, 272–8.

Nelson, R. J., and Demas, G. E. (1996). Seasonal changes in immune function. *Quarterly Review of Biology*, **71**, 511–48.

Nelson, R. J., Demas, G. E., Klein, S. L., and Kriegsfeld, L. J. (2002). *Seasonal Patterns of Stress, Immune Function and Disease*. Cambridge University Press, Cambridge.

Nilsson, J.-Å. (2002). Metabolic consequences of hard work. *Proceedings of the Royal Society B: Biological Sciences*, **269**, 1735–9.

Norberg, U. M. (1996). Energetics of flight. In C. Carey (ed.), *Avian Energetics and Nutritional Ecology*, pp. 199–249. Chapman & Hall, New York.

Obbard, D. J., Welch, J. J., Kim, K. W., and Jiggins, F. M. (2009). Quantifying adaptive evolution in the *Drosophila* immune system. *PLoS Genetics*, **5**, e1000698.

Olsen, B., Munster, V. J., Wallensten, A., Waldenström, J., Osterhaus, D. M. E., and Fouchier, R. A. M. (2006). Global patterns of influenza A virus in wild birds. *Science*, **312**, 384–8.

Palinauskas, V., Valkiûnas, G., Bolshakov, C. V., and Bensch, S. (2011). *Plasmodium relictum* (lineage SGS1) and *Plasmodium ashfordi* (lineage GRW2): the effects of the co-infections on experimentally infected passerine birds. *Experimental Parasitology*, **127**, 527–33.

Palumbi, S. R. (2001). Humans as the world's greatest evolutionary force. *Science*, **293**, 1786–90.

Pap, P. L., Czirjak, G. A., Vagasi, C. I., Barta, Z., and Hasselquist, D. (2010). Sexual dimorphism in immune functions changes during the annual cycle in the house sparrow. *Naturwissenschaften*, **97**, 891–901.

Park, M. G., Yih, W., and Coats, D. W. (2004). Parasites and phytoplankton, with special emphasis on Dinoflagellate infections. *Journal of Eukaryotic Microbiology*, **51**, 145–55.

Pérez-Tris, J., Hellgren, O., Krizanauskiene, A., et al. (2007). Within-host speciation of malaria parasites. *PLoS One*, **2**, e235.

Perrigo, A. L., Romeralo, M., and Baldauf, S. L. (2012). What's on your boots: an investigation into the role we play in protist dispersal. *Journal of Biogeography*, **39**, 995–1003.

Piersma, T. (1997). Do global patterns of habitat use and migration strategics co-evolve with relative investments in immunocompetence due to spatial variation in parasite pressure? *Oikos*, **80**, 623–31.

Piersma, T. (2003). 'Coastal' versus 'inland' shorebird species: interlinked fundamental dichotomies between their life- and demographic histories? *Wader Study Group Bulletin*, **100**, 5–9.

Piersma, T. (2011). Why marathon migrants get away with high metabolic ceilings: towards an ecology of

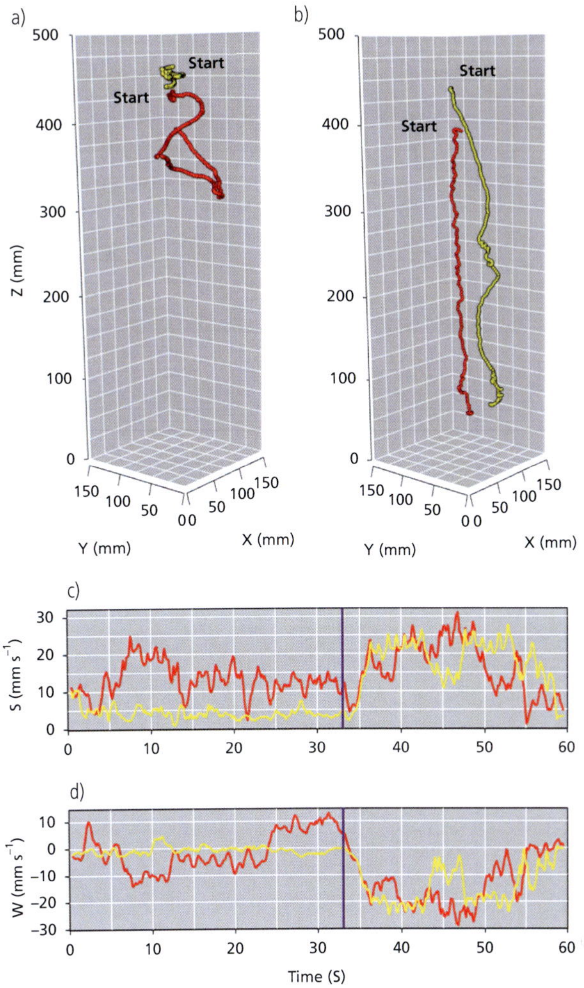

Plate 1 Tracking of the position, speed, and vertical displacement of two *Daphnia magna* individuals in the absence (a) and presence (b) of UV radiation. The animals were marked with yellow and red nanoparticles (quantum dots), and monitored simultaneously to obtain 3D trajectories. Swimming speed (S) for both animals is shown in (c) and vertical speed (w) in (d). From Ekvall et al. (2013). (See Figure 6.5)

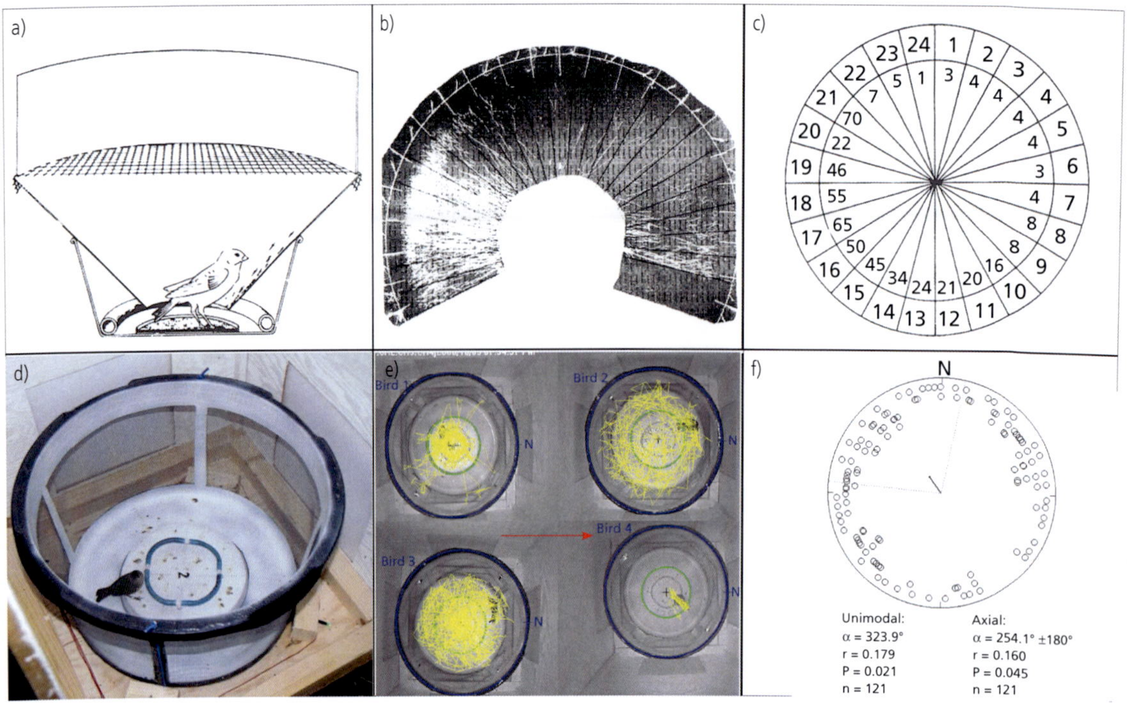

Plate 2 Various types of circular orientation cages have been designed to accurately measure and quantify the orientation preference of caged migratory birds. (a) The original orientation cage was invented by Emlen (Emlen and Emlen 1966) and is referred to as the 'Emlen funnel'. The walls of this funnel-shaped cage were covered with white blotting paper, with an ink pad placed at the bottom of the cage. Thus, a bird that is tested in this experimental setup leaves ink marks on the cage walls whenever trying to fly in a specific direction. (b) This cage design is slightly modified compared to (a), and is typically used as standard method in most orientation studies to date. Here the ink pad is removed, and the walls are covered with typewriter correction paper (e.g. white-out) or thermal paper (Mouritsen et al. 2009) whereon the birds leave scratches whenever touching the cage wall. Shown in the picture is the line-out scratch-sensitive paper (e.g. white-out or thermal paper) after a 90-min orientation experiment, spread on a light table. (c) A circular diagram divided into 24 sectors (covering 15° of the circle) showing the distribution of scratches, hand-counted on the paper shown in (b). (d) More recently, video tracking has been introduced in orientation research. Here, the movements of the birds in cages or funnels are filmed from above or below. Tracking software allows tracking the position of the bird throughout the experiment, and analysing its orientation direction. (e) Digitized orientation data of a 10-min movie interval of a bird filmed in a cage as shown in (d). (f) Orientation of the data shown in (e). Each hop of the bird from the centre (inside of the green inner circle) out to the periphery of the cage and back into the centre again counts as one data point; the position farthest away from the centre is taken as the direction of the hop (round symbols in diagram f). From all these data points (hops), the mean orientation vector with an angle α and a length r (varies between 0 and 1) is calculated. The length *r* can be considered as an inverse analogue of the variance (modified after Helbig 1991; Muheim, unpubl.). (See Figure 9.8)

Plate 3 Polarized light as perceived by the human eye, visible as two faint blue and two faint yellow balloons (Haidinger's brushes), aligned perpendicularly, with the blue axis indicating the axis of polarization. (See Figure 10.5)

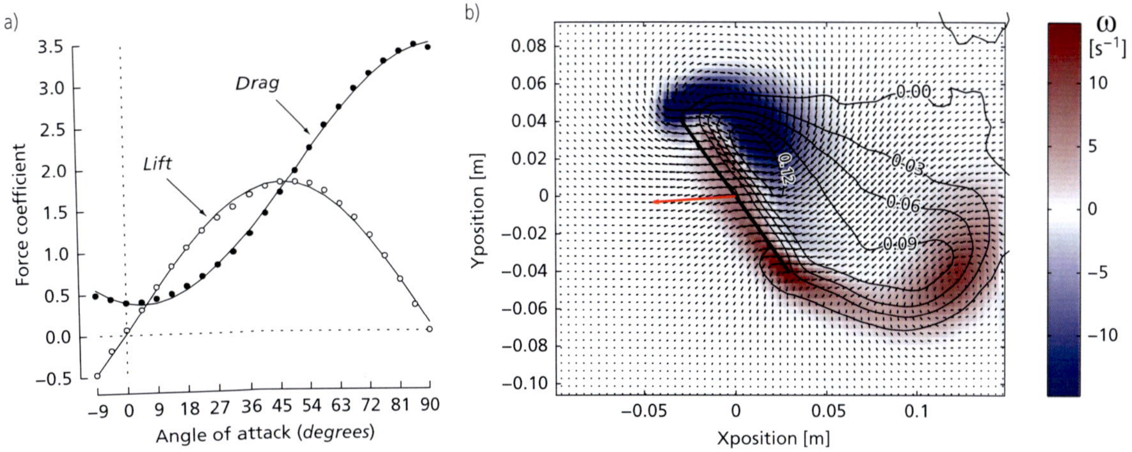

Plate 4 Steady-state lift and drag coefficients for a revolving model fruit fly wing at Reynolds number (Re) ~140. (a) At this low Re, drag forces are relatively high resulting in lift-to-drag ratios (= $1/COT_{mech}$) larger than 1 only at the angles-of-attack range of 9°–45° (from Dickinson et al. 1999). Maximum aerodynamic efficiency occurs at post-stall angles-of-attack, (b) for which an attached leading-edge vortex occurs on the revolving wing. From Poelma et al. (2006). Reproduced with permission from Springer Science + Business Media. (See Figure 13.4)

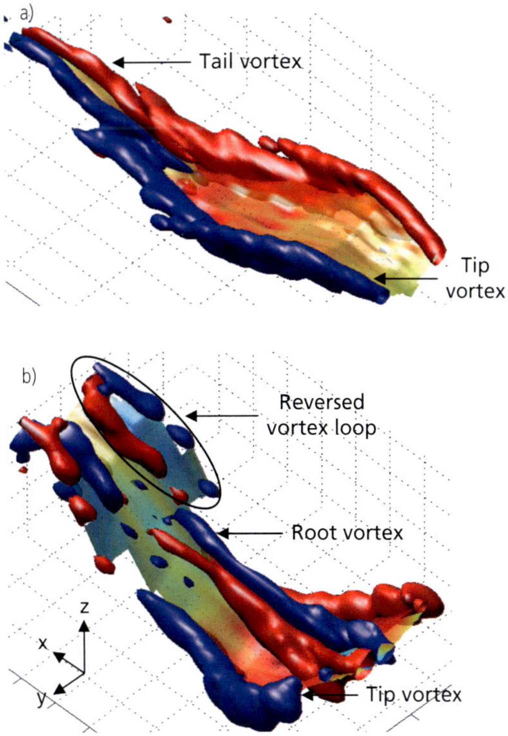

Plate 5 Wake topologies showing the vorticity iso-surfaces of the main vortex structures for one wingbeat of a pied flycatcher, *Ficedula hypoleuca* (a) and a Pallas' long-tongued bat, *Glossophaga soricina*, (b) flying to the left at 7 m/s. The strength of the vortices, measured as circulation, is proportional to the lift generated. The complexity of the wake illustrates how lift is generated by different parts of the animal at different phases of the wingbeat. For example, root vortices seen in the bat wake indicate a lower lift generation by the body than by the wings and the reversed vortex loops generated during the upstroke indicate negative lift being generated. From Muijres et al. (2012b). (See Figure 13.5)

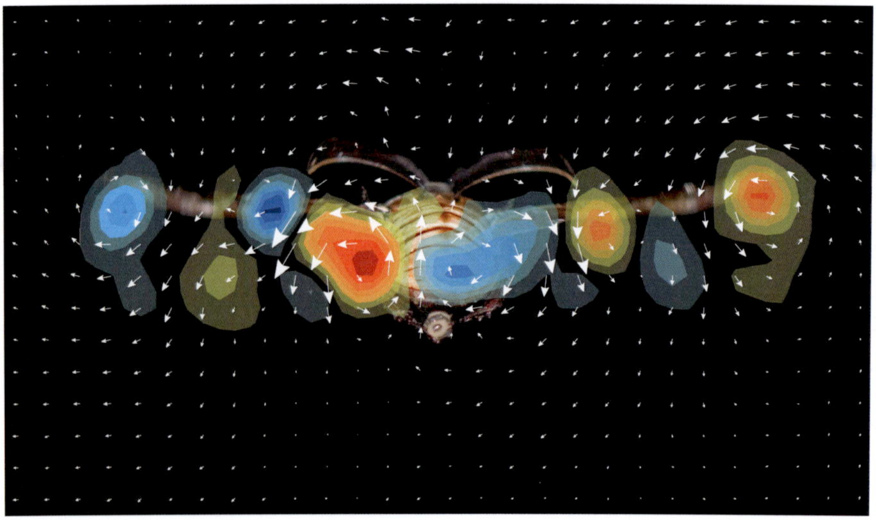

Plate 6 Velocity and vorticity measurements in the wake of a dung beetle during mid-downstroke as seen from behind. The wake contains multiple vortices along the wingspan. Going from the left wingtip to the centre of the body: wing tip vortex, interaction vortex, elytra tip vortex, and body or root vortex. The interaction vortex is the result of a negative interaction between the flow generated by the elytra and the wing, reducing the span efficiency. The strong body/root vortices show that the body does not generate significant lift forces. From Johansson et al. (2012). (See Figure 13.8)

physiological restraint. *Journal of Experimental Biology*, **214**, 295–302.

Pollux, B. J. A. (2011). The experimental study of seed dispersal by fish (ichthyochory). *Freshwater Biology*, **56**, 197–212.

Pollux, B. J. A., Santamaría, L., and Ouborg, N. J. (2005). Differences in endozoochorous dispersal between aquatic plant species, with reference to plant population persistence in rivers. *Freshwater Biology* **50**, 232–42.

Poulin, R., Closs, G. P., Lill, A. W. T., Hicks, A. S., Herrmann, K. K., and Kelly, D. W. (2012). Migration as an escape from parasitism in New Zealand galaxiid fishes. *Behavioral Ecology*, **169**, 955–63.

Prugnolle, F., Manica, A., Charpentier, M., Guegan, J-F., Guernier, V. and Balloux, F. (2005a). Pathogen-driven selection and worldwide HLA class I diversity. *Current Biology*, **15**, 1022–7.

Prugnolle, F., Théron, A., Pointier, J. P., Jabbour-Zahab, R., Jarne, P., Durand, P., de Meeûs, T. (2005b). Dispersal in a parasitic worm and its two hosts: consequence for local adaptation. *Evolution*, **59** (2), 296–303.

Råberg, L., Grahn, M., Hasselquist, D., and Svensson, E. (1998). On the adaptive significance of stress-induced immunosuppression. *Proceedings of the Royal Society B: Biological Sciences*, **265**, 1637–41.

Råberg, L., Nilsson, J.-Å., Ilmonen, P., Stjernman, M., and Hasselquist, D. (2000). The cost of an immune response: vaccination reduces parental effort. *Ecology Letters*, **3**, 382–6.

Råberg, L., Sim, D., and Read, A. F. (2007). Disentangling genetic variation for resistance and tolerance to infectious diseases in animals. *Science*, **318**, 812–14.

Ramey, A. M., Ely, C. R., Schmutz, J. A., Pearce, J. M., and Heard, D. J. (2012). Molecular detection of hematozoa infections in Tundra swans relative to migration patterns and ecological conditions at breeding grounds. *PLoS One*, **7**, e45789.

Schärer, L., Knoflach, D., Vizoso, D. B., Rieger, G., and Peintner, U. (2007). *Thraustochytrids* as novel parasitic protists of marine free-living flatworms: *Thraustochytrium caudivorum* sp. nov. parasitizes *Macrostomum lignano*. *Marine Biology*, **152**, 1095–104.

Schlichting, H. E. J. (1969). The importance of airborne algae and protozoa. *Journal of the Air Pollution Control Association*, **19**, 946–51.

Schlichting, H. E. J. (1974). Ejection of microalgae into the air via bursting bubbles. *Journal of Allergy and Clinical Immunology*, **53**, 185–8.

Sepil, I., Lachish, S., Hinks, A. E., and Sheldon, B. C. (2013). Mhc supertypes confer both qualitative and quantitative resistance to avian malaria infections in a wild bird population. *Proceedings of the Royal Society B: Biological Sciences*, **280**, 20130134.

Skerratt, L. F., Berger, L., Speare, R., et al. (2007). Spread of chytridiomycosis has caused the rapid global decline and extinction of frogs. *EcoHealth*, **4**, 125–34.

Skovgaard, A. (2005). Infection with the dinoflagellate parasite *Balstodinium* spp. in two Mediterranean copepods. *Aquatic Microbial Ecology*, **38**, 93–101.

Skovgaard, A., and Saiz, E. (2006). Seasonal occurrence and role of protistan parasites in coastal marine zooplankton. *Marine Ecology Progress Series*, **327**, 37–49.

Sorensen, A. E. (1986). Seed dispersal by adhesion. *Annual Review of Ecology and Systematics*, **17**, 443–63.

Svensson, E., Råberg, L., Hasselquist, D., and Koch, C. (1998). Energetic stress, immunosuppression, and the costs of an antibody response. *Functional Ecology*, **12**, 912–19.

Szymkowiak, P., Gorski, G., and Bajerlein, D. (2007). Passive dispersal in arachnids. *Biology letters*, **44**, 75–101.

Taylor, D. J., Finston, T. L., and Hebert, P. D. N. (1998). Biogeography of a widespread freshwater crustacean: pseudocongruence and cryptic endemism in the north american *Daphnia laevis* complex. *Evolution* **52**, 1648–70.

Tompkins, D. M., White, A. R., and Boots, M. (2003). Ecological replacement of native red squirrels by invasive greys driven by disease. *Ecology Letters*, **6**, 189–96.

Torchin, M. E., Lafferty, K. D., Dobson, A. P., McKenzie, V. J., and Kuris, A. M. (2003). Introduced species and their missing parasites. *Nature*, **421**, 628–30.

Tschirren, B., Råberg, L., Westerdahl, H. (2011). Signatures of selection acting on the innate immunity gene Toll-like receptor 2 (TLR2) during the evolutionary history of rodents. *Journal of Evolutionary Biology*, **24**, 1232–40.

Valkiûnas, G. (2005). *Avian Malaria Parasites and Other Haemosporidia*. CRC Press, Boca Raton.

van de Meutter, F., Stoks, R., and de Meester, L. (2008). Size-selective dispersal of *Daphnia* resting eggs by backswimmers (*Notonecta maculata*). *Biology letters*, **4**, 494–6.

van Leeuwen, C. H. A., van der Velde, G., van Groenendael, J. M., Klaassen, M., and Traveset, A. (2012). Gut travellers: internal dispersal of aquatic organisms by waterfowl. *Journal of Biogeography*, **39**, 2031–40.

van Riper III, C., van Riper, S. G., Goff, M. L., and Laird, M. (1986). The epizootiology and ecological significance of malaria in Hawaiian land birds. *Ecological Monographs*, **56**, 327–44.

Viana, D. S., Santamaría, L., Michot, T. C., and Figuerola, J. (2012). Migratory strategies of waterbirds shape the continental-scale dispersal of aquatic organisms. *Ecography*, **36**, 430–8.

Vistnes, I. I., Nellemann, C., Jordhøy, P., and Støen, O.-G. (2008). Summer distribution of wild reindeer in relation to human activity and insect stress. *Polar Biology*, **31**, 1307–17.

Waterkeyn, A., Pineau, O., Grillas, P., and Brendonck, L. (2010a). Invertebrate dispersal by aquatic mammals:

a case study with nutria *Myocastor coypus* (Rodentia, Mammalia) in Southern France. *Hydrobiologia*, **654**, 267–71.

Waterkeyn, A., Vanschoenwinkel, B., Elsen, S., et al. (2010b). Unintentional dispersal of aquatic invertebrates via footwear and motor vehicles in a Mediterranean wetland area. *Aquatic Conservation: Marine and Freshwater Ecosystems*, **20**, 580–7.

Webster, M. S., Mara, P. P., Haig, S. M., Bensch, S., and Holmes, R. T. (2002). Links between worlds: unraveling migratory connectivity. *Trends in Ecology and Evolution*, **17**, 76–83.

Westerdahl, H., Asghar, M., Hasselquist, D., and Bensch, S. (2012). Quantitative disease resistance: to better understand parasite-mediated selection on major histocompatibility complex. *Proceedings of the Royal Society B: Biological Sciences*, **279**, 577–84.

Westerdahl, H., Waldenstrom, J., Hansson, B., Hasselquist, D., von Schantz, T., and Bensch, S. (2005). Associations between malaria and MHC genes in a migratory songbird. *Proceedings of the Royal Society B: Biological Sciences*, **272**, 1511–18.

Westerdahl, H., Stjernman, M., Råberg, L., Lannefors, M., and Nilsson, J-Å. (2013). MHC-I Affects Infection Intensity but Not Infection Status with a Frequent Avian Malaria Parasite in Blue Tits. *PLoS One*, **8**(8).

Wilcove, D. S., Rothstein, D., Dubow, J., Phillips, A., and Losos, E. (1998). Quantifying threats to imperiled species in the United States. *Bioscience*, **48**, 607–15.

Wood, M. J., Cosgrove, C. L., Wilkin, T. A., Knowles, S. C. L., Day, K. P., and Sheldon, B. C. (2007). Within-population variation in prevalence and lineage distribution of avian malaria in blue tits, *Cyanistes caeruleus*. *Molecular Ecology*, **16**, 3263–73.

Yeager, M., and Hughes, A. L. (1999). Evolution of the mammalian MHC: natural selection, recombination, and convergent evolution. *Immunological Reviews*, **167**, 45–58.

Yohannes, E., Hansson, B., Lee, R., et al. (2008). Isotope signatures in winter moulted feathers predict malaria prevalence in a breeding avian host. *Oecologia*, **158**, 299–306.

PART III

The Mechanisms and Codes of Navigation and Movement

Chapter 9. Animal navigation
Chapter 10. Sensory mechanisms of animal orientation and navigation
Chapter 11. Movements in the olfactory landscape
Chapter 12. The genetics of migration
Chapter 13. The physics of animal locomotion

CHAPTER 9

Animal navigation

Susanne Åkesson, Jannika Boström, Miriam Liedvogel, and Rachel Muheim

Animals move in their daily lives in search for food, to avoid predators, to find mates, and to transport themselves between areas of importance for their survival. To enable energy-efficient transport, their methods of locomotion and choice of times to move have been under strong selection, shaping a plethora of morphological, physiological, and behavioural adaptations enabling safe and efficient transport (Åkesson and Hedenström 2007; Johansson et al., Chapter 13). However, animals also need to record where they are and have innate information on where to go at a certain instant, and navigational abilities including the use of various external cues have therefore also been of major importance in the evolution of animal movement (Box 9.1). Animal navigation has been studied for many decades and this expanding research field has pushed forward our understanding of the functional characteristics of the biological compasses, which can be based on information from celestial cues, such as the stars and the sun, the related pattern of skylight polarization, as well as the Earth's magnetic field. Furthermore, research on navigation strategies based on information from these compass cues, but also odours (Svensson et al., Chapter 11), landmarks, and the recording of self-generated movements, i.e. path integration, have also been the focus of interest. There has been substantial progress in the field of animal orientation and navigation research, and in this chapter we wish to expand on some of the major findings over the years and highlight some areas of importance for future research where understanding is still lacking. Just as is the case for the book, this chapter is not intended as a review of what is a vast research field, although we hope to point you as a reader to some useful reviews and books, which cover different aspects of animal navigation (e.g. Kramer 1961, Schmidt-Koenig 1975, Emlen 1976, Able 1980, Papi 1982, Baker 1984, Wiltscho and Wiltschko 1995, Wehner 1998, Åkesson 2003, Wallraff 2005). For in-depth information on sensory aspects and recent findings of sensory processing of navigational information in the brain of moving organisms, see also Muheim et al. (Chapter 10).

We may ask ourselves what navigation means to an animal, how the navigation task is perceived and in what circumstances it is challenged. To get an idea of what this can involve we may imagine a juvenile wandering albatross, *Diomedea exulans*, sitting on an isolated island in the Southern Ocean, left behind by its parents and waiting for the appropriate wind to take off for its maiden flight and solo navigation over the open sea. The juvenile albatross will spend the entirety of its first 8 to 10 years of life moving across open sea and thereafter return to the island to breed when it has become sexually mature. How does the young albatross find its way during its first migration and how does it know when it is time to return? How it develops the necessary navigational skills, and what cues are used to navigate back to the island several years after its maiden journey is still a mystery.

In another part of the world a tiny ant leaves the entrance of its underground nest in one of the most extreme habitats of the world, the floor of the sun-baked Sahara desert. The desert ant, *Cataglyphis fortis*, is undertaking its first foraging trip searching for dead insects to bring back to its nestmates

Animal Movement Across Scales. Edited by Lars-Anders Hansson and Susanne Åkesson.
© Oxford University Press 2014. Published 2014 by Oxford University Press.

(Fig. 9.1a). The solo forager spends not more than an hour outside the nest, running on its high stilt-like legs across the hot desert floor and during the trip it covers several hundred metres. The ant's navigational system is pretty well understood, and detailed and careful experimentation has led the scientists to conclude that this system is visually based with input also from proprioceptors (stretch receptors in the legs), but also that advanced neuronal integration of information is needed for the ant to integrate movements in order to successfully return home (e.g. Wehner et al. 1996, Wohlgemut et al. 2000, Wehner 2009). The ant pays attention to its own surroundings while on its journey by memorizing landmarks. To return, it recovers landmark images to reconstruct its way back home. In the hottest central planes in Australia another ant species, the Central Australian desert ant, *Melophorus bagoti*, faces similar navigational challenges in which landmarks, path integration, and spatial memories are also central for navigating the cluttered landscape (Cheng et al. 2009).

In the marine environment, the plaice, *Pleuronectes platessa*, a marine flat fish, aims for its spawning sites in the northeastern part of the English Channel near the Strait of Dover in winter. The plaice rests on the sea floor until the tidal stream changes and then moves in the preferred direction. At the appropriate time the fish then rises in the water column and swims actively, at the same time hitch-hiking with the water currents (e.g. Metcalfe et al. 1993). The plaice will find itself transported many kilometres across the sea floor, sometimes in the darkest part of the night. How does it know when to stop? What information is used to locate the spawning sites and where do the offspring go and by what cues is it steering? In addition to these examples, there are many more of spectacular migrations and navigational achievements across the animal kingdom. To be able to structure these findings and conceptualize the different aspects of navigation we will first need to present some terms and concepts.

9.1 Concepts and terminology

In animal navigation research, as in most other research fields, we need to establish effective communication by using a commonly defined vocabulary. This has been done repeatedly for the field of animal navigation and orientation, but Able (2000), in the most recent attempt, tried to form coherence in

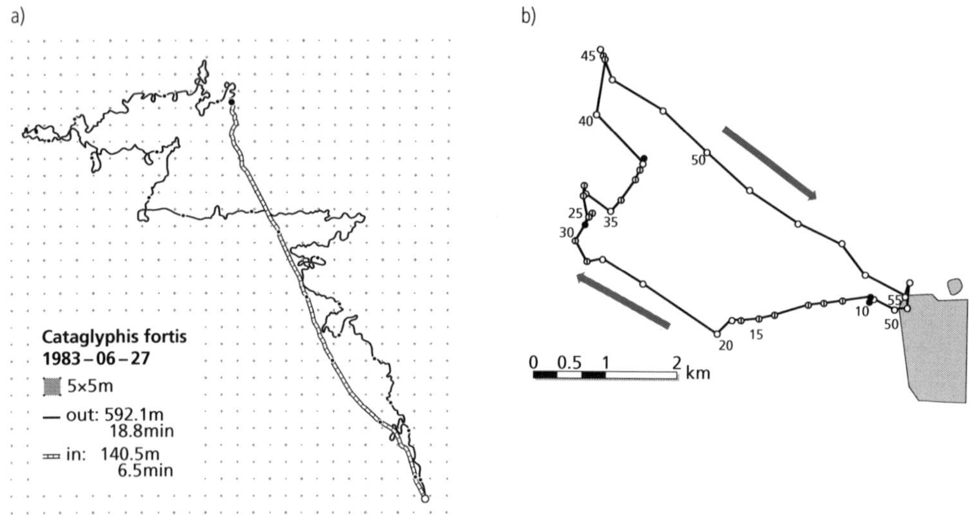

Figure 9.1 Examples of two animals performing path integration during foraging and returning to the nest. (a) A desert ant, *Cataglyphis fortis*, searching for food across the desert floor (modified after Wehner and Wehner 1990 with permission from Taylor and Francis), and (b) a common guillemot, *Uria aalge*, searching for food in the marine environment (modified after Evans et al. 2013). Each circle represents diving events, while the direction of flight is indicated by arrows. Scale bars for each trip are indicated.

the terminology used and came up with some recommendations, which we will mainly adapt to in this chapter. However, other papers, e.g. Papi (2001) and Wiltschko and Wiltschko (2003), provide historical overviews of the development of concepts and terminology highlighting some of the major achievements in this field. In yet another review, Wehner (1998) recommended using a conceptual framework with small, carefully performed experimental steps to unravel the computational strategies of the brain of navigating animals.

9.1.1 Orientation and navigation

Orientation and navigation are the most prominent terms in this field, and over time have been used with somewhat different meaning. However, before we define these terms we need to cover what migration and homing means, as this is central for navigating animals (see also Box 1.1). The different meanings of the term migration have extensively been covered by Dingle (1996), and in its most simple form we refer to animals moving between breeding locations and wintering areas. Homing on the other hand involves movements back to a known site, such as the nest, after displacement or a foraging trip. The term orientation may refer to the orientation of parts of the body with respect to one another. It may also refer to the body orientation in space based on either internal or external information, as well as compass orientation. The latter is perhaps most widely used and here, as suggested by Able (2000), we will use orientation as a synonym for directed movement, or compass orientation. Navigation has also been used in a multitude of ways, but a recommendation by Able (2000) is to restrict the term to the general definition suggested by Schöne (1984, 105) as follows: 'navigation is the theory and practice of charting a course to a remote goal'. We may differentiate orientation behaviour employed in homing behaviour as well as independent from homing. Independent from homing, animals may perform: (a) *non-compass orientation*, (b) *compass orientation*, and (c) *vector navigation* (Able 2000). (a) Non-compass orientation (aka 'piloting') may refer to a young bird flying along a coastline or a line of trees, and thereby temporarily deviating from the intended vector navigation course towards its winter destination. (b) Compass orientation refers to the ability to select a course by using one or several compasses. This course does not have to lead to a known goal, and is not selected based on landmarks, but rather on external compass (i.e. celestial or geomagnetic) information. There are many examples of so-called *non-sense orientation* in birds for which the birds have selected fixed geographic or magnetic courses at free-flying releases or in cage experiments (e.g. Bellrose 1958, Matthews 1961, Sandberg et al. 1988), and which at first have been hard to explain (cf. Sandberg 1994). Further experimentation has revealed that the birds' physiological state may influence their directional choices, especially in coastal areas (Sandberg 2003). Other fixed course responses have been correlated with the shortest distance from land to sea as was observed for Adelie penguins, *Pygoscelis adeliae*, displaced across the Antarctic continent (Emlen and Penney 1964, Penney and Emlen 1967). (c) Vector navigation has originally been referred to as the direction that young birds depart in order to reach their population-specific wintering area (Schmidt-Koenig 1970). Other terms have been used thereafter, but we will keep the original term as this has been used most extensively. The elegant and classical continental-scale experiment by Perdeck (1958) displacing several thousands of ringed European starlings, *Sturnus vulgaris*, from western to central Europe in autumn has revealed a discrepancy in vector navigation expressed by young starlings during their first migration continuing in the expected migratory direction towards southwest after displacement and goal-directed orientation in adults navigating back (northwest) to their normal wintering areas in England and France. The juvenile birds seemed to be unaware of the transport while adults located their position and corrected their flights to again reach the normal wintering area.

To return to a known goal is generally termed homing and will be used by us in this sense here. However, there might be different ways by which an animal is able to return to a goal such as its nest or home range area. The animal might be in direct sensory contact with the goal, seeing it or for instance detecting odours associated with the goal that are transported by winds or currents. It might also integrate several sources of information along the route, or it might have recorded its own movements along the journey by using, e.g. path integration (inertial navigation). Yet another alternative is to memorize a map or use some sort of a positioning system to pin-point the current position relative to home. Systematic or random search is used by animals when information is lacking or all other systems fail. We will cover some of these alternatives later when we discuss route-based navigation and the use of *maps* for navigation. Route-based navigation refers to a situation when the animal is recording its own movements relative to the direction of transport based on information perceived during the outward journey. We may find at least three types of route-based navigation depending on what type of information is used: (a) *piloting*, (b) *beaconing*, and (c) *path integration*. The term piloting has been used in many ways and may partly reflect a lack of empirical understanding of the characteristics of the homing phenomena (Papi 1982, Able 2000). The latest suggestion by Able (2000) is to use piloting when referring to goal-directed orientation based upon following a sequence of familiar landmarks of any type. However, the landmarks do not define the position of the animal in space relative to the goal on a map, but rather provide a route to the goal. With beaconing we refer for instance to a petrel landing on water detecting the odours of dimethyl sulphide emitted by plankton in the sea and transported by the wind. Alternatively odours may play a role when the bird is returning to its nest burrow and thereby is detecting the correct direction to move on land upwind on the basis of nest-specific odours. Male red bollworm moths, *Diparopsis castanea* (Noctuidae), from south-east Africa, that emit sex-pheromones to attract mates are also using beaconing, a topic which is covered in another chapter of this book (Svensson et al., Chapter 11). More examples of beaconing in open field environments are provided by the satellite tracking of homing green turtles at Ascension Island, seemingly relying on information (i.e. chemical cues) transported by wind for the final part of the homing path when the female turtles are heading upwind to reach the island for a new nesting attempt (Luschi et al. 2001, Åkesson et al. 2003, Hays et al. 2003). Recently it has been confirmed that sea turtles are able to detect odours and respond to them with increased activity (Endres et al. 2009, Endres and Lohmann 2012).

Path integration finally has also been termed dead-reckoning, and refers to the situation when the animal is constantly recording and updating its position relative to the starting location, the goal. We may separate allothetic path integration from ideothetic path integration. The former refers to path integration based on external information, while the ideothetic path integration relies on inertial signals, for instance generated by the vestibular system or by counting steps by using proprioreceptors (e.g. Wehner 1998). Inevitable a path-integration system based on allothetic or idiothetic information will generate errors, as distance measurements and compass courses cannot be set precisely by biological sensory systems. Therefore, with distance travelled errors will accumulate, and will be especially problematic for animals moving in fluid media, such as water and air. Path integration has been demonstrated to be used by walking insects (Müller and Wehner 1988), as well as in walking geese during homing (Middlestedt and Middlestedt 1982), but still needs to be investigated for a flying or swimming animal if at all used in these fluid environments.

9.1.2 Maps for navigation

When the animal is defining its position relative to its home solely on information perceived at the distant point of release it is using map-based navigation, i.e. true navigation. We may imagine a homing pigeon transported in a shielded container without access to external navigational clues during transport to be released at a site without direct sensory contact with the home at the release point and from which it has never been exposed. Based solely on local information at the release site the pigeon will make a decision and head off in a direction expected to lead back to its home loft. When performing this task the pigeon is locating its position relative to the goal and then is selecting a course to return home (it is using a map and a compass, see later discussion). There are at least two types of maps: (a) *mosaic* and (b) *gradient maps*. The concept of a mosaic map was first introduced by Wallraff (1974) and refers to a map based on the learning of the spatial relationships of landscape features and their relative position to home. It has been assumed that the landmark positions in a mosaic map are memorized as compass directions and that the size of the map is increasing as the animal is increasing the familiar area during explorations. Therefore the limit of a familiar area map or mosaic map is set by the limit of individual knowledge as a result of explorations (Fig. 9.2a). In this setting a cognitive map has been discussed (Papi 1992), but it does not have to be the same as a mosaic map. A cognitive map, as it was suggested by Tolman (1948) and O'Keefe and Nadel (1978), is the same as a mental representation for which spatial (and perhaps also geometric) relationships among locations are memorized in connection to direction and distance. As a consequence of using a cognitive map, an animal needs to be able to make novel shortcuts

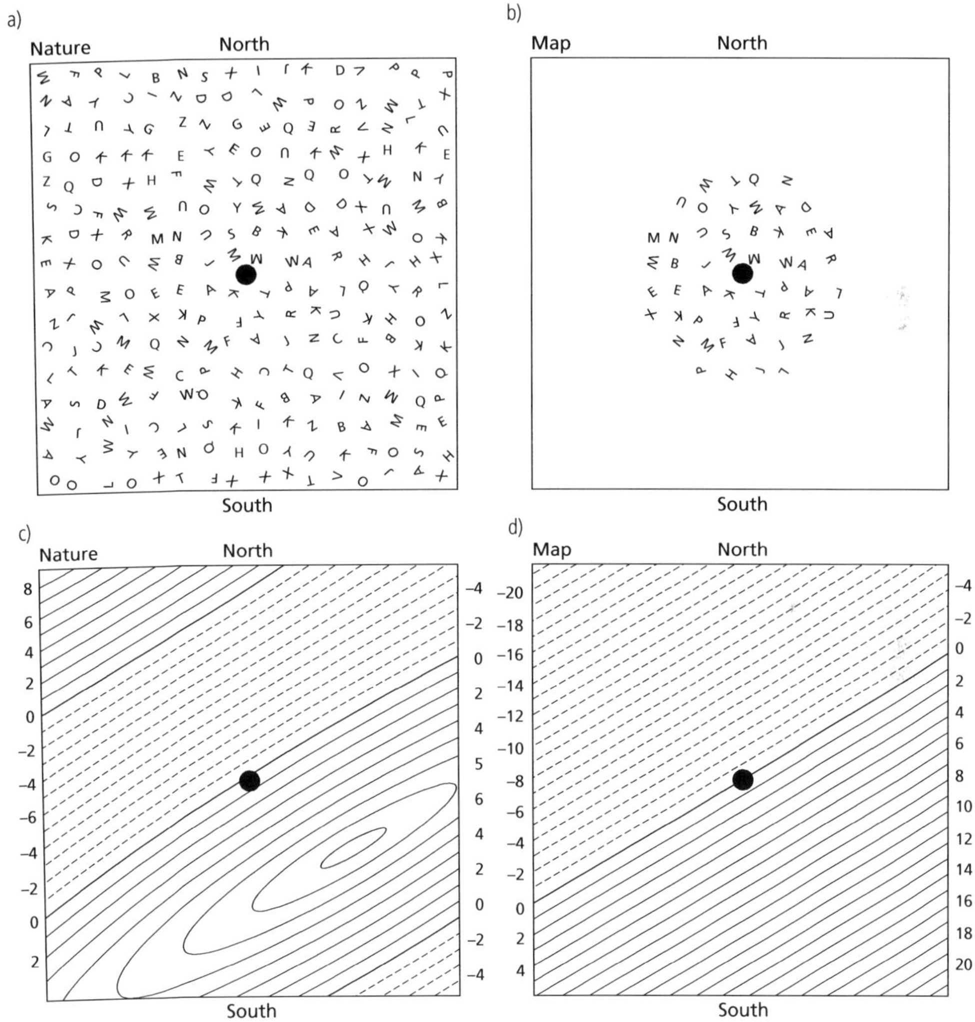

Figure 9.2 (a) Map forming a mosaic of landmarks, symbolized by letters, surrounding the home location (indicated by a filled circle in the centre). (b) The mosaic map experienced by the animal exploring its surroundings is limited in range. (c) Gradient map based on isolines of arbitrary units of a gradient field. In this example the line running through the bird's home is designated as 0, but higher values are illustrated by solid and lower values by broken lines. (d) The bird's corresponding gradient map as established by extrapolation of home site conditions. In (c) and (d) only one gradient is illustrated, but the combination of two gradient fields of sufficiently large intersection is needed for bi-coordinate navigation. Modified after Wallraff (1990) reproduced with permission from Springer Science + Business Media.

between locations. However, this is something which has been difficult to find evidence for (e.g. Bennett 1996, Wehner et al. 2006). Mosaic maps may be based on terrestrial landmarks in for instance a forest or in a city, but may also refer to a mosaic of geographically separated and locally associated odours in terrestrial (e.g. Papi 1972, 1976, 1982; Wallraff and Andreae 2000), as well as marine environments (e.g. Nevitt et al. 1995, Bonnadonna et al. 2003).

A gradient map (in contrast to the mosaic map) is based on one or several gradients, varying in different geographical directions (ideally based on perpendicularly oriented features). A *bi-coordinate gradient map* is based on a combination of two gradients, and the most commonly discussed and with a more global extension is the combination of the gradients of geomagnetic inclination and field strength (Fig. 9.2) (Wallraff 1991, Phillips 1996, Lohmann et al. 2001). Each location will have its unique combinations of parameter values for the two gradients included in the bi-coordinate map, and will provide site-specific knowledge to the animal using it. A clear benefit from using a gradient map, based on gradients which extension may be predictable across space, is that the gradient map is unlimited in its range. In this respect the animal with local knowledge and experience may during exploration memorize the geographical directions in which the gradients are varying. Based on the knowledge in which direction the gradients are decreasing and increasing locally, animals may predict how the same gradients will vary across unknown terrain further away from their range of experience. However, there are possibilities (e.g. Åkesson 1996), but also problems associated with using a bi-coordinate geomagnetic map for global navigation (Åkesson and Alerstam 1998, Boström et al. 2012a).

9.2 The map and compass concepts

Gustav Kramer formulated the concept of homing navigation using a '*map and a compass*' (Kramer 1953) as briefly mentioned earlier. This navigation concept has proven to be one of the most well founded in this area of research, and we still consider the navigational process to involve the two crucial parts relying on a map to define the position and a compass to tell the direction to go home. Kramer (1953) argued that when using the map and compass the navigation process will have three characteristic components: (1) the navigation is a two-component process dependent on the ability to define the position and then use a compass independent on the position-fixing mechanism, (2) the use of a map will provide the direction to go home, and (3) in the last step a compass is used to define the direction to go in order to reach home. The map and compass concept is proven to be a solid conceptual framework of navigation in birds and other animals (Able 2000, Wiltschko and Wiltschko 2003). However, still it is an open question what type of components is providing the map, although e.g. odours, infrasound and geomagnetic field parameters have been recent focus of interest (e.g. Phillips 1996, Wallraff and Andreae 2000, Hagstrum 2013).

9.2.1 Biological compasses

The use of directional information from the sun for orientation and navigation is widespread across the animal kingdom, but has most intensely been studied in insects and birds. How can the sun provide a directional compass cue, when its position changes constantly over the course of a day? Most animals have solved this problem by comparing the directional position of the sun with the time of day, measured by the animal's circadian clock (reviewed by Wiltschko 1980, 1981; Schmidt-Koenig 1990). This concept is called *time-compensated sun compass* (Box 9.1). Animals determine their movement directions by comparing the azimuthal position of the sun in the sky (i.e. the clockwise angle of the sun from North) relative to local time, measured by their internal clock. This is a clever way to avoid the fact that the sun changes 'position' over the day (Fig. 9.3).

9.2.2 Evidence for sun compass orientation in insects

The time-compensated sun compass is the best understood orientation mechanism in insects and it has been demonstrated in several diurnal insect species, such as the honeybee, *Apis mellifera*, desert ants, and several migratory butterfly species.

Box 9.1 Methods for studying animal navigation

Clock-shift experiments: An experimental treatment to test whether animals use a time-compensated sun compass for orientation. By holding animals under altered day–night regimes for a few days, their inner clock can be reset either clockwise or counterclockwise (i.e. 'sunrise' no longer is at the time of natural sunrise, but some hours earlier (counterclockwise), or later (clockwise shift)). This results in a predictable error in the orientation of animals using a time-compensated sun compass, since they think that it is a different time and thus the sun to be in a different direction. At high latitudes the expected shift in orientation can be approximated by a 15° clockwise shift per hour clock shift in the case of a slow shift (clock delayed) and the same amount of shift in the counter-clockwise direction in a fast shift (clock-enhanced). **Cue conflict and cue calibration experiments:** In these two types of experimental manipulations, two orientation cues are tested against each other to assess which cue is dominant over the other. In cue conflict experiments, the animals are exposed to the cue conflict while being tested for orientation, whereas in cue calibration experiments, the animals are exposed to cue conflict not during the actual orientation experiment, but beforehand. The former experimental tool is better suited to test for a preference between cues when the animals determine their orientation direction. The latter experimental tool is used to test which cue calibrates other cues, and has the advantage that the animals can be tested for changes in either cue separately. **Magnetic coil experiments:** Magnetic coils (Helmholtz coils, big enough that experimental cages can be placed in the middle) are the most important experimental tools to test whether an animal uses a magnetic compass, and what type of compass it has (inclination compass vs. polarity compass). With magnetic coils, the magnetic field surrounding the testing apparatus can be artificially changed or cancelled. To test whether an animal uses a polarity or inclination compass, the vertical component of the magnetic field can be artificially inverted which flips the magnetic field vector, i.e. reverses the inclination, but leaves polarity unchanged. Animals responding to such an inversion by changing their orientation by 180° use the inclination angle of the magnetic field to determine the correct side of the axis; animals not responding to an inversion, but only to a shift in the horizontal component of the magnetic field, use the polarity of the magnetic field lines for orientation.

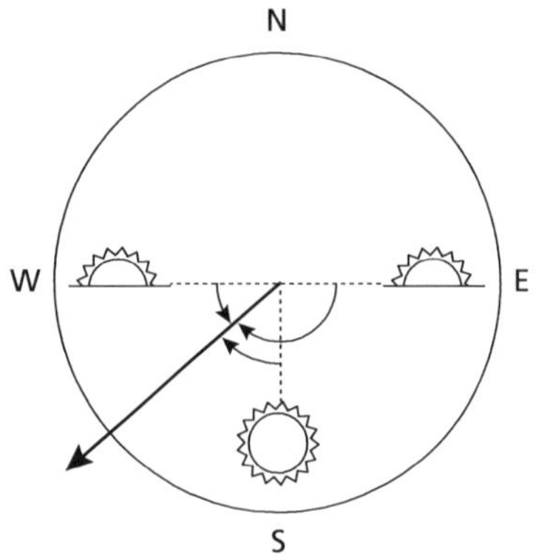

Figure 9.3 Time-compensated sun compass orientation. By comparing the azimuth position of the sun with local time, animals adjust the deviation from the sun's azimuth across the day and thereby determine their movement direction. As an example, the position of the sun at three different times of day, morning , midday, and afternoon are shown, with the thin arrows showing the respective angles that the animals must choose to determine their migratory direction towards southwest (large arrow).

Honeybees are known to use a time-compensated sun compass to perform their waggle dances and direct other bees in the hive to a food resource (von Frisch 1967). Bees are able to choose and keep a compass orientation only based on solar information. In addition, they are able to further use an additional prominent factor, the landscape or skyline panorama. Bees from hives that were translocated from a position next to a treeline to another treeline with a different angular orientation still oriented their dances in the old compass direction when tested under overcast conditions (Dyer and Gould 1981, Dyer 1987, Towne and Moscrip 2008). These bees had apparently learned the movement pattern of the sun in relation to the landscape at the first hive location and remembered that information when performing the dances under overcast conditions at the new location.

Desert ants live in an open, sun-exposed habitat and can use their sun compass both to return to their nests and to return to a food resource after training (Collett et al. 1999). They rely on path integration to keep track of their foraging routes and then vector subtraction (comparison between the known goal and the current state) to determine the direction towards the goal, once a food item has been retrieved (Mittelstaedt 1983). For desert ants the polarized light pattern is more important than the sun itself and their dominant compass cue used for navigation (Wehner and Müller 2006). In addition to the sun and polarized light cues, landmarks or signposts are a further important feature for desert ant orientation (Wehner et al. 1996). The by far best studied species, when it comes to sun compass use in insects, is the North American monarch butterfly, *Danaus plexippus*, which has been shown to use a time-compensated sun compass (for a recent review see Reppert et al. 2010). This species also exhibits the most impressive migration cycle among insects–not only does it cover continents, but also generations (Chapman et al., Chapter 2). Each year, in autumn, millions of butterflies migrate almost 4,000 km to congregate in their wintering area in the Michoacán Mountains in central Mexico. The monarch butterflies start their migration in spring from the wintering sites in Mexico and migrate north to north-east to their first reproduction in the eastern part of the United States each of these butterflies have migrated to the wintering site, but has not yet reproduced meaning that they have just migrated to the wintering site before. The onset of migration is probably triggered by a combination of environmental factors, such as decreasing day length and decreasing temperatures. There also seems to be a hormonal component to the regulation of migratory activity, since summer butterflies do not show oriented flights, whereas migratory butterflies do. The physiological basis for this difference is that the migrants lack a juvenile hormone, an acyclic sesquiterpenoid secreted by endocrine glands associated with the brain. This deficiency leads to reproductive diapause and a prolonged lifespan, enabling long distance migration for this butterfly generation (Herman and Tatar 2001). However, this hormone deficiency does not seem to be required for time-compensated sun compass orientation. The northward migration includes at least two generations of butterflies, as they gradually move north, following the emergence of their larval host plant, eventually ending up in the eastern North America again (Fig. 2.6; Malcolm et al. 1993).

A European counterpart of long-distance migrating butterflies is exemplified by the painted lady, *Vanessa cardui*, occurring in high numbers during irruptive migrations in some years engaging up to five or six generations in the migration cycle covering in total 1,500 km of distance from North and central Africa to northern Scandinavia and back (Stefanescu et al. 2012). Which compasses are used by the painted ladies on migration is still largely an open question, but impressively the small insects select flight altitudes up to 1,000 m during migrations to facilitate long-distance transports (Stefanescu et al. 2012). Recent work on monarch butterflies suggests that this species uses a simple compass system for orientation and follow the topography to reach their wintering sites in autumn (Mouritsen et al. 2013), but also that orientation shifts, from north in spring to south in autumn is triggered by temperature changes (Guerra and Repppert 2013).

9.2.3 Evidence for sun compass orientation in birds

Birds are well known to use directional information from the sun not only during migration and homing, but also for other spatial orientation tasks, for example during food caching and retrieving (Wiltschko 1980, 1981; Schmidt-Koenig 1990). As most animals, the sun compass of birds has been shown to be time-compensated (Fig. 9.3), and was first studied in birds by observing caged migrants in the circular orientation cages and measuring their migratory restlessness (expression of migratory flight in captive birds called *Zugunruhe*). The cages were equipped with windows enabling the bird to view the horizon, and it was noted that European starlings, *Sturnus vulgaris*, were able to maintain their migratory direction throughout the day, despite the movement of the sun across the sky (Kramer 1951). To show that the birds integrate the positional information derived from the sun with their inner clock, Kramer (1951) artificially deflected the position of the sun with mirrors attached to the windows of the orientation cage, either 90° clockwise or 90° counterclockwise, and could show that the birds adjusted their orientation accordingly (Kramer 1951). In conditioning experiments under the natural sky, pigeons and songbirds (including night migrants, for which no sun information is available as reference cue) were trained to relocate a food reward in a fixed compass direction using solely cues from the sun at different times of the day (Kramer and von Saint Paul 1950; von Saint Paul 1953, 1956). By training birds under controlled lab conditions providing an artificial sun, and consequently excluding any other potential directional sources, it was demonstrated that the azimuth position of the sun provided the essential orientation cue (von Saint Paul 1953). Moreover, the fact that the birds were able to find the food reward also at times of day other than the training time demonstrated that the sun compass of these birds took the movements of the sun across the sky during the day into account, which requires the ability to measure time of day and an integration mechanism to compensate for the change in azimuth of the sun over the day. Seed-caching corvids (jays and nutcrackers) also use a time-compensated sun compass in spatial orientation tasks on relatively short distances, such as during food search (Wiltschko and Balda 1989, Balda and Wiltschko 1991, Wiltschko et al. 1999).

Compelling evidence for sun compass orientation has also been provided by homing experiments with pigeons, *Columba livia domesticus* (reviewed by Wiltschko 1980, 1981; Schmidt-Koenig 1990). In clock-shift experiments, the vanishing bearings of clock-shifted homing pigeons were deflected relative to control birds in accordance with the direction and magnitude predicted by the use of a time-compensated sun compass (Schmidt-Koenig 1958a, b; Wiltschko et al. 2000). The accuracy with which pigeons compensate for the apparent movement of the sun across the sky over the course of the day appears to be extremely high and closely tuned to the differential speed of the azimuthal changes of the sun, and specific for both the time of year and testing location of the experimental setup (Wiltschko et al. 2000).

9.2.4 Skylight polarization as a directional compass cue

Skylight polarization arises when the unpolarized skylight that enters the atmosphere gets scattered by hitting particles in the atmosphere, via Rayleigh scattering (Horváth and Varjú 2004). Depending on the scattering angle (i.e. the angle between the unpolarized and the partially polarized light), the degree of polarization is maximal (i.e. 100%) at 90° from the sun, and minimal (0% = unpolarized) in the axial directions of the sun and anti-sun (Brines and Gould 1982). The electric vector (e-vector or direction of vibration) of the polarized light is perpendicular to the scattering angle, and thus, the polarization pattern across the sky provides a clear directional reference component with the sun in the centre (Fig. 9.4). This pattern provides all prerequisites needed in order to provide compass cues for orientation and navigation during migration, or for spatial orientation tasks (Brines and Gould 1982). One clear benefit of the skylight polarization pattern as a reference cue is that this information can also be used when the sun is not visible, and thus not available as compass cue. This mechanism functions as long as some blue, partially polarized sky is still visible, as well as when the sky is partly covered by clouds and the sun itself is obscured (von Frisch 1949, Wehner 1976, Adler and Phillips 1985, reviewed in Horváth and Varjú 2004; Muheim et al., Chapter 10). However, recently it has been shown that low degrees of skylight polarization (5–10%) are transmitted through clouds and fog and the pattern is very similar to that visible in a naturally

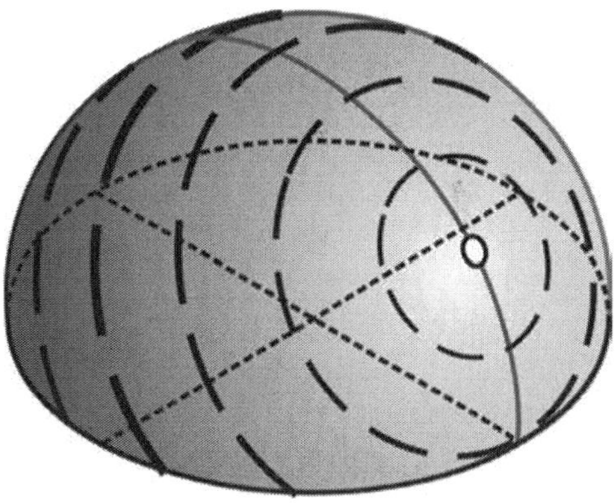

Figure 9.4 Skylight polarization patterns arising from light scattering within the Earth's atmosphere. Skylight becomes linearly polarized if scattered by atmospheric O_2 and N_2 molecules, but remains unpolarized if it reaches the observer directly (scattering angle 0°). Within a theoretical (Rayleigh) atmosphere, the degree of polarization reaches 100% if the scattering angle is 90°. The light is said to be partially linearly polarized when other scattering angles are considered. In the real atmosphere, the degree of polarization is almost always less than 70%. The direction of skylight polarization is indicated by bars, for which the size of bar illustrates the degree of polarization. The position of the sun is indicated by an open circle.

clear sky (Hegedüs et al. 2007a, b). To detect such low degrees of skylight polarization, animals need a very sensitive detection system. Field crickets, *Gryllus campestris*, possess such and have been able to detect degrees of polarization as low as 5% (Herzman and Labhardt 1989), but it is unknown if for example birds have the same capability. Orientation cage experiments performed under natural overcast skies often result in scattered orientation, suggesting a reduced motivation to migrate when overcast, possibly due to an inability to navigate under these conditions (Moore 1987; cf. Åkesson et al. 1996, 2001).

9.2.5 The use of information from skylight polarization for orientation

Evidences demonstrating the use of polarized light cues in migratory birds mainly come from experiments carried out around sunrise and sunset, which usually coincide with departure times of many migrants (Moore 1982, 1987; Able and Able 1993, 1995; Muheim et al. 2006c, 2009; reviewed by Muheim 2011). Migratory birds tested in orientation funnels with polarization filters on top of the cages change their orientation as a response to the shifted polarization axis, clearly indicating that they use the information from the polarized light transmitted though the filters to determine their preferred migratory direction (Able 1982, Helbig and Wiltschko 1989). When the natural polarization pattern is eliminated by depolarizers on top of orientation funnels, songbirds become disoriented in some (Helbig 1990), but not all cases (Åkesson and Bäckman 1999). Information from the pattern of polarized skylight is also used by migratory songbirds for compass calibration. Following experimental exposure to conflicting information between the Earth's magnetic field and polarized light cues, the birds showed a recalibration of the magnetic compass when later tested for magnetic compass orientation or released and followed by radio tracking (Able and Able 1990, 1993, 1995; Cochran et al. 2004; Muheim et al. 2006c, 2009). In contrast, birds did not recalibrate their magnetic compass after being exposed to shifted magnetic field information under depolarized skylight, which further demonstrate that birds use the polarization pattern of the sky to recalibrate their magnetic compass (Able and Able 1993). Experimental compass calibration exposures, which artificially imitated the natural polarization pattern near the horizon at sunrise and sunset with vertically aligned e-vectors, showed that this region of the sky is crucial for magnetic compass calibration. Birds exposed to a shifted polarization pattern near the horizon at sunrise and sunset recalibrated their magnetic compass while birds seeing the overhead polarization pattern only did not recalibrate their magnetic compass (Muheim et al. 2006b, c). To use the information from the skylight polarization pattern at these two times of day has the advantage that the birds can determine the exact timing of local sunrise and sunset irrespective of topography (e.g. mountains blocking the view of the rising/setting sun) and without the need for a time compensation mechanism. Integrating the calibration reference at sunrise and sunset allows the birds to obtain a seasonally and latitudinally independent source of a true geographic reference (Muheim et al. 2006c). However, later studies have not been able to confirm a magnetic compass-recalibration based on polarized skylight (e.g. Wiltschko et al. 2008, Chernetsov et al. 2011, Schmaljohann et al. 2013), and thus the research area needs further attention and repeated experiments in different geographical regions to clarify the reason for the different outcomes. Whether the skylight polarization pattern associated with the sun is part of the time-compensated sun compass remains unclear to date (Muheim 2011; Muheim et al., Chapter 10). Experimental evidence from the times of day other than sunrise and sunset is scarce, and thus, it remains to be demonstrated if polarized light cues provide directional information only at sunrise and sunset, i.e. during maximal polarization values. Possibly polarized light cues are used in a time-compensated manner and function as complementary or backup cue, thereby constituting an integral part of the sun compass during times when the sun disk itself is not visible.

9.2.6 Use of sun board and skylight polarization by Viking navigators

More than a thousand years ago the squared-sailed merchant ships from Scandinavia regularly crossed long stretches of open sea in the North Atlantic when Vikings visited distant trading centres in Europe or exchanged goods in more northerly regions. Archaeological findings support that Vikings established settlements in south Greenland, Newfoundland and recently confirmed also on Baffin Island in northeast Canada several hundred years before Columbus arrived to the continent in 1492. Vikings were further navigating along the rivers in Russia southeast to the Black Sea and Constantinople. The skilful Viking navigators were known to have a number of navigational tools at hand, some of which were based on information from the sun and possibly also the pattern of skylight polarization (Thirslund 2001, Karlsen 2003). Known are the sun board, the sun compass, and the use of drifting lines, birds and waves, but also recognition of local topography (many merchant-ship routes followed coasts for stretches of time) and use of an enigmatic sunstone. The use of a 'sunstone' (Icelandic, sólarsteinn) to pinpoint the position of the sun when it was obscured by clouds was first mentioned in an Icelandic saga where Kung Olaf visited the wise man Sigurd and his sons (Icelandic Saga database). The description of how King Olaf took out his sunstone and pointed it to the sky a cloudy day and read the position of the sun to confirm that Sigurd was right in his estimation of time of day later became food for speculation on how possibly the sunstone was used for navigation purpose on ships and would function also under overcast and foggy skies (Ramskou 1969, Horváth et al. 2011). It was however not mentioned what material the sunstone was made of or how it was used in the original description. Icelandic spaar (i.e. Calcite), from north-east Iceland might be a candidate crystal, characterized by its bi-refringent optical characteristics and large crystal size. By pointing the crystal to the sky, observe the sky through the crystal from below and turning the crystal along the horizontal plane, a double or single image would appear if the crystal contained one single mark on the top (alternatively the turning would cause a change in shading) depending on how the crystal aligned with the pattern of skylight polarization, similar to a linear polarizer.

The optical prerequisite for sunstone navigation under foggy and overcast skies has been confirmed by full sky polarization measurements (Hegedüs et al. 2007a, b; Horváth et al. 2011), and the precision of the crystal compass under similar conditions have been estimated to be sufficient and with high repeatability (Ropars et al. 2012), so there is good theoretical support for its potential use. Still exactly how the crystal was used remains speculative, but a recent discovery of a sunstone on board a Victorian ship dated 1592 from the English Channel near to navigation instruments (Le Floch et al. 2013), further support the potential importance for navigation and use much later than Viking ages. It is therefore likely that the Vikings possessed both the crystals, since Iceland was colonized before Vikings inhabited Greenland and North America, and the knowledge to handle the valuable instrument during sea crossings. As a primary navigation tool a primitive sun compass was used by Viking navigators and to which the crystal was a backup to locate the sun when hidden behind clouds. This instrument, a sun board with a shadow stick (sun dial), needed to be calibrated before start of the journey by indicating a shadow cast from the sun on the board at different times of day and was functioning at a certain latitude and time of year which enabled the Vikings to sail east-west across longitudes (Thirslund 2001, Karlsen 2003). In addition to the technical tools birds and sea mammals were observed and assisted the Vikings to locate coastal areas, but also bright stars guided them when sailing under clear night skies. Just like humans, migratory birds are also known to learn to use stars for orientation, which will be discussed later.

9.3 The star compass

The majority of migratory songbirds are nocturnal migrants, i.e. they migrate at night, when the sun is not available as a reference cue. Celestial cues that are available at night are the stars, and shortly before sunrise and after sunset also polarized skylight is visible. Reed warblers, *Acrocephalus scirpacaeus*, tracked by radio-telemetry on natural migrations select to depart when the sky is sufficiently clear and the stars become visible (Åkesson et al. 2001), suggesting that the stars are important for orientation during their nocturnal migration flights. Soon after the sun compass was described by Kramer (1950), evidence for stars as reference system for compass orientation followed. The first planetarium experiments suggesting that night-migrating passerines use the stars for orientation were performed by Sauer (1957). Sauer was of the opinion that birds could truly navigate to their wintering sites with just the help of stellar cues (Sauer 1957). The definite proof that birds can use the stars for orientation came in 1967, when Steven Emlen published his classical planetarium experiments with indigo buntings, *Passerina cyanea*. His data suggested an imprinting-like learning process, which can only take place during a small sensitive time-window before the onset of migration (Emlen 1967a, b, 1970). In a later planetarium experiment, pied flycatchers, *Ficedula hypoleuca*, and European redstarts, *Phoenicurus phoenicurus*, were exposed to star patterns from different locations on the globe (Rabøl 1998). These data proposed that, in accordance with Sauer (1957), the birds corrected for the simulated displacements, and thus could actively navigate using the stars as orientation cues. There is unfortunately not many experiments addressing the functional characteristics of a star compass in birds, and in contrast to Rabøl (1998), Emlen's data do not suggest true star navigation (Emlen 1967a, b, 1970; Mouritsen and Larsen 2001). Therefore, this field is one which would benefit from further critical experimentation.

Just like the sun, stars are not stationary in the night sky but appear to rotate around the north (and south) sky. Therefore there are several theoretical possibilities on how birds could use the stars to find their way. One possibility, first suggested by Sauer (1957), is that birds deduce their true geographical position from the centre of rotation of the starry sky. Theoretically, the elevation of the centre of rotation above the horizon could provide a measure of latitude, and the rotational phase of the stars, combined with a dual time sense, could provide longitudinal information. However, to detect their longitudinal position, the birds would need to make use of two clocks: one accurately fixed on home-time and one internal clock that quickly adapts to local time (Gwinner 1996, Gwinner et al. 1997), but the time-keeping mechanism needs to be very precise for navigation, which has so far not been demonstrated in any animal (Wehner 1998).

Stellar information could be used by birds to navigate towards a distinct goal travelling along a loxodrome course (i.e. equivalent to a constant compass course route; Fig. 5.2). An alternative approach to explain the use of celestial cues suggests that night-migrating birds use the stars as part of a time-compensated compass (Schmidt-Koenig 1990), but without being able to establish their geographical position from the rotational phase (hour angle) of the stars. Emlen's experiments (Emlen 1967a, b, 1970), however, strongly suggest that indigo buntings do not make use of the rotational phase of the starry sky, neither to identify geographical position nor as part of a time-compensated star compass (working in a similar manner as the sun compass). Instead, his data suggest that the birds only use the location of the rotational point and subsequently the constant geometrical pattern of the stars to define geographical north (Fig. 9.5). Emlen further showed that the star compass is learned and not inherited as mentioned earlier. In his experimental series, hand-raised birds learned the position of the centre of rotation during ontogeny, independent of which star was positioned at the rotational point (Emlen 1970, 1975). Hand-raised birds that had not seen the starry sky (or any artificial light dots) before onset of migration were unable to orient. Even subsequent exposure to the starry sky was not sufficient for gaining star-based orientation (Emlen 1970).

Birds seem to deduce geographical north from the geometrical configurations of the stars independently of current stellar rotation (Emlen 1975), that is, the birds use the stars for time-independent celestial compass orientation. This hypothesis is consistent with the orientation behaviour observed in subsequent experiments (Wiltschko et al. 1987). Garden warblers, *Sylvia borin*, were hand-raised in four corners of one room, where they were exposed to an artificial sky simulated by a set of diodes rotating around the centre of the room. Depending on the birds' position in the room during this imprinting period, they experienced the location of the centre of rotation at different directions. When tested for subsequent star orientation, they oriented in different directions, consistent with an innate strategy, which told them to migrate 'away from the learned centre of rotation' (Wiltschko et al. 1987). This was the direct proof that the birds

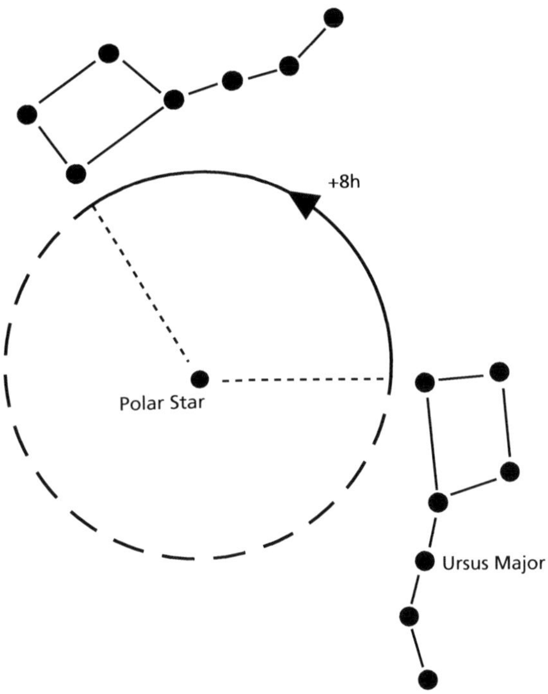

Figure 9.5 Star compass in birds. By observing the rotation of the stars around the centre of rotation, i.e. the Polar star on the Northern hemisphere (on the Southern hemisphere the pendent is Betelgeuse) during a developmental window in their early life, the birds can later during migration determine the direction of true North (or South in the Southern hemisphere) by looking at star patterns on the sky, in a similar way as we do.

perceived the artificial light dots as 'starry sky' and learned the geometrical star configurations to pinpoint and calibrate the rotational point of the starry sky as geographical north (Fig. 9.5). These data suggest that birds do not inherit detailed knowledge about the star pattern, but that they are equipped with an inherited strategy to look for the centre of rotation and interpret this as 'polewards'.

9.4 The magnetic compass

A broad variety of animals across different taxa use directional information from the Earth's magnetic field for orientation (for reviews see Wiltschko and Wiltschko 1995, 2005). In true magnetic compass orientation, animals use directional magnetic information to orient towards a specific direction, which in the case of migratory animals may be the species-specific migratory direction, but may also be a learned compass course towards a food source, shelter, shoreline, etc. (see later). Animals using a magnetic compass can be found across the entire animal kingdom (see also Muheim et al., Chapter 10), and include insects like fruit flies (Phillips and Sayeed 1993), diverse species of social insects like ants and bees (Wajnberg et al. 2010), and representatives from all vertebrate classes, for example, salmon (Quinn and Brannon 1982), eastern red-spotted newts (Phillips 1986), sea turtles (Light et al. 1993), mole rats (Burda et al. 1990), and mice (Muheim et al. 2006a). It has even been speculated that humans have (or maybe once had) a magnetic compass (see e.g. Baker 1980). Magnetic compass orientation is distinctly different from magnetic alignment, where animals passively align relative to the magnetic field in a direction that does not lead to a specific goal. Magnetic alignment is a common feature in insects, which often align along the cardinal directions of the magnetic field (Wiltschko and Wiltschko 1995), but it has been suggested to be important also in mammals, like feeding and resting cows and deer (Begall et al. 2008) and more recently in fish (Hart et al. 2012). Migratory birds are the prime example of animals that use a magnetic compass on their migratory journeys, and were also the first organisms for which a true magnetic compass was demonstrated. In their ground-breaking work, Wolfgang Wiltschko and Friedrich Merkel demonstrated that migratory European robins, *Erithacus rubecula*, changed their migratory orientation when the alignment of the magnetic field around the orientation funnels was horizontally shifted relative to the natural magnetic field (Merkel and Wiltschko 1965). This can be done with electromagnetic coils (Box 9.1), which produce an artificial magnetic field that is superimposed onto the ambient magnetic field. This experimental setup creates an accurate representation of the magnetic field presented to the organism, and allows for alteration of the direction of the resulting vector, intensity and inclination. Furthermore it completely cancel out the magnetic field, thus excluding this variable as a reference cue. So, if magnetic north was shifted either 120° clockwise to geographic southeast or 90° counterclockwise to geographic west, the birds changed their orientation accordingly, demonstrating that

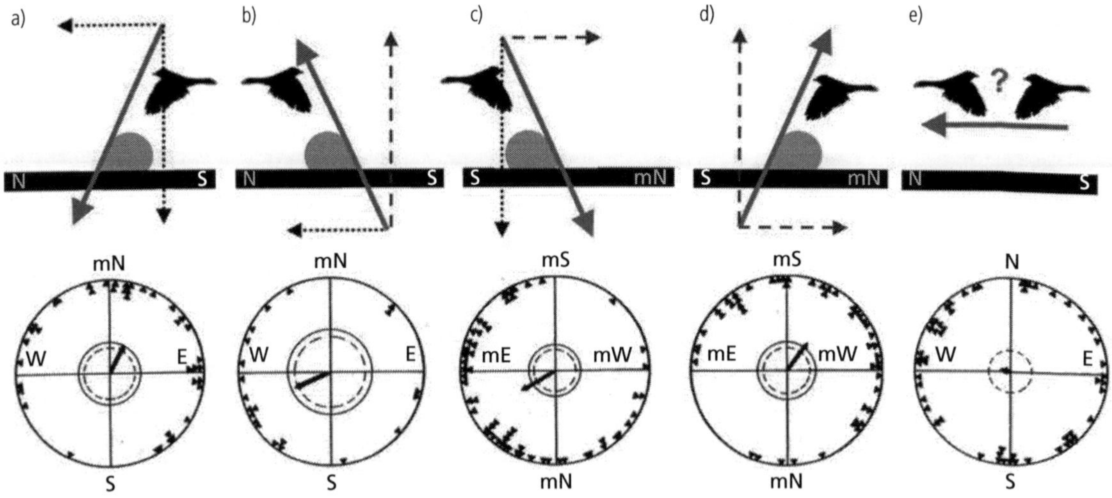

Figure 9.6 The bird's magnetic compass is sensitive to the axis of the magnetic field vector, but not to the polarity, and so birds therefore cannot distinguish between 'south' and 'north' but only between 'poleward' and 'equatorward'. This has been demonstrated in an orientation experiment during spring migrations in European robins, *Erithacus rubecula*. (a) Control condition for robins during spring migration is that they migrate towards the north, as depicted in the upper panel; the orientation diagram in the lower panel orientation data clearly confirms the northerly orientation of all tested birds. Each marking in the compass circles indicates the mean orientation of one bird; the arrow in the centre shows the mean orientation of the group of birds, tested under the given experimental condition. (b) Experimental manipulation of the magnetic field vector. If the vertical component gets inverted, the robins revert their heading compared to (a). In (b), magnetic north (mN) still remains in geographic north, but the course of the field lines, and with it the inclination, has changed. (c) When the horizontal component of the magnetic field gets inverted, this results in a scenario where magnetic north is now located in geographic south. In this case the birds changed their preferred direction correspondingly and change their heading 180° compared to the control situation in (a). If both the horizontal and the vertical component of the magnetic vector get inverted (i.e. inclination angle stays the same, axial direction of the field lines is unchanged, only polarity gets inverted) (d), the birds' behaviour is identical to the pattern shown in the control situation. (e) Depicts the scenario the birds experience at the geomagnetic equator, where the magnetic field lines run horizontal to the Earth's surface (i.e. the magnetic field has only a horizontal component, the length of the vertical field vector is zero, no inclination angle can be deferred). In this test arrangement the birds seemed to be unable to orient. These data clearly demonstrate that the bird's magnetic compass is an inclination compass. (Redrawn from Wiltschko and Wiltschko 1972 with permission from John Wiley & Sons.)

they used the Earth's magnetic field as the primary directional reference (Wiltschko and Merkel 1966). Some years later, it was demonstrated that the magnetic compass of birds is an inclination compass (Wiltschko and Wiltschko 1972; Fig. 9.6). This is in contrast to a polarity compass (see next paragraph) that does not distinguish north from south, but instead between pole-ward and equator-ward directions. This was shown by inverting the vertical component of the magnetic field in their experimental setup by large magnetic coils surrounding the cages where the birds were kept. In this situation the resulting manipulated vector of the magnetic field still has the same polarity as the natural vector pointing towards magnetic north, but is not pointing downwards, but instead upwards. This manipulated situation is thus simulating the situation found in the southern hemisphere (Fig. 9.7). The robins responded by shifting their orientation 180°, which clearly demonstrated that they did not use a conventional polarity compass that senses the direction of the magnetic field (this was unchanged in the treatment; see later in this section), but instead an inclination compass that distinguishes between directions towards the poles and the equator.

The intersection of the magnetic field lines with the surface of the earth (horizon points) always indicates pole-wards towards the closest of both poles, in both the northern and southern hemispheres. In a corresponding way the direction where the inclination angle and the horizon diverge will always point towards the magnetic equator (Fig. 9.7; reviewed by Wiltschko and Wiltschko 1995, 2005). A polarity compass, in contrast, works pretty much

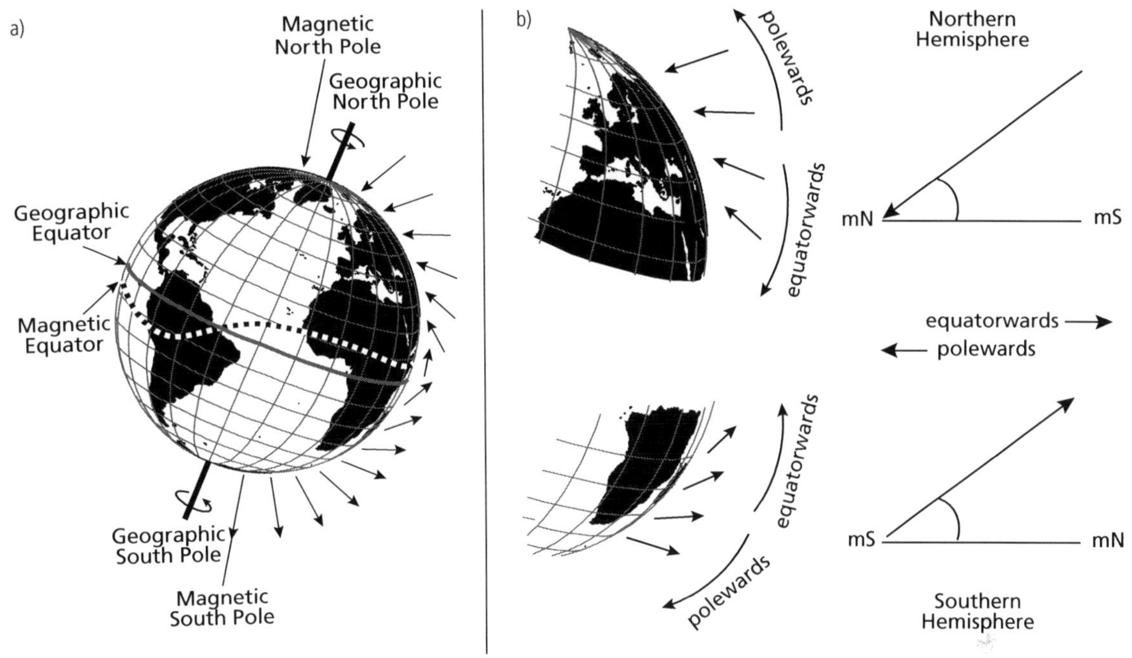

Figure 9.7 The magnetic inclination compass provides animals with information on the axial alignment of the magnetic field (direction along the magnetic north–south meridian). To distinguish the two ends of the axis, the sign of the inclination angle is used: the direction towards the magnetic pole that lies closer coincides with the intersection of the magnetic field lines with the horizon, and the magnetic equator lies towards the direction where the magnetic field lines and the horizon diverge. An inclination compass does not perceive the polarity of the magnetic field vector, like human technical compasses.

like a human technical compass and is in its construction sensitive to the direction of the horizontal component of the magnetic field. Animals using a polarity compass can therefore sense the direction of the magnetic field vector (they know which side the 'arrow' is located) not only its axial orientation. Only a minority of animals, like mole rats, fish, and bats, has been proposed to have a polarity compass; thus, it seems still to be rather the exception than the norm (Wiltschko and Wiltschko 1995, 2005).

An inclination compass seems to be the perfect adaptation for any migrants as the migratory programmes of long-distance migrants living in either the northern or southern hemisphere would then be coded similarly. Birds leaving the northern hemisphere during autumn migration fly equator-wards on their journey to their wintering areas when day length decreases and instead pole-wards towards the breeding grounds when day length increases. A potential problem arises when they approach the equator. What happens when migrants must cross the equator where the magnetic vector is parallel to the Earth's surface and the magnetic inclination is no longer a reliable reference? This situation is tricky in two respects. First, the horizontally aligned magnetic field at the equator does not provide any useful directional information on which side of the axis leads towards the poles or the equator, and secondly, when passing the equator, migratory birds must switch their entire migratory programme from 'fly equator-wards' to 'fly pole-wards' to be able to continue on the intended route. Experimental evidence suggests that exposure to the horizontal magnetic field simulating a situation at the magnetic equator itself may trigger the change in migratory programme (Wiltschko and Wiltschko 1992). A similar mechanism has been confirmed for a North American passerine migrant, the bobolink, *Dolichonyx oryzivorus* (Beason 1992), which also responds to exposure of a horizontal magnetic field by shifting the orientation relative to the angle of inclination. Another interesting property of the magnetic compass in birds is that it operates within a narrow functional window of magnetic field intensity. The working window closely mirrors the intensities found on Earth, but it can artificially be expanded when the birds are pre-exposed to an artificial magnetic field outside the naturally occurring window (Wiltschko 1968, Wiltschko et al. 2006). Nowadays, it is believed that this behavioural peculiarity is a property of the magnetoreception mechanism of the avian magnetic compass (see Muheim et al., Chapter 10).

9.5 Vector navigation in young migratory naïve birds

Imagine a young migratory songbird, the European blackcap, *Sylvia atricapilla*, meeting its first days of life a sunny May morning in a Central European deciduous forest. Just like its nest mates it is hatched with a migration programme inherited from the parents encoding the time, distance (encoded as a time period in relation to internal time sense and annual cycle) and direction to migrate (Helbig 1991). During the first and following autumns it will migrate alone, protected by darkness at night, over a time spanning several weeks. Included in the programme are periods of fuelling and flight involving physiological and behavioural changes (e.g. Åkesson and Hedenström 2007; Hedenström and Lindström, Chapter 5). A population-specific migratory direction, perhaps involving course-changes en route (Gwinner and Wiltschko 1978), will be executed at appropriate times and places in order to reach the intended destination. The endogenously encoded courses have been shown to be stable over time (Helbig 1992). The direction of the migratory restlessness (*Zugunruhe*) can be studied in circular cages, in which the external information may be manipulated and compass mechanisms investigated (Fig. 9.8; see also colour plate section). The first migration in songbirds is thought to be encoded as a simple *clock-and-compass model* (Berthold 1996), while experienced migrants later may incorporate more information in their navigation system and use goal navigation to reach known wintering areas (Perdeck 1967). However, the simple programme has been questioned (Rabøl 1985, Thorup and Rabøl 2001) and the major criticism is directed to the situation birds face on natural migrations, when they actually experience flight conditions in which winds may affect their routes taken in relation to ground as they are exposed to wind drift. Wind forces often reach the same magnitude or stronger than that of the airspeed of the bird itself, having a significant effect on the birds' flight routes. Rabøl (1985) argued that a correction mechanism is more likely to be part of the inherited migration system, executed during migration flights, and that birds are aiming for intermediate goal areas en route. As expected some songbirds have been shown to selectively use intermediate and species-specific stop-over areas during autumn migrations (Fransson et al. 2005). At the same time, the early displacements of starlings suggest that young birds are unaware of translocations, but keep a fixed compass course during migration (Perdeck 1958). It is difficult to explain the discrepancy between studies and how the navigation phenotype is expressed differently during the first and later migrations. Still the solo-migration of young songbirds is truly spectacular and mind-boggling as we still do not fully understand the genetics behind the behaviour and how the expressions of different phenotypic traits are regulated (Liedvogel and Lundberg, Chapter 12). The migration phenotype of songbirds is also under the influence of external information met en route (e.g. Beck and Wiltschko 1988, Fransson et al. 2001), suggesting sophisticated evolution of gene expression and interactions with environmental cues met over evolutionary times. However, there has been some major efforts investigating the genetics of migration and we may see more advances within this field of research within the near future (Liedvogel et al. 2011; Liedvogel and Lundberg, Chapter 12).

Songbirds have been shown to inherit a migratory direction relative to the Earth's magnetic field (e.g. Wiltschko and Gwinner 1974), but they are also dependent on experience from a rotating starry sky pattern in combination with the natural magnetic field to express a population-specific migratory direction as we have just discussed. A rotating star pattern will only generate a migration response away from the rotation centre towards south in autumn for young garden warblers, *Sylvia borin* (Weindler et al. 1996). A migration away from the rotation centre of the night sky in North American songbirds

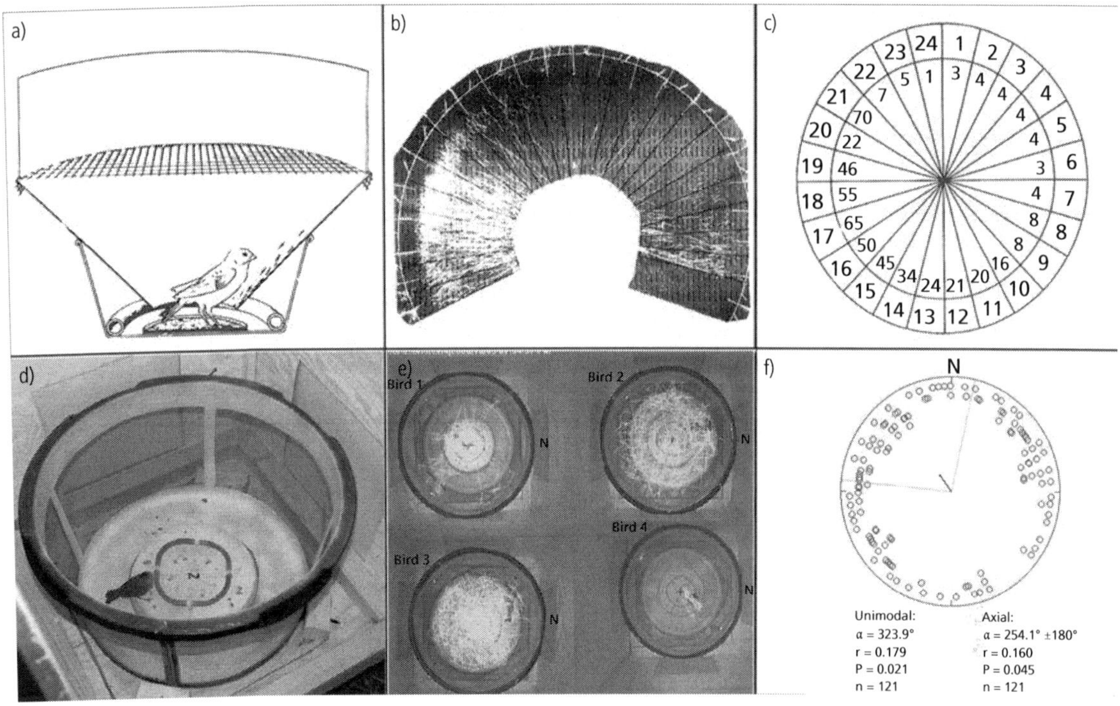

Figure 9.8 Various types of circular orientation cages have been designed to accurately measure and quantify the orientation preference of caged migratory birds. (a) The original orientation cage was invented by Emlen (Emlen and Emlen 1966) and is referred to as the 'Emlen funnel'. The walls of this funnel-shaped cage were covered with white blotting paper, with an ink pad placed at the bottom of the cage. Thus, a bird that is tested in this experimental setup leaves ink marks on the cage walls whenever trying to fly in a specific direction. (b) This cage design is slightly modified compared to (a), and is typically used as standard method in most orientation studies to date. Here the ink pad is removed, and the walls are covered with typewriter correction paper (e.g. white-out) or thermal paper (Mouritsen et al. 2009) whereon the birds leave scratches whenever touching the cage wall. Shown in the picture is the line-out scratch-sensitive paper (e.g. white-out or thermal paper) after a 90-min orientation experiment, spread on a light table. (c) A circular diagram divided into 24 sectors (covering 15° of the circle) showing the distribution of scratches, hand-counted on the paper shown in (b). (d) More recently, video tracking has been introduced in orientation research. Here, the movements of the birds in cages or funnels are filmed from above or below. Tracking software allows tracking the position of the bird throughout the experiment, and analysing its orientation direction. (e) Digitized orientation data of a 10-min movie interval of a bird filmed in a cage as shown in (d). (f) Orientation of the data shown in (e). Each hop of the bird from the centre (inside of the green inner circle) out to the periphery of the cage and back into the centre again counts as one data point; the position farthest away from the centre is taken as the direction of the hop (round symbols in diagram f). From all these data points (hops), the mean orientation vector with an angle α and a length r (varies between 0 and 1) is calculated. The length *r* can be considered as an inverse analogue of the variance (modified after Helbig 1991; Muheim, unpubl.). (See Plate 2).

was also demonstrated by the initial planetarium experiments performed by Emlen (1976). However, the combined experience of a rotating star pattern and the geomagnetic field seems crucial for expression of an adaptive orientation response in autumn migrating young songbirds (Weindler et al. 1996).

9.5.1 Fuelling before migration

Not only compass courses and migration timing are inherited, but also fuelling periods (Hedenström and Lindström, Chapter 5). The necessary physiological responses seem to be under the influence of external information from, e.g. the Earth's magnetic field (Fransson et al. 2001). In laboratory experiments thrush nightingales, *Luscinia luscinia*, were magnetically displaced to a geographic location just north of the Sahara desert in the Mediterranean region during autumn migration, and mass change recorded in cages by electronic balances. The thrush nightingales reacted by increasing their fuelling rate as if they were expecting to soon cross the Sahara desert, despite that the caged birds had not yet experienced a natural migration. In contrast the group of thrush nightingales experiencing only the local geomagnetic field did not put on as much fat as the displaced birds (Fransson et al. 2001). This suggests that fuelling decisions encoded in the endogenous migration programme are dependent on external (and local) geomagnetic information. Later studies investigating the fuelling in other species of songbird migrants have further shown interactions between an external magnetic field and fuelling in birds (e.g. Kullberg et al. 2007; Hensaw et al. 2008; Boström et al. 2010, 2012b), supporting a widespread physiological response to external magnetic fields during this time in the bird's life.

Similar to the young wandering albatross left by its parents on a remote island in the Southern Ocean, discussed in the beginning of this chapter, the young songbird in the European forest faces the same challenge in preparing for its first migration. The albatross needs to find a suitable ocean area where it can forage and spend several years before it becomes sexually mature and returns to the island to breed. What information is used by the young birds during the initial migration and how they find their way during long journeys across the sea is still unexplained. In recent studies using satellite telemetry the initial migrations of young wandering albatrosses have been tracked showing differential migration and preferences for sex-segregated ocean areas in migratory naive albatrosses (Åkesson and Weimerskirch, 2014). The young albatrosses leave the breeding island on prolonged meandering flights towards the north-east when suitable winds from the south-west are hitting the island. Winds from this direction are rare, as north-westerly winds dominate and the young albatrosses are therefore very careful selecting the

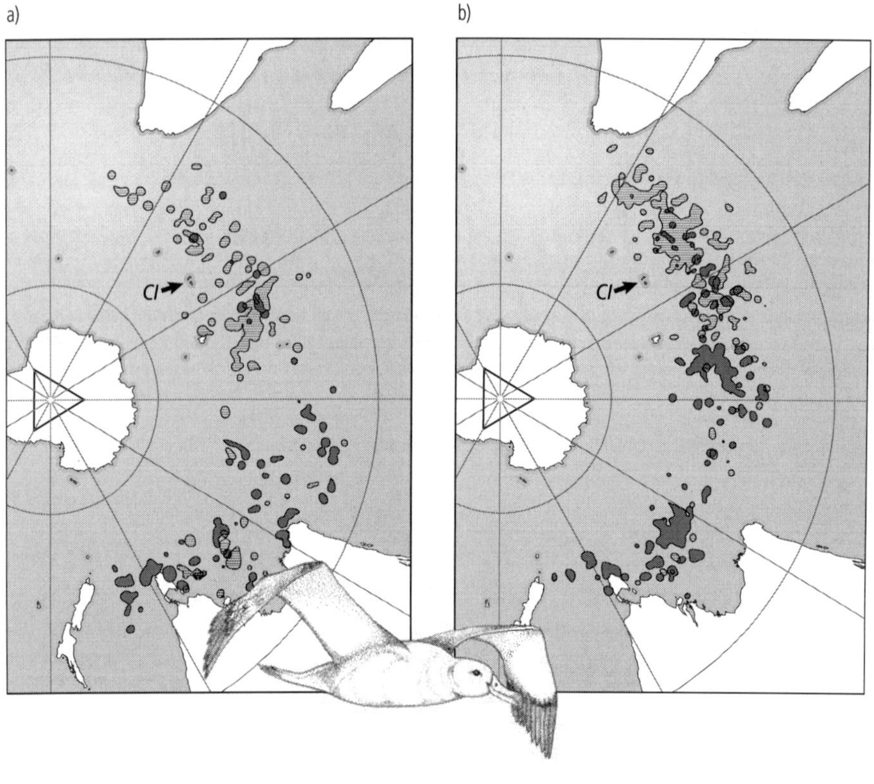

Figure 9.9 Map illustrating post-breeding foraging zones used by juvenile and adult wandering albatrosses, *Diomedea exulans*, breeding at Crozet Islands (indicated with CI) during the post-breeding period as indicated by 75% kernel density zones. (a) Adult females (striped) and males (black). (b) Juvenile females (striped) and males (black). Modified after Åkesson and Weimerskirch (2014).

best time to depart. After the initial flight the individual young albatrosses set out to different ocean areas, females heading to ocean areas south-east of Africa and south of Madagascar located north to north-west of the breeding island, while the young males move more than double the distance to areas south and south-east of Australia (Åkesson and Weimerskirch, 2014). The areas selected by the young wandering albatrosses are very similar to those used by adult birds from the same breeding location at Crozet islands (Fig. 9.9).

It is not possible to explain the initial migration by the young wandering albatrosses by a simple clock-and-compass mechanism, as there must surely be navigational aspects and selective use of winds incorporated in the 'migration programme' used by the different sexes (Åkesson and Weimerskirch, 2014). The timing of migration and flight distance may be encoded as an extended time for dispersal migration in the males which is substantially longer than that for females. However, the courses selected to reach the sex-segregated destination areas selected by female and male albatrosses differ. Still it remains to be shown how they find their way and what navigational mechanism is used, especially when bi-coordinate navigation in this geographical region are expected to be difficult due to large areas with parallel gradients (i.e. inclination and total field intensity, Åkesson and Alerstam 1998). Is the problem of using geomagnetic gradients local and is the situation for magnetic bi-coordinate navigation better in other parts of the ocean? How likely is global bi-coordinate magnetic navigation?

9.6 Bi-coordinate maps—a global perspective of navigation

To perform bi-coordinate gradient map navigation an animal learns how two gradients that vary predictably in time and space vary on a local scale and extrapolates this knowledge during navigation over a larger scale (Wallraff 1974, Wiltschko and Wiltschko 2003). For this navigation method to be successful, the gradients must form a grid, and hence cannot be parallel to each other (Fig. 9.10). Apart from the ability to learn and extrapolate how the gradients vary, the animal also needs a compass to allow it to determine the direction towards the goal and ensure successful navigation (Kramer 1953, 1957). Several different gradients have been suggested as a basis for bi-coordinate maps among animals, e.g. odours (Svensson et al., Chapter 11), the Coriolis force effect (Yeagley and Whitmore 1947) and magnetic field components (total field intensity and its components, inclination and declination; e.g. Wallraff 1991, Gould 1998). Odours have been shown to be an important factor in homing pigeons and seabirds close to their lofts or nests, but in a global perspective, however, the Earth's magnetic field and its components have been suggested to be the most likely basis for a bi-coordinate map sense among migratory animals (first suggested by Viguier 1882; Lohmann et al. 2001).

A bi-coordinate map sense based only on magnetic information could potentially be sufficient for navigation in some areas on the earth, but not all since the magnetic field lines for intensity and inclination are parallel to each other in some areas (Åkesson and Alerstam 1998, Boström et al. 2012a). A possible solution would be to use declination, which would require the ability to determine the difference between true north (geographic north) and magnetic north, and integrate that information into a map (Åkesson and Alerstam 1998, Åkesson et al. 2005). Whether animals are able to use declination to define longitude has been discussed for some time, but it still remains to be investigated (Gould 1998). However, some support of longitude detection by using geomagnetic information has been reported for sea turtles (Putman et al 2011). The most convincing evidence for magnetic bi-coordinate navigation in animals comes from a series of experiments with loggerhead sea turtles, *Caretta caretta* (Lohmann et al. 2001). Hatchling loggerhead sea turtles exposed to altered magnetic fields, simulating displacements to different locations in the Atlantic Ocean, and tested in a circular water-filled arena, choose directions that would keep them in the geographic area of the North Atlantic gyre. Similar experiments with Caribbean spiny lobsters, *Panulirus argus*, and eastern red-spotted newts, *Notophthalmus viridescens*, have also found evidence for magnetic map navigation in these species (Fischer et al. 2001, Phillips et al. 2002, Boles and Lohmann 2003), suggesting the use of magnetic gradient maps over various distances. In homing pigeons,

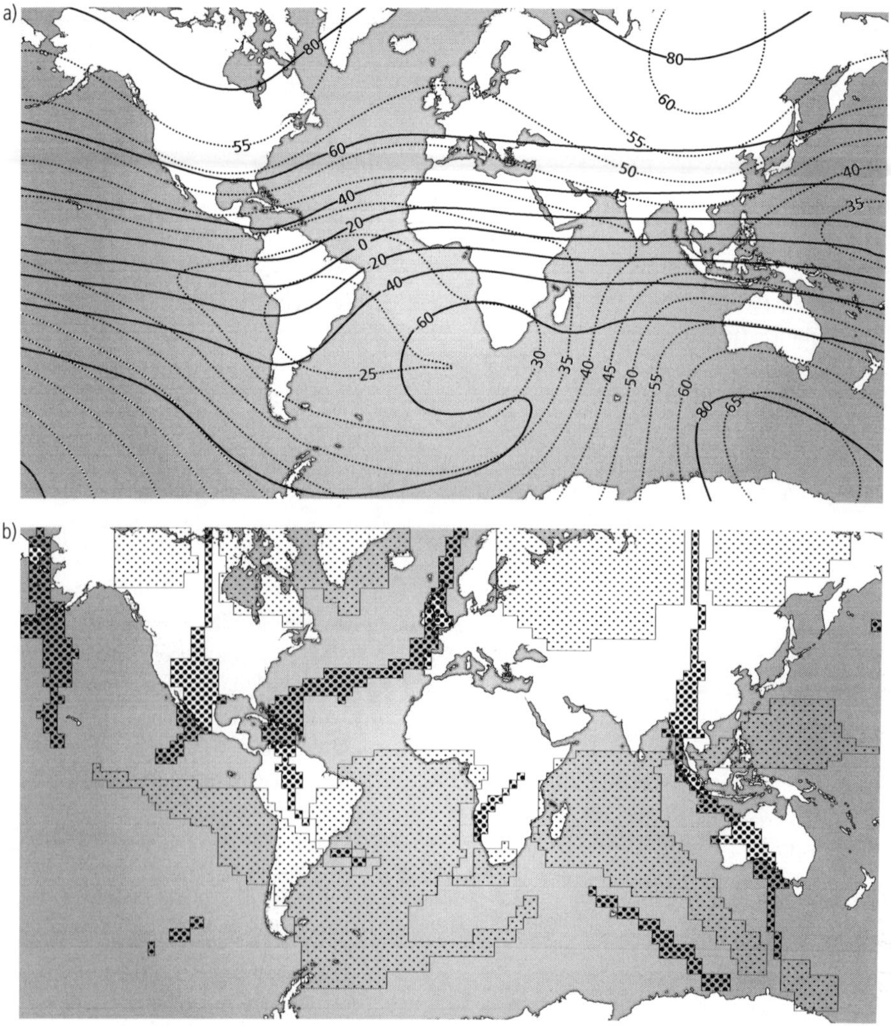

Figure 9.10 (a) Map visualizing the isolines for total field intensity (μ T, dotted line) and inclination (°, filled line). The distance between the isolines is 5 μ T for the total field intensity and 20° for the inclination. The map is in Mercator projection and shows latitudes between 70°N and 70°S. (Modified after Boström et al. 2012.) (b) Map illustrating the possibility to use a bi-coordinate map based on geomagnetic inclination and total field intensity. The map was divided into 3° latitude and 3° longitude squares and the angular difference between isolines of total field intensity and inclination was estimated within each square. Areas with an angular difference of <2° are shown as large dots, 2°–30° as white/grey in between the dotted fields and >30° as small dots (modified after Boström et al. 2012).

Columba livia, there is conflicting evidence regarding the importance of geomagnetic cues, olfactory cues, and landmarks for successful homing (Gagliardo et al. 2009, Wiltschko et al. 2009). All these cues have been demonstrated to play a role during homing events, but their importance seem to differ among lofts and areas. Therefore, further investigations are needed also here to explain the patterns observed and the integration of different types of information.

9.7 Path integration—an egocentric view of navigation

The desert ant explores the sand dunes of the Sahara desert in search for food, and during its trips across the cluttered surface avoiding obstacles and circumventing barriers it is continuously updating a lifeline connecting its current position with a direct route home to its underground nest (Fig. 9.1). It uses path integration to navigate during the several hundred of metres long foraging trip. The path integration is based on updates of distances covered and courses steered with a continuous integration of information from visual compass cues, optic flow detection and movement detectors (Müller and Wehner 1988). For us to resolve the mathematical algorithms describing what desert ants do in minutes is challenging, still the insects have adapted high efficiency and sufficient precision to control their movements to a high degree in order to safely return home. The compass information is read from the skylight polarization pattern, which provides information on the course steered. The apparent movement of the sun across the sky may complicate the use of this dynamic pattern for directional information, but the ants have evolved an ability to approximate the movement and compensate for it (Wehner 1998). In addition the ant has two parallel systems to measure distances, the first using optic flow across the lower part of the eye (measured by observing the speed by which it moves over the ground) and the second by counting of steps (measured by proprioceptors, or stretch receptors, in its own legs; Wehner 1998, 2009).

In an elegant outdoor experiment performed in the natural habitat of desert ants it was discovered that not only the skylight polarization compass information and visual flow measuring speed of the ant is included in the path integration, but also the ant's vertical movement relative to the horizontal plane (Wohlgemuth et al. 2000). By forcing the ant to follow an uphill and downhill track directed by a long channel to a feeder and enabling the ant to only see the sky and the nearest ground under it, as it moved along the channel, the scientists tested its path integration. At the feeder the experimenter captured it and transported it in darkness to a nearby test field, where the ant was released and its preferred distance of the return movement to the expected nest position was measured. With this experiment it was shown that the ants were compensating for the up- and down-movements, presumably assisted by a gravity sensor, and estimated distances covered projected relative to the horizontal plane (Wohlgemuth et al. 2000, Grah et al. 2005). A path integrator taking movements relative to the horizontal plane into account is more safe and useful in a sand dune environment, when also the stability of a landscape environment at this scale may be low over time as sand is easily transported by persistent winds and by occasional sand storms and, in addition, the habitat may change from one day to the next.

The precision of both compass and distance measurements by the neuronal system of the ant is limited, and far from perfect, but apparently sufficiently precise to allow for successful homing over many thousands of body lengths. The ants make turning errors, which are increasing with turning angle (Müller and Wehner 1988), and a path integrator based on optical flow may also generate errors in estimating distances covered. However, as the ant is turning equally many times to the left and right during a forging trip, the angular errors likely cancel out and are therefore reduced (Müller and Wehner 1988). In addition to the path integrator ants also are able to use a systematic search strategy based on the same path integrator (Müller and Wehner 1994) and use landmarks to pinpoint the nest entrance (Wehner and Räber 1979, Collett and Cartwright 1983, Wehner et al. 1996).

9.8 Landmark navigation

Birds (Braithwaite and Guilford 1991, 1995; Burt et al. 1997) and other animals (Collet 1987), including insects, such as desert ants, as seen earlier (Wehner and Räber 1979, Wehner et al. 1996, Collett et al. 2001, Åkesson and Wehner 2002), are able to learn to use landmarks to pinpoint their home, but also to navigate along familiar routes. Insects may memorize two-dimensional landmark images of their home location (e.g. Cartwright and Collett 1983, 1987), for relocation at return journeys rather than memorizing exact positions of landmarks in a 3-dimensional space. Honeybees and ants show the typical turn-back-and-look behaviour during the process of memorizing landmark scenes (Lehrer 1991, 1993). The earliest experiments showed how the apparent size of the landmarks and the position on the retina guided the desert ant back to its nest (Wehner and Müller 1985). The landmark scenes are memorized in the central part of the eye, from the horizontal plane to the dorsal rim area in the top part of the compound eyes. The images memorized in one part of the compound eye are retinotopically fixed to this section of the eye and will later be recovered in the same part when the ant is returning to the same location (Wehner and Müller 1985). Later experiments have shown that the landmark scene is memorized relative to celestial compass information, but that during homing runs the landmark scene and celestial information may temporarily be decoupled, giving initial priority to landmark images suggesting a simplified mechanism in use by the ants in their everyday life (Åkesson and Wehner 2002). Bees also link local landmarks with celestial information (Dickinsson 1994). Homing pigeons returning to their loft following familiar routes seem to combine information from compass information and memorized landmarks (e.g. Biro et al. 2007), further stressing the importance of combining information in navigational tasks, but also with experience use of simplified navigational strategies.

9.9 Future perspectives

Navigation is an important part of the everyday life of many animals, as well as humans. The development of sophisticated biological compasses and navigation strategies used in the natural environment has certainly been important for the evolution of new migration routes and range expansions in animals. Without doubt navigational capacity has been an evolutionary weapon harnessed in competition among different organisms. Thanks to the development of techniques enabling controlled experiments in laboratories and new methods to study animal movements in the natural environment, we now know much about the navigational mechanisms and function of compasses used. However, we still need to explain reasons for contradictory results and investigate the neuronal mechanisms behind navigational output across organisms and scale, a subject further discussed by Muheim et al. (Chapter 10). For instance, whether cognitive maps really exist is still an open question. We further possess limited knowledge about gene expression and the characteristics of the migration phenotype among most organisms (see also Liedvogel and Lundberg, Chapter 12). Hopefully, with continued development of new technology and analytical methods we will be able to answer some of the open questions in the not too distant future.

References

Able, K. P. (1980). Mechanisms of orientation, navigation and homing. In S. A. Gauthreaux, Jr (ed.), *Animal Migration, Orientation, and Navigation*, pp. 283–373. Academic Press, New York.

Able, K. P. (1982). Skylight polarization patterns at dusk influence migratory orientation in birds. *Nature*, **299**, 550–1.

Able, K. P. (2000). The concepts and terminology of bird navigation. *Journal of Avian Biology*, **32**, 174–83.

Able, K. P., and Able, M. A. (1990). Calibration of the magnetic compass of a migratory bird by celestial rotation. *Nature*, **347**, 378–80.

Able, K. P., and Able, M. A. (1993). Daytime calibration of magnetic orientation in a migratory bird requires a view of skylight polarization. *Nature*, **364**, 523–5.

Able, K. P., and Able, M. A. (1995). Manipulations of polarized skylight calibrate magnetic orientation in a migratory bird. *Journal of Comparative Physiology A: Neuroethology, Sensory, Neural, and Behavioral Physiology*, **177**, 351–6.

Adler, K., and Phillips, J. B. (1985). Orientation in a desert lizard (*Uma notata*): time-compensated compass movement and polarotaxis. *Journal of Comparative Physiology A: Neuroethology, Sensory, Neural, and Behavioral Physiology*, **156**, 547–52.

Åkesson, S. (1996). Geomagnetic map used for long-distance navigation? *Trends in Ecology and Evolution*, **11**, 398–400.

Åkesson, S. (2003). Avian long-distance navigation: experiments with migratory birds. In P. Berthold, E. Gwinner, and E. Sonnenschein (eds.), *Avian Migration*, pp. 471–92. Springer Verlag, Berlin.

Åkesson, S., and Alerstam, T. (1998). Oceanic navigation: are there any feasible geomagnetic bi-coordinate combinations for albatrosses? *Journal of Avian Biology*, **29**, 618–25.

Åkesson, S., and Bäckman, J. (1999). Orientation in pied flycatchers: the relative importance of magnetic and visual information at dusk. *Animal Behaviour*, **57**, 819–28.

Åkesson, S., Broderick, A. C., Glen, F., et al. (2003). Navigation by green turtles: which strategy do displaced adults use to locate Ascension Island? *Oikos* **103**, 363–72.

Åkesson, S., and Hedenström, A. (2007). How migrants get there: migratory performance and orientation. *BioScience*, **57**, 123–33.

Åkesson, S., Hedenström, A., and Alerstam, T. (1996). Flight initiation of nocturnal passerine migrants in relation to celestial orientation conditions at twilight. *Journal of Avian Biology*, **27**, 95–102.

Åkesson, S., Morin, J., Muheim, R., and Ottosson, U. (2005). Dramatic orientation shift of displaced birds in response to the geomagnetic field. *Current Biology*, **15**, 1591–7.

Åkesson, S., Walinder, G., Karlsson, L., and Ehnbom, S. (2001). Reed warbler orientation: initiation of nocturnal migratory flights in relation to visibility of celestial cues at dusk. *Animal Behaviour*, **61**, 181–9.

Åkesson, S., and Wehner, R. (2002). Visual navigation in desert ants *Cataglyphis fortis*: are snapshots coupled to a celestial system of reference? *Journal of Experimental Biology*, **205**, 1971–8.

Åkesson, S., and Weimerskirch, H. (2014). Evidence for sex-segregated ocean distributions of first-wintering wandering albatrosses at Crozet Islands. PLoS ONE 9(2): e86779. doi:10.1371/journal.pone.0086779.

Baker, R. R. (1980). Goal orientation by blindfolded humans after long-distance displacement: possible involvement of a magnetic sense. *Science*, **210**, 555–7.

Baker, R.R. (1984). *Bird Migration: The Solution of a Mystery?* Hodder & Stoughton, London.

Balda, R. P., and Wiltschko, W. (1991). Caching and recovery in scrub jays: transfer of sun-compass directions from shaded to sunny areas. *Condor*, **93**, 1020–3.

Beason, R. C. (1992). You can get there from here: responses to simulated magnetic equator crossing by the Bobolink (*Dolichonyx oryzivorus*). *Ethology*, **91**, 75–80.

Beck, W., and Wiltschko, W. (1988). Magnetic factors control the migratory direction of pied flycatchers. In H. Ouellet (ed.), *Acta XIX Congress of International Ornithology*, pp. 1955–62. University of Ottawa Press, Ottawa.

Begall, S., Červený, J., Neef, J., Vojtěch, O., and Burda, H. (2008). Magnetic alignment in grazing and resting cattle and deer. *Proceedings of the National Academy of Sciences USA*, **105**, 13451–5.

Bellrose, F. C. (1958). The orientation of displaced waterfowl in migration. *Wilson Bulletin*, **70**, 20–40.

Bennett, A. T. D. (1996). Do animals have cognitive maps? *Journal of Experimental Biology*, **199**, 219–24.

Berthold, P. (1996). *Control of Bird Migration*. Chapman & Hall, London.

Biro, D., Freeman, R., Meade, J., Roberts, S., and Guilford, T. (2007). Pigeons combine compass and landmark guidance in familiar route navigation. *Proceedings of the National Academy of Sciences USA*, **104**, 7471–6.

Boles, L. C., and Lohmann, K. J. (2003). True navigation and magnetic maps in spiny lobster. *Nature*, **421**, 60–3.

Bonadonna, F., Benhamou, S., and Jouventin, P. (2003). Orientation in 'featureless' environments: the extreme case of pelagic birds. In P. Berthold, E. Gwinner, and E. Sonnenschein (eds.), *Avian Migration*, pp. 367–77. Springer-Verlag, Berlin, Heidelberg.

Boström, J. E., Åkesson, S., and Alerstam, T. (2012a). Where on earth can animals use a geomagnetic bi-coordinate map for navigation? *Ecography*, **35**, 1039–47.

Boström, J. E., Fransson, T., Henshaw, I., Jakobsson, S., Kullberg, C., and Åkesson, S. (2010). Autumn migratory fuelling: a response to simulated magnetic displacements in juvenile wheatears, *Oenanthe oenanthe*. *Behavioral Ecology and Sociobiology*, **64**, 1725–32.

Boström, J. E., Kullberg, C., and Åkesson, S. (2012b). Northern magnetic displacements trigger endogenous fuelling responses in a naive bird migrant. *Behavioral Ecology and Sociobiology*, **66**, 819–21.

Braithwaite, V. A., and Guilford, T. (1991). Viewing familiar landscapes affects pigeon homing. *Proceedings of the Royal Society B: Biological Sciences*, **245**, 183–6.

Braithwaite, V. A., and Guilford, T. (1995). A loft with a view: exposure to the natural landscape during development may encourage adult pigeons to use visual landmarks during homing. *Animal Behaviour*, **49**, 251–3.

Brines, M. L., and Gould, J. L. (1982). Skylight polarization patterns and animal orientation. *Journal of Experimental Biology*, **96**, 69–91.

Burda, H., Marhold, S., Westenberger, T., Wiltschko, R., and Wiltschko, W. (1990). Magnetic compass orientation in the subterranean rodent, *Cryptomys hottentottus* (Bathyergidae). *Experientia*, **46**, 528–30.

Burt, T., Holland, R., and Guilford, T. (1997). Further evidence for visual landmark involvement in the pigeon's familiar area map. *Animal Behaviour*, **53**, 1203–9.

Cartwright, B. A., and Collett, T. S. (1983). Landmark learning in bees: experiments and models. *Journal of Comparative Physiology A: Neuroethology, Sensory, Neural, and Behavioral Physiology*, **151**, 521–43.

Cartwright, B. A., and Collett, T. S. (1987). Landmark maps for honeybees. *Biological Cybernetics*, **57**, 85–93.

Cheng, K., Narendrab, A., Sommera, S., and Wehner, R. 2009. Traveling in clutter: navigation in the Central Australian desert ant *Melophorus bagoti*. *Behavioural Processes*, **80**, 261–8.

Chernetsov, N., Kishkinev, D., Kosarev, V., and Bolshakov, C. V. (2011). Not all songbirds calibrate their magnetic compass from twilight cues: a telemetry study. *Journal of Experimental Biology*, **214**, 2540–3.

Cochran, W. W., Mouritsen, H., and Wikelski, M. (2004). Migrating songbirds recalibrate their magnetic compass daily from twilight cues. *Science*, **304**, 405–8.

Collett, T. S. 1987. The use of visual landmarks by gerbils: reaching a goal when landmarks are displaced. *Journal of Comparative Physiology A: Neuroethology, Sensory, Neural, and Behavioral Physiology*, **160**, 109–13.

Collett, T. S., and Cartwright, B. A. (1983). Eidetic images in insects: their role in navigation. *Trends in Neurosciences*, **6**, 101–5.

Collett, M., Collett, T. S., and Wehner, R. (1999). Calibration of vector navigation in desert ants. *Current Biology*, **9**, 1031–4.

Collett, T. S., Collett, M., and Wehner, R. (2001). The guidance of desert ants by extended landmarks. *Journal of Experimental Biology*, **204**, 1635–9.

Dickinson, J. A. (1994). Bees link local landmarks with celestial compass cues. *Naturwissenschaften*, **81**, 465–7.

Dingle, H. (1996). *Migration: The Biology of Life on the Move*. Oxford University Press, Oxford.

Dyer, F. C. (1987). Memory and sun compensation in honeybees. *Journal of Comparative Physiology A: Neuroethology, Sensory, Neural, and Behavioral Physiology*, **160**, 621–33.

Dyer, F. C., and Gould, J. L. (1981). Honey bee orientation: a backup system for cloudy days. *Science*, **214**, 1041–2.

Emlen, S. T. (1967a). Migratory orientation in the Indigo Bunting, *Passerina cyanea*. Part II: Mechanism of celestial orientation. *Auk*, **84**, 463–89.

Emlen, S. T. (1967b). Migratory orientation in the Indigo Bunting, *Passerina cyanea*. Part I: Evidence for use of celestial cues. *Auk*, **84**, 309–42.

Emlen, S. T. (1970). Celestial rotation: its importance in the development of migratory orientation. *Science*, **170**, 1198–1201.

Emlen, S. T. (1975). The stellar orientation system of a migratory bird. *Scientific American*, **233**, 102–11.

Emlen, S. T. (1976). Migration: orientation and navigation. In D. S. Farner and J. R. King (eds), *Avian Biology*, vol. **5**, pp. 129–219. Academic Press, New York.

Emlen, S. T., and Emlen, J. T. (1966). A technique for recording migratory orientation of captive birds. *Auk*, **83**, 361–7.

Emlen, J. T., and Penney, R. L. (1964). Distance navigation in the Adelie penguin. *Ibis*, **106**, 417–31.

Endres, C. S., and Lohmann, K. J. (2012). Perception of dimethyl sulfide (DMS) by loggerhead sea turtles: a possible mechanism for locating high-productivity oceanic regions for foraging. *Journal of Experimental Biology*, **215**, 3535–8.

Endres, C. S., Putman, N. F., and Lohmann, K. J. (2009). Perception of airborne odors by loggerhead sea turtles. *Journal of Experimental Biology*, **212**, 3823–7.

Evans, T. J., Kadin, M., Olsson, O., and Åkesson, S. (2013). Foraging behaviour of common murres in the Baltic Sea, recorded by simultaneous attachment of GPS and time-depth recorder devices. *Marine Ecology Progress Series*, **475**, 277–89.

Fischer, J. H., Freake, M. J., Borland, S. C., and Phillips, J. B. (2001). Evidence for the use of magnetic map information by an amphibian. *Animal Behaviour*, **62**, 1–10.

Fransson, T., Jakobsson, S., Johansson, P., Kullberg, C., Lind, J., and Vallin, A. (2001). Magnetic cues triggers extensive refuelling. *Nature*, **414**, 35–6.

Fransson, T., Jakobsson, S., and Kullberg, S. (2005). Non-random distribution of ring recoveries from trans-Saharan migrants indicates species-specific stopover areas. *Journal of Avian Biology*, **36**, 6–11.

Gagliardo, A., Ioalè, P., Savini, M., and Wild, M. (2009). Navigational abilities of adult and experienced homing pigeons deprived of olfactory or trigeminally mediated magnetic information. *Journal of Experimental Biology*, **212**, 3119–24.

Gould, J. L. (1998). Sensory basis of navigation. *Current Biology*, **8**, R731–8.

Grah, G., Wehner, R., and Ronacher, B. (2005). Path integration in a three-dimensional maze: ground distance estimation keeps desert ants *Cataglyphis fortis* on course. *Journal of Experimental Biology*, **208**, 4005–11.

Guerra, P. A., and Reppert, S. M. (2013). Coldness triggers northward flight in remigrant monarch butterflies. *Current Biology*, **23**, 419–23.

Gwinner, E. (1996). Circadian and circannual programmes in avian migration. *Journal of Experimental Biology*, **199**, 39–48.

Gwinner, E., and Wiltschko, W. (1978). Endogenously controlled changes in the migratory direction of garden warbler, *Sylvia borin*. *Journal of Comparative Physiology A: Neuroethology, Sensory, Neural, and Behavioral Physiology*, **125**, 267–73.

Gwinner, E., Zeman, M., and Klaassen, M. (1997). Synchronization by low-amplitude light-dark cycles of 24-hour pineal and plasma melatonin rhythms of hatchling European starlings (*Sturnus vulgaris*). *Journal of Pineal Research*, **23**, 176–81.

Hagstrum, J. T. (2013). Atmospheric propagation modeling indicates homing pigeons use loft-specific infrasonic 'map' cues. *Journal of Experimental Biology*, **216**, 687–99.

Hart, V., Kušta, T., Němec, P., et al. (2012). Magnetic alignment in carps: evidence from the Czech Christmas Fish Market. *PLoS One*, **7**(12), e51100. doi:10.1371/journal.pone.0051100.

Hays, G. C., Åkesson, S., Broderick, A. C., et al. (2003). Island finding ability in marine turtles. *Proceedings of the Royal Society B: Biological Sciences*, **270** (suppl.), S5–7.

Hegedüs, R., Åkesson, S., and Horváth, G. (2007b). Polarization patterns of thick clouds: overcast skies have distribution of the angle of polarization similar to that of clear skies. *Journal of the Optical Society of America A*, **24**, 2347–56.

Hegedüs, R., Åkesson, S., Wehner, R., and Horváth, G. (2007a). Could Vikings have navigated under foggy and cloudy conditions by skylight polarization? On the atmospheric optical prerequisites of polarimetric Viking navigation under foggy and cloudy skies. *Proceeding of the Royal Society A: Mathematical, Physical and Engineering Sciences*, **463**, 1081–95.

Helbig, A. J. (1990). Depolarization of natural skylight disrupts orientation of an avian nocturnal migrant. *Experientia*, **46**, 755–8.

Helbig, A. J. (1991). Inheritance of migratory direction in a bird species: a cross-breeding experiment with SE- and SW-migrating Blackcaps (*Sylvia atricapilla*). *Behavioural Ecology and Sociobiology*, **28**, 9–12.

Helbig, A. J. (1992). Ontogenetic stability of inherited directions in a nocturnal bird migrant: comparison between the first and second year of life. *Ethology, Ecology and Evolution*, **4**, 375–88.

Helbig, A. J., and Wiltschko, W. (1989). The skylight polarization pattern at dusk affect the orientation behaviour of blackcaps, *Sylvia atricapilla*. *Naturwissenschaften*, **76**, 227–9.

Henshaw, I., Fransson, T., Jakobsson, S., Lind, J., Vallin, A., and Kullberg, C. (2008). Food intake and fuel deposition in a migratory bird is affected by multiple as well as single-step changes in the magnetic field. *Journal of Experimental Biology*, **211**, 649–53.

Herman, W. S., and Tatar, M. (2001). Juvenile hormone regulation of longevity in the migratory monarch butterfly. *Proceedings of the Royal Society B: Biological Sciences*, **268**, 2509–14.

Herzmann, D., and Labhart, T. (1989). Spectral sensitivity and absolute threshold of polarization vision in crickets: a behavioral study. *Journal of Comparative Physiology A: Neuroethology, Sensory, Neural, and Behavioral Physiology*, **165**, 315–19.

Horváth, G., Barta, A., Pomozi, I., et al. (2011). On the trail of Vikings with polarized skylight: experimental study of the atmospheric optical prerequisites allowing polarimetric navigation by Viking seafarers. *Philosophical Transactions of the Royal Society B: Biological Sciences*, **366**, 772–82.

Horváth, G., and Varjú, D. (2004). *Polarized Light in Animal Vision—Polarization Patterns in Nature*. Springer-Verlag, Heidelberg, Berlin.

Icelandic Saga Database (http://sagadb.org/).

Karlsen, L.K. (2003). *Secrets of the Viking Navigators*. One Earth Press, Seattle.

Kramer, G. (1950). Weitere Analyse der Faktoren, welche die Zugaktivität des gekäfigten Vogels orientieren. *Naturwissenschaften*, **37**, 377–8.

Kramer, G. (1951). Eine neue Methode zur Erforschung der Zugorientierung und die bisher damit erzielten Ergebnisse. *Proceedings of International Ornithological Congress*, **X**, 269–80.

Kramer, G. (1953). Wird die Sonnenhöhe bei der Heimfindeorientierung verwertet? *Journal für Ornithologie*, **94**, 201–19.

Kramer, G. (1957). Experiments on bird orientation and their interpretation. *Ibis*, **99**, 196–227.

Kramer, G. (1961). Long-distance orientation. In A. J. Marshall (ed.), *Biology and Comparative Physiology of Birds*, pp. 341–71. Academic Press, New York.

Kramer, G., and von Saint Paul, U. (1950). Stare (*Sturnus vulgaris* L.) lassen sich auf Himmelsrichtungen dressieren. *Naturwissenschaften*, **37**, 526–7.

Kullberg, C., Henshaw, I., Jakobsson, S., Johansson, P., and Fransson, T. (2007). Fuelling decisions in migratory birds: geomagnetic cues override the seasonal effect. *Proceedings of the Royal Society B: Biological Sciences*, **274**, 2145–51.

Le Floch, A., Ropars, G., Lucas, J., et al. (2013). The sixteenth century Alderney crystal: a calcite as an efficient reference optical compass? *Proceedings of the Royal Society A: Mathematical, Physical and Engineering Sciences*, **469**, 20120651. http://dx.doi.org/10.1098/rspa.2012.0651.

Lehrer, M. (1991). Bees which turn back and look. *Naturwissenschaften*, **78**, 274–6.

Lehrer, M. (1993). Why do bees turn back and look? *Journal of Comparative Physiology A: Neuroethology, Sensory, Neural, and Behavioral Physiology*, **172**, 549–63.

Liedvogel, M., Åkesson, S., and Bensch, S. (2011). The genetics of migration on the move. *Trends in Ecology and Evolution*, **26**, 561–9.

Light, P., Salmon, M., and Lohmann, K. J. (1993). Geomagnetic orientation of loggerhead sea turtles: evidence for an inclination compass. *Journal of Experimental Biology*, **182**, 1–10.

Lohmann, K. J, Cain, S. D., Dodge, S. A., and Lohmann, C. M. F. (2001). Regional magnetic fields as navigational markers for sea turtles. *Science*, **294**, 364–6.

Luschi, P., Åkesson, S., Broderick, A. C., et al. (2001). Testing the navigational abilities of oceanic migrants: displacement experiments on sea turtles (*Chelonia mydas*). *Behavioural Ecology and Sociobiology*, **50**, 528–34.

Malcolm, S. B., Cockrell, B. J., and Brower, L. P. (1993). Spring recolonization of eastern North America by the monarch butterfly: successive brood or single sweep migration? In S. B. Malcolm and M. P. Zalucki (eds), *Biology and Conservation of the Monarch Butterfly*, pp. 235–67. National History Museum of Los Angeles County, CA.

Matthews, G. V. T. (1961). 'Nonsense' orientation in mallard *Anas platyrhynchos* and its relation to experiments on bird navigation. *Ibis*, **103**, 211–30.

Merkel, F. W., and Wiltschko, W. (1965). Magnetismus und Richtungsfinden zugunruhiger Rotkehlchen (*Erithacus rubecula*). *Die Vogelwarte*, **23**, 71–7.

Metcalfe, J. D., Holford, B. H., and Arnold, G. P. (1993). Orientation of plaice (*Pleuronectes platessa*) in the open sea: evidence for the use of external directional clues. *Marine Biology*, **117**, 559–66.

Mittelstaedt, H. (1983). The role of multimodal convergence in homing by path integration. *Fortschritte in der Zoologie*, **28**, 197–212.

Mittelstaedt, H., and Mittelstaedt, M.-L. (1982). Homing by path integration. In F. Papi and H. G. Wallraff (eds), *Avian Navigation*, pp. 290–7. Springer-Verlag, Berlin.

Moore, F. R. (1982). Sunset and the orientation of a nocturnal bird migrant: a mirror experiment. *Behavioural Ecology and Sociobiology*, **10**, 153–5.

Moore, F. R. (1987). Sunset and the orientation behaviour of migrating birds. *Biological Reviews*, **62**, 65–86.

Mouritsen, H., Derbyshire, R., Stalleicken, J., Mouritsen, O. Ø., Frost, B. J., and Norris, D. R. (2013). An experimental displacement and over 50 years of tag-recoveries show that monarch butterflies are not true navigators. *Proceedings of the National Academy of Sciences USA*, **110**, 7348–53.

Mouritsen, H., Feenders, G., Hegemann, A., and Liedvogel, M. (2009). Thermal paper can replace typewriter correction paper in emlen funnels. *Journal of Ornithology*, **150**, 713–15.

Mouritsen, H., and Larsen, O. N. (2001). Migrating songbirds tested in computer-controlled Emlen funnels use stellar cues for a time-compensated compass. *Journal of Experimental Biology*, **204**, 3855–65.

Muheim, R. (2011). Behavioural and physiological mechanisms of polarized light sensitivity in birds. *Philosophical Transactions of the Royal Society B: Biological Sciences*, **366**, 763–71.

Muheim, R., Edgar, N. M., Sloan, K. A., and Phillips, J. B. (2006a). Magnetic compass orientation in C57BL/6J mice. *Learning and Behavior*, **34**, 366–73.

Muheim, R., Moore, F. R., and Phillips, J. B. (2006b). Calibration of magnetic and celestial compass cues in migratory birds—a review of cue-conflict experiments. *Journal of Experimental Biology*, **209**, 2–17.

Muheim, R., Phillips, J. B., and Åkesson, S. (2006c). Polarized light cues underlie compass calibration in migratory songbirds. *Science*, **313**, 837–9.

Muheim, R., Phillips, J. B., and Deutschlander, M. E. (2009). White-throated sparrows calibrate their magnetic compass by polarized light cues during both autumn and spring migration. *Journal of Experimental Biology*, **212**, 3466–72.

Müller, M., and Wehner, R. (1988). Path integration in desert ants, *Cataglyphis fortis*. *Proceedings of the National Academy of Sciences USA*, **85**, 5287–90.

Müller, M., and Wehner, R. (1994). The hidden spiral: systematic search and path integration in desert ants, *Cataglyphis fortis*. *Journal of Comparative Physiology A: Neuroethology, Sensory, Neural, and Behavioral Physiology*, **175**, 525–30.

Nevitt, G. A., Veit, R. R., and Kareiva, P. (1995). Dimethyl sulphide as a foraging cue for Antarctic procellariiform seabirds. *Nature*, **376**, 680–2.

O'Keefe, J., and Nadel, L. (1978). *The Hippocampus as a Cognitive Map*. Oxford Univiversity Press, Oxford.

Papi, F. (1972). Olfaction and homing in pigeons. *Monitore Zoologico Italiano-Italian Journal of Zoology*, **6**, 85–95.

Papi, F. (1976). The olfactory navigation system of the homing pigeon. *Verhandlungen der Deutschen Zoologischen Gesellschaft, Köln*, **69**, 184–205.

Papi, F. (1982). Olfaction and homing in pigeons: ten years of experiments. In *Avian Navigation*, pp. 149–59. Springer-Verlag, Berlin, Heidelberg.

Papi, F. (ed.). 1992. *Animal Homing*. Chapman & Hall, London.

Papi, F. (2001). Animal navigation at the end of the century: a retrospect and a look forward. *Italian Journal of Zoology*, **68**, 171–80.

Penney, R. H., and Emlen, J. T. (1967). Further experiments on distance navigation in the Adelie penguin. *Ibis*, **109**, 99–109.

Perdeck, A. C. (1958). Two types of orientation in migrating Starlings *Sturnus vulgaris* L. and Chaffinches *Fringilla coelebs* L., as revealed by displacement experiments. *Ardea*, **46**, 1–37.

Perdeck, A. C. (1967). Orientation of starlings after displacement to Spain. *Ardea*, **55**, 194–202.

Phillips, J. B. (1986). Two magnetoreception pathways in a migratory salamander. *Science*, **233**, 765–7.

Phillips, J. B. (1996). Magnetic navigation. *Journal of Theoretical Biology*, **180**, 309–19.

Phillips, J. B., Freake, M. J., Fischer, J. H., and Borland, C. S. (2002). Behavioral titration of a magnetic map coordinate. *Journal of Comparative Physiology A: Neuroethology, Sensory, Neural, and Behavioral Physiology*, **188**, 157–60.

Phillips, J. B., and Sayeed, O. (1993). Wavelength-dependent effects of light on magnetic compass orientation in *Drosophila melanogaster*. *Journal of Comparative Physiology A: Neuroethology, Sensory, Neural, and Behavioral Physiology*, **172**, 303–8.

Putman, N. F., Endres, C. S., Lohmann, C. M. F., and Lohmann, K. J. (2011). Longitude perception and bicoordinate magnetic maps in sea turtles. *Current Biology*, **21**, 1–4.

Quinn, T. P., and Brannon, E. L. (1982). The use of celestial and magnetic cues by orienting sockeye salmon smolts. *Journal of Comparative Physiology A: Neuroethology, Sensory, Neural, and Behavioral Physiology*, **147**, 547–52.

Rabøl, J. (1985). The moving goal area and the orientation system of migrant birds. *Dansk Ornitologisk Forenings Tidsskrift*, **79**, 29–42.

Rabøl, J. (1998). Star navigation in pied flycatchers, *Ficedula hypoleuca*, and redstarts, *Phoenicurus phoenicurus*. *Dansk Ornitologisk ForeningsTidsskrift*, **92**, 283–9.

Ramskou, T. (1969). Solstenen. *Skalk*, **2**, 16–17.

Reppert, S. M., Gegear, R. J., and Merlin, C. (2010). Navigational mechanisms of migrating monarch butterflies. *Trends in Neurosciences*, **33**, 399–406.

Ropars, G., Gorre, G., Le Floch, A., Enoch, J., and Lakshminarayanan, V. (2012). A depolarizer as a possible precise sunstone for Viking navigation by polarized skylight. *Proceedings of the Royal Society A: Mathematical, Physical and Engineering Sciences*, **468**, 671–84.

Sandberg, R. (1994). Interaction of body condition and magnetic orientation in autumn migrating robins, *Erithacus rubecula*. *Animal Behavior*, **47**, 679–86.

Sandberg, R. (2003). Stored fat and the migratory orientation of birds. In P. Berthold, E. Gwinner, and E. Sonnenschein (eds), *Avian Migration*, pp. 515–25. Springer Verlag, Berlin, Heidelberg.

Sandberg, R., Pettersson, J., and Alerstam, T. (1988). Shifted magnetic fields lead to deflected and axial orientation of migrating robins, *Erithacus rubecula*, at sunset. *Animal Behaviour*, **36**, 877–87.

Sauer, E. G. F. (1957). Die Sternenorientierung nächtlich ziehender Grasmücken, *Sylvia atricapilla*, borin and curruca. *Zeitung für Tierpsychologie*, **14**, 20–70.

Schmaljohann, H., Rautenberg, T., Muheim, R., Naef-Daenzer, B., and Bairlein F. (2013). Response of a free-flying songbird to an experimental shift of the light polarization pattern around sunset. *Journal of Experimental Biology*, **216**, 1381–7.

Schmidt-Koenig, K. (1958a). Experimentelle Einflussnahme auf die 24-Stunden-Periodik bei Brieftauben und deren Auswirkungen unter besonderer Berücksichtigung des Heimfindevermogens. *Zeitung für Tierpsychologie*, **15**, 301–31.

Schmidt-Koenig, K. (1958b). Der Einfluss experimentell veränderter Zeitschätzung auf das Heimfindevermögen der Brieftauben. *Naturwissenschaften*, **45**, 47.

Schmidt-Koenig, K. (1970). Ein Versuch, theoretisch mögliche Navigationsverfahren zu klassifizieren und relevante sinnesphysiologische Probleme zu umriessen. *Verhandlungen der Deutschen Zoologischen Gesellschaft, Köln*, **1970**, 243–5.

Schmidt-Koenig, K. (1975). *Migration and Homing in Animals*. Springer-Verlag, Berlin.

Schmidt-Koenig, K. (1990). The sun compass. *Experientia*, **46**, 336–42.

Schöne, H. 1984. *Spatial Orientation*. Princeton University Press, Princeton, NJ.

Stefanescu, C., Parámo, F., Åkesson, S., et al. (2012). Multi-generational long-distance migration of insects: studying the painted lady butterfly in the Western Palaearctic. *Ecography*, **35**, 1–14.

Thirslund, S. (2001). *Viking Navigation: Sun-Compass Guided Norsemen First to America*. Gullanders Bogtrykkeri A-S, Skjern, Humlebaek.

Thorup, K., and Rabøl, J. (2001). The orientation system and migration pattern of long-distance migrants: conflict between model predictions and observed patterns. *Journal of Avian Biology*, **32**, 111–19.

Tolman, E. C. (1948). Cognitive maps in rats and men. *Psychological Review*, **55**, 189–208.

Towne, W. F., and Moscrip, H. (2008). The connection between landscapes and the solar ephemeris in honeybees. *Journal of Experimental Biology*, **211**, 3729–36.

Viguier, C. (1882). Le sens de l'orientation et ses organes chez les animaux et chez l'homme. *Revue Philosophque de la France et de l'Etranger*, **14**, 1–36.

von Frisch, K. (1949). Die Polarisation des Himmelslichtes als orientierender Faktor bei den Tänzen der Bienen. *Experientia*, **5**, 142–8.

von Frisch, K. (1967). *The Dance Language and Orientation of Bees*. Harvard University Press, Cambridge, MA.

von Saint Paul, U. (1953). Nachweis der Sonnenorientierung bei nächtlich ziehenden Vögeln. *Behaviour*, **6**, 1–7.

von Saint Paul, U. (1956). Compass directional training of western meadowlarks (*Sturnella neglecta*). *Auk*, **73**, 203–10.

Wajnberg, E., Acosta-Avalos, D., Alves, O. C., De Oliveira, J. F., Srygley, R. B., and Esquivel, D. M. S. (2010). Magnetoreception in eusocial insects: an update. *Journal of the Royal Society Interface*, **7**, S207–25.

Wallraff, H. G. (1974). *Das Navigationssystem der Vögel*. R. Oldenbourg Verlag, Munich.

Wallraff, H. G. (1990). Conceptual approaches to avian navigation systems. *Experientia*, **46**, 379–88.

Wallraff, H. G. (1991). Conceptual approaches to avian navigation systems. In P. Berthold (ed.), *Orientation in Birds*, pp. 128–65. Birkhäuser, Basel.

Wallraff, H. G. (2005). *Avian Navigation: Pigeon Homing as a Paradigm*. Springer-Verlag, Berlin, Heidelberg.

Wallraff, H. G., and Andreae, M. O. (2000). Spatial gradients in ratios of atmospheric trace gases: a study stimulated by experiments on bird navigation. *Tellus B*, **52**, 1138–57.

Wehner, R. (1976). Polarized-light navigation by insects. *Scientific American*, **235**, 106–15.

Wehner, R. (1998). Navigation in context: grand theories and basic mechanisms. *Journal of Avian Biology*, **29**, 370–86.

Wehner, R. (2009). The architecture of the desert ant's navigational toolkit (Hymenoptera: Formicidae). *Myrmecological News*, **12**, 85–96.

Wehner, R., Boyer, M., Loertscher, F., Sommer, S., and Menzi, U. (2006). Ant navigation: one-way routes rather than maps. *Current Biology*, **16**, 75–9.

Wehner, R., Michel, B., and Antonsen, P. (1996). Visual navigation in insects: Coupling of egocentric and geocentric information. *Journal of Experimental Biology*, **199**, 129–40.

Wehner, R., and Müller, M. (1985). Does interocular transfer occur in visual navigation by ants? *Nature*, **315**, 228–9.

Wehner, R., and Müller, M. (2006). The significance of direct sunlight and polarized skylight in the ant's celestial system of navigation. *Proceedings of the National Academy of Sciences USA*, **103**, 12575–9.

Wehner, R., and Räber, F. (1979). Visual spatial memory in desert ants, *Cataglyphis bicolor* (Hymenoptera: Formicidae). *Experientia*, **35**, 1569–71.

Wehner, R., and Wehner, S. (1990) Insect navigation: use of maps or Adriane's thread? *Ethology, Ecology and Evolution*, **2**, 27–48.

Weindler, P., Wiltschko, R., and Wiltschko, W. (1996). Magnetic information affects the stellar orientation of young bird migrants. *Nature*, **383**, 158–60.

Wiltschko, R. (1980). Die Sonnenorientierung der Vögel. 1. Die Rolle der Sonne im Orientierungssystem und die Funktionsweise des Sonnenkompass. *Journal of Ornithology*, **121**, 121–43.

Wiltschko, R. (1981). Die Sonnenorientierung der Vögel. 2. Entwicklung des Sonnenkompass und sein Stellenwert im Orientierungssystem. *Journal of Ornithology*, **122**, 1–22.

Wiltschko, R., Munro, U., Ford, H., and Wiltschko, W. (2008). Contradictory results on the role of polarized light in compass calibration in migratory songbirds. *Journal of Ornithology*, **149**, 607–14.

Wiltschko, R., Schiffner, I., and Wiltschko, W. (2009). A strong magnetic anomaly affects pigeon navigation. *Journal of Experimental Biology*, **212**, 2983–90.

Wiltschko, R., Walker, M., and Wiltschko, W. (2000). Sun-compass orientation in homing pigeons: compensation for different rates of change in azimuth? *Journal of Experimental Biology*, **203**, 889–94.

Wiltschko, R., and Wiltschko, W. (1995). *Magnetic Orientation in Animals*. Springer-Verlag, Heidelberg, Berlin.

Wiltschko, R., and Wiltschko, W. (2003). Avian navigation: from historical to modern concepts. *Animal Behaviour*, **65**, 257–272.

Wiltschko, W. (1968). Über den Einfluß statischer Magnetfelder auf die Zugorientierung der Rotkehlchen (Erithacus rubecula). *Zeitschrift für Tierpsychologie*, **25**, 537–58.

Wiltschko, W., and Balda, R. P. (1989). Sun compass orientation in seed-caching scrub jays (*Aphelocoma coerulescens*). *Journal of Comparative Physiology A: Neuroethology, Sensory, Neural, and Behavioral Physiology*, **164**, 717–21.

Wiltschko, W., Balda, R. P., Jahnel, M., and Wiltschko, R. (1999). Sun compass orientation in seed-caching corvids: its role in spatial memory. *Animal Cognition*, **2**, 215–21.

Wiltschko, W., Daum, P., Fergenbauer-Kimmel, A., and Wiltschko, R. (1987). The development of the star compass in garden warblers, *Sylvia borin*. *Ethology*, **74**, 285–92.

Wiltschko, W., and Gwinner, E. (1974). Evidence for an innate magnetic compass in Garden Warblers. *Naturwissenschaften*, **61**, 406.

Wiltschko, W., and Merkel, F. W. (1966). Orientierung zugunruhiger Rotkehlchen im statischen Magnetfeld. *Verhandlungen der Deutschen Zoologischen Gesellschaft*, **59**, 362–7.

Wiltschko, W., Stapput, K., Thalau, P., and Wiltschko, R. (2006). Avian magnetic compass: fast adjustment to intensities outside the normal functional window. *Naturwissenschaften*, **93**, 300–4.

Wiltschko, W., and Wiltschko, R. (1972). Magnetic compass of European robins. *Science*, **176**, 62–4.

Wiltschko, W., and Wiltschko, R. (1992). Migratory orientation: magnetic compass orientation of garden warblers, *Sylvia borin*, after a simulated crossing of the magnetic equator. *Ethology*, **91**, 70–4.

Wiltschko, W., and Wiltschko, R. (2005). Magnetic orientation and magnetoreception in birds and other animals. *Journal of Comparative Physiology A: Neuroethology, Sensory, Neural, and Behavioral Physiology*, **191**, 675–93.

Wohlgemuth, S., Ronacher, B., and Wehner, B. (2000). Ant odometry in the third dimension. *Nature*, **411**, 795–8.

Yeagley, H. L., and Whitmore, F. C. (1947). A preliminary study of a physical basis of bird navigation. *Journal of Applied Physics*, **18**, 1035–63.

CHAPTER 10

Sensory mechanisms of animal orientation and navigation

Rachel Muheim, Jannika Boström, Susanne Åkesson, and Miriam Liedvogel

Although questions such as 'How do animals find their way, and how do they sense and process this information in the brain?' have been asked for centuries, the field of animal orientation and navigation has seen an immense leap forward in the past few decades. Moreover, our understanding has also expanded considerably regarding the molecular and physiological mechanisms of the different compasses and cues used by animals for orientation and navigation (Åkesson et al., Chapter 9, and Svensson et al., Chapter 11). Most notable are the advances made in our understanding of how animals can sense information provided by the geomagnetic field and use this information for behavioural tasks, for example for compass orientation during migration. But despite interdisciplinary and highly integrative research over recent decades, we do not fully understand how animals perceive the Earth´s magnetic field. We know that animals use geomagnetic information for orientation tasks (see Åkesson et al., Chapter 9), but the receptor(s) remain to be identified. In this chapter, we review current knowledge in this area, outline challenges, and suggest future approaches to elucidate the sensory modalities used by animals for orientation and navigational tasks.

10.1 Magnetic sense

Many hypotheses regarding how animals may sense the Earth's magnetic field have been proposed. Three principally different mechanisms to achieve this could theoretically be used to sense the strength of the Earth's magnetic field, including (1) induction, (2) magnetic particles, and (3) magnetically sensitive biochemical reactions. The latter two possibilities have emerged as the most promising candidate magnetoreceptor mechanisms: A light-dependent process is thought to detect the alignment of the geomagnetic field lines in space. This provides directional information that can be used for a magnetic compass (inclination compass, see Åkesson et al., Chapter 9). The other possibility is a detection process mediated by a ferromineral that reacts to very small changes in the direction and/or intensity of the magnetic field and, thereby, can be used as a magnetic compass and/or a magnetic positioning (map or signpost) sense (for reviews see Wiltschko and Wiltschko 1995a, 2005; Lohmann and Johnson 2000; Mouritsen and Ritz 2005). Both of the latter two mechanisms are supported by behavioural and physiological data in a broad range of organisms (see also Åkesson et al., Chapter 9). In some animals, like newts and birds, the presence of both mechanisms have independently been experimentally demonstrated to be present and used by the animals for different purposes, and thus are believed to be non-exclusive (Phillips 1986, Wiltschko and Wiltschko 1995b, and see 10.1.3). Here, we present the state-of-the-art knowledge of the sensory aspects of the two magnetoreception mechanisms, and highlight recent advances and future challenges.

Animal Movement Across Scales. Edited by Lars-Anders Hansson and Susanne Åkesson.
© Oxford University Press 2014. Published 2014 by Oxford University Press.

10.1.1 Magnetic sense based on ferromagnetic particles

The idea of a ferromineral-based magnetic sense originated from the findings of deposits of ferromagnetic (magnetite and/or maghemite) particles in various organisms, including several animals that use the Earth's magnetic field for behavioural tasks (Kirschvink et al. 1985, Williams and Wild 2001, Falkenberg et al. 2010, but see Treiber et al. 2012). However, we here want to stress that the mere presence of biogenic ferrominerals in any organism is the rule rather than the exception and does not automatically indicate the presence of a magnetic sense. These particles may be involved in many different roles in the physiology of an organism; for example, magnetite synthesis seems to be a general way for organisms to deposit excess iron. With the exception of magnetotactic bacteria, where magnetite crystals act like a compass needle and enable the bacteria to passively align to the geomagnetic field (Blakemore 1975), the physiological mechanisms of magnetic field reception using ferromagnetic minerals remain to be demonstrated for other organisms.

10.1.2 Biophysical and molecular mechanisms of ferromineral-based magnetoreception

Several conditions must be met for a magnetoreception system based on ferromagnetic minerals to work (Kirschvink and Walker 1985, Kirschvink 1989, Winklhofer and Kirschvink 2010; for details see Box 10.1): (1) the ferromagnetic material must be biochemically precipitated by the organism itself, because incorporated external material is usually too contaminated; (2) the ferromagnetic particles must fulfil a variety of magnetophysical properties and have a specific magnetization and size, i.e. be single-domain (SD) or superparamagnetic (SPM) crystals; and (3) they must be mechanically coupled to nerve fibres or sensory organelles; otherwise, they cannot transmit any magnetic information to the brain.

10.1.3 Behavioural and physiological evidence for ferromineral-based magnetoreception

Despite numerous reports of ferromagnetic material in invertebrates, there is little direct behavioural evidence that these particles function as magnetoreceptors. Magnetite, maghemite, or other ferritin-like particles are found in a number of insect species, including bees, ants, and termites, but an actual magnetoreceptor has not been located or described yet (reviewed by Wajnberg et al. 2010). The most convincing data supporting a magnetite-based receptor mechanism come from fish and birds, where there is both behavioural and electrophysiological evidence in favour of a magnetite-based magnetoreception mechanism. Several behavioural studies with fish have reported alignment along magnetic fields or abilities to discriminate magnetic anomalies. For example, rainbow trout, *Oncorhynchus mykiss*, align along an external magnetic field, but become disoriented in a null magnetic field where no magnetic field is present; i.e. all dimensions of the magnetic field are cancelled out and the length of the magnetic vector is zero (Chew and Brown 1989). Trout are also able to discriminate between magnetic anomalies, but only when the conditional response allow movement and when the magnetic fields are spatially distinctive (Walker et al. 1997). Experiments with yellowfin tuna, *Thunnus albacares*, showed that these fish could sense differences in intensity of the magnetic field, but not differences in direction (Walker 1984). Neurophysiological recordings on rainbow trout supported these findings; specifically, single neurons in a side branch of the trigeminal nerve have been shown to respond to changes in intensity, but not to the direction of an imposed magnetic field (Walker et al. 1997). Single-domain magnetite particles have been reported in several locations in different fish species: near the basal lamina of the olfactory epithelium (the area innervated by the trigeminal nerve) and in the dermal bone cartilage of the skull (e.g. Walker et al. 1984, 1997). Based on the findings of iron-rich crystals in the olfactory lamellae in rainbow trout (Walker et al. 1997), Diebel et al. (2000) suggested a vertebrate magnetoreceptor in form of a multi-lobed cell, containing chains of magnetite particles located in the basal lamina of an olfactory lamella. Recently, these findings were supported by a new method developed to detect and characterize candidate vertebrate magnetoreceptor cells (Eder et al. 2012). However, despite this exciting discovery, it remains to be shown how these magnetoreceptor

Box 10.1 Ferromineral-based magnetoreception

Magnetite (Fe_3O_4) and maghemite (Fe_2O_3) are the most frequently occurring of the known biogenic materials that are ferro(i)magnetic at room temperature. Ferro(i)magnets have a spontaneous magnetic moment and are composed of small regions called domains. Crystals composed of only one domain are called single-domain (SD) crystals. They are elongated particles and have a magnetic moment along the long axis directed towards one of the two ends. Magnetite particles which are sufficiently small become superparamagnetic (SP). They have no spontaneous magnetic moment, but align with an external magnetic field. The simplest magnetic compass involving magnetic material is based on SD magnetite particles that align along an external magnetic field like compass needles. The arrangement in chains as found in bacteria (Blakemore 1975) imparts the magnetite crystals an even larger magnetic moment because of additive effects. The rotation or translation of a magnetic particle can in theory translate into a sensory nerve signal in two physically different ways (Kirschvink and Walker 1985, Kirschvink 1989, Winklhofer and Kirschvink 2010): (1) the torque may directly generate elastic resistance or pressure on mechanically coupled physiological structures, like a hair cell, a stretch receptor, or another mechanoreceptor; or (2) the torque may have indirect effects and alter magneto-chemical reactions or interact with nearby superparamagnetic particles (Fig. 10.1). Superparamagnetic ferromagnetic minerals cannot produce a torque, but can theoretically detect magnetic fields by other mechanisms: Clusters of SP magnetite particles dispersed in liquid and enclosed by a biological membrane can theoretically change the shape of these SP clusters under a magnetic field in the order of magnitude of the geomagnetic field, if connected to a mechanoreceptor. Such clusters could provide axial information about the applied field, and the magnetic-field-induced shape of the clusters could be amplified, as well as counterbalanced by osmotic pressure regulation, and thereby provide magnetic intensity information (for recent reviews see Cadiou and McNaughton 2010, Winklhofer and Kirschvink 2010).

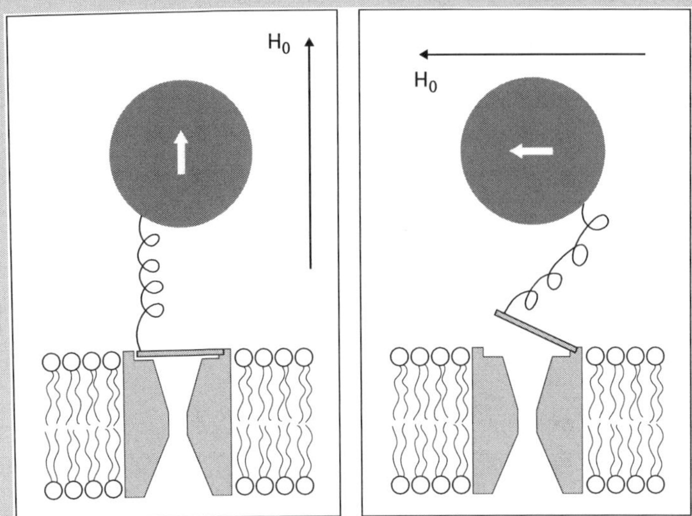

Figure 10.1 In the ferromagnetic transduction model, ion channels in the nerve membrane open or close depending on the magnetic torque acting at magnetite particles connected to the ion channels and thereby let Ca^{2+} ions pass through, which elicits a nerve signal. Redrawn from Kirschvink (1992) with permission.

cells function in detail, and how the information is transferred to a nerve signal.

Both birds and newts are believed to possess a dual magnetoreception mechanism, i.e. a light-dependent magnetic compass and a (putatively independent) magnetite-based magnetic map sense (Phillips 1986, Munro et al. 1997a; Åkesson et al., Chapter 9). Pulse remagnetization experiments and treatment with zinc sulphide of the putative receptor organ (considered as diagnostic tool to test for the involvement of magnetite-based receptor mechanism; see Box 10.2) with migratory songbirds and pigeons, *Columbia livia*, have shown that experienced adult birds, but not inexperienced juveniles, are affected

Box 10.2 Diagnostic tools to distinguish magnetoreceptor models

Diagnostic tools used to test whether a magnetoreceptor based on ferrominerals is involved in a specific behaviour can be grouped into two categories: (1) tools that directly affect the magnetic particles, and (2) tools that affect the sensory receptor or the transduction pathways between the receptor and the brain.

(1) Pulse remagnetization is the application of a brief, strong, directional magnetic pulse (0.5 T for 4–5 ms). This procedure has been widely used to directly affect the ferromagnetic particles in a putative magnetoreceptor (e.g. Wiltschko et al. 1994, 1995B; Beason et al. 1995):

- If SD particles are involved, such a magnetic pulse will permanently remagnetize the particles in the opposite direction.
- If the pulse is applied anti-parallel to the magnetic moment and with an intensity greater than the coercivity of SD particles, i.e. greater than the intensity required to reduce the magnetization of those particles to zero, this should result in a permanent reversal or change in orientation.
- If clusters of SP magnetite are involved, such a strong magnetic pulse is expected to form agglomerations of clusters and impair the magnetoreceptor for a few days.

The downside of pulse remagnetization experiments is that it is difficult to predict the expected outcome of such a treatment as long as the exact structure and function of the receptor are not known. In addition, there is no proper control experiment that allows distinguishing a true effect of the strong pulse on the magnetoreceptor from effects on other unrelated physiological processes.

(2) Tools in the second category aim to disrupt magnetoreceptor function, and include (i) local anaesthetics blockade or treatment with zinc sulphide of the putative receptor organ (i.e. upper beak area in birds; e.g. Holland et al. 2009), and (ii) lesion studies, where either the transmitting nerve (i.e. trigeminal nerve system in birds; cf. Mora et al. 2004) or the putative brain areas involved in the processing of the information are lesioned (via mechanical cut or chemical lesion, e.g. with ibotenic acid; Zapka et al. 2009). Note that studies using local anaesthesia should be treated with caution, as no control experiments exist that can reliably work as treatment control. Further, nothing is known about the time span of efficacy of any such treatment for most migratory animals. A further indirect indication for the involvement of a ferromineral-based mechanism is also directed orientation in total darkness, as this receptor mechanism is independent of light, but the radical-pair-based mechanism requires light to function (but it must be noted that energy for the radical-pair formation could in theory also be taken from chemical energy).

Low-intensity, oscillating radio-frequency magnetic fields (RF fields) have recently been introduced to test whether a radical-pair mechanism is involved in the primary magnetoreception process of an orientation response (Henbest et al. 2004, Ritz et al. 2004). RF fields of distinct frequencies in the lower MHz range (0.1–10 MHz; <1 μT), which

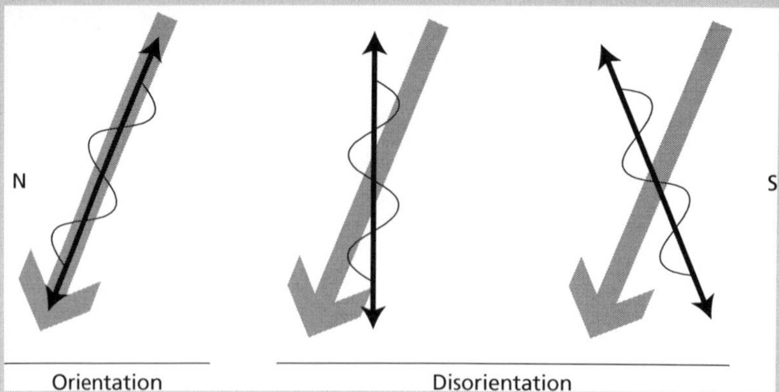

Figure 10.2 European robins, *Erithacus rubecula*, are disoriented when tested in the presence of a radio-frequency (RF) field (black double arrows with snake-line). However, the orientation of the RF field relative to the Earth's magnetic field (large, grey arrows) is crucial. The birds are only disoriented when the RF and the geomagentic fields are aligned non-parallel (right two conditions), but not when the two fields are aligned in parallel (left condition). See Ritz et al. (2004) for details.

Box 10.2 *Continued*

vary with the intensity of the ambient magnetic field and the magnetoreceptor molecule involved, interfere with the interconversion between the singlet and triplet excited states, and mask or alter the magnetic field effects produced by the Earth's magnetic field (Fig. 10.2). Such RF disturbances can lead to either disorientation or change in orientation, depending on the amount and type of change, and how the animals integrate the information into a migratory direction. Magnetite-based magnetoreceptors are not affected by RF fields, since the rotation of magnetite particles as proposed in animal magnetoreceptors is too slow and the ferromagnetic resonance frequency is expected to be in the GHz rather than MHz range (Ritz et al. 2000, 2004; Henbest et al. 2004). This makes the application of RF fields a unique, diagnostic tool for studying the involvement of a radical-pair mechanism in magnetoreception (Fig. 10.2).

by such a treatment (e.g. Wiltschko et al. 1994; Beason et al. 1995, 1997; Munro et al. 1997b; Holland et al. 2009). Presumably, juvenile birds exclusively use their light-dependent magnetic compass to determine their migratory direction during their first migration. Experienced adult birds likely also use their magnetic map sense, which they acquire during their first migration. Blocking the trigeminal nerve with a local anaesthetic eliminates the effects of pulse remagnetization (Beason and Semm 1996).

Where could these putative compass needles (single-domain or superparamagnetic magnetite crystals) be located in birds, and by which pathways could they transmit the magnetic information to the brain? Williams and Wild (2001) found trigeminally innervated structures in the nasal region of pigeons which they interpreted as single-domain magnetite. A few years later, Fleissner and colleagues (2003, 2007) found nerve-cell endings (branches of the trigeminal nerve) in the upper beak of pigeons. Those nerve-cell endings contained dense clusters of ferromagnetic structures at six distinct locations in the beak, which could sense the geomagnetic field and subsequently transmit this information to the brain. Depending on their alignment relative to the magnetic field, these clusters were suggested to become uniquely deformed by the Earth's magnetic field, leading to an opening of mechano-sensitive ion channels and ultimately result in a nerve signal transmitted via the trigeminal nerve to the brain. The birds would thereby be able to measure the intensity, and consequently the polarity, of a magnetic field in all three dimensions, providing them with a '3-D magnetometer'. These structures were later confirmed for several other bird species, including migratory passerines (Falkenberg et al. 2010). The anatomical findings and theoretical assumptions were supported by behavioural experiments which demonstrated that birds are no longer able to detect magnetic stimuli when the trigeminal nerve, and thus the connection from the putative receptor to the brain, was cut (Mora et al. 2004; see also next section). Inactivation of the beak receptor in experiments (by lesioning the ophthalmic branch of the trigeminal nerve) with free-flying birds, both migratory birds and homing pigeons, supported these findings, at least with respect to experienced, adult individuals (Gagliardo et al. 2006, Holland et al. 2009).

Until very recently, such a '3-D magnetometer' in the upper beak of birds was almost regarded as a fact. However, the findings were recently challenged by a study demonstrating that these iron crystals were not located in nerve tissue, but rather in macrophages, and thus could not be the sensors providing the trigeminal nerve, and thereby the brain, with magnetic information (Treiber et al. 2012). Macrophages are immune cells known to contain ferritin proteins (which store iron in the cell), and their presence has been reported throughout the bird's beak and other areas of the body, but nowhere associated to nerve tissue (Treiber et al. 2012). What consequences do these results have on our understanding of how birds may perceive the Earth's magnetic field? They certainly challenge previous interpretations of behavioural experiments that concluded that the trigeminal nerve is the link that makes it possible to transfer information sensed via magnetosensors in the beak to the brain. The new findings add no evidence to support the existence of iron-containing nerve cells at the previously reported locations along the upper beak of birds, but do not exclude the possibility that there still are magnetoreceptors at some yet-to-be-identified location in the bird's upper beak. Consequently, we need to carefully re-evaluate hypotheses and conclusions that were drawn based on behavioural studies that have previously been interpreted to support the function of a magnetite-based receptor located in the upper beak of birds. One plausible scenario could be that the real magnetoreceptors are located in the olfactory tissue, which is adjacent to the area where the supposed magnetic sense system was identified. This region is known to be connected to the nervous tissue, which was shown to transmit magnetic information to the brain in lesion experiments. This would be well supported by the finding of candidate magneto-sensitive cells in olfactory regions in trout (Walker et al. 1997; see earlier in this section).

In conclusion, the hypothesis proposing that magnetic information is perceived by a magnetoreceptor in the beak and transmitted to the brain via the trigeminal nerve might still be valid, as lesioning this link resulted in birds no longer being able to perceive magnetic stimuli. However, we do not understand the detailed function of this putative neuronal link, and we do not know which magnetosensor feeds information into the nerve. Recently, a putative third magnetoreceptor was suggested to be located in the lagena of the inner ear of pigeons and encode information on the direction and intensity of the magnetic field (Wu and Dickman 2011, 2012). However, it remains to be demonstrated whether the reported brain activities as a consequence of changes in the magnetic field originate from an actual new magnetoreceptor, or whether they are the results of the integration of vestibular and magnetic field information, which is necessary for measuring magnetic fields.

10.1.4 Electrophysiological recordings and lesion experiments

The behavioural evidence for a magnetite-based map sensor is also supported by neurophysiological recordings from the ophthalmic branch of the trigeminal nerve of the bobolink, *Dolichonyx oryzivorus*, a migratory songbird from North America. Changes in the rotation of the vertical or horizontal component of the geomagnetic field and changes in intensity of as little as 50–200 nT led to alterations in the action potentials in the trigeminal nerve (Beason and Semm 1987). Magnetic compass responses (choosing and maintaining a direction) were not affected by blocking the ophthalmic branch of the trigeminal nerve (Beason and Semm 1996). All of these findings are suggestive of the involvement of the trigeminal nerve in the processing of magnetic information from the beak to the brain.

10.1.5 Conditioning experiments

Further support for an involvement of the trigeminal nerve system comes from conditioning experiments, which show that pigeons can discriminate between the presence and absence of a strong magnetic anomaly (~100,000 nT; Mora et al. 2004). This ability disappeared when the ophthalmic branch of the trigeminal nerve was cut, but remained when the olfactory nerve was lesioned instead (Mora et al. 2004). Similarly, Pekin ducks, *Anas platyrhynchos domestica*, trained to find a hidden food reward indicated by a magnetic anomaly, were not able to find the reward when the trigeminal nerve was anaesthetized by an injection of lignocaine hydrochloride, strongly indicating a role of the trigeminal nerve in the detection of magnetic anomalies (Freire et al. 2012). In addition, both pigeons (Thalau et al. 2007) and domestic chicks (Denzau et al. 2011) were able to use a small, but strong, anomaly to find a hidden food source, although none of these studies tested for the involvement of the trigeminal nerve. Still, in order for a magnetic 'map sense' to work, birds must be able to detect naturally occurring local changes in the magnetic field strengths that are about five orders of magnitude smaller (~10 nT) than the magnetic anomalies used in the operant conditioning by Mora et al. (2004). Therefore, to function as a putative biologically relevant map or signpost sense, it still remains to be demonstrated that the trigeminal magnetoreception circuit can detect biologically relevant magnetic anomalies.

10.1.6 Chemical magnetoreception based on a radical pair mechanism

In 1978, the physicist Klaus Schulten proposed that magnetic sensing might have spin chemical origins (Schulten et al. 1978, Schulten 1982). In a theoretical paper, he suggested that the yield of a biochemical reaction proceeding via a radical pair might be sensitive to the orientation of an external magnetic field. Electron spins are not strongly coupled to the thermal bath and therefore represent one of only a few molecular features that might plausibly be influenced by the Earth's magnetic field. The suggested mechanism (Box 10.3) involves a light-induced electron transfer between two molecules (note: such processes can also be chemically induced). The electron transfer results in the generation of a radical-pair intermediate that will either exist in a singlet or a triplet excited state, and subsequently decay in chemically different end products. Theoretical calculations and in vitro experiments showed that the ratio between singlet and triplet products from radical-pair reactions can be modulated by an Earth-strength magnetic field during the interconversion step, i.e. when singlet radical intermediates convert to triplets and vice versa (Maeda et al. 2008, 2012). This orientation-dependent interconversion step (and subsequently resulting differences in the ratio of singlet and triplet end products) could theoretically be used to encode directional information. In a revival of this idea, Ritz et al. (2000) proposed that the retina with its almost perfect sphere would be an ideal substrate for such a mechanism. He further speculated that the radical-pair intermediate might be involved in some kind of visual reception system, exploiting the highly efficient visual transduction signalling cascade. According to this suggestion, the reaction yield anisotropy of the receptor radical pair could govern the directional response and thus form the basis of a magnetic compass (for recent reviews see Ritz et al. 2010, Mouritsen and Hore 2012).

Possible receptor candidate molecules involved in the primary magnetoreception process and mediating a light-dependent radical-pair mechanism must meet the following criteria: they need to be light-sensitive, need to be able to form radical pairs that persist long enough so that the radical-pair yields

Box 10.3 Radical-pair-based magnetoreception

A radical is any atom or molecule with one or more unpaired electrons. Electrons possess an angular spin momentum, and radicals are therefore paramagnetic (Fig. 10.3). The radical-pair mechanism starts with an electron transfer from a donor molecule D to an acceptor molecule A upon photo excitation (Fig. 10.3), resulting in a radical pair with both donor and acceptor molecules possessing one unpaired electron (depicted as black dots). The spins of the unpaired electrons can be either opposite (singlet state) or parallel (triplet state) and both states can interconvert. The interconversion step between singlet and triplet intermediates can be affected by an external magnetic field, and therefore modulates the ratio of singlet and triplet transient states depending on the orientation of the molecule within the field. In a final step, singlet, and triplet intermediates will react to distinct products, which may differ in their biochemical properties (modified after Ritz et al. 2000).

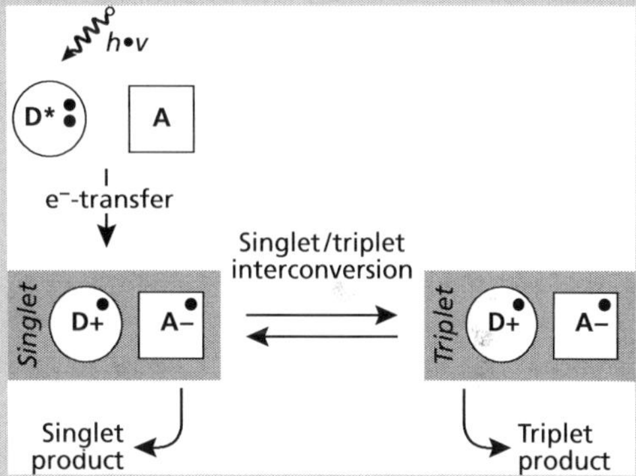

Figure 10.3 Radical pairs are formed in many biochemical transformations as transient (that is: short-lived) reaction intermediates, either as singlet or triplet pairs, depending on the spin correlation of the unpaired electrons. In the singlet state, the total angular momentum of a radical pair is zero and the electron spins are oppositely directed (antiparallel) and paired (↑↓). A radical pair in a singlet state can recombine. In the triplet state, electron spins are parallel (↑↑) and no bonding is possible. Because each electron spin has an associated magnetic moment, the chemical characteristics of the singlet and triplet intermediate states, as well as their interconversion can be influenced by internal and external magnetic fields. Redrawn after Ritz et al. 2000 with permission from Elsevier.

can be modified by an Earth-strength magnetic field, and should be localized in a spatially fixed relationship relative to each other criteria (reviewed by Rodgers and Hore 2009, Liedvogel and Mouritsen 2010, Phillips et al. 2010a, Mouritsen and Hore 2012). Classical photopigments like the opsins do not form radical pairs, and thus are unlikely candidates for magnetoreception. Cryptochromes, a class of blue light photoreceptor molecules, have been suggested to be the most likely molecules involved in magnetoreception. They are the only known photopigments in the vertebrate eye that have the potential to form radical pairs. Cryptochromes form a multi-gene family of blue-green light photoreceptors known to be involved in circadian rhythm regulation (for a recent review see Chaves et al. 2011). They share high sequence homology with photolyases that form radical pair intermediates persisting long enough for magnetic field effects to occur (Weber et al. 2002, Giovani et al. 2003). To date, cryptochromes from several taxa have been shown to also produce persistent, spin-correlated radical species upon photo-excitation in vitro (Liedvogel et al. 2007, Biskup et al. 2009, Schleicher et al. 2009). Cryptochromes are found in a large variety of species known to use a magnetic compass, for example, in retinas of two migratory bird species, European robins, *Erithacus rubecula*, and garden warblers, *Sylvia borin*, and two non-migratory species, chicken, *Gallus gallus*, and zebra finches, *Taeniopygia guttatta* (Liedvogel and Mouritsen 2010). A detailed neuroanatomical study of cryptochrome expression in robins and chicken recently revealed that the outer segments of the avian UV/V cones are most likely the primary sites of the light-dependent magnetoreceptor (Niessner et al. 2011). Niessner and colleagues found cryptochrome expression in virtually every cone across the entire retina, which is one of the requirements for the radical-pair model to work. Still, more research is needed to unambiguously separate observations of cryptochrome expression as a result of circadian rhythmicity and magnetoreception, and to describe the full details of the reception mechanism, including the neuronal pathways mediating the sensory information to the brain.

The development of reliable magnetic compass assays in genetic model organisms, such as *Drosophila* (Phillips and Sayeed 1993, Gegear et al. 2008), laboratory mice (Muheim et al. 2006a), and zebra finches (Voss et al. 2007), has provided access to cryptochrome knockout animals. This has opened up a new and promising avenue in magnetoreception research, allowing direct tests of the involvement of cryptochromes and other molecules possibly involved in the primary magnetoreception process. One recent study with *Drosophila* provided convincing evidence that cryptochromes indeed are involved in light-dependent magnetoreception. In this study, adult *Drosophila* were able to discriminate a magnetic field (although about 10 times stronger than the geomagnetic field) only when the cryptochrome gene was functionally intact, but not when it was genetically engineered to be dysfunctional (Gegear et al. 2008). Also, magnetic fields were shown to influence cryptochrome-mediated effects of blue light on the free-running rhythm of the circadian clock in *Drosophila* (Yoshi et al. 2009). Magnetic field sensitivity could thus be an intrinsic property of cryptochrome-based photo-signalling systems, from which light-dependent magnetoreception might have evolved (Phillips et al. 2010a, b).

From a behavioural perspective, the putative magnetoreceptor must meet the following criteria to be a likely candidate for a light-dependent magnetic compass receptor (Liedvogel and Mouritsen 2010, Phillips et al. 2010a, Ritz et al. 2010; Fig. 10.4): (1) the magnetoreceptor needs to be sensitive to light of specific spectral properties (mainly wavelength), matching the receptor molecule(s) involved in the radical-pair process (see later discussion); (2) the encoded information received by the animal needs to have axial properties and not allow to determine the polarity of the field lines, since the avian magnetic compass generally functions without using polarity, as first described by Wiltschko and Wiltschko (1972); (3) the directional information perceived by the animal must depend on the intensity of the magnetic field; thus exposure to magnetic field intensities never experienced before can lead to disorientation, followed by a slow adaptation to the 'new' intensity (Wiltschko 1968, Wiltschko et al. 2006). A light-dependent magnetic compass has been demonstrated in both invertebrates and vertebrates, including fruit flies, mealworm beetles, bullfrogs, newts, homing pigeons, and several species of migratory birds (reviewed by Wiltschko and Wiltschko 1995a, 2005; Muheim

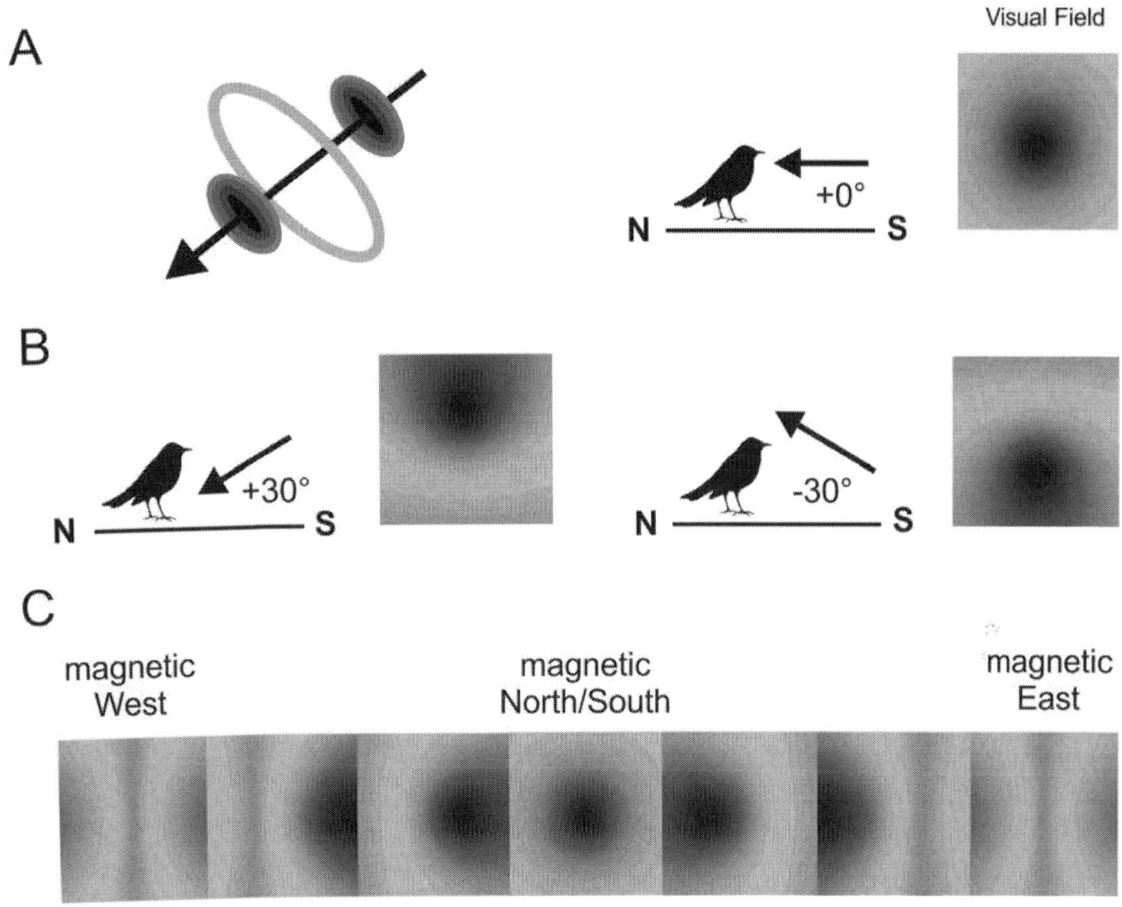

Figure 10.4 Illustration of the visual modulation pattern centred on the Earth's magnetic field (arrow), as suggested to be produced by magnetosensitive photoreceptors in the avian retina, enabling birds to 'see' the magnetic field (freely interpreted after Ritz et al. 2000). (A; left) Three-dimensional visual modulation pattern of the magnetic field, consisting of a dark area on each side of the magnetic field axis and a ring in the centre. (A; right) Visual modulation pattern seen by an animal (bird) looking parallel to the magnetic field at the magnetic equator, where the magnetic field is horizontal. (B) Visual modulation pattern seen by a bird looking towards magnetic south at a geographic latitude, where the magnetic field inclination is about 30°, (left) on the northern and (right) on the southern hemisphere. (C) Visual modulation pattern perceived by a bird looking from the west to the east (on the magnetic equator).

et al. 2002; Phillips et al. 2010a). There is also strong evidence from behavioural experiments for a light-dependent magnetic compass in C57BL/6J mice (Muheim et al. 2006a). A light-independent magnetic compass rather seems to be the exception and has only been demonstrated in a few animals, such as subterranean mole-rats (Marhold et al. 1997) and sea turtles (Lohmann and Lohmann 1993), possibly as a result of an adaptation to the non-terrestrial lifestyle of these organisms.

In birds, the dependence of light of specific wavelengths for magnetic compass orientation has been studied extensively. In general, birds are well oriented under UV to green light, and become disoriented or show shifted orientation under longer wavelengths (e.g. Wiltschko et al. 1993; Wiltschko and Wiltschko 1995b, 2001; Munro et al. 1997b; Muheim et al. 2002). European robins exposed to radio-frequency (RF) fields aligned nonparallel to the geomagnetic field vector become disoriented when tested under either a broadband RF field or distinct single frequencies of 1.375 or 7 MHz (Fig. 10.2; Ritz et al. 2004). The orientation of mole-rats, on the other hand, which is supposed to be mediated by a non-light-dependent magnetoreceptor, is not impaired by an RF field, supporting previous indications that their magnetic compass is based on magnetite (Thalau et al. 2006).

Questions that remain to be answered include: Where in the animals' bodies are the light-dependent magnetoreceptors located, and by which neuronal pathway is the information transmitted from the receptor to the brain? As magnetic fields can transmit through all body material, the receptors could in theory be located just about anywhere. However, behavioural experiments suggest that light is necessary for light-dependent magnetic orientation to function; thus the receptors must be located at a peripheral site of the animals' body. In birds, and also mammals, the only locations where light can reach specialized photoreceptors are the eyes. Already in the 1980s, extracellular recordings provided the first evidence for the involvement of the visual centre in light-dependent magnetoreception in birds. Cells in the nucleus of the basal optic root (nBOR) and in the optic tectum showed light-dependent magnetic responsiveness to changes in the direction of a magnetic field and slow inversions of the vertical component of the magnetic field (Semm et al. 1984). Recent research on the neural basis of magnetoreception has largely substantiated these findings, and provided new insights into the neural pathways and brain areas involved in information transfer and processing in both birds and mammals (e.g. Nemec et al. 2001, Heyers et al. 2007).

In birds, a brain structure in the visual Wulst, named 'Cluster N', and connected to the retina via the thalamofugal pathway, has been identified and suggested to be involved in the processing of light-dependent magnetic information in migratory birds during the night (Mouritsen et al. 2005, Heyers et al. 2007, Zapka et al. 2009). Migratory birds with a (chemically) lesioned Cluster N are disorientated when tested for magnetic compass orientation, but their sunset or star compass remained intact and functional for orientation. These results strongly indicate that Cluster N is involved in processing magnetic compass information at low light levels (Zapka et al. 2009). However, not in all taxa is light-dependent magnetoreception necessarily limited to the retina. In animals that have special structures containing photoreceptors, like the parietal eyes of reptiles or the pineal complex of fish and some reptiles, and the frontal organ of amphibians, light-dependent photoreception could also take place in extra-retinal photoreceptors (for review see Phillips et al. 2010a). This has been nicely demonstrated in Eastern red-spotted newts, *Nothophthalmus viridescens*, which show a 90° shifted response in magnetic compass orientation when tested under red light, compared to tests under blue or full-spectrum white light (Phillips and Borland 1992). Hence, newts trained to learn the shoreward side of a training tank with the top of their head, but not their eyes, covered with a red-light filter, showed the same 90° shifted response as control animals completely illuminated by red light (Deutschlander et al. 1999). Thus, the light-sensitive-magnetosensitive receptors mediating shoreward orientation in newts are most likely located in extraocular photoreceptors in the pineal complex or hypothalamus (Deutschlander et al. 1999). In anuran amphibians and lizards, single photoreceptors with two antagonistic photoreception mechanisms, like those proposed to underlie the light-dependent magnetic compass in newts, have been found in the pineal complex (Eldred and Nolte 1978, Solessio and Engbretson 1993).

10.2 Celestial compasses–sun, polarized light, and star compasses

10.2.1 Physiological evidence for sun and star compass orientation

While the behavioural mechanisms of sun compass orientation are well understood in many species (see Åkesson et al., Chapter 9), we know little about how the sun compass works on the physiological level, especially in higher organisms. Similarly, it remains to be shown what key features from the rotating starry sky the birds use for orientation. Are birds taking 'snapshot pictures' at certain time intervals and overlaying these to identify the centre of rotation? Or are they able to take an overexposed picture, and thus can follow the movement of the stars by seeing their paths? We can currently only speculate on how the star compass in birds functions, and how this information is mediated by the brain and which brain areas are used for processing star compass orientation.

Cleverly designed conditioning experiments could be one approach in order to get further insight into how birds are able to identify the centre of rotation as reference cue for star compass orientation. One possibility for identifying brain areas involved in processing star compass orientation would be 'behavioural molecular mapping', such as the method used to identify Cluster N as the brain area involved in magnetic compass orientation (Mouritsen et al. 2005). This requires an extremely well-designed experiment, including standardized control conditions and exclusion of any alternative variables that could putatively be used for compass orientation.

10.2.2 Behavioural and physiological evidence for polarized light sensitivity

The ability to perceive linearly polarized skylight and use it for orientation is common in the animal kingdom (Åkesson et al., Chapter 9). It has been extensively characterized, on both the behavioural and physiological level, first and foremost in invertebrates, such as arthropods (e.g. honeybees, crickets, ants, spiders, beetles, crabs, shrimp, crayfish, water fleas) and cephalopods (e.g. squids and cuttlefish) (for reviews see Horváth and Varjú 2004, Roberts et al. 2011). However, also vertebrates seem to use this third dimension of light in various behavioural contexts, and polarized light sensitivity has been reported in fish (e.g. Cameron and Pugh 1991, Parkyn et al. 2003), amphibians (Adler and Taylor 1973), reptiles (Adler and Phillips 1985), and birds (e.g. Able 1982, Muheim et al. 2006b, and see below). To date, no evidence for polarization sensitivity in mammals has been published, so mammals may be an exception (but see polarization vision in humans, Box 10.4).

How can animals perceive polarized light? The fundamental prerequisites for a photoreceptor to be able to detect the electric vector of light are inherently polarization-sensitive pigment molecules (dichroic chromophores). Some common chromophores, like retinal, fulfil this requirement. The exceptionally high polarization sensitivity found in many insects, crustaceans, and cephalopods is enhanced by the highly organized microvilli of individual rhabdomeric photoreceptor cells in an ommatidium that are often aligned at 90° to each other (Roberts et al. 2011). In insects, these polarization-sensitive photoreceptors are further concentrated in specialized regions of the eyes (in many cases in an area in the upper field of view,

Box 10.4 Polarization vision in humans

Contrary to the common belief, humans possess the ability to see polarized light, at least in theory. The phenomenon is called Haidinger's Brushes, and is most likely due to an artefact, i.e. a deformation of the macula (yellow spot) in our eyes (Bone 1980). On clear days at times around sunrise and sunset, when the natural polarization of the skylight is most prominent and up to 80–90% of the incoming skylight is polarized and aligned at 90° from the sun, these images can, after some practice, be seen in the sky. The plane of polarization also becomes visible when a sheet polarizer is turned by 90° clockwise and counterclockwise in front of a polarized light source (the natural sky or, for example, an LCD screen). When the two planes of linear polarization are arranged perpendicular to each other, no light is transmitted through the filters and a dark area is produced (Fig. 10.5; see also colour plate section). We can only speculate that humans at some point in their evolutionary history may have used this information for orientation and navigation. In theory, they could have made use of the pattern of skylight polarization as a substitute for the position of the sun when it was covered by clouds or topography, as many insects have been shown to do (see also Åkesson et al., Chapter 9).

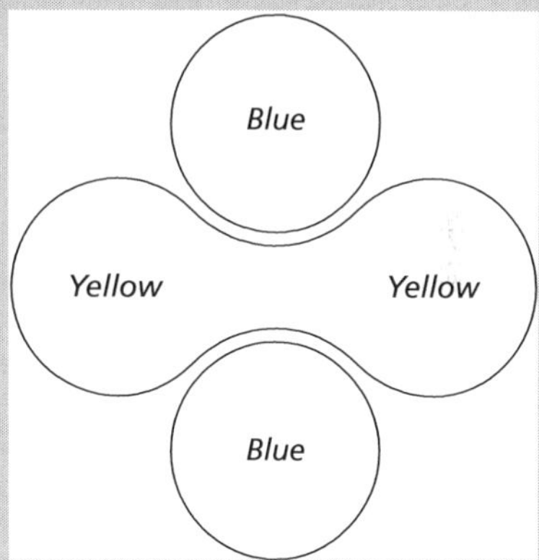

Figure 10.5 Polarized light as perceived by the human eye, visible as two faint blue and two faint yellow balloons (Haidinger's brushes), aligned perpendicularly, with the blue axis indicating the axis of polarization. (See Plate 3).

so-called dorsal rim area), where they are aligned in ordered arrays to measure polarized light from different angles (Wehner 1989).

The insects' compound eyes thus seem to be better suited to perceive polarized light than the vertebrate eye, and the question is then how higher animals can perceive polarized light. When we look up in the sky, we can see the sun and blue patches of sky, but most of us cannot see the e-vector of polarized skylight, and thus the information provided by the skylight polarization pattern. This is mainly due to the fact that we have ciliary photoreceptors (unlike insects that have rhabdomeric photoreceptors). Ciliary photoreceptors do not have the necessary ultrastructure and organization for polarized light reception (Roberts et al. 2011). Thus, the receptor mechanism for perceiving polarized light via ciliary photoreceptors remains one of the big mysteries in sensory biology for most vertebrates. An exception are some fish species, like e.g. anchovies that have axially oriented cone photoreceptors enabling them to distinguish linearly polarized light from different angles, or salmonids that have double cones specialized for polarization sensitivity (Flamarique et al. 1998, Kamermans and Hawryshyn 2011).

In birds, the perception of linearly polarized skylight is still under debate. In his pioneering work, Kenneth Able first demonstrated the importance of polarized light for migratory birds (Able 1982, 1989; Able and Able 1993) by analysing the specific role of polarized light cues in migratory orientation. However, most attempts to demonstrate polarized light sensitivity in birds in indoor settings have failed, mostly due to that carefully controlled conditioning and discrimination experiments are extremely difficult to carry out. One of the problems is the differential reflection of polarized light on surfaces which can lead to light intensity artefacts (for review see Muheim 2011). Two early conditioning experiments successfully demonstrated polarized light sensitivity in homing pigeons (Kreithen and Keeton 1974, Delius et al. 1976), but other studies failed to confirm these findings. Also, discrimination experiments with Japanese quails, *Coturnix coturnix japonica*, and European starlings, *Sturnus vulgaris*, were unsuccessful in demonstrating polarized light sensitivity in birds (Greenwood et al. 2003). Still, there is strong experimental evidence that migratory birds use information from the skylight polarization pattern to determine their departure direction and calibrate their compasses (see Åkesson et al., Chapter 9). However, the (putative) reception mechanism via which birds could perceive the pattern of polarized skylight remains a mystery. There are simply no obvious anatomical structures described in their eyes that could fulfil this job. The avian double cones have been proposed as polarized light receptors, since their orientation forms a cross pattern, with two opposite double cones facing each other, and the other two pointing away from each other (Young and Martin 1984, Cameron and Pugh 1991). However, it is unclear whether they could act as polarized light receptors, and therefore, more research will be necessary to behaviourally characterize and identify the photoreceptors mediating polarized light sensitivity in birds.

10.3 Future perspectives

During the past decade, research on the sensory mechanisms of animal orientation and navigation has become highly integrative and interdisciplinary. The field has progressed from the observation-based study of mainly behavioural aspects of orientation and navigation, to include biophysical, neuroanatomical, and molecular tools. The introduction of new techniques from various neighbouring disciplines has led to significant advances in our understanding of the sensory and cognitive mechanisms underlying animal orientation and navigation. Still, many open questions remain to be answered. Where are the magnetoreceptors located? How is the information processed in the brain? What is the neuronal basis of star compass orientation? Which brain areas are involved in processing celestial compass information from the sun, the stars, or polarized skylight cues? How do birds (and maybe other vertebrates as well) perceive polarized light cues and use this information for compass orientation? Given recent advances in technology and enhanced collaborative effort between disciplines, many of these questions will be addressed, and hopefully answered in the years to come.

References

Able, K. P. (1982). Skylight polarization patterns at dusk influence migratory orientation in birds. *Nature*, **299**, 550–1.

Able, K. P. (1989). Skylight polarization patterns and the orientation of migratory birds. *Journal of Experimental Biology*, **141**, 241–56.

Able, K. P., and Able, M. A. (1993). Daytime calibration of magnetic orientation in a migratory bird requires a view of skylight polarization. *Nature*, **364**, 523–5.

Adler, K., and Phillips, J. B. (1985). Orientation in a desert lizard (*Uma notata*), time-compensated compass movement and polarotaxis. *Journal of Comparative Physiology A: Neuroethology, Sensory, Neural, and Behavioral Physiology*, **156**, 547–52.

Adler, K., and Taylor, D. H. (1973). Extraocular perception of polarized light by orienting salamanders. *Journal of Comparative Physiology A: Neuroethology, Sensory, Neural, and Behavioral Physiology*, **87**, 203–12.

Beason, R. C., Dussourd, N., and Deutschlander, M. E. (1995). Behavioural evidence for the use of magnetic material in magnetoreception by a migratory bird. *Journal of Experimental Biology*, **198**,141–6.

Beason, R. C., and Semm, P. (1987). Magnetic responses of the trigeminal nerve system of the bobolink, *Dolichonyx oryzivorus*. *NeuroScience Letters*, **80**, 229–34.

Beason, R. C., and Semm, P. (1996). Does the avian ophthalmic nerve carry magnetic navigational information? *Journal of Experimental Biology*, **199**, 1241–4.

Beason, R. C., Wiltschko, R., and Wiltschko, W. (1997). Pigeon homing: effects of magnetic pulses on initial orientation. *Auk*, **114**, 405–15.

Biskup, T., Schleicher, E., Okafuji, A., et al. (2009). Direct observation of a photoinduced radical pair in a cryptochrome blue-light photoreceptor. *Angewandte Chemie Internationale Edition*, **48**, 404–7.

Blakemore, R. P. (1975). Magnetotactic bacteria. *Science*, **190**, 377–9.

Bone, R. A. (1980). The role of the macular pigment in the detection of polarized light. *Vision Research*, **20**, 213–20.

Cadiou, H., and McNaughton, P. A. (2010). Avian magnetite-based magnetoreception: a physiologist's perspective. *Journal of the Royal Society Interface*, **7**, S193–S205.

Cameron, D. A., and Pugh, E. N. (1991). Double cones as a basis for a new type of polarisation vision in vertebrates. *Nature*, **353**, 161–4.

Chaves, I., Pokorny, R., Byrdin, M., et al. (2011). The cryptochromes: blue light photoreceptors in plants and animals. *Annual Review of Plant Biology*, **62**, 335–64.

Chew, G. L., and Brown, G. E. (1989). Orientation of rainbow trout (*Salmo gairdneri*) in normal and null magnetic fields. *Canadian Journal of Zoology*, **67**, 641–3.

Delius, J. D., Perchard, R. J., and Emmerton, J. (1976). Polarized light discrimination by pigeons and an electroretinographic correlate. *Journal of Comparative Physiology and Psychology*, **90**, 560–71.

Denzau, S., Kuriakose, D., Freire, R., Munro, U., and Wiltschko, W. (2011). Conditioning domestic chickens to a magnetic anomaly. *Journal of Comparative Physiology A: Neuroethology, Sensory, Neural, and Behavioral Physiology*, **197**, 1137–41.

Deutschlander, M. E., Borland, S. C., and Phillips, J. B. (1999). Extraocular magnetic compass in newts. *Nature*, **400**, 324–5.

Diebel, C. E., Proksch, R., Green, C. R., Neilson, P., and Walker, M. M. (2000). Magnetite defines a magnetoreceptor. *Nature*, **406**, 299–302.

Eder, S. H. K., Cadiou, H., Muhamad, A., McNaughton, P. A., Kirschvink, J. L., and Winklhofer, M. (2012). Magnetic characterization of isolated candidate vertebrate magnetoreceptor cells. *Proceedings of the National Academy of Sciences USA*, **109**, 12022–7.

Eldred, W. D., and Nolte, J. (1978). Pineal photoreceptors: evidence for a vertebrate visual pigment with two physiologically active states. *Vision Research*, **18**, 29–32.

Falkenberg, G., Fleissner, G., Schuchardt, K., et al. (2010). Avian magnetoreception: elaborate iron mineral containing dendrites in the upper beak seem to be a common feature of birds. *PLoS One*, **5**, e9231.

Flamarique, I. N., Hawryshyn, C. W., and Hárosi, F. I. (1998). Double-cone internal reflection as a basis for polarization detection in fish. *Journal of the Optical Society of America A*, **15**, 349–58.

Fleissner, G., Holtkamp-Rötzler, E., Hanzlik, M., et al. (2003). Ultrastructural analysis of a putative magnetoreceptor in the beak of homing pigeons. *Journal of Comparative Neurology*, **458**, 350–60.

Fleissner, G., Stahl, B., Thalau, P., Falkenberg, G., and Fleissner, G. (2007). A novel concept of Fe-mineral-based magnetoreception: histological and physicochemical data from the upper beak of homing pigeons. *Naturwissenschaften*, **94**, 631–42.

Freire, R., Dunston, E., Fowler, E. M., McKenzie, G. L., Quinn, C. T., and Michelsen, J. (2012). Conditioned response to a magnetic anomaly in the Pekin duck (*Anas platyrhynchos domestica*) involves the trigeminal nerve. *Journal of Experimental Biology*, **215**, 2399–404.

Gagliardo, A., Ioale, P., Savini, M., and Wild, J. M. (2006). Having the nerve to home: trigeminal magnetoreceptor versus olfactory mediation of homing in pigeons. *Journal of Experimental Biology*, **209**, 2888–92.

Gegear, R. J., Casselman, A., Waddell, S., and Reppert, S. M. (2008). Cryptochrome mediates light-dependent magnetosensitivity in *Drosophila*. *Nature*, **454**, 1014–18.

Giovani, B., Byrdin, M., Ahmad, M., and Brettel, K. (2003). Light-induced electron-transfer in a cryptochrome blue-light photoreceptor. *Nature Structural Biology*, **10**, 489–90.

Greenwood, V. J., Smith, E. L., Church, S. C., and Partridge, J. C. (2003). Behavioural investigation of polarisation sensitivity in the Japanese quail, *Coturnix coturnix japonica*, and the European starling, *Sturnus vulgaris*. *Journal of Experimental Biology*, **206**, 3201–10.

Henbest, K. B., Rodgers, C. T., Hore, P. J., and Timmel, C. R. (2004). Radio frequency magnetic field effects on a radical recombination reaction: a diagnostic test for the radical pair mechanism. *Journal of the American Chemical Society*, **126**, 8102–3.

Heyers, D., Manns, M., Luksch, H., Güntürkün, O., and Mouritsen, H. (2007). A visual pathway links brain structures active during magnetic compass orientation in migratory birds. *PLoS One*, **2**, e937.

Holland, R. A., Thorup, K., Gagliardo, A., et al. (2009). Testing the role of sensory systems in the migratory heading of a songbird. *Journal of Experimental Biology*, **212**, 4065–71.

Horváth, G., and Varju, D. (2004). *Polarized Light in Animal Vision: Polarization Patterns in Nature*. Springer-Verlag, New York.

Kamermans, M., and Hawryshyn, C. (2011). Teleost polarization vision: how it might work and what it might be good for. *Philosophical Transactions of the Royal Society B: Biological Sciences*, **366**, 742–56.

Kirschvink, J. L. (1989). Magnetite biomineralization and geomagnetic sensitivity in higher animals: an update and recommendations for future study. *Bioelectromagnetics*, **10**, 239–59.

Kirschvink, J. L. (1992). Comment on 'Constraints on biological effects of weak extremely-low-frequency electromagnetic fields'. *Physical Review A*, **46**, 2178–84.

Kirschvink, J. L., and Walker, M. M. (1985). Particle-size considerations for magnetite-based magnetoreceptors. In K. L. Kirschvink, D. S. Jones, and B. J. Mcfadden (eds), *Magnetite Biomineralization and Magnetoreception in Organisms: A New Biomagnetism*, pp. 243–54. Plenum Press, New York/London.

Kirschvink, J. L., Walker, M. M., Chang, S.-B. R., Dizon, A. E., and Petersen, K. A. (1985). Chains of single-domain magnetite particles in chinook salmon, *Oncorhynchus tshawytscha*. *Journal of Comparative Physiology A: Neuroethology, Sensory, Neural, and Behavioral Physiology*, **157**, 375–81.

Kreithen, M. L., and Keeton, W. T. (1974). Detection of polarized light by the homing pigeon, *Columba livia*. *Journal of Comparative Physiology A: Neuroethology, Sensory, Neural, and Behavioral Physiology*, **89**, 83–92.

Liedvogel, M., Maeda, K., Henbest, K. B., et al. (2007). Chemical magnetoreception: bird cryptochrome 1a is excited by blue light and forms long-lived radical-pairs. *PLoS One*, **2**, e1106.

Liedvogel, M., and Mouritsen, H. (2010). Cryptochromes—a potential magnetoreceptor: what do we know and what do we want to know? *Journal of the Royal Society Interface*, **7**, S147–62.

Lohmann, K. J., and Johnson, S. (2000). The neurobiology of magnetoreception in vertebrate animals. *Trends in Neuroscience*, **23**, 153–9.

Lohmann, K. J., and Lohmann, C. M. (1993). A light-independent magnetic compass in the leatherback sea turtle. *Biological Bulletins*, **185**, 149–51.

Maeda, K., Henbest, K. B., Cintolesi, F., et al. (2008). Chemical compass model of avian magnetoreception. *Nature*, **453**, 387–90.

Maeda, K., Robinson, A. J., Henbest, K. B., et al. (2012). Magnetically sensitive light-induced reactions in cryptochrome are consistent with its proposed role as a magnetoreceptor. *Proceedings of the National Academy of Sciences USA*, **109**, 4774–9.

Marhold, S., Wiltschko, W. and Burda, H. (1997). A magnetic polarity compass for direction finding in a subterranean mammal. *Naturwissenschaften*, **84**, 421–3.

Mora, C. V., Davison, M., Wild, J. M., and Walker, M. M. (2004). Magnetoreception and its trigeminal mediation in the homing pigeon. *Nature*, **432**, 508–11.

Mouritsen, H., Feenders, G., Liedvogel, M., Wada, K. and Jarvis, E. D. (2005). Night-vision brain area in migratory songbirds. *Proceedings of the National Academy of Sciences USA*, **432**, 8339–44.

Mouritsen, H., and Hore, P. (2012). The magnetic retina: light-dependent and trigeminal magnetoreception in migratory birds. *Current Opinion in Neurobiology*, **22**, 343–52.

Mouritsen, H., and Ritz, T. (2005). Magnetoreception and its use in bird navigation. *Current Opinion in Neurobiology*, **15**, 406–14.

Muheim, R. (2011). Behavioural and physiological mechanisms of polarized light sensitivity in birds. *Philosophical Transactions of the Royal Society B: Biological Sciences*, **366**, 763–71.

Muheim, R., Bäckman, J., and Åkesson, S. (2002). Magnetic compass orientation in European robins is dependent on both wavelength and intensity of light. *Journal of Experimental Biology*, **205**, 3845–56.

Muheim, R., Edgar, N. M., Sloan, K. S., and Phillips, J. B. (2006a). Magnetic compass orientation in C57BL/6 mice. *Learning and Behaviour*, **34**, 366–73.

Muheim, R., Phillips, J. B., and Åkesson, S. (2006b). Polarized light cues underlie compass calibration in migratory songbirds. *Science*, **313**, 837–9.

Munro, U., Munro, J. A., and Phillips, J. B. (1997a). Evidence for a magnetite-based navigational 'map' in birds. *Naturwissenschaften*, **84**, 26–8.

Munro, U., Munro, J. A., Phillips, J. B., and Wiltschko, W. (1997b). Effect of wavelength of light and pulse magnetisation on different magnetoreception systems in a migratory bird. *Australian Journal of Zoology*, **45**, 189–98.

Nemec, P., Altmann, J., Marhold, S., Burda, H., and Oelschläger, H. A. (2001). Neuroanatomy of magnetoreception: the superior colliculus involved in magnetic orientation in a mammal. *Science*, **294**, 366–8.

Niessner, C., Denzau, S., Gross, J. C., et al. (2011). Avian ultraviolet/violet cones identified as probable magnetoreceptors. *PLoS One*, **6**, e20091.

Parkyn, D. C., Austin, J. D. and Hawryshyn, C. W. (2003). Acquisition of polarized light orientation in salmonids under laboratory conditions. Animal Behaviour **65**, 893–904.

Phillips, J. B. (1986). Two magnetoreception pathways in a migratory salamander. *Science*, **233**,765–7.

Phillips, J. B., and Borland, S. C. (1992). Behavioural evidence for use of light-dependent magnetoreception mechanism by a vertebrate. *Nature*, **359**, 142–4.

Phillips, J. B., Jorge, P. E., and Muheim, R. (2010a). Light-dependent magnetic compass orientation in amphibians and insects: candidate receptors and candidate molecular mechanisms. *Journal of the Royal Society Interface*, **7**, S241–56.

Phillips, J. B., Muheim, R., and Jorge, P. E. (2010b). A behavioural perspective on the biophysics of the light-dependent magnetic compass: a link between directional and spatial perception? *Journal of Experimental Biology*, **213**, 3247–55.

Phillips, J. B., and Sayeed, O. (1993). Wavelength-dependent effects of light on magnetic compass orientation in *Drosophila melanogaster*. *Journal of Comparative Physiology A: Neuroethology, Sensory, Neural, and Behavioral Physiology*, **172**, 303–8.

Ritz, T., Adem, S., and Schulten, K. (2000). A model for photoreceptor-based magnetoreception in birds. *Biophysical Journal*, **78**, 707–18.

Ritz, T., Ahmad, M., Mouritsen, H., Wiltschko, R., and Wiltschko, W. (2010). Photoreceptor-based magnetoreception: optimal design of receptor molecules, cells, and neuronal processing. *Journal of the Royal Society Interface*, **7**, S135–46.

Ritz, T., Thalau, P., Phillips, J. B., Wiltschko, R., and Wiltschko, W. (2004). Resonance effects indicate a radical-pair mechanism for avian magnetic compass. *Nature*, **429**, 177–80.

Roberts, N. W., Porter, M. L., and Cronin, T. W. (2011). The molecular basis of mechanisms underlying polarisation vision. *Philosophical Transactions of the Royal Society B: Biological Sciences*, **366**, 627–37.

Rodgers, C. T., and Hore, P. J. (2009). Chemical magnetoreception in birds: the radical pair mechanism. *Proceedings of the National Academy of Sciences USA*, **106**, 353–60.

Schleicher, E., Bittl, R., and Weber, S. (2009). New roles of flavoproteins in molecular cell biology: blue-light active flavoproteins studied by electron paramagnetic resonance. *FEBS Journal*, **276**, 4290–303.

Schulten, K. (1982). Magnetic field effects in chemistry and biology. *Advanced Solid State Physics*, **22**, 61–83.

Schulten, K., Swenberg, C., and Weller, A. (1978). A biomagnetic sensory mechanism based on magnetic field modulated coherent electron spin motion. *Zeischrift für Physikalische Chemie* NF, **111**, 1–5.

Semm, P., Nohr, D., Demaine, C., and Wiltschko, W. (1984). Neural basis of the magnetic compass: interactions of visual, magnetic and vestibular inputs in the pigeon's brain. *Journal of Comparative Physiology A: Neuroethology, Sensory, Neural, and Behavioral Physiology*, **155**, 283–8.

Solessio, E., and Engbretson, G. A. (1993). Antagonistic chromatic mechanisms in photoreceptors of the parietal eye of lizards. *Nature*, **364**, 442–5.

Thalau, P., Holtkamp-Rötzler, E., Fleissner, G., and Wiltschko, W. (2007). Homing pigeons (*Columba livia f. domestica*) can use magnetic cues for locating food. *Naturwissenschaften*, **94**, 813–19.

Thalau, P., Ritz, T., Burda, H., Wegner, R. E., and Wiltschko, R. (2006). The magnetic compass mechanisms of birds and rodents are based on different physical principles. *Journal of the Royal Society Interface*, **3**, 583–7.

Treiber, C. D. Salzer, M. C., Riegler, J., et al. (2012). Clusters of iron-rich cells in the upper beak of pigeons are macrophages not magnetosensitive neurons. *Nature*, **484**, 367–70.

Voss, J., Keary, N., and Bischof, H.-J. (2007). The use of the geomagnetic field for short distance orientation in zebra finches. *NeuroReport*, **18**, 1053–7.

Wajnberg, E., Acosta-Avalos, D., Alves, O. C., De Oliveira, J. F., Srygley, R. B., and Esquivel, D. M. S. (2010). Magnetoreception in eusocial insects: an update. *Journal of the Royal Society Interface*, **7**, S207–25.

Walker, M. M. (1984). Learned magnetic field discrimination in yellowfin tuna, *Thunnus albacares*. *Journal of Comparative Physiology A: Neuroethology, Sensory, Neural, and Behavioral Physiology*, **155**, 673–9.

Walker, M. M., Diebel, C. E., Haugh, C. V., Pankhurst, P. M., Montgomery, J. C., and Green, C. R. (1997). Structure and function of the vertebrate magnetic sense. *Nature*, **390**, 371–6.

Walker, M. M., Kirschvink, J. L., Chang, S.-B. R., and Dizon, A. E. (1984). A candidate magnetic sense organ in the yellowfin tuna, *Thunnus albacares*. *Science*, **224**, 751–3.

Weber, S., Kay, C. W. M., Mogling, H., Mobius, K., Hitomi, K., and Todo, T. (2002). Photoactivation of the flavin cofactor in *Xenopus laevis* (6–4) photolyase: observation of a transient tyrosyl radical by time-resolved electron paramagnetic resonance. *Proceedings of the National Academy of Sciences USA*, **99**, 1319–22.

Wehner, R. (1989). Neurobiology of polarization vision. *Trends in Neurosciences*, **12**, 353–9.

Williams, M. N., and Wild, J. M. (2001). Trigeminally innervated iron-containing structures in the beak of homing pigeons, and other birds. *Brain Research*, **889**, 243–6.

Wiltschko, W. (1968). Über den Einfluss statischer Magnetfelder auf die Zugorientierung der Rotkehlchen, *Erithacus rubecula*. *Zeitschrift für Tierpsychologie*, **25**, 537–58.

Wiltschko, W., Munro, U., Beason, R. C., Ford, H., and Wiltschko, R. (1994). A magnetic pulse leads to a temporary deflection in the orientation of migratory birds. *Experientia*, **50**, 679–700.

Wiltschko, W., Munro, U., Ford, H., and Wiltschko, R. (1993). Red light disrupts magnetic orientation of migratory birds. *Nature*, **364**, 525–7.

Wiltschko, W., Stapput, K., Thalau, P., and Wiltschko, R. (2006). Avian magnetic compass: fast adjustment to intensities outside the normal functional window. *Naturwissenschaften*, **93**, 300–4.

Wiltschko, W., and Wiltschko, R. (1972). Magnetic compass of European robins. *Science*, **176**, 62–4.

Wiltschko, R., and Wiltschko, W. (1995a). *Magnetic Orientation in Animals*. Springer, New York.

Wiltschko, W., and Wiltschko R. (1995b). Migratory orientation of European robins is affected by the wavelength of light as well as by a magnetic pulse. *Journal of Comparative Physiology A: Neuroethology, Sensory, Neural, and Behavioral Physiology*, **177**, 363–9.

Wiltschko, W., and Wiltschko, R. (2001). Light-dependent magnetoreception in birds: the behaviour of European robins, *Erithacus rubecula*, under monochromatic light of various wavelengths and intensities. *Journal of Experimental Biology*, **204**, 3295–302.

Wiltschko, W., and Wiltschko, R. (2005). Magnetic orientation and magnetoreception in birds and other animals. *Journal of Comparative Physiology A: Neuroethology, Sensory, Neural, and Behavioral Physiology*, **191**, 675–93.

Winklhofer, M., and Kirschvink, J. L. (2010). A quantitative assessment of torque-transducer models for magnetoreception. *Journal of the Royal Society Interface*, **7**, S273–89.

Wu, L.-Q., and Dickman, J. D. (2011). Magnetoreception in an avian brain in part mediated by inner ear lagena. *Current Biology*, **21**, 418–23.

Wu, L.-Q., and Dickman, J. D. (2012). Neural correlates of a magnetic sense. *Science*, **336**, 1054–7.

Yoshii, T., Ahmad, M., and Helfrich-Förster, C. (2009). Cryptochrome mediates light-dependent magnetosensitivity of drosophila's circadian clock. *PLoS Biology*, **7**, e1000086.

Young, S. R., and Martin, G. R. (1984). Optics of retinal oil droplets: a model of light collection and polarization detection in the avian retina. *Vision Research*, **24**, 129–37.

Zapka, M., Heyers, D., Hein, C.M., et al. (2009). Visual but not trigeminal mediation of magnetic compass information in a migratory bird. *Nature*, **461**, 1274–7.

CHAPTER 11

Movements in the olfactory landscape

Glenn P. Svensson, Maria Strandh, and Christer Löfstedt

The ability to trace distant odours is crucial for many animals in search for mates or food sources. Some of these long-distance olfactory-mediated behaviours are among the most impressive examples of animal navigation observed, such as the migration of salmon over hundreds of kilometres as they return from the open ocean to their natal stream to reproduce, or the return flights of procellariiform seabirds to their nesting islands after travelling thousands of kilometres on the wing in search for food in the open sea. On a smaller scale, one cannot but marvel when observing the dense traffic of ants as they move back and forth along a narrow chemical trail between a food source and the nest (Fig. 1.1). The ability to locate an aquatic or airborne odour plume, or a substrate-borne odour trail, and to follow it all the way to its source requires many adaptations, such as an efficient locomotion system (Johansson et al., Chapter 13; Hedenström and Lindström, Chapter 5), an optimized search strategy, and an efficient sensory apparatus (Muheim et al., Chapter 10) to accurately resolve the spatiotemporal distribution of the olfactory stimulus. It is reasonable to assume that such traits have been subject to strong natural selection over evolutionary time to optimize the way animals find remote odour sources, which are often patchily distributed and ephemeral.

Chemical communication, where the sender emits a substance that elicits a response in the receiver, is the most ancient mode of communication and occurs in more or less all organisms, from unicellular bacteria to elephants, and over a wide range of spatial scales. Chemical signals most likely evolved from emitted substances (cues) that were exploited, to chemical stimuli shaped by the producer to increase their information-transmitting function. Despite the fact that chemical communication is the most potent and widespread form of information transfer in nature, the research field of chemical ecology is historically young, and took off in the 1950s when the first identification of a sex pheromone was published (Butenandt et al. 1959). There are probably several reasons for this slow start. As humans, we find it easier to study how animal behaviours are influenced by visual and auditory cues, because we are familiar with these senses, but due to our poorly developed sense of olfaction, we are in many cases not able to perceive the chemical cues influencing animal behaviours. In addition, olfactory signals have been hard to analyse because they are often complex, released in minute amounts, and without any apparent change in the behaviour of the producer. During the last decades, however, the advancement of analytical, electrophysiological, behavioural, and molecular assays have enabled the field of chemical ecology to grow rapidly, and today behaviourally active compounds have been identified for all major taxonomic groups of animals (Symonds and Elgar 2008).

The Pherobase website (www.pherobase.com) is a freely available database of chemical compounds known to mediate intra- or interspecific interactions. A visitor exploring this database soon realizes that the vast majority of published compounds have been identified from insects. Their accessible and relatively simple olfactory system, in combination with their often very stereotypic behavioural response to olfactory cues, has made insects a major research model for any aspect of chemical ecology.

Animal Movement Across Scales. Edited by Lars-Anders Hansson and Susanne Åkesson.
© Oxford University Press 2014. Published 2014 by Oxford University Press.

In addition, their economic importance as pest species in agriculture and forestry, as vectors for various human diseases, and as major pollinators of crops, has triggered extensive research on insect olfaction with the aim of developing methods to manipulate this sensory system. In contrast, information about the roles and mechanisms of olfactory communication in other taxonomic groups of animals are still rather fragmentary. Extensive progress has been made recently for some groups, but the chemical identity of most compounds mediating specific interactions is still unknown, and the lack of synthetic lures for laboratory and field experiments is a major limitation for the advancement of chemical ecology research in these taxa. As a consequence of this asymmetry in research effort between insects and other animals, published data on odour-guided movement, i.e. how animals locate and follow an odour plume or trail to its source, are strongly biased toward insects, especially moths.

This chapter will focus on the diversity of navigational strategies, with olfaction as a prerequisite, that different animals use to find remote odour sources. We have tried to select interesting model systems of olfactory search from throughout the animal kingdom to highlight similarities and differences in navigational performance, and the sensory machinery responsible for odour tracking in various groups. Two large taxonomic groups—amphibians and reptiles—are not covered here. Although the use of short-range chemosensory signals is widespread in both groups (Woodley 2010, Mason and Parker 2010), information regarding olfactory navigation over large distances is still limited.

11.1 The olfactory landscape

In terrestrial, as well as in aquatic environments, chemosensory signals originating from various sources produce an olfactory landscape through which animals must navigate to efficiently locate important resources and avoid enemies. Many species have a remarkable capacity to detect and discriminate among such signals. Olfactory signals are detected by specific sensory neurons of the antennae of invertebrates, and the nasal cavity of vertebrates, respectively. Because olfactory cues (excluding chemical trails) are transmitted via the current flow, they can travel over large distances, but at a slower speed compared to visual and auditory cues, and the directional control of emission is poor. However, the most important features of olfactory signals with regard to animal movement are their highly intermittent structure and chemical complexity. Thus, in order to understand how animals track odour plumes, information about the physical and chemical properties of such plumes is needed. The insight that an odour plume is not a smooth Gaussian distribution of odour intensity, but instead a highly discontinuous signal of rapidly fluctuating concentrations of odour, has been essential for our understanding of how olfactory and locomotory systems in animals have evolved for the efficient location of remote odour sources. Visualization of odour plumes has enabled detailed analyses of their spatial distribution as well as the relationship between stimulus arrival and the resulting behavioural reaction in the target organism.

11.2 Structure of odour plumes

Odours disperse like smoke in air, in characteristic plume formations (Fig. 11.1) that we will briefly describe here. For more detailed information about various aspects of plume dispersal in aquatic and terrestrial habitats, see Murlis et al. (1992), Riffell et al. (2008), Webster and Weissburg (2009), and Atema (2012). After being emitted from the source, molecules are influenced by two physical forces that will affect their dispersal at different spatial scales: (i) molecular diffusion and (ii) turbulence and advection (Murlis et al. 1992, Vickers 2000, Riffell et al. 2008). At the micro-scale (<1 mm), molecules will diffuse due to Brownian motion, but this process is very slow and has probably limited influence on odour-induced behaviours in organisms larger than bacteria. In contrast, when molecules are carried away by fluid in the form of a chemical plume, turbulence (the dynamic mixture of odour with clean fluid), and advection (the bulk transport of odour in the average fluid direction) will cause the plume to stretch out, compress, and mix, to produce a highly discontinuous signal. Individual filaments containing high concentrations of odour molecules will be interspersed with the surrounding fluid (Fig. 11.1).

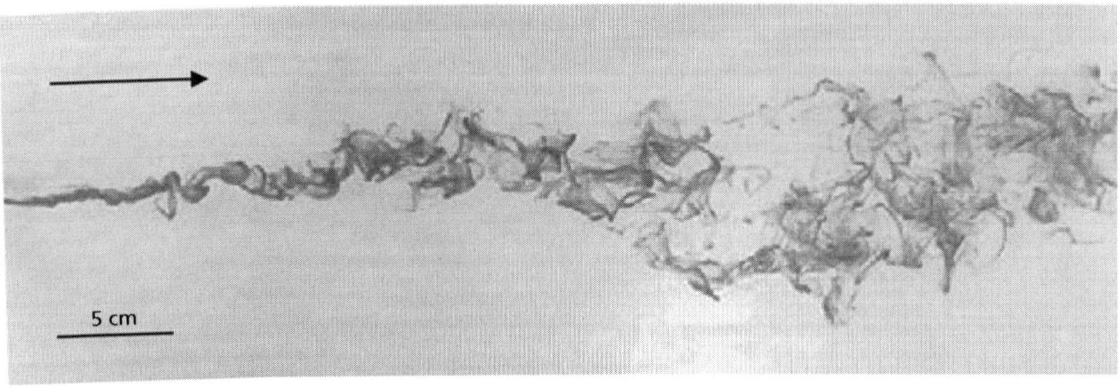

Figure 11.1 An aquatic odour plume visualized with a fluorescent dye (Rhodamine 6G). The arrow indicates the direction of flow (velocity set to 5 cm s^{-1}). At a distance from the source, the plume breaks up into smaller units (filaments) and becomes intermittent, with areas of high concentrations interspersed with areas of low or zero concentration. Modified with permission from Page et al. (2011a).

Thus, turbulence is the major force in plume formation in both terrestrial and aquatic environments. When the plume has travelled further away from its source, changes in the speed and direction of the fluid will cause it to meander and break up even more. As the distance from the source increases, the probability of encountering an odour filament at a fixed point decreases, although individual filaments can still be rather undiluted. Due to the intermittency of the signal, which is calculated as the proportion of time for which a signal is present, a moving animal will encounter filaments only infrequently during its tracking of an odour plume. For example, Murlis (1997) stated that the intermittency of terrestrial odour plumes is usually between 0.1 and 0.4, which means that the signal is absent for 60–90% of the time.

To directly measure the structure of odour plumes has been a complicated task. Various tracers, e.g. surrogate molecules, have traditionally been used to build models about plume structure and dynamics (Murlis 1997). For a tracer system to be relevant for olfactory research, the tracer itself has to be passive and conservative, and the method for its detection has to be both sensitive and active at sufficient time resolution. In terrestrial systems, both gases and ions have been used as tracers, although both systems have limitations. Recently, proton transfer reaction mass spectrometry (PTRMS) and photoionization detection (PID) have been applied in studies on plume structure (Riffell et al. 2008). Methods for analysing plume structure in aquatic systems have included rhodamine dye (Johnson et al. 2012), dopamine-sensitive carbon fibre microelectrodes (Moore and Atema 1991), and laser imaging techniques (Page et al. 2011a). The general conclusion from these measurements at rapid temporal scales is that terrestrial and aquatic plumes are highly intermittent, with bursts of high odour concentration interspersed with zero concentration or concentration below the detection threshold. However, the most reliable way of analysing plume structure in the context of odour tracking would be to use the chemosensory system of the target organism as a biological sensor. For insects, two different methods have been widely used to measure responses from olfactory receptor neurones (ORNs) upon stimulation. With electroantennography (EAG), the combined response from stimulated ORNs on the antenna can be measured, whereas single-sensillum recording (SSR) allows for monitoring the activity of individual ORNs. The SSR technique has the greatest potential for real-time analysis of odour plume structure, because ORNs typically respond to fluctuations in concentration within a few milliseconds, thus matching the rapid odour flux occurring inside plumes. In addition, recordings from single ORNs are not influenced by background levels of plant volatiles, or factors like temperature, moisture, or wind, which is the case for EAG. The most detailed analysis of odour plume structure using the SSR technique is probably by Valeur et al. (1999), in which the spike frequencies of individual ORNs of males of the turnip moth, *Agrotis segetum*, were measured when exposed to plumes from sex pheromone sources. Measurements were done 2 m downwind from the source and relative pheromone concentrations could be measured with a time resolution of 20 s. As predicted, the spatiotemporal distribution of odour filaments was found to be highly fluctuating.

11.3 Chemical features of odour plumes and trails

Communication signals come in a wide variety of chemical forms, and they mediate a wide variety of interactions both within and between species. Box 11.1 provides information about various types of chemicals used for information transfer between organisms, as well as their definitions. For more detailed overviews about the terminology of semiochemicals, see Wyatt (2010, 2014). Key features of semiochemicals are their molecular weight and polarity, which are the major factors determining their volatility in air, and solubility in water, respectively. In terrestrial systems, compounds used for long-range airborne communication are typically small and volatile. For example, female-produced sex pheromones of moths normally consist of chemical compounds that are 10–18 carbons long. In contrast, compounds used as trail pheromones of ants are generally less volatile because these signals should last for longer time periods on the substrate to enable other members of the community to detect, follow, and reinforce the trail if needed (Fig. 1.1; Morgan 2009).

The specificity of chemical signals is often achieved by blends of compounds rather than by single constituents, and the same compound can be part of the sex pheromone blend of many species, even those of distantly related taxonomic groups. A classic example is (Z)-7-dodecen-1-yl acetate, which has been identified as a female-produced sex pheromone for the Asian elephant, *Elephas maximus* (Rasmussen et al. 1996). This compound is also known to be a sex pheromone component for several hundred species of moths (www.pherobase.com). In aquatic systems, compounds of higher molecular weight, such as bile acids, soluble steroids, peptides and proteins, are commonly used for communication (Johnson and Li 2010, Woodley 2010). Regardless of structure, substances mediating interactions in terrestrial as well as in aquatic systems are generally released in minute amounts, and the receiver often demonstrates very high sensitivity to such signals.

Box 11.1 Definition of chemicals used for communication

Semiochemical is the broad term for compounds used for communication between individuals (Law and Regnier 1971). These can be divided into two main categories: *pheromones*, which mediate intraspecific interactions (Karlson and Lüscher 1959), and *allelochemicals*, which mediate interspecific interactions (Nordlund 1981). Pheromones are assumed to be of mutual benefit for both the producer and the receiver, and they can be further classified according to context, e.g. sex pheromones for mate finding, or trail pheromones for guiding colony members to a food source. A major limitation when classifying all chemical signals that mediate intraspecific interactions as pheromones is that many social interactions among conspecifics, e.g. recognition of kin or nest mates, are mediated by highly variable cues consisting of mixtures of many molecules, which are learned by other conspecifics to recognize a particular individual or colony. Such cues are thus fundamentally different from e.g. sex pheromones, which are the same within males or females across the species (though quantities of the pheromone may differ between individuals). Wyatt (2010) proposed the term *signature mixtures* to describe such learned chemical cues.

In contrast to pheromones, allelochemicals are classified according to the costs and benefits to producer and receiver. *Kairomones* define signals, which harm the producer but benefit the receiver. Such 'eavesdropping' is common among insect predators and parasitoids exploiting the sex pheromone channel of their prey or host. In contrast, *allomones* define signals, which benefit the producer, but harm the receiver. Plants mimicking the scent of a mate or food source of the pollinator are typical examples of such chemical deceit. Finally, *synomones* define signals, which are beneficial for both the producer and the receiver, and include any compound mediating a mutualistic interaction. Note that if a chemical signal is mediating more than one interaction, it can be classified as more than one type of semiochemical. For instance, a host-marking pheromone will also function as a kairomone if it is used by a female parasitoid to locate its host. Hence, we may conclude that the message is in the nose of the beholder.

A calling female moth releases her sex pheromone at a concentration of 10^{-9} g or less to attract males, and the male sex pheromone of the sea lamprey, *Petromyzon marinus*, triggers upstream movement in ovulated females at concentrations as low as 10^{-14} M (Johnson et al. 2009). In both cases, attraction of potential mates occurs over distances of several hundreds of metres.

11.4 The evolution of olfaction and olfactory-guided movement

The olfactory ability varies tremendously, both regarding quantity and quality of detectable olfactory cues, across the animal kingdom. What could explain such differences in the olfactory sense between taxa? The interplay during evolutionary time between an animal's particular olfactory environment, and the animal's genetic architecture will set the frame for the rate of dependence on the olfactory sense (Table 11.1). The visual and auditory senses always handle the same type of stimuli, a distribution of light or sound wavelengths, whereas olfaction has to adapt to a moving target of arising cues, similar to what the immune system has to deal with (Bargmann 2006). The complexity and amounts of useful olfactory cues (from the environment, including conspecifics and heterospecifics) could be hypothesized to be important in determining the complexity and sensitivity of the olfactory sense in a given species. External, physical factors like medium (water or air), temperature and the chemical structure and size of odour molecules present will further shape the olfactory sense to the environment (Table 11.1). However, cost–benefit trade-offs of having a good sense of smell certainly will influence how much an animal will invest in it. If e.g. other senses are more useful (i.e. give higher fitness) depending on the lifestyle of the animal, olfaction will not be selected for. The ability to use long-range olfactory cues will depend on the evolution of an appropriate number of olfactory receptors (ORs) coupled to complex enough neural circuits so that relevant odour cues can be integrated and used.

Table 11.1 Factors determining the evolution of olfaction and olfactory-guided movement in animals. External (physical) and internal (constraints of animal) factors are grouped separately.

External (physical) factors	Internal factors of animal
Life medium (air/water)	Spatial range
Temperature	Number of genes/genome size
Complexity of olfactory cues	Placement of olfactory organ
Amount of olfactory cues	Number of olfactory receptors
Volatility/size of olfactory cues	Complexity of neural circuits

Discrimination between different odour cues is achieved by binding of odour molecules to olfactory receptors, specific to certain molecules, whereupon signalling to the brain follows. Potentially, the more different olfactory receptors are present in a species, the more scent molecules can be distinguished. Insects have a relatively small but 'efficient' OR repertoire ranging from around 60 to 260 functional genes and with a low proportion of pseudogenes (Robertson et al. 2003, Robertson and Wanner 2006, Engsontia et al. 2008; Fig. 11.2). In crustaceans, no homologs of insect ORs have been found whereas other types of insect homologous chemoreceptors (gustatory and ionotropic receptors) probably are involved in olfactory reception (Peñalva-Arana et al. 2009, Corey et al. 2013). The rapid expansion of the insect OR gene family has been predicted to coincide with the evolution of terrestrial insects from their aquatic arthropod ancestors (Robertson et al. 2003). For the crustaceans the total number of ORs is not known. Inferring the numbers from their closest relatives, the insects, is probably risky since crustaceans mainly live in water and not in air.

Among vertebrates, the lowest numbers of ORs are found in the aquatic spotted green pufferfish, *Tetraodon nigriviridis*, with only 44 functional genes (Niimura and Nei 2005), whereas rat, *Rattus norvegicus*, has 1,284 functional ORs (Zhang et al. 2007). In birds, the repertoire of functional olfactory receptor genes is in the order of 300–400 genes (Steiger et al. 2009), which is comparable to the human repertoire of 388 functional ORs (Niimura and Nei 2003). The switch from aquatic to terrestrial life in vertebrates some 400 million years ago, coincides with a fast expansion of their ORs (Glusman et al. 2000). This fast expansion of a subset of the fish ORs most probably is a result of adaptation to airborne rather than waterborne odorants (Niimura and Nei 2005, Bargmann 2006) similar to the predicted cause

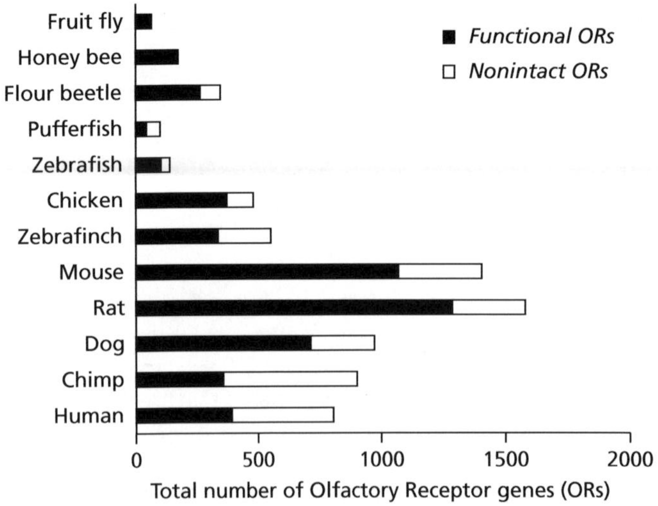

Figure 11.2 Distribution of olfactory receptor (OR) genes found in the genomes of a selection of insects and vertebrates. The proportion of functional genes (in black) and nonintact genes (truncated or pseudogenes, in white) are shown for the following taxa: fruit fly, *Drosophila melanogaster* (Robertson et al. 2003), honey bee, *Apis mellifera* (Robertson and Wanner 2006), red flour beetle, *Tribolium castaneum* (Engsontia et al. 2008), spotted green pufferfish, *Tetraodon nigriviridis*, zebrafish, *Danio rerio* (Niimura and Nei 2005), chicken, *Gallus gallus*, zebrafinch, *Taeniopygia guttata* (Steiger et al. 2009), mouse, *Mus musculus* (Zhang et al. 2004), rat, *Rattus norwegicus* (Zhang et al. 2007), dog, *Canis lupus familiaris* (Olender et al. 2004), chimpanzee, *Pan troglodytes* (Gilad et al. 2005), and human (Niimura and Nei 2003).

of the insect OR gene family expansion (Robertson et al. 2003).

11.5 Odour tracking

Different animals show surprisingly similar paths of olfactory navigation, regardless of their phylogenetic relationships, whether they walk, fly or swim, or whether they live in aquatic or terrestrial habitats. Comparing the search trajectories of a fly, a freshwater eel, a seabird, and a rat (Fig. 11.3), all share some common features, such as meandering or zigzagging patterns of movement in the horizontal plane. It is thus tempting to suggest that these paths are the result of a shared neuroethological pathway for odour navigation. However, the sensory organs and neural systems differ widely between taxonomic groups, and different animals have solved the problem of processing the information provided by odour plumes or trails for efficient navigation in different ways, supporting the convergent evolution of odour navigation systems. Although our understanding of how animals move in the olfactory landscape has advanced a lot in recent years, a potential problem is that this knowledge has mainly been derived from laboratory experiments using various bioassays, such as wind tunnels, Y-mazes, and flumes, where the medium (air or water) often has a constant speed and direction, producing continuous and relatively straight odour plumes. Much can be learnt about basic odour navigation from these controlled experiments, but the full picture of how animals navigate under natural conditions when the odour environment is much more turbulent may be difficult to deduce from laboratory experiments.

All animals use some kind of taxis for location of remote odour sources. Taxis is defined as the ability of an organism to direct its movement in relation to the source of stimulation. A key feature of olfactory search in many animals is to incorporate visual cues and mechanical assessment. In terrestrial systems, wind-induced movement is defined as anemotaxis, whereas movement in response to water flow in aquatic systems is referred to as rheotaxis. Animals use mechanosensory information when tracking an odour plume or following a scent trail. However, in contrast to invertebrates, for which information about such mechanisms has been derived from detailed experimentation in both terrestrial and aquatic environments, the modalities involved in odour tracking in other animals are not as well understood, because such information has mainly been derived from observational rather than experimental data.

11.5.1 Odour tracking in insects

Insects may be equipped with brains smaller than the size of a pinhead, but they can nevertheless

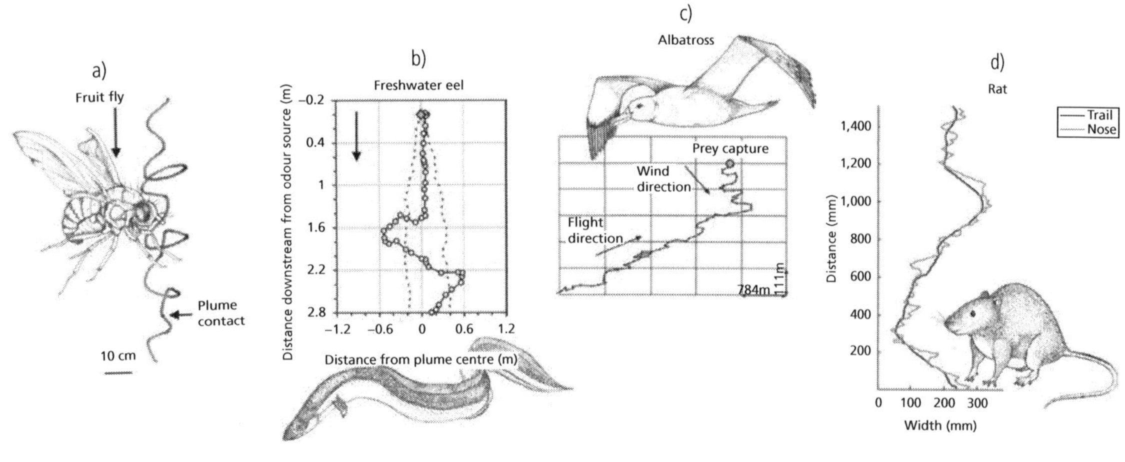

Figure 11.3 Odour tracking in different animals. (A) Flight track of a fruit fly, *Drosophila melanogaster*, following an odour plume of fermented banana. The flow direction and plume contact by the fly are indicated by arrows. Figure modified with permission from Budick and Dickinson (2006). (B) Search path of a freshwater eel, *Anguilla* spp., navigating towards a source of salmon extract (grey square). Dashed lines indicate the average distribution of the odour plume, and the arrow shows the direction of flow. Figure modified from Carton and Montgomery 2003. Reproduced with permission of John Wiley & Sons. (C) Flight path of a wandering albatross, *Diomedea exulans*, equipped with a Geographical Position System (GPS) logger, and a stomach temperature recorder to detect feeding events. The initial crosswind direction changes to meandering upwind flight followed by a feeding event (grey circle). Figure modified from DeBose and Nevitt 2008. Reproduced with permission from Springer Science + Business Media. (D) Nose trajectory (grey) of a walking rat following a meandering odour trail (black) applied on a sheet of paper attached to a treadmill. Movement is from bottom to top. Figure modified from Khan et al. (2012). Reproduced with permission from Nature Publishing Group.

perform tasks such as accurately detecting remote odour sources hundreds of metres away. In fact, chemosensory cues are crucial in most events in the life of insects, e.g. for mate finding, host and prey location, trail following, and the structuring of societies in social species. Olfactory cues are mainly detected by the antennae, where the chemosensors are organized as sensilla, i.e. cuticular extensions housing ORNs, which are directly exposed to the surrounding air. These ORNs project into the antennal lobes, which constitute the primary olfactory brain centre in insects (Schachtner et al. 2005). The easy access of these sensory neurons for electrophysiological recordings has been a very important factor for the many successful semiochemical identifications in insects. The by far most well-studied system regarding odour-guided navigation is the pheromone-induced flight behaviour of male moths. Baker and Vickers (1997) and Cardé and Willis (2008) have provided excellent overviews of this research field, and a brief summary is given here. Over several decades of wind tunnel experimentation, the flight behaviour of male moths was analysed in detail in all kinds of situations: with and without air flow, with and without pheromone stimulation, and with various kinds of pheromone exposure regimes, ranging from homogenous clouds of pheromone to narrow pheromone plumes with continuous, turbulent, or pulsed structures. Based on the early studies, the male flight response was explained as a combination of two mechanisms: (i) optomotor anemotaxis, i.e. a visually based, wind-steering mechanism, and (ii) a self-regulated counterturning program.

However, the realization that the intermittency of odour plumes is a basic requirement for males to sustain upwind flight (they generally fail to progress upwind in a homogenous cloud of pheromone, but see Justus and Cardé 2002), and the observation that males could respond behaviourally to single odour filaments, as well as to the gaps in between filaments, resulted in an alternative explanation (Baker 1990). The new model postulated that male pheromone-mediated flight relies on a combination of two distinct mechanisms: a phasic response, in which the insect proceeds upwind after contact with an odour filament, and a tonic

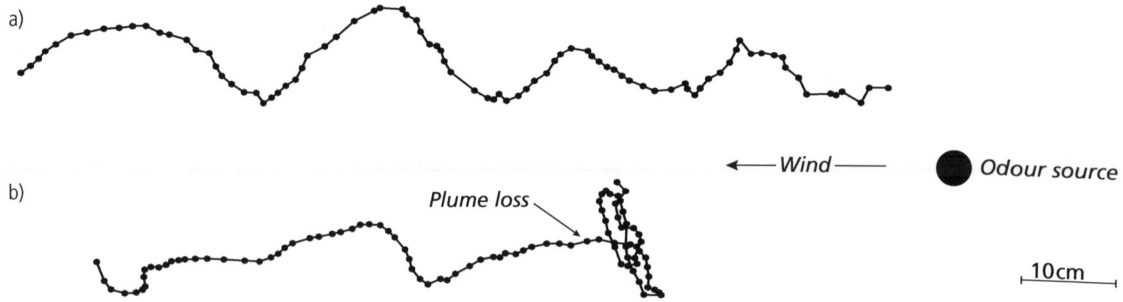

Figure 11.4 (a) Conceptual flight track showing the casting behaviour of a male moth during olfactory search in an odour plume. (b) After a short latency time following the loss of a pheromone plume (indicated by thin arrow) the male ceases upwind flight and instead initiates extended flights back and forth across the wind-line with no net progress upwind.

response, in which the insect initiates casting, i.e. lateral flight excursions with no net upwind progress, when exposed to clean air in between two filaments (Fig. 11.4). Such a reiterative process, with each 'template' consisting of upwind surge and casting flight, will make the insect sustain upwind progress in a turbulent and highly intermittent odour plume. The straightness of the flight trajectory will then be determined by how frequently the insect will encounter odour filaments in relation to gaps with clean air, i.e. how well the phasic mechanism would suppress the tonic one (Fig. 11.4; Baker 1990).

To test the specific predictions derived from this model, the male pheromone-modulated flight of the tobacco budworm, *Heliothis virescens*, was dissected using plumes with different pulse rates (Vickers and Baker 1994, 1996). As predicted, flight trajectories were straighter, and showed a unimodal distribution of track angles in relation to the wind, when males were exposed to the higher encounter rate of pheromone (10 filaments s^{-1}), whereas exposure to the lower encounter rate (4 filaments s^{-1}) resulted in meandering flights and a bimodal distribution of track angles. The upwind surge response after interception of a single filament was ≈0.30 s, and males reverted to casting after ≈0.27 s in absence of a filament, revealing a very rapid mechanism for processing the dynamic flux of odour concentration within the plume to maintain progress toward the source. In addition, electroantennograms were recorded from a third antenna, carried as a 'back pack', of free-flying males navigating in a turbulent pheromone plume, to correlate filament interception with flight behaviour. Upwind flight following filament contact was ≈0.23 s, and casting was initiated after ≈0.30 s if no more filaments were encountered, nicely matching the results from the pulsed plume trials (Vickers and Baker 1994). The same rapid surge-cast template has been described for males of the almond moth, *Cadra cautella* (Mafra-Neto and Cardé 1994, 1996), manoeuvring in pheromone plumes, and rapid initiation of casting following plume loss (Fig. 11.4) has been observed for males and females of the navel orangeworm, *Amyelois transitella*, navigating toward sources of sex pheromone and host plant odours, respectively (Haynes and Baker 1989). The accumulated data from these dissections of the odour-modulated flight in moths demonstrate the generality of the surge-cast template as a mechanism by which both sexes navigate in odour plumes. The casting flight has been suggested to increase the ability of moths to scan the spatial distribution of odour filaments across the plume to facilitate navigation, as well as to enhance the probability of relocating a meandering plume in situations where it has been lost (see below).

Although the flight behaviour of male moths in response to female sex pheromone has been the major model for understanding odour-guided navigation in insects, recent studies on fruit flies (Budick and Dickinson 2006) and mosquitoes (Geier et al. 1999, Dekker and Cardé 2011) have been important to allow for comparisons between taxonomic groups in search for similarities and differences in odour tracking mechanisms. Females of the yellow fever mosquito, *Aedes aegypti*, show a similar surge-cast behaviour when encountering a CO_2 filament as male moths do when encountering a pheromone filament (Dekker and Cardé 2011). Moreover, a homogenous cloud of CO_2 will not trigger sustained upwind flight in female mosquitoes, similar to what has been observed for male moths exposed to a homogenous fog of pheromones (Baker et al. 1985, Geier et al. 1999). On the other hand, higher levels of upwind flight and source contact were found for mosquitoes when exposed to homogenous versus turbulent plumes of skin odours (Dekker and Cardé 2011). In *Drosophila melanogaster*, upwind surge was observed following contact with a plume of fruit odour (Budick and Dickinson 2006). However, similar to the mosquitoes, the flies responded most strongly to a homogenous cloud of odour, thus departing from the Baker model for odour-guided navigation in moths. In addition, flies require sensory inputs from both antennae to accurately track an odour plume (Duistermars et al. 2009), whereas male *H. virescens* can perform such a task with a single intact antenna (Vickers and Baker 1991). There is clearly no single model covering all flight behaviours associated with odour tracking in insects, and data for additional taxonomic groups are needed to get a better insight into the diversity of navigational strategies employed by insects to find remote odour sources.

In contrast to the thorough understanding of how insects orient in odour plumes, much less is known about the search strategies they employ to find odour plumes in the first place, known as 'appetitive' flight or 'ranging' flight. This is presumably because such flight behaviours occur over much longer distances and have thus been harder to analyse, especially for nocturnal species. It is well known that male moths, after being released in a wind tunnel without exposure to pheromones, exhibit casting flight (Fig. 11.4), and such behaviour has been proposed to be the most optimal search strategy to establish contact with an odour plume. However, the very stable wind conditions in such a bioassay are rarely found in a natural setting, and whether crosswind flight is the optimal search strategy in situations with more turbulent wind has been much debated. Several models have been produced to explain how insects should use information about the current wind direction and its fluctuation over time to optimize their search for odour plumes. Depending on the specific assumptions given for any particular model, the most favourable strategy to maximize the likelihood of encountering an odour plume would be to fly crosswind, upwind or downwind in relation to the odour source (Sabelis and Schippers 1984; Dusenbery 1989, 1990; Cardé and Willis 2008). A crosswind strategy is optimal when the plume is stretched out along the direction of the wind, which is the case when the wind direction is fairly stable. On the other hand, if the wind direction is shifting more than 60°, an upwind or downwind strategy seems to be optimal, because the distribution of the plume will then be crosswind-biased.

Unfortunately, few 'landscape scale' studies are available regarding appetitive flights by insects in search for odour plumes, mainly due to the technical problems associated with tracking such small animals over large distances with high spatiotemporal resolution. Direct visual observations of search flights have been performed on two diurnally active moth species, the gypsy moth, *Lymantria dispar* (Elkinton and Cardé 1983), and the bog moth, *Virbia lamae* (Cardé et al. 2012). The observations were conducted over rather restricted areas, and no directional flight preference in relation to the average wind direction was found in any of the studies. Field studies on onion flies, *Delia antiqua*, also revealed a random pattern of initial flight direction in females in the absence of host plant odour (Judd and Borden 1988). However, because crosswind navigation encompasses twice the possibility of either upwind or downwind navigation, Cardé et al. (2012) argued that the random pattern of navigational directions observed in these studies is in agreement with a crosswind strategy.

A limitation with the above-mentioned studies addressing optimal plume finding is that insect flights are normally not straight paths, but instead more or less irregular trajectories including various turns and straight flights of different lengths, and more complex models are needed to describe such patterns of movement. Traditionally, search paths in animals have been quantified using models of correlated random walk (Kareiva and Shigesada

1983, Bovet and Benhamou 1988), which were developed from observations of movements occurring over relatively small scales in space and time. More recent analyses of animal movement over larger areas and over longer time periods have suggested that such paths are sometimes best described by Lévy flight models (Reynolds and Rhodes 2009). Lévy flights are a special class of random walks, and consist of frequently occurring clusters of short segments connected with relatively longer segments occurring less frequently, which are in turn connected with even longer segments occurring at even lower frequency, and so on (Fig. 11.5). The first report of Lévy flight in animals was based on an analysis of prey search of wandering albatrosses, *Diomedea exulans* (Viswanathan et al. 1996), but the conclusion from that study later turned out to be wrong (Edwards et al. 2007). This pioneering work nevertheless triggered extensive research on patterns of animal movement, and although flight trajectories classified as Lévy flight have since then been suggested for several insect species (Reynolds and Frye 2007, Reynolds et al. 2009), there is still a controversy regarding the importance of this type of flight in animal foraging (James et al. 2011).

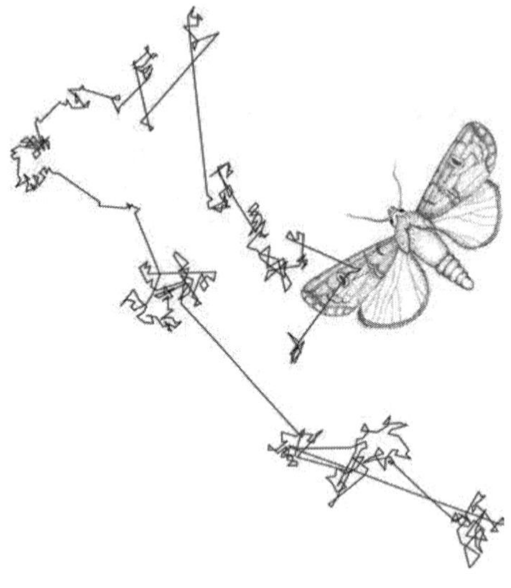

Figure 11.5 A conceptual example of Lévy flight, consisting of frequently occurring short-distance movements that are randomly alternated with more occasionally occurring movements over longer distances, which in turn are alternated with even longer movements occurring even less frequently, and so on. This type of flight behaviour has been suggested for, e.g., pheromone-mediated mate finding in moths.

With the recent development of a scanning harmonic radar system (Chapman et al. 2011) it is now possible to track individual flying insects over large geographic scales with high resolution (Box 11.2). This technique has been used to study the foraging flights of bees (Capaldi et al. 2000), bumblebees (Osborne et al. 1999), and butterflies (Cant et al. 2005) at a landscape scale. The first attempt to use harmonic radar in the context of olfactory search in nocturnal moths was made by Riley et al. (1998) using males of *A. segetum* as a model. A few years later, a more extensive study on the same species was performed (Svensson et al. 2001) with the aim of comparing the flight tracks of males in areas with and without high doses of synthetic pheromones, to gain insights into possible mechanisms involved in pheromone-based mating disruption. A re-analysis of these radar tracks was later done with the specific aim of studying appetitive flights in moths

Box 11.2 Harmonic radar for tracking low-altitude flights of insects

Several radar techniques have been applied to study high altitude movements of insects, e.g. during migration (Riley 1975, Beerwinkle et al. 1994, Alerstam et al. 2011). In contrast, radar studies of low altitude flights have been seriously constrained due to background clutter obscuring the received signal from the target insect. This is a major problem because almost all non-migratory flight behaviours, e.g. those associated with foraging or mating, occur at low altitudes. However, this problem can be overcome by using a scanning harmonic radar system. With this technique the low altitude flights of medium-sized insects can be monitored over distances of several hundreds of metres. The insect to be tracked is equipped with a miniature (16 mm) transponder. When the emitted radar signal reaches the target, the transponder absorbs energy from the beam and re-transmits some of it at a harmonic of the incoming signal. Because the radar receiver is selectively tuned to this new frequency, echoes from other objects are suppressed, i.e. the signal can be extracted from noise. All energy required for re-transmission of the signal is used from the emitted radar beam, and the tagging device needs no battery and can thus be made very small. The position of the target insect can be estimated every third second with a spatial resolution of ±3 m, allowing for high-resolution recordings of flight tracks.

(Reynolds et al. 2007), and some interesting features of the resulting paths of movement were identified. After being released, most males flew downwind. However, when close to a pheromone source, males that initially flew in a downwind direction changed to crosswind flight, and those that initially flew in a crosswind direction changed to upwind flight. In addition, the movement patterns for some of the longer flight tracks were described as an optimal Lévy flight search strategy (Fig. 11.6), and subsequent harmonic radar studies on honeybees have shown that these insects also employ this search strategy to locate distant odour sources (Reynolds et al. 2009). These pioneering studies demonstrate the high potential of using harmonic radar for analyses of olfactory search on a landscape level in both nocturnal and diurnal insects.

As mentioned earlier, odour plumes in a natural setting show a more or less discontinuous and meandering pattern at a distance from the source, due to changes in the speed and direction of the air, and landscape features such as high vegetation.

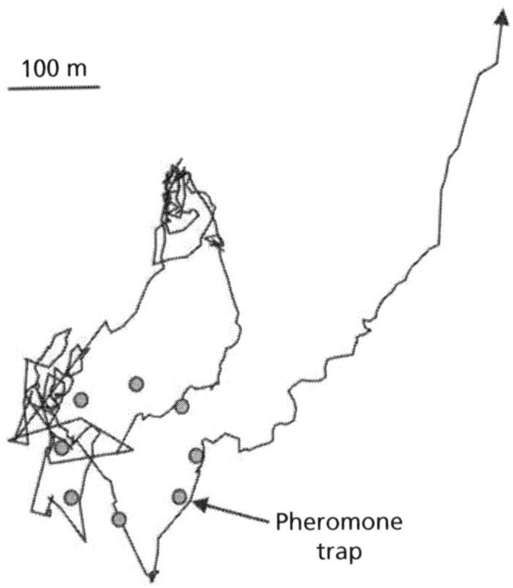

Figure 11.6 Flight track of a male turnip moth, *Agrotis segetum*, monitored by scanning harmonic radar. The moth was released from the centre of a circle of pheromone traps at 10.58 pm and could be followed for 65 min, corresponding to a track length of 4,745 m. The transponder weighed 8 mg, which was about 6% of the body weight of a male. The observed flight pattern is consistent with the adoption of an optimal Lévy-flight strategy for locating pheromone sources. The flight track is modified from Reynolds et al. (2007) and reproduced with permission from Elsevier.

As a consequence, an insect flying upwind in such highly intermittent odour plume is expected to lose contact with it frequently. There are two principal ways in which plume loss can occur: (1) odour filaments within a plume are detected at too low frequency by the insect, which will cause cessation of movement toward the source; (2) due to the meandering pattern of the plume, moving upwind may not lead along the plume's path, but instead cause the animal to exit the plume. For moths of both sexes, plume loss results in casting flight (Kennedy 1983, Haynes and Baker 1989, Kuenen and Cardé 1994), and similar crosswind movements in the absence of odour stimulation has been reported for fruit flies (Budick and Dickinson 2006) and mosquitoes (Dekker and Cardé 2011). Thus, the casting response seems to be shared by dipterans and lepidopterans, and is predicted to occur also in other insect groups facing the common problem of frequently losing contact with a meandering odour plume when tracking it upwind.

11.5.2 Odour tracking in crustaceans

Decapod crustaceans, such as lobsters, crabs, and crayfish, rely heavily on olfactory cues for various tasks, e.g. conspecific interactions, predator avoidance, and prey location. Extensive progress has been made recently regarding the mechanisms these animals use to navigate in odour plumes, especially for spiny lobsters, *Panulirus interruptus* and *P. argus*, and blue crabs, *Callinectes sapidus* and *C. similis* (Grasso and Basil 2002, Derby and Zimmer 2012, Weissburg 2012). These crustaceans live in the benthic zone, which is at the interface of solid and fluid environments, and the high degree of turbulence in this region puts specific requirements on their olfactory system. Similar to their terrestrial cousins, aquatic crustaceans have their chemical sensors organized as sensilla, which are in direct contact with the surrounding fluid. Such sensors can be found on most body surfaces, but are concentrated to the appendages, especially the antennules, antennae, legs and mouthparts (Horner et al. 2004). The chemosensilla can be unimodal, housing only chemosensory neurons, or bimodal, housing both chemosensory and mechanosensory neurons. Thus, chemical and flow information are provided by the antennule, and both senses are important for successful navigation to distant odour sources in crustaceans, similar to what has been documented for insects (Weissburg and Zimmer-Faust 1993). For the detection of relevant odours, the antennules have been considered the main appendage for chemoreception and experiments on both lobsters and crayfish, where different chemosensory populations on the antennules have been ablated or lesioned, have confirmed this hypothesis (Kraus-Epley and Moore 2002, Horner et al. 2004). These ORNs project to the olfactory lobes, i.e. the primary olfactory brain centre in crustaceans (Schachtner et al. 2005).

In addition to the chemoreceptors found on the antennules, numerous chemoreceptors are also present on the walking legs of decapod crustaceans, and these have been assumed to regulate navigation in odour plumes (Moore et al. 1991, Keller et al. 2003). This view was confirmed in recent studies on *C. sapidus*, where properties of odour plumes at the antennules and legs could be directly related to upstream navigation (Page et al. 2011a, b). In these experiments, odour plumes originating from shrimp chemical exudates were used, and a three-dimensional laser-induced fluorescence system was applied to analyse the fine-scale structure of the plume, providing the most detailed examination of signal input-behavioural output relationships with respect to odour tracking in a non-insect model. Similar to what was known from previous studies on lobsters and crayfish (Kraus-Epley and Moore 2002, Horner et al. 2004), odour-triggered upstream movement in *C. sapidus* is mainly dictated by information received by the antennules, but the crabs also use chemosensory information from the widely spaced legs to detect differences in filament encounter rates across the plume's width, and move cross-stream if needed to accurately track a meandering plume. Blue crabs could respond to shifts in the position of the plume's centre by less than 1 cm, and the reaction time after such shift was 2–4 s (Page et al. 2011b). Thus, input from antennule and leg chemoreceptors are processed in parallel during odour tracking in *C. sapidus*, and this is probably the case for other decapod crustaceans as well. The time taken for crabs to increase their speed to above average following the detection of an odour filament with the antennules was estimated to be 0.4–0.7 s (Page et al. 2011a), revealing rapid processing of the incoming signal, and corresponded well to the latency times observed for ORNs in lobsters following odour stimulation (Gomez and Atema 1996).

The cumulative body of research on decapod crustacean chemosensory navigation indicates that these animals rely primarily on odour-conditioned rheotaxis for odour plume tracking. In this mechanism, the animal moves upstream as long as sufficient encounters with odour filaments is maintained, but if odour contact is lost for long enough, it will move cross-stream until odour contact is re-established or the search is terminated (Grasso and Basil 2002, Page et al. 2011a). The recent finding that blue crabs use leg chemosensors for navigation suggests that they combine odour-conditioned rheotaxis with chemotropotaxis, i.e. spatially separated chemosensors compare odour encounter rates at both sides of the body to sample the spatial distribution of odour across the main plume axis (Grasso and Basil 2002, Page et al. 2011b). Thus, the mechanism responsible for the turning behaviour in blue crabs during olfactory search is distinct from the zigzagging flight observed by male moths approaching a pheromone source, which is driven by an internal program rather than sampling of odour filament flux across the plume. The counter-turning in insects has been suggested as an alternative mechanism for these small animals to increase the sampling area within a plume and thereby increase the efficiency of navigation (Page et al. 2011b), although they cannot quantify the spatial distribution of the plume as a whole.

11.5.3 Odour tracking in fish

Living in the aquatic environment, fish rely heavily on chemosensory signals for various behaviours, such as foraging, predator avoidance, reproduction, kin recognition, homing, and migration. However, detailed analyses of the behavioural mechanisms involved in odour plume tracking in fish are still scarce (but see Gardiner and Atema 2010). Most studies of olfactory search have been conducted under laboratory or flume conditions, focusing on behaviours triggered by chemical cues at short range. Odour-mediated behaviours occurring over larger spatial scales (hundreds of metres up to hundreds of kilometres) have mainly focused on homing at various scales, i.e. whether displaced fish can return to their natal river or even to a specific part of such river. Although chemosensory cues are known to mediate long-range migratory behaviours in fish, little is known about how these animals use the information provided from remote odour sources. In addition, in most cases the chemical identity of the attractant has not been revealed. The sensory organs involved in fish olfaction have been extensively studied, and considerable differences in size and structure of the organs across taxonomic groups have been documented (see Hara 1975 for a detailed overview). Teleost fish have paired olfactory organs located on the dorsal side of the head, each including an olfactory cavity. The cavity is connected to the outside environment via the incurrent and recurrent nares, which direct the flow of water through the organ as the fish swims. Each cavity has a series of olfactory lamellae (the olfactory rosette), where sensory epithelia are located, and the ORNs are situated within such epithelium. The ORNs project directly to the olfactory bulb, i.e. the structure of the vertebrate brain that is involved in olfaction (Laberge and Hara 2001).

The by far most studied taxa in the context of fish migration are the salmonids. Their dependence on olfactory cues during their migratory journey from the oceanic feeding sites to the coastal waters, and finally to their home rivers, is well known, and several theories have been put forward to explain how they navigate back from the sea to their spawning grounds. The oceanic and freshwater phases of the migration route in salmonids likely involve different sets of sensory cues, with olfactory cues presumably being of increasing importance the closer the fish are to home (Dittman and Quinn 1996). Studies on the migration behaviour of Atlantic salmon, *Salmo salar*, in coastal fjord systems in Norway have provided important insights into how salmonids use olfactory cues over large spatial scales. In fjords, the water column is stratified with different layers differing in temperature, salinity, and other features. Studies on the vertical movements of salmon in such stratified water columns, using passive sonar tracking, have shown that fish with an intact olfactory system restricted their vertical oscillatory movements to specific layers at the upper 30 m of the water column, whereas anosmic individuals did not show any preference and oscillated between the surface and 80 m depth (Døving et al. 1985). In addition, recordings made from single olfactory bulb neurons upon stimulation with water samples from different depths demonstrated that the olfactory system in salmon is capable of discriminating among different layers in the stratified water column (Døving et al. 1985). The combined results from this study suggest that salmon move vertically in the water column in search for chemosensory information provided by specific water layers, and use this information to guide them closer to the natal river.

The freshwater phase of salmonid homeward migration includes upstream movement to the spawning site. Imprinting to the olfactory environment during a sensitive period during the juvenile stage may play an important role in how salmonids accomplish this last part of their migration journey (Dittman et al. 1996). In experiments applying 'artificial' odour cues during the juvenile stage, the adult fish migrated upstream in response to the imprinting cue, also into previously unfamiliar rivers (Scholz et al. 1976, Johnsen and Hasler 1980, Dittman et al. 1996). In addition, these fish display a high degree of site fidelity within rivers, and are known to occupy specific 'home areas'. Many studies have shown that if individuals are displaced within their river system, they are able to return to their home area, and site-specific odours have been suggested to play a key role in such behavioural responses. The strongest evidence for an olfactory basis of residence behaviour in salmonids is probably the study on brown trout (Halvorsen and Stabell 1990), in which fish were tagged and displaced 200 m upstream or downstream in a river system. Before being displaced, fish were either made anosmic by heat treatment or sham-operated. Nine weeks later, fish from all four groups were captured and their position in the river recorded. Most fish with a non-functional olfactory organ remained close to the area of release, regardless if they had been released upstream or downstream in relation to their home area. In contrast, almost half (47%) of the fish with an intact olfactory system were recaptured either in the section of original capture or its neighbouring section (Halvorsen and Stabell 1990). Recapture rates were high for all groups (69–82%), and the results strongly indicate an olfactory basis for the residence behaviour of brown trout. Further support of odour-evoked residence behaviour in this species was provided from two-choice tests in the laboratory (Arnesen and Stabell 1992). Fish were able to discriminate between water from their home stream versus that from a neighbouring stream, and in most cases they were also able to discriminate between water scented with fish from the home area versus water scented with fish from other sections of the same stream.

Despite the recent progress in research on olfactory-mediated behaviours in salmonids, the chemical structures of the cues involved in migration and residence behaviours in these fish are still unknown. In contrast, the sea lamprey has emerged as an important fish model system, for which the olfactory signals guiding adults to suitable spawning sites as well as to mates have been chemically characterized. This jawless fish invaded the Laurentian Great Lakes of North America during the first half of the 20th century. It rapidly became a problem due to its aggressive parasitism on key predator species and game fish (Smith and Tibbles 1980), which prompted extensive research efforts to find methods to control it, including detailed studies on its chemical communication system. The chemosensory cues involved in the terminal stage of the sea lamprey's life cycle have been well investigated. In spring, adults migrate into streams to spawn, and location of suitable streams is mediated by a pheromone released by resident larvae (Sorensen et al. 2003). Males arrive first and build nests in parts of the stream where the flow of water is fairly high (0.5–1.5 m s^{-1}), and the presence of a long-range male-produced sex pheromone to attract ovulating females had been suggested for a long time. With the identification of the bile acid, 7α, 12α, 24-trihydroxy-5α-cholan-3-one 24-sulfate (3kPZS), from male secretions (Li et al. 2002), and subsequent experiments in the laboratory (Siefkes et al. 2005) and in the field (Johnson et al. 2009) demonstrating that ovulating females were highly attracted to synthetic 3kPZS, this prediction could be confirmed. The pheromone is extremely potent, and when released downstream at picomolar concentration, it elicited strong attraction of females over distances of several hundreds of metres (Johnson et al. 2009). In fact, females could be lured into traps baited with 3kPZS, and the sea lamprey may thus represent the first vertebrate for which pheromone-based methods may be used successfully for population control, similar to what has been achieved for many pest species of insects.

Sea lampreys scan the olfactory landscape by transporting water via the nostril into the olfactory capsule and out again with each respiratory cycle. The mechanism only allows for a few sampling events per second, thus restricting the ability to obtain high-resolution information regarding the dynamic structure of odour plumes. The study by Johnson et al. (2009) gave important insights into how sea lampreys navigate in odour plumes. Females were equipped with either radio tags or passive integrated transponder (PIT) tags, and their upstream movements toward traps baited with 3kPZS were recorded. Tracking data revealed that the pheromone-mediated behaviours in female lampreys show many similarities with those observed in insects. Upstream movement appears to occur as long as contact with the pheromone plume is achieved, and when plume contact is lost, females move downstream and initiate a typical pattern of cross-stream movements, similar to the casting behaviour observed in moths (Fig. 11.4; Johnson et al. 2009). In contrast to male moths, for which the casting behaviour is expressed within less than a second following plume loss (Vickers and Baker 1994, Mafra-Neto and Cardé 1996), it takes several seconds for female lampreys to switch to cross-stream zigzagging in absence of odour stimulation. Thus, the rate in which these fish sample their odour environment is much lower than that for insects.

The establishment of a robust system for odour tracking has allowed for hypothesis testing regarding the navigational strategies used by female lampreys to locate distant odour sources under natural conditions. The mean flow is a critical component of odour tracking in this species, and observations of females navigating in odour plumes of synthesized 3kPZS under different flow conditions were combined with engineering algorithms to test which type of taxis they employed during odour tracking (Johnson et al. 2012). Three algorithms were tested with specific assumptions for rheotaxis, odour-conditioned rheotaxis, or klinotaxis, i.e. movement in the direction of increased odour concentration without integration of flow direction. Under flow conditions, females generally moved fast and straight to the odour sources, but in reduced flow conditions, many females failed to move upstream and fewer located the source of pheromone. These results clearly support the view that odour-conditioned rheotaxis is the main mechanisms used by females to track odour plumes, similar to what has been observed for crustaceans.

11.5.4 Odour tracking in birds

Olfaction and olfactory-guided behaviours in birds has been intensively debated over the years. Birds were previously regarded as close to anosmic and therefore not relying on chemosensory signals, and avian chemical communication received little attention compared to acoustic and visual communication (Hagelin and Jones 2007, Caro and Balthazart 2010). However, evidence that birds indeed are able to use olfactory signals has accumulated since. Avian olfaction has been documented in foraging (Grubb 1972, Nevitt et al. 2008) as well as in homing and navigation (Grubb 1974, Papi 1990, Bonadonna et al. 2004). Olfactory-guided foraging and navigation may well have paved the way for chemical communication between birds, exploiting present sensitivity to odorants. Emitted olfactory cues have been proposed to influence avian social behaviour, such as the tangerine-scented body odour that is recognized among conspecifics in the crested auklet, *Aethia cristatella* (Hagelin et al. 2003). Recent studies show that birds have individual body odours (Karlsson et al. 2010, Mardon et al. 2010), are able to discriminate familiar from unfamiliar odours (Caspers and Krause, 2011), and are also able to recognize their partner or kin by odour (Bonadonna and Nevitt 2004, Krause et al. 2012). Taken together, these studies of signature mixtures suggest that there is a large potential for odour-mediated movement to be important also in birds.

The lifestyle of most birds, on the wing moving through turbulent airflow, with sometimes extremely long migrations or foraging routes, sets the frame for their olfactory ability. The chemosensory system in birds is similar to other vertebrates; an internal structure with air contact, but the vomeronasal organ (VNO) is absent (reviewed in Roper 1999). Two nares or nostrils on the beak are connected to a series of chambers, where the third chamber houses the olfactory epithelium containing the ORNs. These neurons project to the olfactory bulbs that are connected to other parts of the brain, e.g. the limbic system (Roper 1999). The relative size of the avian olfactory bulbs to the whole brain is highly variable between species, which may reflect variation in odour reliance in different taxa (Bang and Cobb 1968, Zelenitsky et al. 2011). As expected, nocturnal species, that cannot depend as much on visual cues, tend to have larger relative olfactory bulb sizes than diurnal species within families (Healy and Guilford 1990). The sensitivity of the olfactory neurons in birds were found, at least in some cases, to be as high as in rats, in an early electrophysiological study (Tucker 1965). Roper (1999) summarizes the experimental thresholds found in various studies for avian odour detection to be in the range 10^{-5} to 10^{-7} M, which is close to the human olfactory spectrum, whereas e.g. rats often are in the range 10^{-8} to 10^{-12} M.

Perhaps the best-studied group of birds regarding olfactory-guided behaviours are the tube-nosed seabirds (order Procellariiformes), including petrels, albatrosses and shearwaters. These birds only spend a few months on land per year for breeding, and the remaining part in flight over the ocean, where they forage over hundreds and thousands of square kilometres. They have well-developed olfactory systems (Bang and Cobb 1968) and their olfactory behaviours have been extensively investigated (for a review, see Nevitt 2008). One example is the blue petrel, *Halobaena caerulea*, a burrow nesting species with a circumpolar distribution in Antarctic and Subantarctic seas (Brooke 2004). Many aspects of the daily life of the blue petrel is facilitated by olfaction: food search is odour-mediated (Bonadonna et al. 2006) and underground nests and partners can be identified solely by olfactory cues (Bonadonna et al. 2004, Mardon and Bonadonna 2009, Mardon et al. 2010). In addition, partners were found to have dissimilar major histocompatibility complex (MHC) genotypes (Strandh et al. 2012; see also Westerdahl et al., Chapter 8), suggesting that these birds may choose mates based on MHC-related odour cues similar to what has been described for some mammals (Penn and Potts 1999, Penn 2002).

Tube-nosed seabirds are able to find patchily distributed food sources over long distances using olfaction alone (Grubb 1972). One of the odorants related to zooplankton grazing on phytoplankton, dimethyl sulphide (DMS), is highly attractive to both petrels and storm petrels that feed on zooplankton (Nevitt et al. 1995, Bonadonna et al. 2006). Observations of foraging procellariiforms in the open ocean have shown that they often approach food-related odour sources from downwind and that low-flying birds make broad zigzag turns progressing to smaller and smaller turns when approaching the odour source (Hutchison and Wenzel 1980). In a study by Nevitt et al. (2008), wandering albatrosses, *Diomedea exulans*, were equipped with global positioning system (GPS) tags and sensors to monitor changes in stomach temperature during food intake of foraging routes. In nearly half of the tracks preceding prey capture, the bird changed from crosswind flight to upwind flight upon detection of an odour plume (maximum 5 km from the source), showing that anemotaxis is an important mechanism for olfactory search in this seabird (Fig. 11.3C). The more or less featureless landscape of the open ocean may in many cases not provide these birds with sufficient visual information to forage efficiently, at least not for the first individual to arrive at a food source, and instead they seem to rely on olfactory cues for prey detection.

Evidence for long-range kairomonal response to DMS at sea has also been reported for penguins, the procellariiforms' closest relatives. Experiments on African penguins, *Spheniscus demersus*, which capture live prey underwater, have revealed that these birds are able to detect the prey-related compound DMS over large distances. Odour slicks released 500 m upwind from penguin colonies and treated with DMS attracted significantly more individuals than control slicks (Wright et al. 2011). The maximum number of birds at odour slicks occurred about half an hour after the odour was applied. An assumed travel speed of 1.2 m s^{-1} for these penguins suggests that they were attracted to the DMS patch from a distance of 2 km or more. This penguin is also attracted to DMS on land (Wright et al. 2011). Another sulphur-containing compound, ethyl mercaptan, which is associated with animal decay, has been shown to be a potent long-range attractant for the turkey vulture, *Cathartes aura*, which is one of few birds with a terrestrial lifestyle that are assumed to have a good olfactory sense (reviewed by DeBose and Nevitt 2008). The behavioural mechanisms these vultures use to locate remote odour sources are, however, not well studied. Another major model for bird olfaction is the homing pigeon. Olfaction was claimed in the 1970s to be responsible for long-range navigation in these birds (Benvenuti et al. 1973, Papi et al. 1973). Hereafter, research on navigational mechanisms in homing pigeons took off and the hypothesis on olfaction as the overruling navigation cue has been both supported and disputed (reviewed in Able 1996). One of the important concepts in this work is the 'olfactory map theory' postulating that animals navigate primarily based on the spatial distribution of natural odours (Wallraff 2004; see Åkesson et al., Chapter 9, and Muheim et al., Chapter 10, for a more comprehensive description of navigation).

11.5.5 Odour tracking in mammals

Mammals often have good olfactory abilities; the relatively mediocre human olfactory sense should be seen as an exception (though see Shepherd 2004). Chemical communication occurs in many aspects of mammalian life, such as kin and individual recognition (Hepper 1983, Hurst 1993, Sun and Müller-Schwarze 1998), mate choice (Yamazaki et al. 1976, Singer et al. 1997), and territorial behaviour (Ralls 1971, Sillero-Zubiri and Macdonald 1998). Many of these behaviours involve detection of odour signals at relatively small spatial scales. In contrast to odour plume navigation in air and water, which have been extensively studied for several groups of animals, the mechanisms involved in long-range odour tracking in mammals have hardly been investigated, apart from scent-trail tracking.

In mammals, odour sampling often involves active sniffing to concentrate the stimuli. Air containing the odorants is inhaled through paired nostrils and then the odour molecules reach the upper nasal cavity where the main olfactory epithelium (MOE) covers structured cartilaginous flaps, 'turbinates', that increase the MOE area. The MOE is filled with ORNs (in the range of 1,000 different types) that project to glomeruli in the olfactory bulb (reviewed in Firestein 2001). Most mammals also have a separate VNO that project to the accessory olfactory bulb. Traditionally, the VNO has been considered to handle only pheromones and the MOE all other odours, but recent research shows that this view is overly simplistic and that there is overlap between the two systems (reviewed in Martinez-Marcos 2009).

Although rodents are known to have a well-developed olfactory system, detailed studies of the mechanisms involved in odour trail tracking in this group have mainly been performed in a single species, the rat. As for most animals, rats use several sensory modalities when navigating. The hierarchy of preferred cues when homing from foraging trips was suggested to be visual > olfactory > self-movement cues (the animals movement is recorded and integrated by the brain which enables 'back-tracking' of the trip), but there is a lot of flexibility in switching between modalities (Maaswinkel and Whishaw 1999). However, immature rats (12-days-old) navigate using olfactory cues and are unable to navigate on visual cues, whereas when they mature, olfactory cues enables the use of visual cues, and when they are adult they can navigate on visual cues alone (Rossier and Schenk 2003). Rats can discriminate between different kinds of odour trails, even trails left by themselves or unknown conspecifics (Wallace et al. 2002). The sniffing/breathing rate for odour sampling was found to be 6–9 Hz compared to normal respiration at 2–4 Hz, whereas anticipation of a water reward was found to elicit sniffing at 9–12 Hz (Kepecs et al. 2007). In rats, more than 90% of the responsive neurons in the olfactory bulb responded differentially to odours presented from the left or right, and only one sniff during a time period as short as 125 ms was enough to distinguish from what side the odour came, implying that the brain makes a simultaneous comparison rather than a sequential (Rajan et al. 2006).

In a study by Khan et al. (2012), individual female rats were trained to follow a narrow meandering odour trail, which was applied onto a sheet of paper attached to a treadmill. All rats could accurately track surface-borne odours by moving their noses in a zigzag pattern over the trail (Fig. 11.3D). When losing contact with the trail, some rats initiated extended, sideway movements, similar to the casting behaviour observed in moths following plume loss (Fig. 11.4), indicating that they have an efficient behavioural strategy for relocation of a scent trail. The distance between the rat's nose and the trail in the horizontal plane was used as a measure of tracking accuracy, and this deviation from the trail was shown to decrease significantly with the time period after the training trials were initiated, i.e. that rats seemed to learn to track the trails more efficiently. Rats are known to have stereo olfaction (Rajan et al. 2006), and treatment rats with one nostril closed by stitching could follow the odour trail, but with larger deviation from the trail when compared with control rats, showing that stereo sampling of olfactory information improved the ability to track a scent trail. Sniffing rate was significantly higher during scent tracking (11 Hz) than during non-tracking (10 Hz), and much higher than rates observed in other odour-based tests in rats (see above). Thus, the combination of a high sniffing rate and zigzag nose movement seems to be an optimal strategy for scent trail tracking in rats. Stereo olfaction was recently documented also for blind eastern American moles, *Scalopus aquaticus* (Catania 2013), indicating that such a mechanism for scent tracking may be widespread among mammals.

The ancient ability of carnivores to odour track potential prey is retained in many modern dogs, where breeding and training has refined this ability to serve human purposes. Dogs are well known for their marvellous ability to e.g. track missing persons using odour cues. When confronted with a human-laid odour trail, trained trace dogs follow the trail in the correct direction (Mackenzie and Schultz 1987). There is large variation between dogs in ability to find the direction of odour trails, where male dogs and young dogs (< 2 years) perform better (Wells and Hepper 2003). The decision on what direction to take seems to rely on the patchiness of odour sources, the steps since the dog failed to determine the correct direction when contiguous odour trails were tested (Steen and Wilsson 1990). The odour tracking can be divided into three phases: (1) the searching phase, where the dog moves quickly and sniffs 10–20 times per breath to find the track, (2) the decision phase, moving slowly and sniffing 2–5 steps during 3–5 s, and finally (3) the tracking phase, moving quickly following the track (Thesen et al. 1993).

As mentioned earlier, humans have a rather poorly developed sense of smell compared to many other mammalian species. We extract spatial information from smell by comparing the olfactory input across our nostrils, either by time or concentration of the odour, along the same principle as we use our ears for the localization of acoustic stimuli. Nostril-specific receptive fields of the olfactory cortex have been shown in humans during localization of odour stimuli, showing that spatial smell localization relies on spatially different reception in the brain (Porter et al. 2005). In scent-trail tracking tests in humans, the typical zigzagging trajectories were found and one-nostril tracking was less accurate and slower than using two nostrils (Porter et al. 2007), similar to results obtained on rats (Khan et al. 2012).

Finally, a fascinating example of underwater olfactory search in mammals was recently documented. Although many mammalian species spend a considerable time underwater foraging, researchers have traditionally assumed that they cannot use olfactory information in the aquatic medium because sniffing is required for odorant molecules to reach the olfactory epithelium. However, a study by Catania (2006) has shown evidence of underwater sniffing as a mechanism in olfactory search in two species of semi-aquatic mammals: the star-nosed mole, *Condylura cristata*, and the water shrew, *Sorex palustris*. To detect olfactory cues in the underwater environment, these species emit air bubbles through their nostrils onto a surface and then re-inhale them as a way to capture odour molecules from that surface and carry them to the olfactory epithelium. In tracking trials with moles, the test animal was allowed to 'sniff' underwater trails (earthworm or fish scent) covered under a steel mesh, which prevented the animal from touching the surface with its star-like nose. Accurate tracking of scent trails were observed when using a mesh that allowed air bubbles to pass through, capture odour molecules from the trail, and then be re-inhaled by the mole. However, when a mesh with a grid size preventing air bubbles to reach the scent trail, the mole failed to track the trail. Water shrews also performed well in scent tracking trials. No mesh covered the scent trail in these tests, because water shrews use their whiskers instead of their nose for making contact with the substrate. However, when a blocking grid was used, accuracy of tracking in test shrews dropped. These experiments reveal a novel mechanism for odour tracking which may be widespread in semi-aquatic mammals relying on olfactory information both on land and in water.

11.6 Future perspectives

Over recent decades we have gained increased insight into the various navigational strategies employed by animals, from flies to humans, to find remote odour sources. These insights have mainly been derived from experimental studies using various laboratory bioassays, such as wind tunnels and flumes, which typically limit the spatial scale of observation to just a few metres. The odour-evoked behaviours observed under such small-scale, controlled conditions might not be representative for how animals perform under field conditions with a much more turbulent odour environment. The lack of data on odour tracking in natural settings is a general problem for most taxonomic groups. As an example, a review of 400 papers on chemical communication in crustaceans and fish found that only ≈20% of the studies were performed in the field (Johnson and Li 2010). This constraint is mainly because of the technical difficulties in studying animal movement over large geographical scales at high spatiotemporal resolution, and we still lack answers to many fundamental questions concerning odour tracking, such as the distances over which odour signals can be attractive under natural conditions.

New techniques for tracking animal movement over large distances with high precision, e.g. harmonic radar (Box 11.2; Riley et al. 1996) and GPS loggers (Weimerskirch et al. 2002), can give unique insights into long-range olfactory search in a wide range of taxa, and provide a basis for the development of general principles of odour tracking on a broad landscape scale. Moreover, the emergence of new theoretical models, such as the Lévy flight concept, to explain the complex patterns of movement observed has broadened our knowledge about the search strategies animals employ during foraging and mate finding. Information on how individual species move in the landscape will be crucial for future conservation strategies, and analysing olfactory-guided search behaviours in e.g. insects may help in predicting how the current decline and fragmentation of natural habitats will affect their ability to find new resources. Moreover, understanding how altered agricultural practices in a changing climate will affect olfactory search in pollinators and pests will be important to predict future changes in ecosystem services and crop damage (Smith et al., Chapter 4). Implementing odour-based methods when tracking the movements of threatened insects can provide important information about their dispersal capacity (Svensson et al. 2011).

In addition, the identification of chemosensory stimuli in non-insect taxa is urgently needed. The application of synthetic compounds in field experiments would give insights into the importance of single compounds versus blends in mediating olfactory search, and more precise estimates of what concentrations are needed to elicit long-range odour-evoked behaviours. The studies on sea lampreys (Johnson et al. 2009, Johnson et al. 2012) and procellariiform seabirds (Nevitt et al. 1995) clearly show how important it is to have a potent synthesized attractant available for controlled experiments on odour tracking to enable hypothesis testing regarding olfactory search strategies, as well as potential applications for attractants, such as population monitoring or direct control of pest species.

Finally, recent fascinating innovations derived from research on odour-mediated navigation in animals have included developments of various algorithms to be implemented into autonomous sensor-driven tracking of turbulent chemical plumes. The odour-driven search strategies employed by moths and crustaceans have been models for such algorithms, and several of these have been tested in autonomous underwater vehicles for tracking turbulent chemical plumes in natural settings (reviewed by Kang and Li 2012). Some of these tracking devices show very high precision of navigation in turbulent odour plumes (Webster et al. 2012), indicating that the performance of the algorithm is on the same level as the performance of the model organism. These studies highlight the potential of implementing bio-inspired search algorithms into robotics (see also Johansson et al., Chapter 13) to be used for e.g. localization of hazardous chemicals in near shore areas, and perhaps similar systems can be used to solve related problems in terrestrial habitats.

References

Able, K. P. (1996). The debate over olfactory navigation by homing pigeons. *Journal of Experimental Biology*, **199**, 121–4.

Alerstam, T., Chapman, J. W., Bäckman, J., et al. (2011). Convergent patterns of long-distance nocturnal migration in noctuid moths and passerine birds. *Proceedings of the Royal Society B: Biological Sciences*, **278**, 3074–80.

Arnesen, A. M., and Stabell, O. B. (1992). Behaviour of stream-dwelling brown trout towards odours present in home stream water. *Chemoecology*, **3**, 94–100.

Atema, J. 2012. Aquatic odour dispersal fields: opportunities and limits of detection, communication, and navigation. In C. Brönmark and L. A. Hansson (eds), *Chemical Ecology in Aquatic Systems*. Oxford University Press, Oxford.

Baker, T. C. 1990. Upwind flight and casting flight: complementary phasic and tonic systems used for location of sex pheromone sources by male moths. In K. B. Doving (ed.), *Proceedings of the Tenth International Symposium on Olfaction and Taste*. Graphic Communication System, Oslo.

Baker, T. C., and Vickers, N. J. 1997. Pheromone-mediated flight in moths. In R. T. Cardé and A. K. Minks (eds), *Insect Pheromone Research: New Directions*. Chapman & Hall, New York.

Baker, T. C., Willis, M. A., Haynes, K. F., and Phelan, P. L. (1985). A pulsed cloud of sex pheromone elicits upwind flight in male moths. *Physiological Entomology*, **10**, 257–65.

Bang, B. G., and Cobb, S. (1968). The size of the olfactory bulb in 108 species of birds. *Auk*, **85**, 55–61.

Bargmann, C. I. (2006). Comparative chemosensation from receptors to ecology. *Nature*, **444**, 295–301.

Beerwinkle, K. R., Lopez, J. D., Witz, J. A., Schleider, P. G., Eyster, R. S., and Lingren, P. D. (1994). Seasonal radar and meteorological observations associated with nocturnal insect flight at altitudes to 900 meters. *Environmental Entomology*, **23**, 676–83.

Benvenuti, S., Fiaschi, V., Fiore, L., and Papi, F. (1973). Homing performances of inexperienced and directionally trained pigeons subjected to olfactory nerve section. *Journal of Comparative Physiology A: Neuroethology, Sensory, Neural, and Behavioral Physiology*, **83**, 81–92.

Bonadonna, F., Caro, S., Jouventin, P., and Nevitt, G. A. (2006). Evidence that blue petrel, *Halobaena caerulea*, fledglings can detect and orient to dimethyl sulphide. *Journal of Experimental Biology*, **209**, 2165–9.

Bonadonna, F., and Nevitt, G. A. (2004). Partner-specific odor recognition in an Antarctic seabird. *Science*, **306**, 835.

Bonadonna, F., Villafane, M., Bajzak, C., and Jouventin, P. (2004). Recognition of burrow's olfactory signature in blue petrels, *Halobaena caerulea*: an efficient discrimination mechanism in the dark. *Animal Behaviour*, **67**, 893–8.

Bovet, P., and Benhamou, S. (1988). Spatial analysis of animal's movement using a correlated random walk model. *Journal of Theoretical Biology*, **131**, 419–33.

Brooke, M. (2004). *Albatrosses and Petrels across the World*. Oxford University Press, New York.

Budick, S. A., and Dickinson, M. H. (2006). Free-flight responses of *Drosophila melanogaster* to attractive odors. *Journal of Experimental Biology*, **209**, 3001–17.

Butenandt, A., Beckmann, R., Stamm, D., and Hecker, E. (1959). Über den sexuallockstoff des seidenspinners *bombyx mori*. Reindarstellung und konstitution. *Zeitschrift für Naturforschung*, **14**b, 283–4.

Cant, E. T., Smith, A. D., Reynolds, D. R., and Osborne, J. L. (2005). Tracking butterfly flight paths across the landscape with harmonic radar. *Proceedings of the Royal Society B: Biological Sciences*, **272**, 785–90.

Capaldi, E. A., Smith, A. D., Osborne, J. L., et al. (2000). Ontogeny of orientation flight in the honeybee revealed by harmonic radar. *Nature*, **403**, 537–40.

Cardé, R. T., Cardé, A. M., and Girling, R. D. (2012). Observations on the flight paths of the day-flying moth *Virbia lamae* during periods of mate location: do males have a strategy for contacting the pheromone plume? *Journal of Animal Ecology*, **81**, 268–76.

Cardé, R. T., and Willis, M. A. (2008). Navigation strategies used by insects to find distant, wind-borne sources of odor. *Journal of Chemical Ecology*, **34**, 854–66.

Caro, S. P., and Balthazart, J. (2010). Pheromones in birds: myth or reality? *Journal of Comparative Physiology A: Neuroethology, Sensory, Neural, and Behavioral Physiology*, **196**, 751–66.

Carton, A. G., and Montgomery, J. C. (2003). Evidence of a rheotactic component in the odour search behaviour of freshwater eels. *Journal of Fish Biology*, **62**, 501–16.

Caspers, B. A., and Krause, E. T. (2011). Odour-based natal nest recognition in the zebra finch (*Taeniopygia guttata*), a colony-breeding songbrid. *Biology Letters*, **7**, 184–6.

Catania, K. C. (2006). Underwater 'sniffing' by semi-aquatic mammals. *Nature*, **444**, 1024–5.

Catania, K. C. (2013). Stereo and serial sniffing guide navigation to an odour source in a mammal. *Nature Communications*, **4**, 1441.

Chapman, J. W., Drake, V. A., and Reynolds, D. R. (2011). Recent insights from radar studies of insect flight. *Annual Review of Entomology*, **56**, 337–56.

Corey, E. A., Bobkov, Y., Ukhanov, K., and Ache, B. W. (2013). Ionotropic crustacean olfactory receptors. *PLoS One*, **8**, e60551.

Debose, J. L., and Nevitt, G. A. (2008). The use of odors at different spatial scales: comparing birds with fish. *Journal of Chemical Ecology*, **34**, 867–81.

Dekker, T., and Cardé, R. T. (2011). Moment-to-moment flight manoeuvres of the female yellow fever mosquito (*Aedes aegypti*) in response to plumes of carbon dioxide and human skin odour. *Journal of Experimental Biology*, **214**, 3480–94.

Derby, C. D., and Zimmer, R. K. (2012). Neuroecology of predator-prey interactions. In C. Brönmark and L. A. Hansson (eds), *Chemical Ecology in Aquatic Systems*. Oxford University Press, Oxford.

Dittman, A. H., and Quinn, T. P. (1996). Homing in Pacific salmon: mechanisms and ecological basis. *Journal of Experimental Biology*, **199**, 83–91.

Dittman, A. H., Quinn, T. P., and Nevitt, G. A. (1996). Timing of imprinting of natural and artificial odors by coho salmon (Oncorhynchus kisutch). *Canadian Journal of Fisheries and Aquatic Sciences*, **53**, 434–42.

Døving, K. B., Westerberg, H., and Johnsen, P. B. (1985). Role of olfaction in the behavioral and neuronal responses of Atlantic salmon, *Salmo salar*, to hydrographic stratification. *Canadian Journal of Fisheries and Aquatic Sciences*, **42**, 1657–67.

Duistermars, B. J., Chow, D. M., and Frye, M. A. (2009). Flies require bilateral sensory input to track odor gradients in flight. *Current Biology*, **19**, 1301–7.

Dusenbery, D. B. (1989). Optimal search direction for an animal flying or swimming in a wind or current. *Journal of Chemical Ecology*, **15**, 2511–19.

Dusenbery, D. B. (1990). Upwind searching for an odor plume is sometimes optimal. *Journal of Chemical Ecology*, **16**, 1971–6.

Edwards, A. M., Phillips, R. A., Watkins, N. W., et al. (2007). Revisiting Lévy flight search patterns of wandering albatrosses, bumblebees and deer. *Nature*, **449**, 1044–8.

Elkinton, J. S., and Cardé, R. T. (1983). Appetitive flight behavior of male gypsy moths (Lepidoptera: Lymantridae). *Environmental Entomology*, **12**, 1702–7.

Engsontia, P., Sanderson, A. P., Cobb, M., Walden, K. K. O., Robertson, H. M., and Brown, S. (2008). The red flour beetle's large nose: an expanded odorant receptor gene family in *Tribolium castaneum*. *Insect Biochemistry and Molecular Biology*, **38**, 387–97.

Firestein, S. (2001). How the olfactory system makes sense of scents. *Nature*, **413**, 211–18.

Gardiner, J. M., and Atema, J. (2010). The function of bilateral odor arrival time differences in olfactory orientation of sharks. *Current Biology*, **20**, 1187–91.

Geier, M., Bosch, O. J., and Boeckh, J. (1999). Influence of odour plume structure on upwind flight of mosquitoes towards hosts. *Journal of Experimental Biology*, **202**, 1639–48.

Gilad, Y., Man, O., and Glusman, G. (2005). A comparison of the human and chimpanzee olfactory receptor gene repertoires. *Genome Research*, **15**, 224–30.

Glusman, G., Bahar, A., Sharon, D., Pilpel, Y., White, J., and Lancet, D. (2000). The olfactory receptor gene superfamily: data mining, classification, and nomenclature. *Mammalian Genome*, **11**, 1016–23.

Gomez, G., and Atema, J. (1996). Temporal resolution in olfaction: stimulus integration time of lobster chemoreceptor cells. *Journal of Experimental Biology*, **199**, 1771–9.

Grasso, F. W., and Basil, J. A. (2002). How lobsters, crayfishes, and crabs locate sources of odor: current perspectives and future directions. *Current Opinion in Neurobiology*, **12**, 721–7.

Grubb, T. C. J. (1972). Smell and foraging in shearwaters and petrels. *Nature*, **237**, 404–5.

Grubb, T. C. J. (1974). Olfactory navigation to the nesting burrow in Leach's petrel (*Oceanodroma leucorrhoa*). *Animal Behaviour*, **22**, 192–202.

Hagelin, J. C., and Jones, I. L. (2007). Bird odors and other chemical substances: a defense mechanism or overlooked mode of intraspecific communication? *Auk*, **124**, 741–61.

Hagelin, J. C., Jones, I. L., and Rasmussen, L. E. L. (2003). A tangerine-scented social odour in a monogamous seabird. *Proceedings of the Royal Society B: Biological Sciences*, **270**, 1323–9.

Halvorsen, M., and Stabell, O. B. (1990). Homing behaviour of displaced stream-dwelling brown trout. *Animal Behaviour*, **39**, 1089–97.

Hara, T. J. (1975). Olfaction in fish. *Progress in Neurobiology*, **5**, 271–335.

Haynes, K. F., and Baker, T. C. (1989). An analysis of anemotactic flight in female moths stimulated by host odour and comparison with males' response to sex pheromones. *Physiological Entomology*, **14**, 279–89.

Healy, S., and Guilford, T. (1990). Olfactory-bulb size and nocturnality in birds. *Evolution*, **44**, 339–46.

Hepper, P. G. (1983). Sibling recognition in the rat. *Animal Behaviour*, **31**, 1177–91.

Horner, A. J., Weissburg, M. J., and Derby, C. D. (2004). Dual antennular pathways can mediate orientation by Caribbean spiny lobsters in naturalistic flow conditions. *Journal of Experimental Biology*, **207**, 3785–96.

Hurst, J. L. (1993). The priming effects of urine substrate marks on interactions between male house mice, *Mus musculus domesticus* Schwarz & Schwarz. *Animal Behaviour*, **45**, 55–81.

Hutchison, L. V., and Wenzel, B. M. (1980). Olfactory guidance in foraging by Procellariiforms. *Condor*, **82**, 314–19.

James, A., Plank, M. J., and Edwards, A. M. (2011). Assessing Lévy walks as models of animal foraging. *Journal of the Royal Society Interface*, **8**, 1233–47.

Johnsen, P. B., and Hasler, A. D. (1980). The use of chemical cues in the upstream migration of coho salmon, *Oncorhynchus kisutch* Walbaum. *Journal of Fish Biology*, **17**, 67–73.

Johnson, N. S., and Li, W. (2010). Understanding behavioral responses of fish to pheromones in natural freshwater environments. *Journal of Comparative Physiology A: Neuroethology, Sensory, Neural, and Behavioral Physiology*, **196**, 701–11.

Johnson, N. S., Muhammad, A., Thompson, H., Choi, J., and Li, W. (2012). Sea lamprey orient toward a source of a synthesized pheromone using odor-conditioned rheotaxis. *Behavioral Ecology and Sociobiology*, **66**, 1557–67.

Johnson, N. S., Yun, S. S., Thompson, H. T., Brant, C. O. and Li, W. (2009). A synthesized pheromone induces upstream movement in female sea lamprey and summons them into traps. *Proceedings of the National Academy of Sciences USA*, **106**, 1021–6.

Judd, G. J. R., and Borden, J. H. (1988). Long-range host-finding behaviour of the onion fly *Delia antiqua* (Diptera: Anthomyiidae): ecological and physiological constraints. *Journal of Applied Entomology*, **25**, 829–45.

Justus, K. A., and Cardé, R. T. (2002). Flight behaviour of males of two moths, *Cadra cautella* and *Pectinophora gossypiella*, in homogenous clouds of pheromone. *Physiological Entomology*, **27**, 67–75.

Kang, X., and Li, W. (2012). Moth-inspired plume tracing via multiple autonomous vehicles under formation control. *Adaptive Behavior*, **20**, 131–42.

Kareiva, P. M., and Shigesada, M. (1983). Analyzing insect movement as a correlated random walk. *Oecologia*, **56**, 234–8.

Karlsson, A. C., Jensen, P., Elgland, M., et al. (2010). Red junglefowl have individual body odors. *Journal of Experimental Biology*, **213**, 1619–24.

Karlson, P., and Lüscher, M. (1959). 'Pheromones': a new term for a class of biologically active substances. *Nature*, **183**, 55–6.

Keller, T. A., Powell, I., and Weissburg, M. J. (2003). Role of olfactory appendages in chemically mediated orientation of blue crabs. *Marine Ecology Progress Series*, **261**, 217–31.

Kennedy, J. S. (1983). Zigzagging and casting as a programmed response to wind-borne odour: a review. *Physiological Entomology*, **8**, 109–20.

Kepecs, A., Uchida, N., and Mainen, Z. F. (2007). Rapid and precise control of sniffing during olfactory discrimination in rats. *Journal of Neurophysiology*, **98**, 205–13.

Khan, A. G., Sarangi, M., and Bhalla, U. S. (2012). Rats track odour trails accurately using a multi-layered strategy with near-optimal sampling. *Nature Communications*, **3**, 703.

Krause, E. T., Krüger, O., Kohlmeier, P., and Caspers, B. A. (2012). Olfactory kin recognition in a songbird. *Biology Letters*, **8**, 327–9.

Kraus-Epley, K. E., and Moore, P. A. (2002). Bilateral and unilateral lesions alter orientation abilities of the crayfish, *Orconectes rusticus*. *Chemical Senses*, **27**, 49–55.

Kuenen, L. P. S., and Cardé, R. T. (1994). Strategies for recontacting a lost pheromone plume: casting and upwind flight in the male gypsy moth. *Physiological Entomology*, **19**, 15–29.

Laberge, F., and Hara, T. J. (2001). Neurobiology of fish olfaction: a review. *Brain Research Reviews*, **36**, 46–59.

Law, J. H., and Regnier, F. E. (1971). Pheromones. *Annual Review of Biochemistry*, **40**, 533–48.

Li, W., Scott, A. P., Siefkes, M. J., et al. (2002). Bile acid secreted by male sea lamprey that acts as a sex pheromone. *Science*, **296**, 138–41.

Maaswinkel, H., and Whishaw, I. Q. (1999). Homing with locale, taxon, and dead reckoning strategies by foraging rats: sensory hierarchy in spatial navigation. *Behavioural Brain Research*, **99**, 143–52.

Mackenzie, S. A., and Schultz, J. A. (1987). Frequency of back-tracking in the tracking dog. *Applied Animal Behaviour Science*, **17**, 353–9.

Mafra-Neto, A., and Cardé, R. T. (1994). Fine-scale structure of pheromone plumes modulates upwind orientation of flying moths. *Nature*, **369**, 142–4.

Mafra-Neto, A., and Cardé, R. T. (1996). Dissection of the pheromone-modulated flight of moths using single-pulse response as a template. *Experentia*, **52**, 373–9.

Mardon, J., and Bonadonna, F. (2009). Atypical homing or self-odour avoidance? Blue petrels (*Halobaena caerulea*) are attracted to their mate's odour but avoid their own. *Behavioral Ecology and Sociobiology*, **63**, 537–42.

Mardon, J., Saunders, S. M., Anderson, M. J., Couchoux, C., and Bonadonna, F. (2010). Species, gender, and identity: cracking petrel's sociochemical code. *Chemical Senses*, **35**, 309–321.

Martinez-Marcos, A. (2009). On the organization of olfactory and vomeronasal cortices. *Progress in Neurobiology*, **87**, 21–30.

Mason, R. T., and Parker, M. R. (2010). Social behaviour and pheromonal communication in reptiles. *Journal of Comparative Physiology A: Neuroethology, Sensory, Neural, and Behavioral Physiology*, **196**, 729–49.

Moore, P. A., and Atema, J. (1991). Spatial information in the three-dimensional fine structure of an aquatic odor plume. *Biological Bulletin*, **181**, 408–18.

Moore, P. A., Scholz, N., and Atema, J. (1991). Chemical orientation of lobsters, *Homarus americanus*, in turbulent odour plumes. *Journal of Chemical Ecology*, **17**, 1293–1307.

Morgan, E. D. (2009). Trail pheromones of ants. *Physiological Entomology*, **34**, 1–17.

Murlis, J. 1997. Odor plumes and the signal they provide. In R. T. Cardé and A. K. Minks (eds), *Insect Pheromone Research: New Directions*. Chapman & Hall, New York.

Murlis, J., Elkinton, J. S., and Cardé, R. T. (1992). Odor plumes and how insects use them. *Annual Review of Entomology*, **37**, 505–32.

Nevitt, G. A. (2008). Sensory ecology on the high seas: the odor world of procellariiform seabirds. *Journal of Experimental Biology*, **211**, 1706–13.

Nevitt, G. A., Losekoot, M., and Weimerskirch, H. (2008). Evidence for olfactory search in wandering albatross, *Diomedea exulans*. *Proceedings of the National Academy of Sciences USA*, **106**, 4576–81.

Nevitt, G. A., Veit, R. R., and Kareiva, P. (1995). Dimethyl sulphide as a foraging cue for Antarctic procellariiform seabirds. *Nature*, **376**, 680–2.

Niimura, Y., and Nei, M. (2003). Evolution of olfactory receptor genes in the human genome. *Proceedings of the National Academy of Sciences USA*, **100**, 12235–40.

Niimura, Y., and Nei, M. (2005). Evolutionary dynamics of olfactory receptor genes in fishes and tetrapods. *Proceedings of the National Academy of Sciences USA*, **102**, 6039–44.

Nordlund, D. A. 1981. Semiochemicals: a review of the terminology. In D. A. Nordlund, R. L. Jones, and W. J. Lewis (eds), *Semiochemicals: Their Role in Pest Control*. Wiley, New York.

Olender, T., Fuchs, T., Linhart, C., et al. (2004). The canine olfactory subgenome. *Genomics*, **83**, 361–72.

Osborne, J. L., Clark, S. J., Morris, R. J., et al. (1999). A landscape-scale study of bumble bee foraging range and constancy, using harmonic radar. *Journal of Applied Ecology*, **36**, 519–33.

Page, J. L., Dickman, B. D., Webster, D. R., and Weissburg, M. J. (2011a). Getting ahead: context-dependent responses to odorant filaments drive along-stream progress during odor tracking in blue crabs. *Journal of Experimental Biology*, **214**, 1498–1512.

Page, J. L., Dickman, B. D., Webster, D. R., and Weissburg, M. J. (2011b). Staying the course: chemical signal spatial properties and concentration mediate cross-stream motion in turbulent plumes. *Journal of Experimental Biology*, **214**, 1513–1522.

Papi, F. (1990). Olfactory navigation in birds. *Experentia*, **46**, 352–63.

Papi, F., Fiore, L., Fiaschi, V., and Benvenuti, S. (1973). An experiment for testing the hypothesis of olfactory navigation of homing pigeons. *Journal of Comparative Physiology A: Neuroethology, Sensory, Neural, and Behavioral Physiology*, **83**, 93–102.

Peñalva-Arana, D. C., Lynch, M., and Robertson, H. M. (2009). The chemoreceptor genes of the waterflea *Daphnia pulex*: many Grs but no Ors. *BMC Evolutionary Biology*, **9**, 79.

Penn, D. J. (2002). The scent of genetic compatibility: sexual selection and the major histocompatibility complex. *Ethology*, **108**, 1–21.

Penn, D. J., and Potts, W. K. (1999). The evolution of mating preferences and major histocompatibility genes. *American Naturalist*, **153**, 145–64.

Porter, J., Anand, T., Johnson, B., Khan, R. M., and Sobel, N. (2005). Brain mechanisms for extracting spatial information from smell. *Neuron*, **47**, 581–92.

Porter, J., Craven, B., Khan, R. M., et al. (2007). Mechanisms of scent-tracking in humans. *Nature Neuroscience*, **10**, 27–9.

Rajan, R., Clement, J. P., and Bhalla, U. S. (2006). Rats smell in stereo. *Science*, **311**, 666–70.

Ralls, K. (1971). Mammalian scent marking. *Science*, **171**, 443–9.

Rasmussen, L. E. L., Lee, T. D., Roelofs, W. L., Zhang, A., and Daves Jr, G. D. (1996). Insect pheromone in elephants. *Nature*, 379, 684.

Reynolds, A. M., and Frye, M. A. (2007). Free-flight odor tracking in *Drosophila* is consistent with an optimal intermittent scale-free search. *PLoS One*, **2**, e358.

Reynolds, A. M., Reynolds, D. R., Smith, A. D., Svensson, G. P., and Löfstedt, C. (2007). Appetitive flight patterns of male *Agrotis segetum* moths over landscape scales. *Journal of Theoretical Biology*, **245**, 141–9.

Reynolds, A. M., and Rhodes, C. J. (2009). The Lévy flight paradigm: random search patterns and mechanisms. *Ecology*, **90**, 877–87.

Reynolds, A. M., Swain, J. L., Smith, A. D., Martin, A. P., and Osborne, J. L. (2009). Honeybees use a Lévy flight search strategy and odour-mediated anemotaxis to relocate food sources. *Behavioral Ecology and Sociobiology*, **64**, 115–23.

Riffell, J. A., Abrell, L., and Hilderbrand, J. G. (2008). Physical processes and real-time chemical measurement of the insect olfactory environment. *Journal of Chemical Ecology*, **34**, 837–53.

Riley, J. R. (1975). Collective orientation in night-flying insects. *Nature*, **253**, 113–14.

Riley, J. R., Smith, A. D., Reynolds, D. R., et al. (1996). Tracking bees with harmonic radar. *Nature*, **379**, 29–30.

Riley, J. R., Valeur, P., Smith, A. D., Reynolds, D. R., Poppy, G. M., and Löfstedt, C. (1998). Harmonic radar as a means of tracking the pheromone-finding and pheromone-following flight of male moths. *Journal of Insect Behavior*, **11**, 287–96.

Robertson, H. M., and Wanner, K. W. (2006). The chemoreceptor superfamily in the honey bee, *Apis mellifera*: expansion of the odorant, but not gustatory, receptor family. *Genome Research*, **16**, 1395–1403.

Robertson, H. M., Warr, C. G., and Carlson, J. R. (2003). Molecular evolution of the insect chemoreceptor gene superfamily in *Drosophila melanogaster*. *Proceedings of the National Academy of Sciences USA*, **100 Suppl**, 14537–42.

Roper, T. J. (1999). Olfaction in birds. *Advances in the Study of Behavior*, **28**, 247–332.

Rossier, J., and Schenk, F. (2003). Olfactory and/or visual cues for spatial navigation through ontogeny: olfactory cues enable the use of visual cues. *Behavioural Neuroscience*, **117**, 412–25.

Sabelis, M. W., and Schippers, P. (1984). Variable wind directions and anemotactic strategies of searching for an odour plume. *Oecologia*, **63**, 225–8.

Schachtner, J., Schmidt, M., and Homberg, U. (2005). Organization and evolutionary trends of primary olfactory brain centers in Tetraconata (Crustacea + Hexapoda). *Arthropod Structure and Development*, **34**, 257–99.

Scholz, A. T., Horrall, R. M., Cooper, J. C., and Hasler, A. D. (1976). Imprinting to chemical cues: the basis for home stream selection in salmon. *Science*, **192**, 1247–9.

Shepherd, G. M. (2004). The human sense of smell: are we better than we think? *PLoS Biology*, **2**, 572–5.

Siefkes, M. J., Winterstein, S. R., and Li, W. (2005). Evidence that 3-keto petromyzonol sulphate specifically attracts ovulated female sea lamprey, *Petromyzon marinus*. *Animal Behaviour*, **70**, 1037–45.

Sillero-Zubiri, C., and Macdonald, D. W. (1998). Scent-marking and territorial behaviour of Ethiopian wolves *Canis simensis*. *Journal of Zoology*, **245**, 351–61.

Singer, A. G., Beauchamp, G. K., and Yamazaki, K. (1997). Volatile signals of the major histocompatibility complex in male mice urine. *Proceedings of the National Academy of Sciences USA*, **94**, 2210–14.

Smith, B. R., and Tibbles, J. J. (1980). Sea lamprey (*Petromyzon marinus*) in Lake Huron, Michigan, and superior: history of invasion and control, 1936–1978. *Canadian Journal of Fisheries and Aquatic Sciences*, **37**, 1780–1801.

Sorensen, P. W., Vrieze, L. A., and Fine, J. M. (2003). A multi-component migratory pheromone in the sea lamprey. *Fish Physiology and Biochemistry*, **28**, 253–7.

Steen, J. B., and Wilsson, E. (1990). How do dogs determine the direction of tracks? *Acta Physiologica Scandinavica*, **139**, 531–4.

Steiger, S. S., Kuryshev, V. Y., Stensmyr, M. C., Kempenaers, B. and Mueller, J. C. (2009). A comparison of reptilian and avian olfactory receptor gene repertoires: Species-specific expansion of group gamma genes in birds. *BMC Genomics*, **10**, 446.

Strandh, M., Westerdahl, H., Pontarp, M., et al. (2012). Major histocompatibility complex class ii compatibility, but not class i, predicts mate choice in a bird with a highly developed olfaction. *Proceedings of the Royal Society B: Biological Sciences*, **279**, 4457–63.

Sun, L., and Müller-Schwarze, D. (1998). Anal gland secretion codes for family membership in the beaver. *Behavioral Ecology and Sociobiology*, **44**, 199–208.

Svensson, G. P., Sahlin, U., Brage, B., and Larsson, M. C. (2011). Should I stay or should I go? Modelling dispersal strategies in saproxylic insects based on pheromone capture and radio telemetry: a case study on the threatened hermit beetle *Osmoderma eremita*. *Biodiversity and Conservation*, **13**, 2883–902.

Svensson, G. P., Valeur, P. G., Reynolds, D. R., et al. (2001). Mating disruption in *Agrotis segetum* (Lepidoptera: Noctuidae) monitored by harmonic radar. *Entomologia Experimentalis et Applicata*, **101**, 111–21.

Symonds, M. R. E., and Elgar, M. A. (2008). The evolution of pheromone diversity. *Trends in Ecology and Evolution*, **23**, 220–8.

Thesen, A., Steen, J. B., and Døving, K. B. (1993). Behaviour of dogs during olfactory tracking. *Journal of Experimental Biology*, **180**, 247–51.

Tucker, D. (1965). Electrophysiological evidence for olfactory function in birds. *Nature*, **207**, 34–36.

Valeur, P. G., Hansson, B. S., and Löfstedt, C. (1999). Real-time measurement of pheromone release from individual female moths and synthetic dispensers in a wind tunnel by recording of single receptor-neurone responses. *Physiological Entomology*, **24**, 240–50.

Vickers, N. J. (2000). Mechanisms of animal navigation in odor plumes. *Biological Bulletin*, **198**, 203–212.

Vickers, N. J., and Baker, T. C. (1991). The effects of unilateral antennectomy on the flight behaviour of male *Heliothis virescens* in a pheromone plume. *Physiological Entomology*, **16**, 497–506.

Vickers, N. J., and Baker, T. C. (1994). Reiterative responses to single strands of odor promote sustained upwind flight and odor source location by moths. *Proceedings of the National Academy of Sciences USA*, **91**, 5756–60.

Vickers, N. J., and Baker, T. C. (1996). Latencies of behavioral response to interception of filaments of sex pheromones and clean air influence flight track shape in *Heliothis virescens* (f.) males. *Journal of Comparative Physiology A: Neuroethology, Sensory, Neural, and Behavioral Physiology*, **178**, 831–47.

Viswanathan, G. M., Afanasyev, V., Buldyrev, S. V., Murphy, E. J., Prince, P. A. and Stanley, H. E. (1996). Lévy flight search patterns of wandering albatrosses. *Nature*, **381**, 413–15.

Wallace, D. G., Gorny, B. and Whishaw, I. Q. (2002). Rats can track odors, other rats, and themselves: implications for the study of spatial behavior. *Behavioural Brain Research*, **131**, 185–92.

Wallraff, H. G. (2004). Avian olfactory navigation: its empirical foundation and conceptual state. *Animal Behaviour*, **67**, 189–204.

Webster, D. R., Volyanskyy, K. Y., and Weissburg, M. J. (2012). Bioinspired algorithm for autonomous sensor-driven guidance in turbulent chemical plumes. *Bioinspiration and Biomimetics*, 7, 036023 (11 pp).

Webster, D. R., and Weissburg, M. J. (2009). The hydrodynamics of chemical cues among aquatic organisms. *Annual Review of Fluid Mechanics*, **41**, 73–90.

Weimerskirch, H., Bonadonna, F., Bailleul, F., Mabille, G., Dell'omo, G., and Lipp, H. P. (2002). GPS tracking of foraging albatrosses. *Science*, **295**, 1259.

Weissburg, M. J. (2012). Death from downstream: chemosensory navigation and predator-prey processes. In C. Brönmark and L. A. Hansson (eds), *Chemical Ecology in Aquatic Systems*. Oxford University Press, Oxford.

Weissburg, M. J., and Zimmer-Faust, R. K. (1993). Life and death in moving fluids: hydrodynamics effects on chemosensory-mediated predation. *Ecology*, **74**, 1428–43.

Wells, D. L., and Hepper, P. G. (2003). Directional tracking in the domestic dog, *Canis familiaris*. *Applied Animal Behaviour Science*, **84**, 297–305.

Woodley, S. K. (2010). Pheromonal communication in amphibians. *Journal of Comparative Physiology A: Neuroethology, Sensory, Neural, and Behavioral Physiology*, **196**, 713–27.

Wright, K. L. B., Pichegru, L., and Ryan, P. G. (2011). Penguins are attracted to dimethyl sulphide at sea. *Journal of Experimental Biology*, **214**, 2509–11.

Wyatt, T. D. (2010). Pheromones and signature mixtures: defining species-wide signals and variable cues for identity in both invertebrates and vertebrates. *Journal of Comparative Physiology A: Neuroethology, Sensory, Neural, and Behavioral Physiology*, **196**, 685–700.

Wyatt, T. D. 2014. *Pheromones and Animal Behavior: Chemical Signals and Signature Mixtures*, 2nd edn., Cambridge University Press, Cambridge.

Yamazaki, K., Boyse, E. A., Miké, V., et al. (1976). Control of mating preferences in mice by genes in the major histocompatibility complex. *Journal of Experimental Medicine*, **144**, 1324–35.

Zelenitsky, D. K., Therrien, F., Ridgely, R. C., Mcgee, A. R. and Witmer, L. M. (2011). Evolution of olfaction in non-avian theropod dinosaurs and birds. *Proceedings of the Royal Society B: Biological Sciences*, **278**, 3625–34.

Zhang, X., Rodriguez, I., Mombaerts, P. and Firestein, S. (2004). Odorant and vomeronasal receptor genes in two mouse genome assemblies. *Genomics*, **83**, 802–11.

Zhang, X., Zhang, X., and Firestein, S. (2007). Comparative genomics of odorant and pheromone receptor genes in rodents. *Genomics*, **89**, 441–50.

CHAPTER 12

The genetics of migration

Miriam Liedvogel and Max Lundberg (The authors contributed equally to this chapter)

Food, water, and protective cover are basic to the survival of all animals. But the changing seasons can transform a comfortable environment into an unlivable one: the food and water supply can dwindle or disappear, plant cover can vanish, and competition with other animals may increase. Many animals therefore face the problem of occupying a habitat that is suitable for only a portion of the year. Animals that possess the ability to perform movements or migrate are in the favourable position that they can exploit seasonally benign habitats at all times of their annual cycle. In many taxa migration is the rule rather than the exception and the most well-known example here is probably birds. Hence, although we will in this chapter use examples from several taxa, we will mainly focus on bird migration, since the genetic basis of seasonal migration is well known in birds. As outlined in the introductory chapter (Box 1.1), we define seasonal migration as the oriented directional movement of individuals of a population between distinct locations including a return journey (e.g. breeding and wintering grounds). The timescale over which these migratory cycles occur can span hours, days, months, or years, and can even be multigenerational (Chapman et al., Chapter 2). Migration requires remarkable navigation abilities in order to combine inherited information on timing and direction into a spatiotemporal orientation programme that allows the animal to return to the same area on a regular basis (Åkesson et al., Chapter 9). The phenomenon of migration includes a complex suite of behavioural, sensory, morphological, and physiological traits. Key migratory traits are (i) migratory direction and orientation skills using compass reference systems to keep this direction and orient or navigate during the migratory journey (Åkesson et al., Chapter 9); (ii) an endogenous time programme that determines both the onset and termination of migratory behaviour (thus also defining the duration of migration), as well as precise timing of physiological adaptations necessary for a successful migratory journey, such as fuelling (hyperphagia) and moult; and (iii) the intensity of migratory activity, which may vary between different populations and during different migratory seasons (i.e. spring and autumn migration; Lindström et al., Chapter 3). Even when kept in captivity, migratory birds express these traits in a very characteristic behavioural repertoire—so-called *Zugunruhe* or migratory restlessness behaviour (Kramer 1949). The inherited time programme equips birds with information on when to moult and start depositing fat in order to cope with the challenge of the migratory journey, when to start their migratory journey, how long to fly, and when to stop. Migratory restlessness behaviour can easily be quantified in nocturnally migrating birds, as it can clearly be distinguished from other movement behaviours. Migratory restlessness behaviours are in very good accordance with the onset, termination, directional orientation, and intensity of migratory activity in wild conspecifics, and thus their precise quantification under controlled conditions in the lab serves as a good proxy for these migratory traits (Berthold 1996).

Animal Movement Across Scales. Edited by Lars-Anders Hansson and Susanne Åkesson.
© Oxford University Press 2014. Published 2014 by Oxford University Press.

12.1 How do we know that migratory traits are innate?

When aiming at understanding the genetic architecture of migratory traits, it is important to choose a study species that exhibits a highly diverse repertoire of migratory phenotypes. Ideally the species' repertoire includes a continuum of migratory distances ranging from long-distance migrants to fully sedentary (non-migratory) populations, as well as populations exhibiting a migratory divide, i.e. neighbouring populations that follow different migratory directions. One of these extremely well-suited 'migratory model organisms' is the blackcap, *Sylvia atricapilla*. Most of our current understanding of the genetics of migration is based on common garden experiments with passerines that allow us to disentangle the genetic and environmental origin of phenotypic differences (reviewed by van Noordwijk et al. 2006). These kinds of experiments include a series of large-scale crossbreeding and selection experiments on blackcaps (e.g. Berthold et al. 1992, Helbig 1996; see Fig. 12.1), and displacement experiments in some other species of passerines (e.g. Perdeck 1958, Thorup et al. 2007). These experiments have suggested a considerable genetic component for several migratory traits, and further showed that these traits (when under strong artificial selection) can drastically change within a few generations.

In addition to experimental approaches, there are also 'unmanipulated' examples in the wild, which clearly show that migratory traits are inherited. The common cuckoo, *Cuculus canorus*, is an excellent example where we know that young birds never interact with their parents, but nevertheless know when they should migrate to sub-Saharan Africa. Another fascinating example is the multigenerational migration cycle of monarch butterflies, *Danaus plexippus* (see Box 12.1 and Fig. 2.6). North American monarchs carry the genetic machinery to migrate, but the programme only gets initiated in migratory populations. This scenario clearly demonstrates that the amazing orientation mechanisms involved in multigenerational migration are not learnt, but must be inherited, since there are at least two non-migratory generations between the migratory ones (Chapman et al., Chapter 2). It further highlights the most likely involvement of interactions between environmental cues and the genetic programme, which can differ between populations or even vary among individuals.

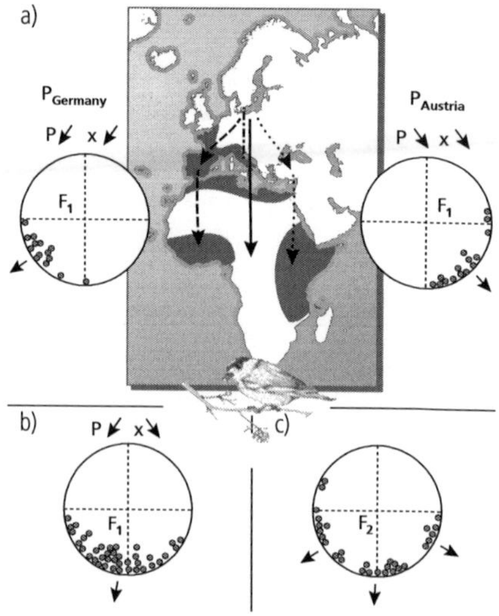

Figure 12.1 Inheritance of blackcap, *Sylvia atricapilla*, migratory direction in a selection- and crossbreeding experiment (modified from Helbig 1996). (a) Orientation diagrams (see also Åkesson et al., Chapter 9) of selectively bred first-generation (F_1) blackcaps during the autumn of parental (P) blackcap populations west (left, dashed) and east (right, dotted) of the central European migratory divide (each dot in the circle indicates the migratory direction of one bird; the orientation vector of all birds as a group is indicated by the arrow outside the circle). (b) F_1 offspring of mixed pairs crossbred in aviaries from parents of either side of the divide show an intermediate orientation phenotype. (c) The scatter of orientation phenotype in the second generation of crossbred offspring (F_2) segregates (compared to F_1) and clusters around both the intermediate and parental phenotypes. This pattern is indicative of a scenario where modulation of the migratory phenotype could be regulated by variation in only a small number of genes with large effects.

12.2 Quantitative genetic assessment of migratory traits

Quantitative genetics (see Box 12.2) allows us to assess the relative roles of genetic variation and environment in explaining the variation of phenotypic traits. For a single trait this relationship, and the response to selection of the trait, is quantified as heritability (h^2) (see Box 12.2 for more information). Migration-related traits that have been investigated

Box 12.1 Genomics of migration: next-generation sequencing approaches

Recent advances in next-generation sequencing (NGS) technology will most likely revolutionize our understanding of the genetics of migration. These resources add a completely new level to the field of migration genetics. This toolbox can now be applied to non-model migratory species and will facilitate the identification of genes or genomic regions that harbour variation with relevant consequences for the migratory phenotype, ranging from the generation of expressed sequence tag (EST) libraries for transcriptional comparison between different migratory phenotypes, up to fully sequenced and annotated genomes of non-model migratory species. However, the basis of migratory phenotype differences might not only lay in the DNA sequence. It is plausible that a limited number of genetic changes can result in large and widespread differences in gene expression downstream in a signalling cascade, either through sequence differences or governed by epigenetic processes. Here we can make use of the fact that migratory animals don't migrate all year round, but the migratory phenotype is only expressed during certain migratory periods. Transcriptomics or gene expression profiling (via RNAseq or microarray technology) allows us to characterize and compare gene expression profiles. This is a promising approach to uncover new gene candidates for further characterization without a priori knowledge of specific genes, and to identify gene networks that are differentially regulated between different phenotypic groups and thus likely involve in modulating the migratory phenotype. Gene expression profiling is typically based on mRNA (transcripts from protein-coding genes) extracts from different tissues (e.g. brain, muscle, fat, liver), and makes it possible to compare gene expression between phenotypically divergent populations at different time points. This approach is expected to be especially powerful in species with populations that exhibit a continuum of migratory phenotypes, i.e. species that include long-distance migrants, short-distance migrants, populations following different migratory directions, or even sedentary populations. The expression of at least some migratory traits is also regulated by external cues, and a lot of phenotypic plasticity could be explained by epigenetic processes. Changes in day length can trigger onset of migratory restlessness, and alterations of the magnetic field have been shown to modulate fuel deposition in birds. The most dramatic example is probably the transition to the migratory phenotype in the desert locust, *Locusta migratoria*, which is triggered by increased serotonin levels owing to social interactions (Anstey et al. 2009). How could epigenetic processes regulate migratory phenotype expression? The migratory cycle of North American monarch butterflies, *Danaus plexippus*, is exceptional in the way that it involves successive generations, including migratory and sedentary populations (reviewed in Brower 1996). All butterfly generations are equipped with the same core genetic material, and it is possible that epigenetic mechanisms are triggering changes in gene expression or function that underlie the migratory phenotype. This means that there is a genetic basis to the migratory process, but the generational divergence in phenotype might be epigenetically regulated by environmental cues that yet remain to be identified.

Another approach to find new potential candidates is to compare the association between sequence polymorphisms and a phenotypic trait. Quantitative trait locus (QTL) mapping requires genetic markers (e.g. single nucleotide polymorphisms (SNPs) or microsatellites) on a linkage map derived from pedigrees of wild populations or from captive populations selectively bred to generate a F_2 or backcross population. Constructing sufficiently large pedigrees (several hundreds to thousands of individuals) of crossbred captive populations is extremely labour- and cost-intensive for many organisms. Genome-wide association studies (GWAS) offer an alternative approach that will be facilitated by NGS technology. GWAS rely on historical recombination events between genetic variants in a population. These events will have a randomizing effect that allows genetic variants to be individually associated with phenotypic traits in a population. This approach can be used in any population but typically requires a large number of individuals and genetic markers.

Using next generation sequencing technology, variation in a large number of phenotypically characterized individuals may be screened to establish associations between genotype and phenotype in non-model organisms. One recent genome-wide association study focused on understanding the genetic basis of migration in two wild populations of migratory steelhead and resident rainbow trout, *Oncorhynchus mykiss*, from the Pacific Northwest (USA). This study identified several genetic loci associated with migratory life-history traits, and suggests a complex multi-genic basis (with several loci of small effect distributed throughout the genome) contributing to migration in this species (Hecht et al. 2013).

Ultimately, we want to assess functionality of any identified candidate gene; i.e. we want to test the effect of the gene on the phenotype expression of the focal trait. To do this it is necessary to modify the genotype (i.e. knockout and rescue mutants). This has, for example, been successfully applied for the monarch butterfly using zinc-finger nucleases, which are a class of synthetic DNA binding proteins that allow targeted genome manipulation (Reppert et al. 2010). However, when working with wild caught migratory animals, generating genetically modified organisms is usually not an option, due to methodological reasons and ethical concerns. In this case the use of small interfering RNA (RNAi)-mediated post-transcriptional gene silencing might be an alternative technological approach.

Box 12.2 Mendelian traits and quantitative traits

Mendelian traits usually have a few qualitative states which are governed by variation at a single locus. The allele can be either dominant or recessive. If the dominant allele is inherited, the dominant phenotype is always expressed—no matter if heterozygous or homozygous for the dominant allele. If the recessive allele is inherited, the recessive phenotype is only expressed if the recessive allele is inherited by the offspring from both the father and the mother, i.e. homozygous recessive. Only a minority of traits are inherited in a purely Mendelian way because dominance may be incomplete or phenotypic traits are controlled by co-dominance (phenotypic expression of both alleles), or variation in traits arise from variation at several loci. Examples of purely Mendelian traits are melanism or the ability to smell hydrogen cyanide (like bitter almond), which are both examples of recessiveness and the immunity to poison ivy, which is an example of dominance.

Quantitative traits are often continuous traits that are influenced by variation at several loci, for example height in humans, wing length in birds, and migration distances across different populations. The environment will often have a large effect on the phenotype. Quantitative traits are commonly measured in terms of their variances. In its simplest form the phenotypic variance can be partitioned into its environmental and genetic component. Of particular interest is the additive genetic variance, which results from the additive effects of alleles at different loci. The ratio between the additive genetic variance and the phenotypic variance within populations is termed *heritability* (h^2). This is a (population) measurement of how much of the phenotypic variance could be explained by the (additive) genetic variance and determines how strongly this trait responds to selection. Note that these measurements are population-specific and are among other things dependent on the environment (e.g. Hoffmann and Merilä 1999, Wilson et al. 2006, Visscher et al. 2008). A constant environment, for example a lab environment, may inflate heritability and, in cases where individuals experience large differences in environmental conditions (e.g. access to food resources), the ordinary heritability of the trait might be underestimated. Since relatives are expected to share a more similar genetic background, heritability could be estimated from comparing a trait among close relatives, for example by linear regression or more sophisticated models such as the animal model, which could take into account more complicated relationships among relatives (Kruuk 2004).

in crossbreeding and selection experiments have generally shown moderate to high heritabilities, which may facilitate rapid evolutionary responses. There is little information on natural variation of migratory traits in wild populations, but in general heritability estimates calculated from quantified behaviour of migratory birds in the lab appear to be similar to heritabilities in the wild (Pulido and Berthold 2003). Values can be estimated by different methods, such as parent–offspring regressions and by full-sibling correlations, or using more sophisticated methods, such as the animal model (Lynch and Walsh 1998, Kruuk 2004). Depending on the modelling approach and the size of pedigree data, estimated heritabilities for migratory activity in blackcaps range from 0.37 to 0.46, and for the timing of (autumn) migration between 0.34 and 0.45 (Berthold and Pulido 1994, Pulido and Berthold 2010). The spread of heritability estimates increases if other species are considered as well (Pulido and Berthold 2003). These quantitative genetic analyses suggest that variation observed within migratory traits has a strong genetic basis. Quantitative genetics also allows us to predict the correlated selective responses of phenotypic traits that are genetically correlated with any focal migratory traits. For example, in blackcaps, genetic correlations have been observed between various timing aspects of migratory restless behaviours such as amount of activity, intensity, and timing (Pulido and Berthold 2003). This suggests that variation in one trait is not necessarily independent of variation of another trait and that traits could be seen as components of a migratory 'gene package' (Berthold 1999). Consequently, selection on a specific trait expressed in the migratory phenotype could ultimately cause indirect selection on genetically correlated traits of the migratory phenotype. Some of these traits might also have consequences on other life-history events throughout the annual cycle. For example, correlations between the timing of migration and moult (Pulido and Coppack 2004), and the timing of migration and breeding (Teplitsky et al. 2011) have been demonstrated.

A quantitative genetics approach is, however, of limited use if we want to understand to what extent the 'migratory gene package' is similar between species across different phylogenetic scales, or to gain insight into the genetic architecture of movement within species. For such comparative analyses, the genes underlying the different components need to be identified. Quantitative genetics analyses have clearly shown that migratory animals have an *inherent time schedule* (Gwinner 1967, 1996; Pulido et al. 2001; O'Malley et. al. 2010; Anderson and Beer 2009; O'Malley and Banks 2008) and at least an inherited initial migratory direction (reviewed in Helbig 1996), which they combine into a spatiotemporal migration programme that leads them to their species- or population-specific winter quarters. We will focus on our current understanding of the genetics of these two key behavioural adaptations.

12.3 The genetics of migratory direction

Small migratory songbirds migrate at night on their own, which means that first-year migrants are heading towards an area they have never been to before. They travel completely by themselves and directional information is thus crucial for their survival (Åkesson et al., Chapter 9). Most of our current understanding of the genetics of directional information comes from crossbreeding and displacement experiments (see Fig. 12.1 for further details and illustration). Crossbreeding experiments of blackcaps from two sides of the central European migratory divide suggest an intermediate mode of inheritance (at least in blackcaps), as first-generation (F_1) offspring of crossbred birds show a directional preference that is intermediate to the parental (P) phenotypes (Fig. 12.1). These experiments also show that genetically determined migratory direction is susceptible to microevolution, i.e. small evolutionary changes that can be detected at the population level. The particular segregation pattern of migratory direction observed in crossbred birds of the second generation (F_2) might suggest a scenario where the modulation of the migratory phenotype is governed by variation in a small number of genes of considerable importance (Fig. 12.1). A different mode of inheritance has been suggested from a study of pied flycatchers, *Ficedula hypoleuca*, and collared flycatchers, *Ficedula albicollis*. Using stable isotope analyses of feathers grown on their wintering grounds as a proxy for wintering location and thus migratory direction, Veen et al. (2007) suggested that the migratory direction of the pied flycatcher, which winters in West Africa, is dominantly expressed in hybrids between the two species. A similar pattern has also been found in the great reed warbler, *Acrocephalus arundinaceus*, and the clamorous warbler, *Acrocephalus stentoreus*, where hybrids appear to follow the migratory route of great reed warblers (Yohannes et al. 2011). Not only birds, but also many fishes perform seasonal mass migrations. A study on Atlantic eels provides indirect evidence for a genetic basis of migratory direction in fish. Both the American *Anguilla rostrata* and the European eel, *Anguilla anguilla*, start their migratory journey in the Sargasso Sea, but their routes differ in distance and direction: American eels migrate towards the North American coast and European eels follow a slightly longer northeasterly route towards Europe, whereas Icelandic eels follow an intermediate migratory direction. A proportion of Icelandic eels could be identified as hybrids based on their genotype, which, again, suggests an intermediate mode of inheritance in migratory direction (Albert et al. 2006). There are only a few studies that have explored the genetics of migratory direction, and too little is known to draw any general conclusions about the mode of inheritance of migratory direction. In general, sex-specific inheritance of migratory traits seems to be negligible, but maternal effects have been found to be significant for the onset of migratory activity in the blackcap (Pulido and Berthold 2010).

12.4 The genetics of migratory timing

In migratory birds, the timing aspects of the migratory route appear to be encoded by the circannual clock of the bird. The circannual clock is a biochemical oscillator that controls the biological rhythm, underlying the timing of behavioural and physiological processes, for example related to breeding, moult, and migration (extensively reviewed by Gwinner 2003, Wikelski et al. 2008, Visser et al. 2010). As the name suggests this rhythm is only approximately annual and must be synchronized with external cues, of which the most important is day length, in order for the bird to time its behaviours to the natural year. Different components of the endogenous timing programme (i.e. amount, intensity, onset and termination, duration) have been extensively studied in several migratory passerines (reviewed e.g. in Pulido and Berthold 2003). Heritability values for onset of migratory behaviour are medium to high (0.34–0.45) in blackcaps, and even higher estimates have been observed in garden warblers, *Sylvia borin* (0.67 for onset of both autumn and spring migratory activity). Heritability estimates for termination of migratory activity are slightly lower, and in blackcaps they range from 0.16 to 0.44. Values for migratory activity range from 0.36 to 0.52 (for an overview of heritability (h^2) estimates for various traits and species see Pulido and Berthold 2003).

Fish are also well suited for studying spatiotemporal aspects of migration. Recent studies have focused in particular on a genus of trouts and salmons, *Oncorhynchus*, which shows high variability in both spatial orientation and temporal return patterns (Weitkamp 2010). Fish are of great economic value as well, and hatchlings are often tagged with coded wire tags when released from hatcheries. This provides a massive data set on population movement pattern and overall spatial distribution. A candidate gene approach led to the identification of a latitudinal cline in the *CLOCK* gene, a central and molecularly well-characterized component of the circadian clock in animals. *CLOCK* gene variation corresponded with spawning time in the Chinook salmon, *Oncorhynchus tshawytscha*, and identified day length as a primary cue for migrating and spawning in fish (O'Malley and Banks 2008, O'Malley et al. 2010). Genetic data from a pink salmon, *Onchorhynchus gorbuscha*, population further indicate a genetic change for earlier migration timing (Kovach et al. 2012). These findings are in line with a latitudinal cline in *CLOCK* gene variability reported for blue tits, *Cyanistes caeruleus* (Johnsen et al. 2007), which might suggest a general adaptation to ecological factors correlated with latitude, such as breeding or moult phenology (Liedvogel et al. 2009, Caprioli et al. 2012, and Saino et al. 2013 in barn swallows, *Hirundo rustica*). These data suggest that *CLOCK* may be involved in controlling seasonal adaptation and influences geographical variation in timing of reproduction in at least some migratory fish and bird species. However, this picture is far from being consistent, and several studies in fish and birds actually report the lack of significant correlation between the candidate gene *CLOCK* and latitude or the timing of phenological events (Johnsen et al. 2007 (blue throats); Liedvogel and Sheldon 2010; Dor et al. 2012a, b; O'Brien et al. 2013; Peterson et al. 2013). These studies highlight that caution is advised when interpreting variation in candidate genes in the context of seasonal activities, particularly migration, or in relationship to photoperiodism along geographical gradients. One possible explanation may be that this discrepancy in data from various species indicates that this particular 3'UTR polymorphism of the candidate gene is less significant in modulating the expression of *CLOCK* in other species. These studies further highlight how more detailed understanding of the genetic architecture of migratory traits can shed light on other life-history traits, such as the timing of reproduction. One likely consequence of a connection between the timing of migration and breeding might, for example, result in reproductive isolation between sympatric populations, and direct or indirect selection on migratory timing could result in different breeding times (allochrony), reproductive isolation, and eventually speciation.

12.4.1 The threshold model of migration

The results from crossbreeding and selection experiments (Fig. 12.1) not only show that there is considerable genetic variation in migratory traits, but they further suggest that it is possible that many, if not all animals have the genetic machinery to migrate. With appropriate selection pressure, migratory or sedentary behaviour could dominate within a population. This switch between alternate behaviours can occur within very few generations, particularly in populations that include a continuum of migratory strategies, for example in blackcaps (Berthold et al. 1992, Outlaw and Voelker 2006, Rolshausen et al. 2009, Pulido and Berthold 2010).

The evolutionary potential of migratory behaviour is also evident in wild populations. In birds it is quite common to find closely related species that show very different migratory strategies. Furthermore, in species with a wide latitudinal breeding distribution, more northern populations often show a higher frequency of migrants, a reduced sensitivity to environmental cues, and increased migratory distances (Berthold and Querner 1981, Pulido 2011). There are also several instances when a change in migratory behaviour has been recorded over a very short timescale. German blackcaps were first observed to winter in England in the 1960s and the wintering population had increased from 0% to 7% before 1960 and then to 11% in 1992 (Berthold et al. 1992). Similarly, North American house finches, *Carpodacus mexicanus*, have recently expanded their range from the North American east coast, where they were introduced in the 1940s, to most of North America. While the house finch was sedentary over its traditional range in western North America, the newly introduced eastern populations now show migratory behaviour (Able and Belthoff 1998). The mode of inheritance of migratory behaviour has been suggested to be a threshold trait (Pulido et al. 1996; Fig. 12.2). This is a special case of a quantitative trait in which there are two (or several) distinct phenotypic states. The state of an individual depends on a liability variable that is assumed to show a normal distribution in the population. When the combined effect of multiple genes in an individual reaches a threshold, the individual expresses migratory behaviour, and conversely, if the combined effect of the genes is below the threshold, the individual does not express migratory behaviour (reviewed in Pulido 2011; Fig. 12.2: white area under the distribution curve). One important evolutionary consequence of the threshold model is that genetic variation for the alternative state could be hidden from selection in a population. For example, if a population becomes sedentary from an ancestrally migratory state, genetic variation for the expression of migratory behaviour could still be present but be hidden in the resident individuals. Conversely, if selection would favour migratory behaviour in this now sedentary population (for example, due to climate change), this genetic variation could be selected for and the phenotypic trait may subsequently be expressed again (Fig. 12.2, black area under the curve). The threshold model could also be extended to explain a continuum of migratory strategies, such as residency, facultative migration, and obligate migration (Pulido 2011). This could be accommodated by assuming different environmental influences on the threshold. In some populations, such as in obligate long-distance migrants, the threshold might be largely insensitive to environmental cues and whether the individual should migrate is dependent on the genotype. Among more facultative migrants, the genotype might prepare for migration, but the ultimate decision to migrate is also dependent on environmental cues and social interactions (Nilsson et al., Chapter 6). Even though this is a simple and plausible genetic model to explain different migratory strategies, it needs to be verified and explored at a molecular level. Of particular interest is also the interactions between genes and environment that

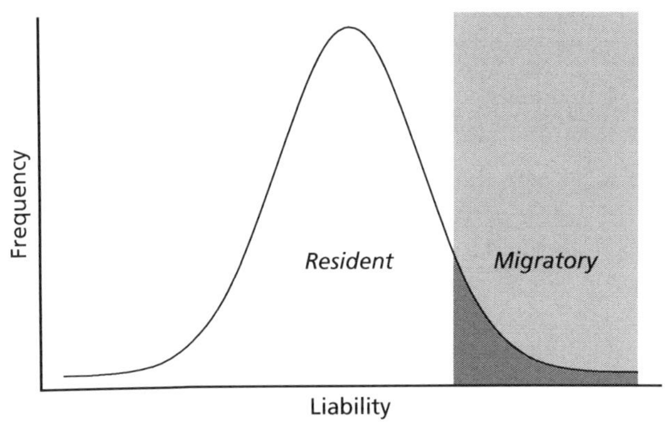

Figure 12.2 The threshold model of migration describes the distribution of migration propensity (i.e. liability) in a partially migratory population comprising both resident and migratory individuals. Individuals with migration propensities below the threshold are residents (white area under the curve); individuals with propensities above the threshold (black area) are migrants. The position of the threshold is not fully fixed and can be shifted to either side by environmental factors (see Pulido 2011 for an in-depth discussion of the model).

could potentially be mediated by epigenetic effects (for an overview on epigenetic consequences on behavioural traits see Champagne 2012, Ledón-Rettig et al. 2012). Epigenetic effects are changes in gene expression that occur without any modification in the genic DNA sequence. Epigenetic factors that govern these expression changes are chemically stable and potentially reversible, and can be modulated or induced by environmental factors and different developmental or physiological stages. Thus, epigenetic effects could, for example, explain some differences between adult migrants and young inexperienced animals on their first migration. In addition, we cannot fully explain how a most likely complex multi-gene adaptation such as migration can repeatedly be lost and gained in the course of evolution until we understand its underlying genetic architecture.

The molecular genetics of the transition from a resident to a migratory phenotype have been explored in trout. In these species of fish, all individuals hatch and are raised in freshwater, but some individuals later undergo a change to a sea-living migratory phenotype through a process termed smoltification (Chapman et al.; Chapter 2). The use of both QTL mapping (Hecht et al. 2012a) and genome-wide association (Hecht et al. 2013; Box 12.1) has discovered the involvement of several genomic regions in the smoltification of the rainbow trout, *Oncorhynchus mykiss*.

12.5 Morphological and behavioural adaptations for migration

Besides behavioural adaptations, the migratory phenotype also includes morphological and physiological adaptations. A well-studied morphological adaptation in migrating animals is wing morphology. Long-distance migrants among birds (Winkler and Leisler 1992) and insects (Lockwood et al. 1998, Alizer and Davis 2010) tend to have longer and more pointed wings than short-distance migrants and sedentary species. A longer wing is beneficial for long-distance flight, but is also thought to decrease manoeuvrability when foraging in dense habitats (Winkler and Leisler 1992; Johansson et al., Chapter 13). A recent scan for genetic variation associated with wing length in great reed warblers, *Acrocephalus arundinaceus*, identified a significant quantitative trait locus (QTL) explaining a substantial part of wing length variability (Tarka et al. 2010). This may be a first important step towards understanding the genetic architecture of wing length and shed light on our understanding of the genetics of associated migratory traits. The genetics of wing morphology has also been studied in the sand cricket, *Gryllus firmus* (reviewed in Roff and Fairbairn 2007). In this species wing length has been characterized as a threshold trait with a strong genetic basis ($h^2 = 0.65$).

Other adaptive traits likely to be genetically controlled are hyperphagia, fat deposition (birds start to eat almost continuously in order to store excess fat to serve as energy during the journey), and organ plasticity (regression of reproductive organs) as preparation for migration, as well as the choice of food, including seasonal shifts from insectivorous to a fruit-biased diet in many migrants before and during the migratory period (e.g. Biebach and Bauchinger 2003, Pulido and Berthold 2003, Piersma et al. 2005).

12.6 Genetics of migration: the molecular toolbox

Many migratory species show a strong phenotypic differentiation between populations (Chapman et al., Chapter 2). Given the data showing that migratory traits have a genetic basis, an obvious question is to ask whether it is possible to link these phenotypic differences to specific genetic differences among populations. A first step is to explore the overall genetic structure between the populations. Several tools have traditionally been used in population genetics to assess genetic structure between populations, for example mitochondrial DNA (mtDNA) or microsatellites, which are often assumed to be selectively neutral and reflect population processes. So far only a few studies have investigated overall genetic structure in migratory species, but the general pattern from various taxa (including birds, fish, and insects) is that intra-specific differences in migratory traits only weakly, or not at all, correlate with the overall genetic differentiation (e.g. Bensch et al. 1999, 2002; Pérez-Tris et al. 2004). This suggests that migratory phenotypes are not associated with the overall genetic differentiation, but on the other hand, many of the traditional markers only have very limited coverage of the genome. It could also be that variation in migratory traits is mainly modulated by variation at relatively few loci (regions in the genome), and that these subtle but targeted genetic differences may result in dramatic phenotypic effects if they alter the expression levels or up- or downstream signalling cascades of many genes (for example, as transcription complexes). A whole genome AFLP scan in willow warblers, *Phylloscopus trochilus*, a migratory songbird exhibiting a central migratory divide in Scandinavia, lead to the identification of one promising AFLP marker exhibiting a strong genetic differentiation between the subspecies and matches the differences in migratory behaviour across the divide (Bensch et al. 2009; Fig. 12.3). Future characterization and mapping of the chromosomal location of derived AFLP markers on the genome may lead to the identification of the gene(s) under selection. An alternative approach is to limit the study to candidate genes that have shown to have a similar or otherwise interesting function in other species. This approach was used in a study of blackcaps in which longer alleles of a locus with the *ADCYAP1* gene correlated with higher migratory activity (quantified via migratory restlessness activity in night-migratory songbirds) as a proxy for migratory distance (Müller et al. 2011, but see Peterson et al. 2013). As highlighted earlier, results from candidate gene studies must be interpreted with caution and in order to draw general conclusions future studies (e.g. Peterson et al. 2013) are needed to investigate whether this locus is also associated with migratory traits in other birds as well.

For both the AFLP and the candidate gene approach it is further important to bear in mind that genetic variation is not always independently segregating. This non-random association of variation is referred to as linkage disequilibrium and could potentially stretch over large distances of a chromosome. In this case a genetic difference might not be functionally relevant to the trait in question, but is co-inherited and thus strongly correlated with nearby causative functional variation. This highlights that detection of interesting variation in a gene should be accompanied by screening the genetic variation in the nearby chromosome region.

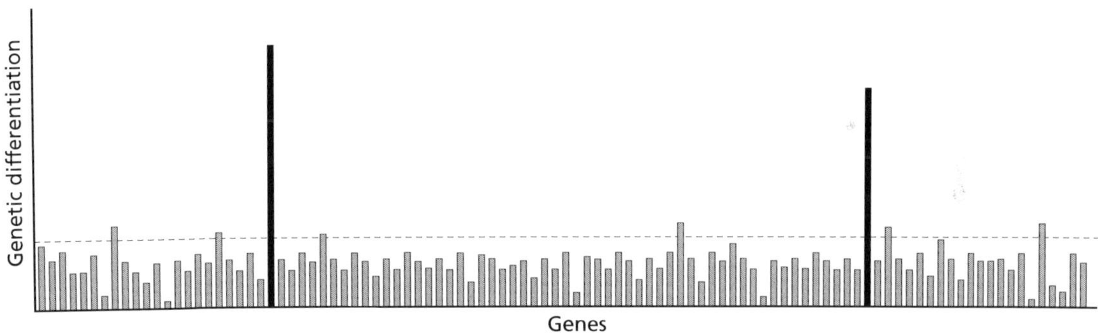

Figure 12.3 A hypothetical example showing genetic structure between two populations exhibiting different migratory phenotypes. The genetic structure is measured as genetic differentiation of several genes. Most genes (grey bars) only show little genetic differentiation between populations around the background level (dotted line). The low differentiation of most genes could be explained by that the populations separated recently and that genetic differences have not had time to accumulate between them, and/or that gene flow between them has homogenized the gene pool. Genes that are under divergent selection between the populations (represented by black bars) are expected to have different variants favoured in each of the two populations and therefore show higher genetic differentiation than other genes. In this particular example, these genes may be involved in adaptations associated with the different migratory phenotypes of each population. However, it is important to keep in mind that the highly differentiated genes could be associated with other adaptations in the populations than those associated with migratory phenotypes. Furthermore, populations are expected to become increasingly genetically differentiated over time due to neutral processes and accumulation of new adaptations. As a consequence, in more genetically separated species, migration genes might be harder to detect because of an overall higher genetic differentiation (dashed line). A suitable species for studying the genetics of migration should thus show very small overall differentiation between migratory phenotypes. In this case, genes involved in migratory phenotypes could be expected to show a more pronounced genetic structure between the populations than is seen with neutral markers.

12.7 Limitations and future perspectives in identifying 'the migratory gene'

Studying the genetics of migration is challenging because migratory traits are complex and often difficult to clearly define, measure, and experimentally manipulate. The complex nature of traits makes it likely that the adaptations may be driven by a large number of genetic variants of which most are expected to have a minor effect (for a general reference see Lynch and Walsh 1998). Behavioural traits have generally been shown to be controlled by genes with small effects (e.g. Flint et al. 2010, Hecht et al. 2013), but genes with major impacts on specific behaviour do exist, as has been successfully demonstrated for gaits in horses and mice (DMRT3, Andersson et al. 2012). Thus, the investigation of genetic variants with large effect size on the migratory phenotype (as expected by the phenotype distribution in crossbreeding experiments with blackcaps, see Fig. 12.1) is of particular interest.

General conclusions about migration genes among taxa is still not possible, because studies of different taxa have frequently focused on different aspects of the migratory phenotype. While some studies have specifically looked at genes involved in e.g. phenological events (*CLOCK* studies in birds and fish), others have focused on genes influencing migration and breeding on a broader scale, such as those responsible for a transition into a migratory phenotype (e.g. Hecht et al. 2012a). Direct comparison is further complicated by the fact that different technological approaches have been used. For example, a study on gene expression patterns may detect genes that are differentially expressed between migrating and breeding individuals. The expression changes of these particular genes that are picked up by gene expression pattern comparison could be due to a downstream effect of an important regulatory sequence variant that may not be detected in this specific data set, but might be detected in a genome-wide association study. In addition, different methods of assessing the same phenotype could potentially explain some differences in gene sets found between different study populations (Hecht et al. 2013). Even in cases where the same candidate gene has been investigated in populations of the same (or a different species), an association established in one of the populations has often been difficult to replicate in another population. This does not rule out the fact that this gene may be of (maybe exclusive) importance in this particular environment, but not in the other populations investigated. However, it is also possible that the same phenotypic effect in other populations could, for example, be produced by different regulatory mechanisms or genetic variants of the same gene, or by other genes in the same pathway. These studies highlight the architecture of complex traits, for which phenotypic variation is likely produced by complex interactions between many genes. Thus taking gene ontology information into account when interpreting these data, such as molecular functions and involvement in a particular biological process of the genes, and attempting to draw general conclusions is extremely important.

The traditional molecular methods that have previously been used in the field of migration research have several limitations. The candidate gene approach is limited by genetic resources available from other populations or species. As highlighted earlier, the same polymorphism in a candidate gene might be of less importance in the new study population, and even if it is important, its particular association with a phenotypic trait should ideally be corroborated with an independent set of genome-wide markers. Similarly, the number of markers is relatively low if one uses microsatellites and AFLPs. This means that only a relatively small fraction of the genome can be screened simultaneously, which clearly limits the resolution achieved. The annotation of these markers also relies on the availability of genomic resources from other closely related species. A general problem of molecular studies of migration has until recently been that none of the model species used as references for genomic analyses of migratory species show a distinct migratory phenotype (e.g. *Drosophila*, zebra finch *Taeniopygia guttata*). Hence, we cannot be sure that the 'migration genes' being looked for are expressed, are functional, or even occur in the model organisms used as default references.

Due to the recent advances in technology, more and more genomic resources of migratory species are becoming available (e.g. monarch butterfly, Zhan et al. 2011; *Ficedula* flycatchers, Ellegren et al.

2012; and willow warbler, Lundberg et al. 2013), which will ultimately start an entirely new era and allow us to validate the function of identified gene candidates. These resources will be invaluable for large-scale analyses, such as high-throughput genotyping and quantification of gene expression or 'next generation sequencing' (NGS; Box 12.1). The detection of migration genes is likely to also be aided by parallel advances in tracking technologies (Åkesson et al., Chapter 9, and Nilsson et al., Chapter 6), which allow for more precise phenotyping of migration in wild populations. These phenotyping improvements will also serve to enhance our understanding of environmental effects on migration and subsequently also genotype versus environment interactions. Once differently expressed gene or variable genomic regions have been identified, it will be possible to test how this variation relates to the phenotypes of individuals, and how epigenetic changes contribute to variation in migratory traits. Consequently, this will lead to an understanding of the genetic architecture of complex traits and ultimately the association between the variation in the environment and the genotype of an individual.

Key questions that could be answered once migration genes have been identified are, for example: Have the same genes been used in migratory traits across distantly related taxa such as insects and birds? How many independent 'genetic solutions' to migratory behaviour have there been? What behavioural adaptations has migration evolved from? What molecular, behavioural, or physiological processes have these groups of genes been involved with in the past before they specialized to shape migratory traits? The future for solving these questions is very exciting and given recent advances, the field of migration genetics certainly looks bright.

References

Able, K. P. and Belthoff, J. R. (1998). Rapid 'evolution' of migratory behaviour in the introduced house finch of eastern North America. *Proceedings of the Royal Society B: Biological Sciences*, **265**, 2063–71.

Albert, V., Jónsson, B., and Bernatchez, L. (2006). Natural hybrids in Atlantic eels (*Anguilla anguilla A. rostrata*): evidence for successful reproduction and fluctuating abundance in space and time. *Molecular Ecology*, **15**, 1903–16.

Altizer, S., and Davis, A. K. (2010). Populations of monarch butterflies with different migratory behaviors show divergence in wing morphology. *Evolution*, **64**, 1018–28.

Anderson, J. J. and Beer, W. N. (2009). Oceanic, riverine, and genetic influences on spring chinook salmon migration timing. *Ecological Applications*, **19**, 1989–2003.

Anstey, M. L., Rogers, S. M., Ott, S. R., Burrows, M., and Simpson, S. J. (2009). Serotonin mediates behavioural gregarization underlying swarm formation in desert locusts. *Science*, **323**, 627–39.

Andersson, L. S., et al (2012). Mutations in DMRT3 affect locomotion in horses and spinal circuit function in mice. *Nature*, **488**, 642–646.

Bensch, S., Åkesson, S., and Irwin, D. E. (2002). The use of AFLP to find an informative SNP: genetic differences across a migratory divide in willow warblers. *Molecular Ecology*, **11**, 2359–66.

Bensch, S., Andersson, T., and Åkesson, S. (1999). Morphological and molecular variation across a migratory divide in willow warblers. *Phylloscopus trochilus. Evolution*, **53**, 1925–35.

Bensch, S., Grahn, M., Müller, N., Gay, L., and Åkesson, S. (2009). Genetic, morphological, and feather isotope variation of migratory willow warblers show gradual divergence in a ring. *Molecular Ecology*, **18**, 3087–96.

Berthold, P. (1996). *The Control of Bird Migration*. Chapman & Hall, London.

Berthold, P. (1999). A comprehensive theory for the evolution, control and adaptability of avian migration. *Ostrich*, **70**, 1–11.

Berthold, P., Helbig, A. J., Mohr, G., and Querner, U. (1992). Rapid microevolution of migratory behavior in a wild bird species. *Nature*, **360**, 668–70.

Berthold, P., and Pulido, F. (1994). Heritability of migratory activity in a natural bird population. *Proceedings of the Royal Society B: Biological Sciences*, **257**, 311–15.

Berthold, P., and Querner, U. (1981). Genetic basis of migratory behavior in European warblers. *Science*, **212**, 77–9.

Biebach, H., and Bauchinger, U. (2003). Energetic savings by organ adjustment during long migratory flights in garden warblers (*Sylvia borin*). In P. Berthold, E. Gwinner, and E. Sonnenschein (eds), *Avian Migration*, pp. 269–80. Springer-Verlag, Berlin.

Brower, L. P. (1996). Monarch butterfly orientation: missing pieces of a magnificent puzzle. *Journal of Experimental Biology*, **199**, 93–103.

Caprioli, M., Ambrosini, R., Boncoraglio, G., et al. (2012). *Clock* gene variation is associated with breeding phenology and maybe under directional selection in the migratory barn swallow. *PLoS One*, **7**(4), e35140.

Champagne, F. A. (2012). Interplay between social experiences and the genome: epigenetic consequences for behavior. *Advances in Genetics*, **77**, 33–57.

Dor, R., Cooper, C. B., Lovette, I. J., et al. (2012b). *Clock* gene variation in *Tachycineta* swallows. *Ecology and Evolution*, **2**. 95–105.

Dor, R., Lovette, I. J., Safran, R. J., et al. (2012a). Low variation in the polymorphic *Clock* gene Poly-Q region despite population genetic structure across barn swallow (*Hirundo rustica*) populations. *PLoS One*, **6**, e28843.

Ellegren, H, Smeds, L., Burri, R., et al. (2012). The genomic landscape of species divergence in *Ficedula* flycatchers. *Nature*, **491**, 756–60.

Flint, J., Greenspan, R. J., and Kendler, K. S. (2010). *How Genes Influence Behavior*. Oxford University Press, Oxford.

Gwinner, E. (1967). Circannuale Periodik der Mauser und der Zugunruhe bei einem Vogel. *Naturwissenschaften*, **54**, 447.

Gwinner, E. (1996). Circadian and circannual programmes in avian migration. *Journal of Experimental Biology*, **199**, 39–48.

Gwinner, E. (2003). Circannual rhythms in birds. *Current Opinions in Neurobiology*, **13**, 770–8.

Hecht, B. C., Campbell, N. R., Holecek, D. E., and Narum, S. R. (2013). Genome-wide association reveals genetic basis for the propensity to migrate in wild populations of rainbow and steelhead trout. *Molecular Ecology*, **22**, 3061–76.

Hecht, B. C., Thrower, F. P., Hale, M. C., Miller, M. R., and Nichols, K. M. (2012). Genetic architecture of migration-related traits in rainbow and steelhead trout, *Oncorhynchus mykiss*. *G3*, 2, 1113–27.

Helbig, A. J. (1996). Genetic basis, mode of inheritance and evolutionary changes of migratory directions in Palearctic warblers (Aves: Sylviidae). *Journal of Experimental Biology*, **199**, 49–55.

Hoffman, A. A., and Merila, J. (1999). Heritable variation and evolution under favourable and unfavourable conditions. *Trends in Ecology and Evolution*, **14**, 96–101.

Johnsen, A., Fidler, A. E., Kuhn, S., et al. (2007). Avian *Clock* gene polymorphism: evidence for a latitudinal cline in allele frequencies. *Molecular Ecology*, **16**, 4867–80.

Kovach, R. K., Gharrett, A. J., and Tallmon, D. A. (2012). Genetic change for earlier migration timing in a pink salmon population. *Proceedings of the Royal Society B: Biological Sciences*, **279**, 3870–978.

Kramer, G. (1949). Über Richtungstendenzen bei der nächtlichen Zugunruhe gekäfigter Vögel. In E. Mayr and E. Schüz (eds), *Ornithologie als biologische Wissenschaft*, Winter, Heidelberg, p. 269.

Kruuk, L. E. B. (2004). Estimating genetic parameters in natural populations using the 'animal model'. *Philosophical Transactions of the Royal Society B: Biological Sciences*, **359**, 873–90.

Ledón-Rettig, C. C., Richards, C. L., and Martin, L. B. (2012). Epigenetics for behavioral ecologists. *Behavioral Ecology*, **24**, 311–24.

Liedvogel, M., and Sheldon, B. C. (2010). Low variability and absence of phenotypic correlates of *Clock* gene variation in a great tit *Parus major* population. *Journal of Avian Biology*, **41**, 543–50.

Liedvogel, M., Szulkin, M., Knowles, S. C. L., Wood, M., and Sheldon, B.C. (2009). Phenotypic correlates of variation at the *Clock* gene in a wild blue tit population: evidence for a role in seasonal timing of reproduction. *Molecular Ecology*, **18**, 2444–56.

Lockwood, R., Swaddle, J. P., and Rayner, M. V. (1998). Avian wingtip shape reconsidered: wingtip shape indices and morphological adaptations to migration. *Journal of Avian Biology*, **29**, 273–92.

Lundberg, M., Boss, J., Canbäck, B., et al. (2013). Characterisation of a transcriptome to find sequence differences between two differentially migrating subspecies of the willow warbler *Phylloscopus trochilus*. *BMC Genomics*, **14**, 330.

Lynch, M., and Walsh, B. (1998). *Genetics and Analysis of Quantitative Traits*. Sinauer Assoc., Sunderland, MA.

Mueller, J. C., Pulido, F., Kempenaers, B. (2011). Identification of a gene associated with avian migratory behaviour. *Proceedings of the Royal Society B*, **278**, 2848–2856.

O'Brien, C., Unruh, L., Zimmerman, C., Bracshaw, W. E., Holzapfel, C. M., and Cresco, W. A. (2013). Geography of the circadian gene clock and photoperiodic response in western North American populations of the three-spined stickleback *Gasterosteus aculeatus*. *Journal of Fish Biology*, **82**, 827–39.

O'Malley, K. G., and Banks, M. A. (2008). A latitudinal cline in the Chinook salmon *Oncorhynchus tshawytscha Clock* gene: evidence for selection on PolyQ length variants. *Proceedings of the Royal Society B: Biological Sciences*, **275**, 2813–21.

O'Malley, K. G., Ford, M. J., and Hard, J. J. (2010). *Clock* polymorphism in Pacific salmon: evidence for variable selection along a latitudinal gradient. *Proceedings of the Royal Society B: Biological Sciences*, **277**, 3703–14.

Outlaw, D. C., and Voelker, G. (2006). Phylogenetic tests of hypotheses for the evolution of avian migration: a case study using the Motacillidae. *Auk*, **123**, 455–66.

Perdeck, A. C. (1958). Two types of orientation in migrating starlings *Sturnus vulgaris* L. and chaffinches *Fringilla coelebs* L., as revealed by displacement experiments. *Ardea*, **46**, 1–37.

Pérez-Tris, J., Bensch, S., Carbonell, R., Helbig, A. J., and Tellería, J. L. (2004). Historical diversification of migration patterns in a passerine bird. *Evolution*, **58**, 1819–32.

Peterson, M. P., Abolins-Abols, M., Atwell, J. W., Rice, R. J., Milá, B., and Ketterson, E. D. (2013). Variation in candidate genes *CLOCK* and *ADCYAP1* does not consistently predict differences in migratory behavior in the songbird genus *Junco* [v1; ref status: indexed, http://f1000r.es/11p] *F1000Research*, 2,115.

Piersma, T., Pérez-Tris, J., Mouritsen, H., Bauchinger, U., and Bairlein, F. (2005). Is there a 'migratory syndrome' common to all migrant birds? *Annals of the New York Academy of Sciences*, **1046**, 282–93.

Pulido, F. (2011). Evolutionary genetics of partial migration—the threshold model of migration revis(it) ed. *Oikos*, **120**, 1776–83.

Pulido. F., and Berthold, P. (2003). Quantitative genetic analysis of migratory behavior. In P. Berthold, E. Gwinner, and E. Sonnenschein (eds), *Avian Migration*, pp. 53–77, Springer-Verlag, Berlin.

Pulido, F., and Berthold, P. (2010). Current selection for lower migratory activity will drive the evolution of residency in a migratory bird population *Proceedings of the National Academy of Sciences USA*, **107**, 7341–6.

Pulido, F., Berthold, P., Mohr, G., and Querner, U. (2001). Heritability of the timing of autumn migration in a natural bird population. *Proceedings of the Royal Society B: Biological Sciences*, **268**, 953–9.

Pulido, F., Berthold, P., and van Noordwijk, A. J. (1996). Frequency of migrants and migratory activity are genetically correlated in a bird population: Evolutionary implications. *Proceedings of the National Academy of Sciences USA*, **93**, 14642–7.

Pulido, F., and Coppack, T. (2004). Correlation between timing of juvenile moult and onset of migration in the blackcap, *Sylvia atricapilla*. *Animal Behaviour*, **68**, 167–73.

Reppert, S. M., Gegear, R. J., and Merlin, C. (2010). Navigational mechanisms of migrating monarch butterflies. *Trends in Neurosciences*, **33**, 399–406.

Roff, D. A., and Fairbairn, D. J. (2007). The evolution and genetics of migration in insects. *BioScience*, **57**, 155–64.

Rolshausen, G., Segelbacher, G., Hobson, K. A., and Schaefer H. M. (2009). Contemporary evolution of reproductive isolation and phenotypic divergence in sympatry along a migratory divide. *Current Biology*, **19**, 2097–101.

Saino, N., Romano, M., Caprioli, M., et al. (2013). Timing of molt of barn swallows is delayed in a rare *Clock* genotype. *PeerJ* 1:e17 http://dx.doi.org/10.7717/peerj.17.

Tarka, M., Åkesson, M., Beraldi, D., et al. (2010). A strong quantitative trait locus for wing length on chromosome 2 in a wild population of great reed warblers. *Proceedings of the Royal Society B: Biological Sciences*, **277**, 2361–9.

Teplitsky, C., Mouawad, N. G., Balbontin, J., De Lope, F., and Møller, A. P. (2011). Quantitative genetics of migration syndromes: a study of two barn swallow populations. *Journal of Evolutionary Biology*, **24**, 2025–39.

Thorup, K., Bisson, I.-A., Bowlin, M. S., et al. (2007). Evidence for a navigational map stretching across the continental U.S. in a migratory songbird. *Proceedings of the National Academy of Sciences USA*, **104**, 18115–19.

van Noordwijk, A., Pulido, F., Helm, B., et al. (2006). A framework for the study of genetic variation in migratory behaviour. *Journal of Ornithology*, **147**, 221–33.

Veen, T., Svedin, N., Forsman, J.R., et al. (2007). Does migration of hybrids contribute to post-zygotic isolation in flycatchers? *Proceedings of the Royal Society B: Biological Sciences*, **274**, 707–12.

Visscher, P. M., Hill, W. G., and Wray, N. R. (2008). Heritability in the genomics era—concepts and misconceptions. *Nature Reviews Genetics*, **9**, 255–66.

Visser, M. E., Caro, S. P., van Oers, K., Schaper, S. V., and Helm, B. (2010). Phenology, seasonal timing and circannual rhythms: towards a unified framework. *Philosophical Transactions of the Royal Society B: Biological Sciences*, **365**, 3113–27.

Weitkamp, L. A. (2010). Marine distributions of Chinook salmon from the west coast of North America determined by coded wire tag recoveries. *Transactions of the American Fisheries Society*, **139**, 147–70.

Wikelski, M., Martin, L. B., Scheuerlein, A., et al. (2008). Avian circannual clocks: adaptive significance and possible involvement of energy turnover in their proximate control. *Philosophical Transactions of the Royal Society B: Biological Sciences*, **363**, 411–23.

Wilson, A. J., Pemberton, J. M., Pilkington, J. G., et al. (2006). Environmental coupling of selection and heritability limits evolution. *PLoS Biology*, **4**(7), e216.

Winkler, H., and Leisler, B. (1992). On the ecomorphology of migrants. *Ibis*, **134**, Suppl. 21–8.

Yohannes, E., Lee, R. W., Jochimsen, M. C., and Hansson, B. (2011). Stable isotope ratios in winter-grown feathers of Great Reed Warblers *Acrocephalus arundinaceus*, Clamorous Reed Warblers *A. stentoreus* and their hybrids in a sympatric breeding population in Kazakhstan. *Ibis*, **153**, 502–8.

Zhan, S., Merlin, C., Boore, J. L., and Repper, S. M. (2011). The Monarch butterfly genome yields insights into long-distance migration. *Cell*, **147**, 171–85.

CHAPTER 13

The physics of animal locomotion

Christoffer L. Johansson, Florian T. Muijres, and Anders Hedenström

When a bar-tailed godwit, *Limosa lapponica*, takes off from the shores of Alaska on its autumn migration, it starts the longest direct flight known in the animal kingdom. When the bird lands 8 days later on the shores of New Zealand, it has been flying 11,000 km non-stop at an average speed of 16 m/s (Gill et al. 2009, Hedenström 2010). Such flights highlight two of the factors under selection in animal locomotion, in particular in relation to migration, namely being able to move fast and at a low energy cost. The ability to minimize costs and maximize movement speed is part of the biomechanical physics of animal locomotion. For any type of active movement, the animal is required to produce forces to overcome resistance and in many cases also gravity. For terrestrial locomotion, this is achieved by generating ground reaction forces, while in air and water by generating fluid dynamic forces. Due to scaling laws in physics, the speed that can be achieved and the cost of transport (COT) are correlated with the size of the animal. Moving across scales thus has consequences for our expectations regarding animal movement, including the occurrence of seasonal migrations, which may be limited by the speed and cost of locomotion. In this chapter we will discuss the physics behind animal movement, and focus on adaptations found in locomotion to generate forces and on mechanisms to reduce the cost of transport.

13.1 Scaling of speed and cost of transport

The speed and power required for locomotion are the two main factors determining animal behaviour during locomotion (Tokić and Yue, 2012). For animals moving from one location to another, speed determines the duration spent in transition and power determines the energetic cost of the journey. Being able to move fast will allow for seasonal migrations, such as among birds, whales, and fish (Chapman et al., Chapter 2; Lindström et al., Chapter 3), where different climates or resources separated by large distances can be utilized. A lower speed may make it impossible to complete a long-distance migration within an annual cycle, but may still allow for migratory life strategies, for example where juveniles and adults are separated by large distances (e.g. salmons, eels, and squids). At some point, when the speed of travel becomes too low, migration will not be feasible and these animals have to cope with their environment or use alternative strategies, such as hibernation or resting stages.

13.1.1 Cost of transport (COT)

Since all forms of movement come with costs for the organism, it is of considerable importance to understand the main components of these costs. Hence, animals move according to a number of different modes of locomotion (e.g. running, walking, crawling, swimming, and flying), characterized by their power consumption and speed. In order to compare the performance of different modes of locomotion, taxa, and sizes, we may consider the cost of transport (COT), which is the energy cost of transporting a unit weight of the animal over a unit distance, defined as

$$\mathrm{COT} = P/(mgU_{\mathrm{loc}}), \qquad (1)$$

where P is the power required for locomotion (mechanical or metabolic), mg is the weight of the

Animal Movement Across Scales. Edited by Lars-Anders Hansson and Susanne Åkesson.
© Oxford University Press 2014. Published 2014 by Oxford University Press.

animal (body mass, m, times acceleration due to gravity, g), and U_{loc} is the speed of locomotion. If P is the mechanical power required to push the animal along at speed U_{loc}, the ratio between P_{mech}/U_{loc} is equal to the average thrust (T), which is equal to the time-averaged drag in steady locomotion (D, see later discussion). Therefore COT_{mech} is the ratio between the horizontal D and the vertical force, mg, which is a ratio between two forces and therefore dimensionless (Pennycuick 1992). In aeronautics, the inverse of this ratio is the well-known lift to drag ratio (L/D), where lift (L) is the vertical aerodynamic force required for weight support. The mechanical forces involved in locomotion are often hard to obtain, which is one reason why researchers often measure metabolic rate of locomotion at certain speeds instead of determining COT_{mech}. The metabolic rate includes the cost of contracting the muscles involved in locomotion and the basal metabolic rate (BMR). Assuming that BMR is an insignificant proportion of the total power and that the (often unknown) energy conversion efficiency is constant across animal size and speed these measurements can be used to evaluate how the mechanical power of locomotion varies with speed. Schmidt-Nielsen (Schmidt-Nielsen 1972) collated data from the literature to create his famous graph shown in Fig. 13.1, where he plotted the minimum COT_{met} across the size of the animals. The graph shows a decreasing COT_{met} with increasing size, where, for a given size, swimming is the cheapest followed by flying and running. However, these results are influenced by BMR and in mammals COT has been suggested to be the same across different locomotory modes (Williams 1999).

In a series of isometrically scaled flying animals the mechanical power required to fly scales as proportions to $m^{7/6}$ (where m is the mass) and the cruising flight speed scales as proportional to $m^{1/6}$, which in accordance with Eq. (1) results in a mass-independent COT scaling (Pennycuick 1992). If COT, according to isometric scaling, should be independent of body mass, what causes this negative scaling found in real animals (Fig. 13.1)? One reason might be that the power required to fly does not increase as fast with body mass as prescribed by biomechanics (Hedenström 2008), but with a mass-exponent of about 0.78. Characteristic flight speed does not rise as fast as predicted either, but with a mass-exponent of about 0.1 (Alerstam et al. 2007).

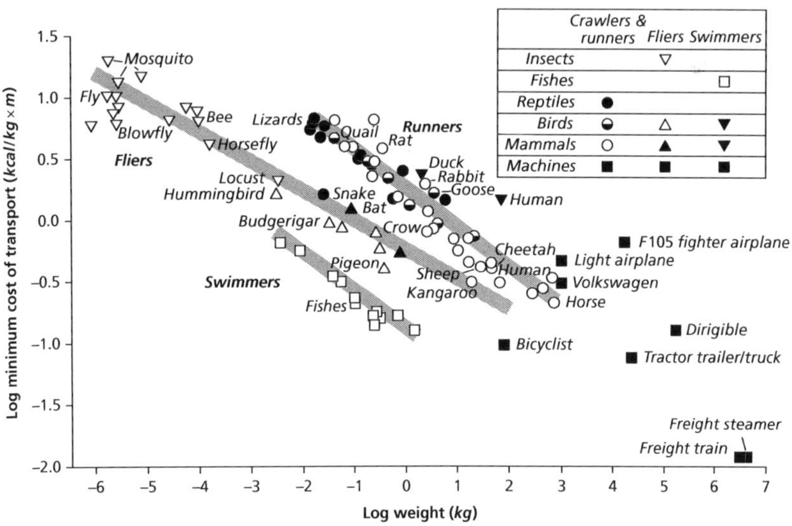

Figure 13.1 Scaling of cost of transport (COT) with weight for running, swimming, and flying animals. All groups follow a negative scaling with highest COT when walking/running/crawling and lowest when swimming. Examples are given of COT for specific taxa, as well as for some human ways of transport, such as cars, boat, trains, and airplanes. The table to the right shows different modes of transport in different taxa. Modified from Schmidt-Nielsen (1972). Reprinted with permission from the American Association of the Advancement of Sciences.

The net effect of these deviations from the expected slopes is a negative slope between COT and body mass in flying animals. For running and swimming animals scaling analysis also suggests that cruising speed scales proportional to $m^{1/6}$, which fits relatively well with measurements of locomotory speeds in diverse groups of animals (Bejan and Marden 2006).

The relationship of COT to body size is potentially important for our understanding of the development and evolution of migratory behaviour (Fig. 13.1). The lower the COT, the longer the expected potential migration distance is. A survey of migration distances, measured as number of body lengths, among animals that swim, fly and walk/run during migration indeed showed that swimmers and flyers perform the longest migrations (Alerstam et al. 2003), where in some cases, such as the arctic tern, *Sterna paradisaea*, the annual migration distance is limited only by the size of planet Earth (Egevang et al. 2010).

13.1.2 Speed of migration

It is, however, not only the energy cost (COT) that determines the propensity for long-distance migration, but also the time required to perform the round-trip between consecutive seasons. If the overall speed of migration, U_{migr}, i.e. the reciprocal of time required for migration of a specific distance, is the main selective agent we get a slightly different picture than for the COT. The overall migration speed can be written as

$$U_{migr} = U_{loc} P_{dep} / (P_{loc} + P_{dep}), \qquad (2)$$

where P_{loc} is power required for locomotion and P_{dep} is the rate of energy accumulation at stopovers. U_{migr} as a measure of migration performance accounts for the need of fuelling and actual movement, where a typical long-distance migration consists of several such fuelling-transport cycles. In animals using powered flight (i.e. flapping flight), migration speed is favoured by being small, while in soaring-flight migration, as well as in swimming and walking/running migration, there is an advantage of being large (Hedenström 2003). Also this result seems to be consistent with what we find in nature. The longest distance swimmers are found among large mammals such as whales and large sharks (Bonfil et al. 2005, Mate et al. 2007), while long-distance migrating self-powered flyers are mostly found in small to medium sized birds (Hedenström 2010; Hedenström and Lindström, Chapter 5).

Swimming and flying power can be divided into (a) maintenance power (basal metabolic rate, BMR), (b) power required to move the body through the fluid (parasite power), (c) power to move the wings/fins through the fluid (profile power), (d) power loss due to generating fluid dynamic forces (induced power) and, (e) power for oscillating the wings (inertial power). The different power components will affect the relation between locomotion speed and COT differently. A low basal metabolic rate (BMR) reduces the COT, but result in a reduced optimal speed. A low parasite and profile power reduces COT and result in a higher optimal speed. More efficient propulsion and low induced power, reduces COT, but in flying animals will result in a reduced optimal flight speed since lowering the induced power will lower the power curve mainly at low speeds and therefore shift the minimum of the P/U curve towards lower speeds.

In terrestrial locomotion, such as walking and running, the power can be divided into external power, which is related to changes in potential and kinetic energy, and internal power, which is related to the acceleration and rotation of elements within the body (e.g. legs) (Cavagna and Kaneko 1977). The relationship with speed will depend on energy storing mechanisms, such as tendons and exchange between kinetic and potential energy, and adaptations to reduce the cost of moving the limbs. In addition, different gaits, patterns of movement, often show distinct optima at different speeds and many species show a decreasing COT with increasing speed. In the following section we will dissect different modes of locomotion in more detail with respect to adaptations that affects speed and minimize the cost of transport.

13.2 Moving across a surface

In human engineering wheels are by far the most preferred terrestrial locomotory system, while in nature wheels for locomotion are essentially absent (Biewener 2003), except among fantasy animals like, for example, the yellow toy duck on the back cover. There are two main explanations for this: first rotary systems are difficult to construct in biological systems (only found in bacterial flagella, see later discussion) and second, wheels are a bad choice when moving in uneven terrain. Moving across a surface may be done in a multitude of ways, including pushing against a layer of self-generated slime (e.g. snails; Chan et al. 2005), crawling in a serpentine motion using ground friction to push forward (e.g. snakes) and pulling against attachments to the substrate (e.g. earthworms). However, if moving relatively far or fast the most common way of travel involve legged locomotion such as walking, running and hopping. Since natural selection seems to have favoured legged locomotion, especially for faster transport, we will focus on this aspect of terrestrial locomotion.

As soon as animals evolved a terrestrial lifestyle the limbs needed to support their body weight, which called for a more robust skeletal support system. These first explorers of land had a rather sprawled posture (Pierce et al. 2012), i.e. with the legs extended laterally, which is reflected in contemporary amphibians and reptiles. Locomotion with a sprawled posture in contemporary tetrapods involves a substantial lateral motion of the body. As the demand for larger weight or higher speed increased a more upright leg posture was favoured by natural selection, since it reduces the moments of the ground reaction force (see later discussion) and allows for a unidirectional swing of the legs during the swing phase.

The relative moments exerted by the limb extensor muscles compared to the moments generated by the ground reaction force is known as the effective muscle advantage (EMA; Biewener 1989). As mass increases, scaling laws dictate that the cross sectional area of the muscle, which is proportional to the force that can be generated, increases as $m^{2/3}$, which suggests that relatively less muscle force is available as size of the animal increases. Smaller animals generally have a more crouched i.e. bent (with smaller angles between the leg bones) or sprawling posture than larger animals (Fig. 13.2). This crouched posture results in a constant EMA, which suggests that the animal needs to compensate the increased force demands with larger muscle mass, as size increases (Fig. 13.2) (Reilly et al. 2007). Larger animals, and particularly large mammals, are instead thought to have evolved higher EMA in response to the otherwise higher force requirements as size increases (Biewener 1989). A higher EMA is achieved by straighter limbs, which reduce the moments generated by the ground reaction force. A notable exception to this scaling is found in the hopping kangaroos, which show a constant EMA with increasing size (Bennett and Taylor 1995). Although this results in higher force demands, a bent leg posture is considered necessary to allow for effective hopping (Bennett and Taylor 1995) and the energy storing tendons reduce the work required by the muscles (see later discussion).

When increasing speed, the ground reaction forces increase and a selection towards a higher EMA is therefore expected. Interestingly, the largest land animal, the African elephant, decreases EMA when increasing speed (Ren et al. 2010). EMA increases until the leg is completely straight, which seem to happen at the size of a horse (Ren et al. 2010). The reason behind the decreasing EMA in running elephants is not clear, but a likely explanation to the crouched position when running is that it reduces the vertical fluctuation in the centre of mass (CoM). However, running in a crouched position has a price in the form of increased muscle power requirements, which is manifested as high metabolic rate in fast walking/running elephants (Langman et al. 2012). This is linked to the scaling results showing that above a certain size, terrestrial animals do not increase their speed further, but may instead decrease their speed due to higher ground reaction forces as speed increases (Bejan and Marden 2006).

13.2.1 Factors influencing speed on land

It is widely known that the fastest animal is the cheetah, *Acinonyx jubatus*, with top speeds of 104 km/h, but it is less well known that it rapidly gets exhausted (Sharp 1997). When it comes to endurance in running, only one animal stands out: the pronghorn antelope, *Antilocapra americana*, of North America. The antelope has a top speed of 100 km/h, nearly matching the cheetah, and can perform sustained speeds above 65 km/h (Lindstedt et al. 1991). The speed of a walking, running, or hopping animal is determined by the product of the stride length and the stride frequency. Maximum stride frequency is ultimately determined by leg swing speed, which is limited by the muscle contraction speed or the maximum force required to accelerate the mass of the appendage (Hudson et al. 2012).

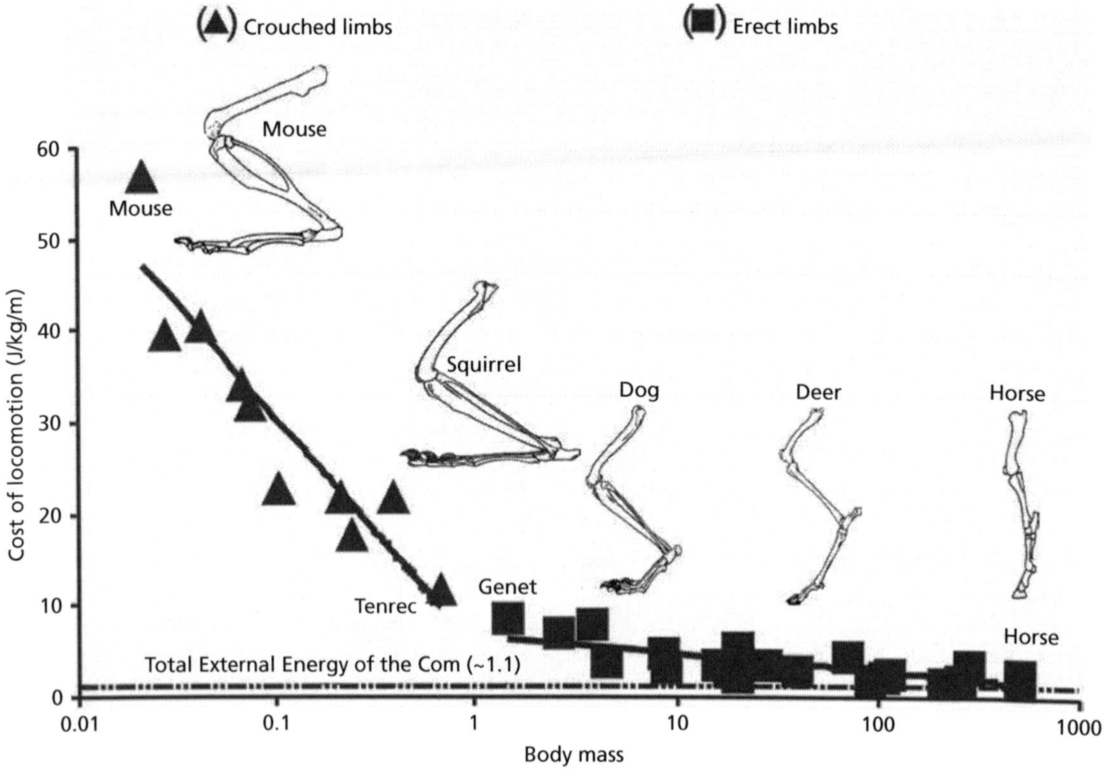

Figure 13.2 Cost of transport (COT) against body mass (kg) for walking and running animals, showing the effect of limb posture, crouched (plantigrade) versus erect, as shown by the limb posture above the graph. From Reilly et al. (2007), reproduced with permission from Elsevier.

However, the force a muscle can generate decreases with the contraction speed of the muscle (the highest force is during extension) (Edman 1988) and a faster contraction thus requires more muscle mass. In addition, the moment arm the muscle acts on at a joint is inversely proportional to the force required to move the leg. At the same time the output speed of the leg is inversely proportional to the muscle moment arm. In other words, to reduce the force required by the muscle to perform a specific motion, the moment arm should be long, but this results in a lower output speed. Hence, not surprisingly, to be fast requires more muscle output. Furthermore, when increasing the locomotory speed the duty factor (the time a limb is in contact with the ground) decreases, which results in higher force requirements for a given change in momentum of the body.

When walking, the step length is essentially determined by the leg length and the angular excursion of the leg. When running, which unlike walking contains an aerial phase, the step length can be increased by extending the aerial phase of the stride, but in quadrupeds also by incorporating the spine in the effective leg length by flexing the spine in synchrony with the stride. Some of the fast running animals, such as the cheetah, show substantial flexion of the spine during fast running (Hildebrand 1959). This may also function as a spring allowing for transfer of energy between the impact of the front legs to the hind leg (Alexander 1988). The horse, and probably the pronghorn, use relatively little flexion of the spine during running (Hildebrand 1959). The stride length is instead almost entirely dependent on the leg length and the length of the aerial phase of the stride. Also in hopping, which is a gait employed by, for example, kangaroos and birds, speed enhancement is achieved at almost constant frequency, by extending the aerial phase (Kram and Dawson 1998).

13.2.2 Factors influencing COT on surfaces

If the speed is determined by how fast the animal can swing its legs back and forth and how long strides it can take, which factors are then determining the cost of transport? When moving on land, a main energy consumer is to overcome the gravity, whereas aerodynamic costs are generally ignored. Animals moving on a surface under water have much lower ground reaction forces, but have to overcome a substantial drag due to the higher density of water compared to air (Martinez et al. 1998). In addition to overcoming gravity and drag, power may be required for accelerating the limbs back and forth during the stride. In other words,

$$W_{tot} = W_{mech} + W_{int}, \quad (3)$$

where W_{tot} is the total work and the mechanical work, W_{mech}, is the work associated with changes in potential and kinetic energy in the centre of mass (CoM), and W_{int} is the internal work associated with movements of limbs, flexion of the spine, etc. The swinging of the leg can be likened to a pendulum motion where the natural frequency of the pendulum is inversely proportional to its length. The length of the pendulum is defined as the length between the CoM of the pendulum and the point of rotation. Swinging the leg at the natural frequency (resonance frequency) or below this frequency does not require any energy (except for compensating for fluid dynamic forces), but any increase in frequency above the natural frequency requires work to slow and accelerate the leg during the swing (Srinivasan and Ruina 2006, Srinivasan 2011). Operating at a high natural frequency selects for having the mass of the leg as close to the hip/shoulder as possible (Myers and Steudel 1985, 1997). This is commonly found in cursorial animals, for example horses, where the distal parts of the legs are often light and without muscles, while the proximal parts have big muscles.

Generally speaking, walking and running animals minimize mechanical energy losses by exchanging potential and kinetic energy and/or by storing and recovering energy in elastic materials (Srinivasan 2011). Walking has been modelled as an inverted pendulum, where the animal converts kinetic energy of the CoM at the beginning of the stance to potential energy at mid stance and back to kinetic energy at the end of stance as the CoM 'vaults' over the stiff leg (Dickinson et al. 2000). When walking the CoM should thus be at its lowest position in the beginning and end of the stance and at its highest position during mid-stance. In a perfect motion the energy losses would be zero. However, bipedal walking does not show the predicted vertical motion of the CoM. Nor do force plate measurements (registering the force and direction of force) during bipedal walking show the simple relationship predicted by the inverted pendulum model. Measurements show a double peak in the ground reaction force instead of the peak during mid-stance predicted by the model (Geyer et al. 2006). Consequently, it has been recognized that walking is better modelled as legs with springs, alternatively as legs having knees. Such models capture the double peak in ground reaction force found in bipedal walking (Geyer et al. 2006, Srinivasan 2011). Having legs with springs allow for energy storing during walking, making the models more similar to the ones used for running.

Running is often modelled as a mass-spring system, where the kinetic energy is used to load the spring by deforming it during the first half of the stance phase after the initial impact of the leg. The stored energy is then released during the second phase of the stance, assisting the acceleration of the CoM (Srinivasan 2011). When running, CoM is thus at its lowest position at mid stance. Energy storage in tendons may reach a high efficiency (97% in resilin; Weis-Fogh 1960, Chen et al. 2012) allowing for efficient running with energy savings approaching 60–70% (Reilly et al. 2007).

An alternative view to the traditional models of terrestrial locomotion has recently been explored by various researchers (Ruina et al. 2005, Srinivasan and Ruina 2006). The proposed model is based on the minimization of collision effects during the impact between the legs and the ground. The model of Lee and colleagues (Lee et al. 2011) suggests large differences in collision, which correspond to the COT_{mech}, between different gaits. The model is attractive since it gives a physical basis for the COT_{mech} in walking and running animals. The power required to move is calculated as the scalar product of the force acting on the CoM and the velocity of the CoM. However, COT_{mech} shows low correlation with the COT_{met}, but since their model only accounts for the cost of moving the CoM, they suggest that adaptations in the legs, with muscles and tendons elastically storing energy, could explain the discrepancy with COT_{met} data (Lee et al. 2011).

A realistic model of terrestrial legged locomotion should take into account the mechanical properties of tendons and muscles to properly account for the work and energy savings that can be achieved. Springs in terrestrial locomotion may have three main functions (Roberts and Azizi 2011): firstly, to store energy during one phase of the stride to release it at a later phase, and secondly, to allow for a higher power output of the muscle–tendon unit by allowing the muscle to work at a lower power, storing energy in the tendon that is later used to generate a higher power output for a shorter time period. This mechanism is particularly important during jumping. A final function may be to allow for brief periods of high loads, for example during impact after landing, where stretching of a tendon allows for a lower contraction rate of the muscles responsible for the slowing down of the body. This allows the muscles to operate with a slower contraction rate, which increases the maximum force that can be generated. This in turn lowers the necessary muscle recruitment and ultimately the energy consumption of the muscles and hence the COT. Muscles and tendons work in series and their function is therefore closely linked. In fact, models suggest that the optimal range of stiffness characteristics of tendons is rather narrow and, for example, having too compliant tendons may actually increase the power required to move (Srinivasan 2011). When considering energy storing in animals we often think of tendons with elastin in vertebrates and resilin in insects, but the capacity of the muscles as springs, allowing for a higher level of control, is becoming increasingly appreciated (Roberts and Azizi 2011). In addition, recent findings suggest that the mechanical properties of the tendons may be altered by the contraction of the muscle, which opens up for a dynamic control of the stiffness of tendons and hence control of the energetics and mechanics of locomotion (Azizi and Roberts 2009).

Energy savings by tendons and vaulting exchange of potential and kinetic energy of centre of mass (CoM) are smaller relative to the total cost of locomotion in smaller animals compared to walking gaits in larger animals (Reilly et al. 2007). The reduction in COT with increasing size in crouched animals (Fig. 13.2) is seen as a result of a decrease in stride frequency (Reilly et al. 2007) and an increase in stride length, since COT is inversely correlated to the stride length (Kram and Taylor 1990). There is consequently weak selection against a crouching/sprawling leg posture in small animals, and if there is positive selection for crouching, associated with for example manoeuvring, this would be likely to maintain a crouching posture despite the lower effective muscle advantage (EMA). In large animals elastic storing and kinetic to potential energy storing makes up a significant proportion of the gross COT, and it has been suggested that without these savings, COT would increase rather than decrease with size in terrestrial locomotion (Reilly et al. 2007). However, there may be an upper limit in size to benefit from these savings since recent estimates of scaling of COT in elephants show that COT is constant across size, suggesting that contrary to what one might predict based upon interspecific scaling, larger elephants do not show a lower COT (Langman et al. 2012).

One characteristic of legged locomotion is that the footfall sequence and relative timing of support by the limbs change with speed. These gaits, which are discrete combinations of speed, duty factor, frequency and stride length, or time lags between limbs (e.g. Abourachid 2003), result in different COT across speed (Fig. 13.3; e.g. Nudds et al. 2011). The transition from one gait to another, for example from walking to running, in bipeds has been motivated by the result that the inverted pendulum model requires higher than achievable centripetal accelerations when speed increases (Srinivasan 2011). However, this does not explain why animals do not use other non-flight phase gaits (e.g. horizontal CoM motion) at high speeds (Srinivasan 2011). Instead the walk-run transition is likely to be motivated by energetic considerations, which is supported by results from springless models and models with knees that are able to predict the gait transition well based on power requirements (Srinivasan 2011).

Switching between gaits has experimentally been shown to result in more beneficial COT when changing speed, and animals seem to prefer to move at energetically optimal speeds in nature (Watson et al. 2011). The optimal gait, minimizing COT, is size

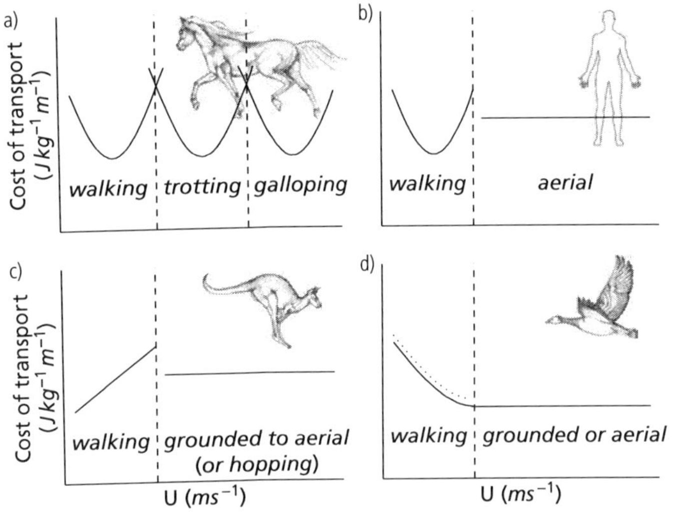

Figure 13.3 Gait transitions at increasing speed often lead to reduced cost of transport (COT), and a variety of gait-related COT patterns across speed (U) can be found in animals. Examples are (a) horse; (b) human, donkey, and camel; (c) ostrich and kangaroo; and (d) barnacle goose (dotted line), and platypus and small mammals (solid line). In (c), grounded (duty factor above 0.5, i.e. no aerial phase) and aerial running refers to the ostrich, and hopping refers to the kangaroo. In (d), the type of running (grounded or aerial) is not known. From Nudds et al. (2011), reproduced with permission.

dependent with smaller animals having an optimum in running gaits, while larger animals have an optimum gait when walking (Reilly et al. 2007, Nudds et al. 2011). At some size between a dog and a horse, walking becomes cheaper than running. In human locomotion, walking minimizes COT (Fig. 13.3), while large ratites, such as ostrichs, *Struthio camelus*, of similar size, show the lowest COT when running at the highest recorded speeds (Watson et al. 2011). One explanation for this may be that these large birds elastically store more energy in their legs than most other animals (Rubenson et al. 2011) making running more efficient. Unlike in most running animals, power during hopping stays almost constant over a large range of speeds (Baudinette et al. 1992, Kram and Dawson 1998), resulting in a decreasing COT with increasing speed, although these results may underestimate the effect of air resistance (Baudinette et al. 1992). This remarkable result is likely due to a much larger energy storage capacity of kangaroo tendons than in other mammals (Bennett and Taylor 1995).

13.3 Moving through fluids

Flying and swimming animals need to generate fluid dynamic forces to propel themselves forward. Since fluids, such as air and water (Box 13.1), have a much lower resistance to deformations than solids, fluid dynamic propulsion differs fundamentally from propulsion at a solid surface (e.g. terrestrial locomotion). However, within the speed range of animal locomotion, both air and water behave as an incompressible fluid making the fluid dynamics of flying and swimming strikingly similar. Here, we will discuss how fluid dynamic forces are generated and what propulsion mechanisms are used for locomotion. We will also discuss factors influencing force and COT in flying and swimming animals.

The most important difference between air and water with respect to animal locomotion is the difference in the relative fluid density with respect to that of the animal. In water the ratio between animal and fluid density is in the order of one, making most aquatic animals close to neutrally buoyant, while for flying animals is in the order of 10^3. Consequently, to swim steadily, a neutrally buoyant animal needs to generate only a horizontal force, thrust, to counteract fluid dynamic resistance, drag, on the body. In steady flight, on the other hand, the animal needs to generate an upward force equal to its weight, which is often significantly larger than the drag force experienced by the flying animal. The three main fluid dynamic force generators in nature are the flagella and cilia as found in the smallest swimming animals, the jet propulsion system as seen in jellyfish and squids, and fluid foils, such as wings and fins.

13.3.1 Flagella and cilia

The smallest actively swimming animals operate at very low Reynolds number, Re, in the order of 1 to 10 (Box 13.1). Due to the relatively high viscosity,

Box 13.1 Fluid forces and Reynolds number

A fluid is a substance that deforms plastically when subjected to a force. An animal that moves through a fluid is subject to and may utilize fluid dynamic forces of different origin; viscous forces, pressure forces, and added mass forces. Viscous forces relate to fluid viscosity, which is a measure for the resistance of a fluid to deformation. When an object deforms a fluid by moving through it, the fluid will exert a force, viscous friction drag, on that body. Due to fluid viscosity, an object moving through a fluid will drag along part of that fluid. The amount of entailed fluid is expressed in the boundary layer, as it is defined as the fluid layer around the object that moves at a speed of at least 95% of the animal's speed. For a steady moving object, dragging along a boundary layer of fluid does not result in an added mass force, but when the object accelerates, the entailed fluid needs to be accelerated too. This results in the added mass force

$$F_a = \rho\, V_a\, a, \qquad (1)$$

where V_a is the volume of fluid entailed by the body and a is the acceleration of the body (or boundary layer fluid).

Pressure forces arise from velocity changes in the fluid surrounding an object. This can best be explained using Bernoulli's principle: A change in the velocity of an inviscid fluid, results in a change in the pressure within that part of the fluid such that $p + \frac{1}{2}\rho U^2 = constant$, where p is local fluid pressure, ρ is fluid density, and U is the fluid velocity. The *constant* is equal to the total pressure. Thus, by altering the speed of a fluid relative to the body, an animal can manipulate the pressure distribution around its body, which results in a net pressure force on that body.

Whether pressure and viscous forces are dominant for a flying or swimming animal depends on both the size of the animal and on the characteristics of the fluid. How the ratio between the pressure force and the viscous force scale on a body is expressed by the Reynolds number (Re),

$$Re = Ul / \upsilon, \qquad (2)$$

where U is the speed at which the animal moves through the fluid, l is the length of the animal (or wing/fin) in the direction of movement, and υ is the kinematic viscosity of the fluid, a measure of the viscosity relative to the density of the fluid.

Under normal conditions υ of both water and air can be assumed constant and Re increase with both the size and the speed of the animal. For a small, slow-moving animal, viscous forces govern the force dynamics, while for large and fast-moving animals, pressure forces govern the force dynamics. This means that a small animal experiences the same fluid as much thicker and stickier (higher relative viscosity) than a larger, faster-moving animal. The Re range relevant for animal locomotion ($Re < 1 \times 10^8$) can roughly be divided into three sections. At low Re ($Re < 30$), it is virtually impossible to produce inertia forces as viscosity directly dampens any significant fluid acceleration. This is closely linked to the fact that the smallest stable fluid eddy (vortex) has a Re ~ 30 (Saffman 1995).

At intermediate Re ($30 < Re < 10^6$), animals are able to generate significant inertial forces, which results in the generation of patches of rotating fluid, called eddies or vortices. These inertial forces become relatively more dominant with increasing Re, but throughout the intermediate Re regime viscosity has still a strong effect on the overall flow characteristics. Friction forces are not negligible, and viscosity has a stabilizing effect on the flow dynamics making the flow around bodies largely laminar at intermediate Re (Anderson 2007). For $Re > 10^6$, viscous forces and thus viscous drag can largely be ignored. As a result also the stabilizing effect of the viscous forces on the flow is less and flow separation is more easily triggered, i.e. the flow cannot follow the contours of the animal, but instead forms large vortices, resulting in high pressure drag. In addition, the flow around the animal may become turbulent, which generally increases pressure drag, but interestingly it also functions as an alternative source for fluid flow stabilization, reducing detrimental characteristics such as flow separation (Anderson 2007). At the transition region between the medium and high Re regime ($Re \sim 5{*}10^5$), it is notoriously difficult to predict flow dynamics and the resulting force production, as small changes in flow conditions can trigger relatively large changes in the flow character (Spedding and McArthur 2006). Still, most flying and swimming animals operate within this Re regime, and possibly they can exploit the benefits of both the relatively stable flow dynamics at low Re and the increased efficiency as a result of reduced friction drag at the high Re regime. The Re thus determines the relative usefulness of different force generating mechanisms.

a fluid at these Re scales behaves very different from how we experience water and air, and animals such as spermatozoids experience their surrounding fluid more similar to how we would experience moving through mud. As a result, these animals cannot use fluid inertia to propel themselves, but rely exclusively on viscosity-based friction forces produced by flagella and cilia (Box 13.1). Flagella and cilia are hair-like structures that are moved in a wave-like pattern, i.e. the alternating extension as in most cilia (power and recovery stroke), a planar sinusoidal wave as in the flagella of spermatozoids, or a corkscrew motion as in the flagella of bacteria (Chia et al. 1984). The viscous drag acting on the moving cilia or flagellum results in a relatively inefficient net thrust production, as only a small portion of the total friction force can be used for propulsion (Brennen and Winet 1977, Dillon and Fauci 2000).

13.3.2 Jet propulsion

In swimming animals, jet propulsion has independently evolved several times during evolution, but is not found in flying animals. Well-known examples are jellyfish (Dabiri et al. 2010), squid (Bartol et al. 2009a, Anderson and Grosenbaugh 2005), salps (Sutherland and Madin 2010), and scallops (Carsen et al. 1996). Jet propulsion functions by accelerating fluid through a narrow orifice or siphon in the opposite direction to the desired swim direction, resulting in the generation of a fluid jet, and a thrust force (T). Virtually all natural jet propelling animals have a pulsed jet, with a fluid intake phase where a chamber is filled with fluid, and an exit phase in which the fluid is exerted at high speed through a siphon to produce thrust. During the intake phase the fluid decelerates from some free-stream velocity to zero velocity (with respect to the animal), and then accelerates to a jet speed U_{jet} during the propulsion phase. The thrust produced during the jet phase, is similar to that of a rocket engine as in both systems the jet fluid accelerates from zero speed to the jet speed. Therefore, thrust force (T) in natural jets can be modelled as

$$T = \rho\, A\, U_{jet}^{2}\,, \qquad (4)$$

where A is the cross sectional area of the jet and U_{jet} is the velocity of the jet relative to the animal. To model the complete pulsed cycle jet dynamics of animals, one must incorporate also the intake phase, and when assuming that the intake phase does not produce significant thrust (or drag), the whole-cycle net thrust is

$$T_{net} = \tau\, \rho\, A\, U_{jet}^{2}\,, \qquad (5)$$

where the duty cycle $\tau = t_{jet}/(t_{in} + t_{jet})$ is defined as the temporal ratio between the jet phase (t_{jet}) and the total cycle period ($t_{in} + t_{jet}$). These steady-state thrust estimates are based on average velocities and due to the temporal dynamics of the propulsion mechanism (Anderson and DeMont 2000) and fluid dynamics losses in the jet wake (Gharib et al. 1998), real jet performance in swimming animals is lower.

13.3.3 Wings and fins

A foil is a relatively flat solid object that, when moved through a fluid, can be used as a propulsion system. The fluid foil is the most commonly used propulsion system in nature, as virtually all actively flying animals use fluid foils, where they are called airfoils or wings, and many swimming animals use so-called hydrofoils. Fins, as used by most fish and marine mammals, as well as webbed feet of birds and frogs are all hydrofoils. When a foil is moved through a fluid, e.g. by flapping or beating it, it generates two forces, a lift force (L) and a drag force (D). Drag is defined as the force parallel and opposite to the direction of motion of the fluid foil, while lift is defined as the force perpendicular to the direction of motion. Lift consists almost solely of pressure forces (Box 13.1), as it is a result of a fluid pressure difference between the two sides of the foil perpendicular to the motion. Drag, on the other hand, consists of pressure drag due to a pressure difference between the fluid in front and behind the foil and viscous friction drag as a result of the fluid moving along the surface of the foil. In addition, when a fluid foil produces lift, air is deflected causing a downwash, which tilts the lift force vector backwards, resulting in streamwise directed lift vector component. This force vector is therefore by definition a drag force, and is called the lift-induced drag (Spedding and McArthur 2010). Lift and drag in fluid foils are generally modelled as

$$L = \frac{1}{2}\rho AU^2 C_L, \quad (6)$$

$$D = \frac{1}{2}\rho AU^2 C_D, \quad (7)$$

respectively, where A is a characteristic surface area of the foil; C_L and C_D are the dimensionless lift and drag coefficient, respectively, both depending strongly on the orientation of the foil. Foil orientation is primarily expressed in the angle-of-attack (α), defined as the angle between the foil surface and the movement vector. Generally, both lift and drag increase linearly with increasing α up to a critical angle of about 20° (Fig. 13.4a; see also colour plate section). For angles-of-attack beyond this critical angle the fluid flow cannot follow the contours of the foil and flow separation occurs. The onset of flow separation, referred to as stall, results in a sudden increase in drag, and depending on the shape of the foils lift can either continue to increase or decrease. For all airfoils, though, when approaching an angle-of-attack of 90° lift will go down to 0, while drag is maximal (Fig. 13.4a). Details about the C_L–α and C_D–α curves depend on Re as well as various wing shape factors such as thickness, camber (curvature of the foil profile), and aspect ratio (the ratio between span and cord length) (Anderson 2007). Natural flyers and swimmers use fluid foils throughout the complete range of angles-of-attack to produce forces, ranging from pure drag based propulsion at α~90° as in frog paddling (Johansson and Lauder 2004), to lift based flight with low drag at small α.

13.3.4 Factors influencing speed when moving in air and water

The fastest actively flying bird recorded so far, the common swift (*Apus apus*), is capable of flying at 30 m/s (Henningsson et al. 2010), while the fastest swimming animal is probably the sailfish or blue marlin, *Makaira nigricans*, swimming faster than 20 m/s (Block et al. 1992). It is not a coincidence that both species use lift-based foil systems. The maximum speed of animals using fluid dynamic propulsion is limited by the drag of the animal and the thrust the animals can generate. To be able to move fast thus requires low drag and an ability to generate large propulsive forces. However, the relationship between swim and flight speed and the amount of thrust that can be generated varies across Re, and between the different propulsion systems.

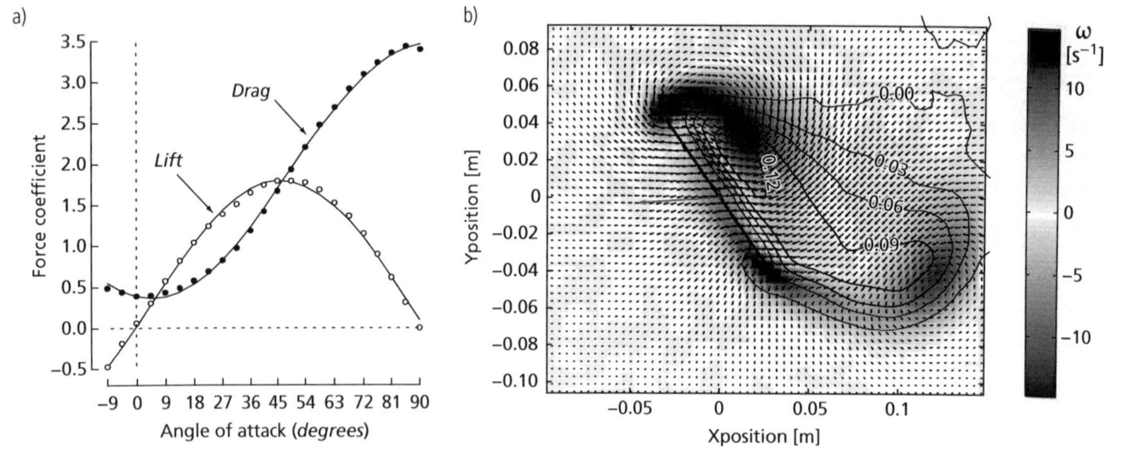

Figure 13.4 Steady-state lift and drag coefficients for a revolving model fruit fly wing at Reynolds number (Re) ~140. (a) At this low Re, drag forces are relatively high resulting in lift-to-drag ratios (= $1/COT_{mech}$) larger than 1 only at the angles-of-attack range of 9°–45° (from Dickinson et al. 1999). Maximum aerodynamic efficiency occurs at post-stall angles-of-attack, (b) for which an attached leading-edge vortex occurs on the revolving wing. From Poelma et al., 2006 reproduced with permission from Springer Science + Business Media. (See Plate 4).

The drag on a body can be modelled using Eq. (7), and the corresponding drag coefficient varies both with body shape and Re, and consequently with speed, due to change in the relative contribution of viscous and pressure forces (Box 13.1). At low Re, friction drag is the primary source of drag and reducing the wetted area, i.e. the body area in contact with the fluid, is the best strategy to minimize drag.

At the intermediate Re range the relative contribution of friction and pressure drag depends strongly on the combination of Re and body shape. One reason is that with increasing Re the boundary layer can change from laminar to turbulent (at Re ~ $5*10^5$), which will cause abrupt changes in the drag coefficient. The effect of these changes is strongly shape dependent; circular shapes experience a decrease in drag since a turbulent boundary layer will have a smaller separated flow region behind the animal than a laminar boundary layer. On the other hand, flat plates in line with the flow experience an increase in drag due to the increased friction drag resulting from a turbulent boundary layer. For streamlined objects this transition is more similar to the intermediate between these extreme cases, resulting only in a plateau or a local drag minimum at Re ~ 10^6 (Hoerner 1965).

Streamlining is a mechanism for reducing drag at both medium and high Re, as it reduces the pressure difference between the front and rear of an object, and is a commonly found adaptation in both flying and swimming animals. The amount of streamlining for an object is described by the fineness ratio (length/diameter ratio) with optimal streamlining thought to occur at a fineness ratio of ~4.5 (Hoerner 1965). However, an analysis of fineness ratio in whales have suggested a positive relationship between optimal fineness ratio and Re, with the largest whales having an fineness ratio of 8 (Ahlborn et al. 2009). The reason is that the relative contribution of pressure compared to friction drag increases with Re making a relatively longer body more beneficial since it increases friction drag but reduces pressure drag (Ahlborn et al. 2009). When looking at flying animals moving at high Re, such as most birds, they are generally streamlined. Other large birds, for example swans, *Cygnus* spp., however have long necks, making their bodies less streamlined, indicating a higher drag. Interestingly though, one of the few studies that have measured the drag of bird bodies suggested that at high Re a long neck actually results in a lower drag coefficient (Lovvorn et al. 2001). In addition to adaptations in body shape, drag can also be reduced by manipulating the material properties of the wing and body surface. One example is the scales of sharks, which reduce drag (Bechert et al. 1997, 2000; Chen et al. 2012). Shark scales have small ridges in line with the body axis that control the spreading of turbulence in the boundary layer and thereby reduce the drag (Bechert et al. 2000). Whether feathers and fur can have a similar function is still an open question.

The other speed-limiting factor is the propulsive system. The way animals can increase thrust production depends strongly on the type of locomotion system. Animals operating at low Re use flagella and cilia as propulsion systems, and are thus restricted to using viscous forces for thrust production (Brennen and Winet 1977, Dillon and Fauci 2000). Due to the relatively high fluid viscosity at low Re, the boundary layer around these animals is relatively thick. This results in large added mass forces (Box 13.1), and causes the flagella and cilia to be submerged in the boundary layer. Both effects reduce the ratio between thrust and drag, and thus the maximum attainable speed. At higher Re, jet propulsion becomes a feasible alternative to generate thrust. Natural jet propulsors can be divided into three groups: animals that use the same siphon for intake and exit (e.g. jellyfish; Dabiri et al. 2010), animals that have a separate intake and exit siphon that are oriented in a similar direction (e.g. squid; Bartol et al. 2009a, Anderson and Grosenbaugh 2005) and animals that have separate intake and exit siphons facing opposite directions (e.g. salps (Sutherland and Madin 2010) and scallops (Carsen et al. 1996)). In the first two groups filling the fluid chamber during the intake phase results in negative thrust, reducing the cycle average thrust. For the third group, actively filling the fluid chamber results in net positive thrust. Alternatively, animals with forward facing intake siphon can use the momentum resulting from forward swimming to passively fill the fluid chamber. In both these cases the net thrust is increased relative to the first two groups, which would suggest higher relative swim speeds. Since jet propulsion in animals involves fluid accelerations that are relatively high and the fluid mass used for jet propulsion is often similar if not greater than the mass of the animal, added mass effects reduce the effective jet thrust significantly in animals such as jellyfish (Dabiri 2005), thereby reducing the maximum attainable speed.

To produce thrust using a fluid foil, most animals flap their wings or fins up and down or side to side, which enables adjusting the lift and drag forces into the direction of movement. The flapping motion causes the foil to constantly accelerate and decelerate, and consequently added mass forces can be a significant part of the overall fluid dynamic forces (Dabiri 2005). Added mass forces are relatively large in most swimming animals due to the high density of water, and in most small flying animals such as insects, because of their high flap frequencies and relative thick boundary layers at low Re (Dabiri 2005). Added mass forces can, however, often be ignored in larger, faster moving animals, such as birds and bats in cruising flight (Hedenström et al. 2006). Where relevant, added mass forces are mostly considered a cost factor that reduces net force production, but in some cases they increase thrust forces, such as in swimming beetles (Nachtigall 1960, Daniel 1984).

Fluid foil propulsion can roughly be divided into drag-based and lift-based propulsion, although many species use both, sometimes even within the same propulsive stroke (Johansson and Norberg 2003). Pure drag-based propulsion, also called paddling or rowing, is mainly found in secondary swimmers such as frogs (Johansson and Lauder 2004) and muskrats (Fish 1984). Similar to jet propulsion in animals, paddling can provide relatively high propulsion forces at low swimming speeds, making this mechanism useful for fast accelerating from rest, such as in escape manoeuvres. However, it is difficult to sustain the fast steady swimming required for migration using drag-based paddling. In addition, similarly to jet propulsion in many animals, the paddle stroke is followed by a recovery stroke that mostly generates negative thrust. Hence, drag-based paddling is rarely seen in animals moving large distances.

To produce weight support, flying animals have an asymmetric wingbeat, where the downstroke and upstroke are functionally different. In all flying animals, the downstroke contributes to weight support, but depending on the relative thrust requirements the contribution of the upstroke to both thrust and weight support varies. As a result the aerodynamic function of the upstroke differs between species, but also throughout the flight speed range of an individual flying animal (Fig. 13.4; see also colour plate section). At low flight speeds thrust requirements are low (in hovering flight no thrust is required), and most insects and hummingbirds are specialized to operate at these low flight speeds. They have an almost symmetric, horizontal, wingbeat in which they flip their wings upside-down during the upstroke (supination), generating weight support during both the downstroke and upstroke. Thrust is a result of drag differences between downstroke and upstroke (Dudley 2002, Altshuler and Dudley 2003, Warrick et al. 2005). To a lesser extent, this mechanism is found in both birds and bats when flying slowly (Hedenström et al. 2007, Johansson et al. 2008, Crandell and Tobalske 2011).

Small birds that are able to fly at a large range of flight speeds from hovering to migratory relevant speeds and have an inclined stroke plane generally spread their wing feathers during the upstroke, enabling the air to flow through the gaps between the feathers (Norberg 1975, 1985). This causes the wing to be largely aerodynamically inactive, and thus the wing upstroke is effectively a recovery stroke (Muijres et al. 2012a). In steady forwards flight, most birds can use the upstroke to add to weight support, which comes at the cost of negative thrust production (Pennycuick 2008; Johansson and Hedenström 2009). Steadily flying bats, on the other hand, move their wings upwards at a negative angle-of-attack during part of the upstroke, resulting in a downward and forward directed flight force (Fig. 13.5; Hedenström et al. 2007; Johansson et al. 2008; Hubel et al. 2010, 2012). The forward force adds to thrust, while the downward force is assumed to be a cost factor resulting from the thrust production (Muijres et al. 2011a).

Flying animals that operate at low range of intermediate Re (Re = 100–1000) are almost exclusively insects, although small hummingbirds can also be added to this group as 'honorary insects' (Alexander 2004). At this Re range, the high viscosity causes the flow around an object to be relatively

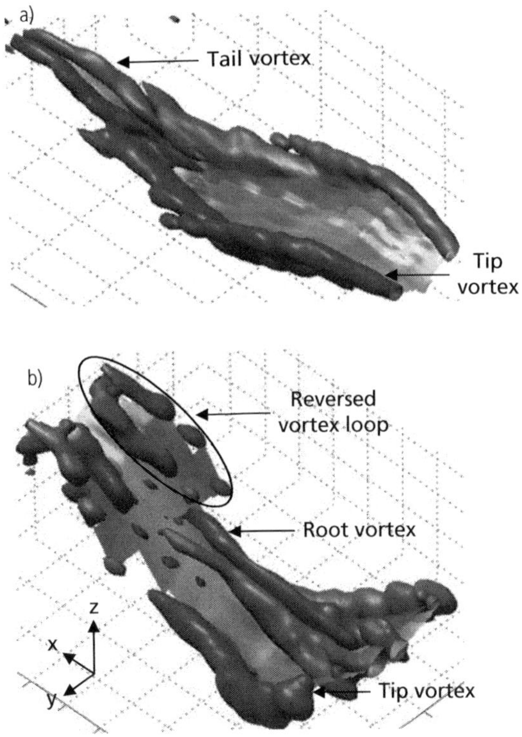

Figure 13.5 Wake topologies showing the vorticity iso-surfaces of the main vortex structures for one wingbeat of a pied flycatcher, *Ficedula hypoleuca* (a) and a Pallas' long-tongued bat, *Glossophaga soricina*, (b) flying to the left at 7 m/s, starting with a downstroke followed by an upstroke. The strength of the vortices, measured as circulation, is proportional to the lift generated. The complexity of the wake illustrates how lift is generated by different parts of the animal at different phases of the wingbeat. For example, root vortices seen in the bat wake indicate a lower lift generation by the body than by the wings and the reversed vortex loops generated during the upstroke indicate negative lift being generated. From Muijres et al. (2012b). (See Plate 5).

stable (Box 13.1). This enables insects to operate their wings at post stall angles-of-attack, and to use a range of unsteady aerodynamic mechanisms to boost flight forces (Sane 2003, Lehmann 2004), including leading edge vortices (LEVs) (Ellington et al. 1996, Dickinson and Götz 1993), wake capture (Birch and Dickinson 2003), rotational circulation (Dickinson et al. 1999), and clap-and-fling (Bennett 1970, Weis-Fogh 1973, Maxworthy 1979). Arguably one of the most important mechanisms is the LEV, which is a vortex positioned on top of and along the leading edge of wings with separated flow (Fig. 13.4b; see also colour plate section). LEVs enhance lift due to their low pressure core in proximity of the wing surface increasing lift by as much as 65% (van den Berg and Ellington 1997), although at a cost of increased drag. For small insects, the LEV stays stably attached to the wing throughout the complete wingstroke (Birch and Dickinson 2001). A recent study on the LEV dynamics in larger and free flying insects, though, shows a more complex structure (Johansson et al. 2013) and stabilizing mechanisms have been suggested to allow the use of LEVs at these higher Re (Lentink and Dickinson 2009).

Until recently LEVs were assumed to be purely a low Re phenomenon only used by insects, due to the LEV stabilization required at higher Re (Birch and Dickinson 2001, Birch et al. 2004). The use of LEVs outside insects was first demonstrated in slow flying bats (Fig. 13.6; Muijres et al. 2008) and subsequently in hummingbirds (Warrick et al. 2009) and pied flycatchers (Muijres et al. 2012c). LEVs at these intermediate Re are possibly stabilized by rotational accelerations caused by the flapping wing (Lentink and Dickinson 2009). Also, the relative contribution and structure of the LEVs varies significantly between groups, suggesting differences in how the LEVs are controlled, for example by wing twist and camber (Muijres et al. 2008, 2012c; Nakata and Liu 2012). Up to now, LEVs in vertebrates have only been found in slow and hovering flight, although model bird studies suggest the potential use at higher flight speeds (Videler et al. 2004; Hubel and Tropea 2010). A study using a mechanical bat wing flapper suggests that bats are probably able to use LEVs throughout their complete natural flight speed range, but that these are not used in steady cruising flight (Koekkoek et al. 2012). During fast steady flight only low C_L is required, and thus LEVs with their associated high drag and thus high COT are not needed. LEVs could still be used in certain fast flight manoeuvres for which large lift and/or drag forces are required. This was also suggested for swifts (Videler et al. 2004), although Lentink et al. (2007) showed that, at least when minimizing turn radius, swifts should not use LEVs.

Among the flying animals in the intermediate Re range we find birds, bats and large insects. From an aerodynamic perspective the intermediate Re range is particularly interesting as it allows the animals

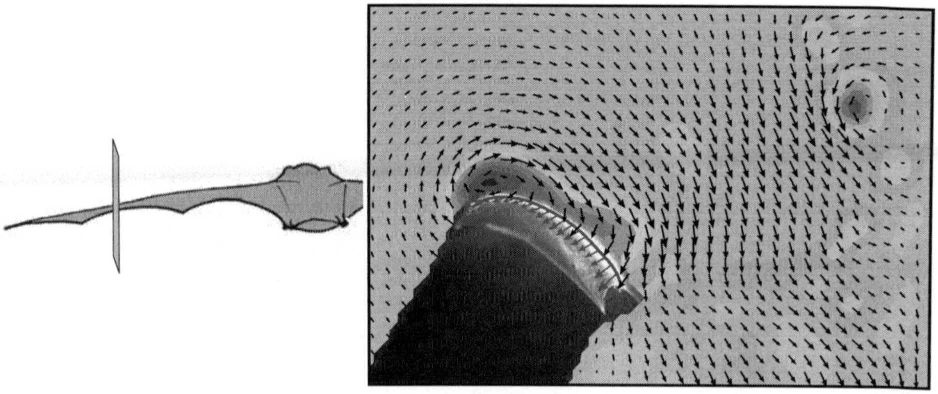

Figure 13.6 Velocity and vorticity fields around a bat wing in slow forward flight (1 m/s) showing the disturbance caused by the wing with the uniform mean flow (of 1 m/s) removed. A leading edge vortex (LEV), which increases lift, is seen above the wing. The spanwise position of the measurement is indicated on the bat silhouette to the left. The flight direction is from right to left. The local wing profile and its relative motion are shown with a curve and arrows. From Muijres et al. (2008).

to use primarily efficient lift-based flight force following attached flow with low COT_{mech} during cruising flight, while allowing for low Re mechanisms such as LEVs to be used to boost flight forces at low flight speed or during landing/take off and during certain manoeuvres (Koekkoek et al. 2012). However, due to sensitivity to perturbations in the flow operating at this Re range this may also result in increased demands of passive and active control mechanisms to respond to variation in the forces and moments generated.

Flying animals operating at high Re, e.g. at the fast flight speeds relevant for migration and commuting, almost exclusively use lift-based foil forces, as it is the most efficient means of propulsion at this Re range. Due to the relatively high speed of the flow, only a small deflection of the fluid (a small lift coefficient) is necessary to generate high forces. The drag of the propulsor is at the same time minimized, resulting in a low COT_{mech}. However, the high Re also means that the flow is sensitive to perturbations and the foils can easily stall resulting in a sudden increase in drag that works effectively as an airbrake. This significantly limits the flight speed envelope compared to low Re flyers. Birds can increase their flight speed envelope by wing morphing, where wing feathers slide over each other resulting in a change of wing area (Pennycuick 1968a, b, Lentink et al. 2007). Compared to birds, bats can do this to a lesser extent, due to the stronger limitations on the morphing capabilities of membranous wings (Pennycuick 1971).

Many fish swim with flapping motions of the pectoral fins (Fricke and Hissmann 1992, Combes and Daniel 2001, Walker and Westneat 2002) and the ocean sunfish propels itself with lateral oscillations of the dorsal and anal fins (Watanabe and Sato 2008). Also squids (Stewart et al. 2010), turtles (Davenport 1987, Wyneken 1997), wing propelled diving birds (Clark and Bemis 1979, Hui 1988, Johansson and Wetterholm Aldrin 2002), and fur seals (Feldkamp 1987) use lift-based propulsion by flapping fins. These animals operate in the Re range 10^3–$5*10^5$ and use fluid dynamic lift for locomotion. These 'underwater flyers' are close to neutrally buoyant, requiring only thrust to swim. As a result the kinematics are, unlike in most flying animals, mostly symmetric, cancelling the fluid dynamic force perpendicular to the swimming direction during the semi-strokes. This results in a ladder wake pattern with vortices of opposite sense being shed at each half-stroke (Lentink et al. 2008).

Perhaps less intuitively, lift-based propulsion can also be found in animals using paddling motions of their legs (Johansson and Lindhe Norberg 2000, 2001; Johansson and Norberg, 2003). Grebes, *Podiceps* spp., use a sweeping motion of the feet moving them from underneath the body to above the back of the body with the lobed toes meeting the water at low angles of attack (Johansson and Lindhe

Norberg 2000, 2001). Also cormorants, *Phalacrocorax* spp., sweep their feet upwards, at low angle of attack during part of the power stroke. A start vortex is seen to be left in the wake, indicating a lift generating mechanism, which is aided by LEVs on the suction side of the feet (Johansson and Norberg 2003). Unlike the flapping described earlier, these modes of locomotion have a recovery phase in the same way as in drag-based paddling. However, the use of a lift-based mechanism increases the maximum attainable swimming speed compared to drag-based paddling. In general, lift based propulsion is less limited by the flapping speed of the foils, since the flapping motion is perpendicular to the swimming direction and the foil therefore will be able to generate substantial forces also when the forward velocity supersedes the flapping velocity (Vogel 1994).

13.3.5 Factors influencing COT when moving in air and water

When considering the power required to move through a fluid, the drag of the animal stands out as the most important predictor (see earlier discussion). However, also the efficiency of the propulsion system is important, as reflected in the induced and metabolic power. Since we discussed shape factors reducing the drag in the previous section we will here mainly focus on behavioural issues that may reduce COT here. The change in drag coefficient, C_D, associated with the transition between laminar and turbulent boundary layer has been used to explain the preferred swim speed (determined by minimum COT) of both fish and birds. Theoretical studies of fish swimming and experimental studies of bird swimming have suggested that animals adjust their swimming speed to stay on the 'right side' of the Re transition even though the predicted velocity based on their size is different. For example, swimming birds have been found to be faster than their counterpart mammals, which is suggested to be a result of a sharp decrease in drag coefficient with increasing Re (Watanabe et al. 2011). By swimming faster than predicted by their size (based on the mammal scaling) the birds would 'outswim' the high drag and thus lower COT. Although an interesting hypothesis, it needs further exploration considering the drag coefficient relationship with Re is most likely not as stated by the authors (i.e. drag coefficients of 2D instead of 3D streamlined bodies). Moreover, also large birds, that would normally be above the critical Re, tend to swim faster than predicted (Watanabe et al. 2011). A more likely reason for the higher speed in birds compared to mammals is a higher basic metanbolic rate (BMR) caused by higher thermoregulation costs in birds compared to mammals (Watanabe et al. 2011) resulting in higher optimal swim speed, but probably also higher COT (see earlier discussion).

The motion of the body can be used to reduce drag and thereby COT, as indicated from studies on robotic fish, showing that the power required during undulating motions is lower than the power needed to move the same fish when gliding with a stiff body (Barrett et al. 1999, Shen et al. 2003, Dong and Lu 2005, Lu and Yin 2005, Yao et al. 2011). When the speed of the undulation is 1.25 times the free-stream speed, power is minimized, which is due to the growth of the boundary layer. With a stiff body, the boundary layer grows along the length of the body, whereas with a body undulation moving rearwards over the body at a greater speed than the swimming speed, the growth of the boundary layer and the formation of separation bubbles, are suppressed and drag is thereby reduced (Shen et al. 2003).

Swimming close to the surface introduces an additional drag due to the generation of waves (Byquist 1973), which may be highly relevant for migratory swimming animals. Wave drag has been suggested to directly influence the swimming behaviour of salmonid fish during spawning migration (Hughes 2004). When swimming upstream in rivers the flow speed is often dependent on water depth and distance from shore, with the highest speeds at the deepest location and furthest from the shore. Large migrating fish prefer the deeper section, which results in a higher flow speed and higher drag. However, since the wave drag depends on the relative depth (body heights to the surface) larger fish will experience higher wave drag if swimming in the shallower and slower flowing sections of the river (Hughes 2004). Since wave drag can be a substantial fraction of the total drag, it seems as if large fish try to avoid it. Most long distance migrating animals can avoid swimming in direct proximity to the surface. However, air breathing animals need to spend a relatively higher proportion of their time at the surface (compared to fish), suggesting that their overall COT is higher.

The propulsive efficiency is strongly influenced by the type of propulsion system and by Re. At the lower end of the Re scale, where cilia and flagella provide friction-based locomotion, COT is notoriously high, but no better alternative exists at these low Re. In combination with the low movement velocities at these small scales, migration through active movement of flagella or cilia is not realistic. Instead, large-scale movements in these smallest animals is primarily by drifting within currents. Jet propulsion in animals is commonly modelled by adapting standard rocket efficiency,

$$\eta_r = 2U_\infty U_{jet} \ / \ (U_\infty^2 + U_{jet}^2), \qquad (8)$$

to periodic propulsion in animals. This results in a whole-cycle jet efficiency for animals with rear-facing intake of

$$\eta_{rear} = 2U_\infty U_{jet} \ / (2U_R U_\infty + 3U_\infty^2 + U_{jet}^2), \qquad (9)$$

where U_R is the fluid speed in the intake siphon. For forward facing intakes, and assuming a passive intake phase, the efficiency is

$$\eta_{forward} = 2U_\infty U_{jet} \ / (3U_\infty^2 + U_{jet}^2) \qquad (10)$$

Thus, jet-propelling animals with a forward facing intake can generate both higher thrust forces and a higher efficiency than animals with rearward facing intakes. Squid use two alternative strategies when swimming, each with its own benefits regarding the formation of the jet; firstly, short pulsed jets forming stacked vortex rings in the wake and secondly elongated plugs of water used at high swimming speeds and at Re < 6, where shear forces dominate the jet fluid dynamics. At Re > 6, the shear layer between the jet and the undisturbed fluid rolls up into a leading-edge vortex (LEV) ring. Throughout the jet phase, this LEV grows until it reaches a maximum sustainable size, when it eventually separates from the siphon and convects into the wake. After this point, the pinch point, a periodic formation and shedding of vortex rings from the siphon occurs. Jet propulsion is most efficient at the pinch point and less so before and after. The time of the pinch point, within the vortex ring formation time, can be estimated using the formation time $F = l/d$, where l is the length of the jet wake and d is the diameter of the jet. Regardless of the temporal dynamics of a jet and even for many biological systems where the jet siphon area varies throughout the jet phase, the pinch point occurs generally at a formation time of $F_{pinch} = 4.0 \pm 0.5$ (Gharib et al. 1998, Dabiri and Gharib 2005), although higher values have been reported in natural jet propulsion systems (Bartol et al. 2009a, b). In the case of the formation of a fluid plug, the efficiency of the force generation increases with decreasing difference between the ejected speed and the swim speed (the slip) (Anderson and Grosenbaugh 2005) and efficiency of the jetting phase may be as high as 0.93. Although efficiency during the propulsive phase may be high, the whole-cycle jet efficiency is rather low (max 0.58; Anderson and Grosenbaugh 2005). Due to this low efficiency, few animals use jet propulsion to actively move over large distances.

The efficiency of a fluid foil depends on the size of the vortex ring generated relative to the force. The larger the vortex ring, the lower the induced velocities, and thus the higher the efficiency (Rayner 1979). The vortex rings of drag-based paddlers are approximately the size of the foil. This means that drag-based foils, which produce relatively small vortex rings, compared to lifting foils (where the rings approximate to the size swept by the foil), are less efficient, and the efficiency of lifting foils increases with forward speed (although the pressure drag of the whole system increases simultaneously). Lifting foils differ in their efficiency depending on their shape and comparative measurements of flight efficiency in different groups of animals based on quantitative measurements are now available. A recent study compared span efficiency (the ratio of the minimum induced drag for a given force to the actual induced drag) and COT_{mech} for two bird and two bat species and showed that the birds were more efficient than bats (Fig. 13.7; Muijres et al. 2012b). The wake analysis suggests a higher body drag and lower lift generation of the body in bats increasing the COT_{mech} (Muijres et al. 2012b). A likely cause for this result is that the ears and nose leaf of bats, which are used for echolocation, disrupt the flow over the body causing it to generate less lift and more drag than the bird's body. The bats thus require relatively more thrust, which may, at least partially, explain the differences in upstroke function between birds and bats (Fig. 13.5, and see also colour plate section; Hedenström et al. 2007; Johansson et al. 2008; Hubel et al. 2009, 2010, 2012; Muijres et al. 2011a). Higher body drag should result in both higher COT_{mech}, and potentially a higher COT_{met} and a lower speed at minimum COT, resulting in a lower migration potential for bats compared to birds (Muijres et al. 2011a; Fig. 13.7). This is in line with ecological data on birds and bats, where bats are generally found to prefer a lower flight speed and migrate less and shorter distances than birds (Hedenström et al. 2009).

Detailed quantitative fluid dynamics measurements from free flying insects, to the best of our knowledge, does not exist, and only a few species have been studied during tethered flight in such detail that the results can be compared with vertebrate data (Johansson et al. 2012; Henningsson and Bomphrey 2012, 2013). The span efficiency estimated in these species suggests that insects may have a lower span efficiency than vertebrates (Johansson et al. 2012, Henningsson and Bomphrey 2013). A lower span efficiency is consistent with results from flow visualizations and computational fluid dynamics studies in, for example, hawkmoths, bumblebees and model fruitflies, suggesting that the wing on each side of the animal generates a discrete vortex ring, indicating relatively little lift production by the body (Bomphrey et al. 2009), resulting in reduced span efficiency (Muijres et al. 2011b).

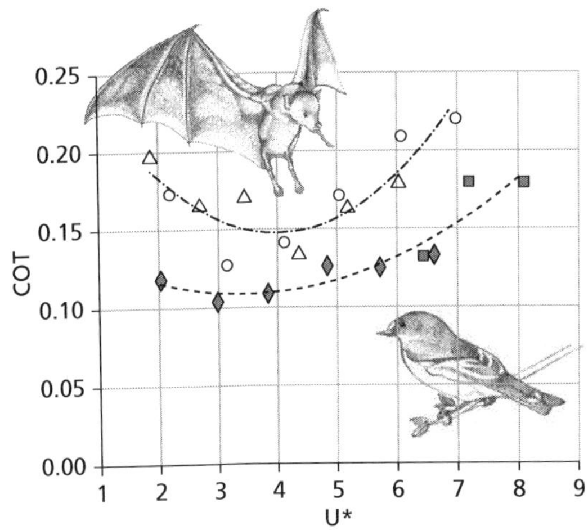

Figure 13.7 Normalized mechanical cost-of-transport COT_{mech} against normalized flight speed, for the pied flycatcher, *Ficedula hypoleuca* (filled diamonds); blackcap, Sylvia atricapilla (filled squares); Pallas' long-tongued bat, *Glossophaga soricina* (open circles); and lesser long-nosed bat, *Leptonycteris yerbabuenae* (open triangles). The trend lines are for birds (lower) and bats (upper). The estimates are based on wake measurements of animals flying in a wind tunnel. From Muijres et al. (2012b).

The energetics of insect flight has been modelled by Ellington (Ellington 1984), by defining and modelling the various aspects that contribute to insect flight COT, i.e. lift-induced drag costs, profile drag costs, inertial costs as a result of wing flapping, as well as various efficiency factors such as aerodynamic efficiency and muscle efficiency. Although the exact estimates of this model are partially outdated, the model itself is still the basis for most insect flight performance studies. One of the most relevant improvements to the model is the incorporation of unsteady aerodynamic mechanisms (Sane 2003, Lehmann 2004). Also the empirical estimates of various flight performance factors have been improved, such as the drag coefficients of insect wings (Dickinson et al. 1999) and span efficiency of insect flight (Henningsson and Bomphrey 2012, Johansson et al. 2012, Henningsson and Bomphrey 2013). The currently most detailed flight cost estimates for a single insect species is in the fruit fly, *Drosophila melanogaster* (Dickson et al. 2008). Studies based on respirometry measurements (Lehmann and Dickinson 1997), computational fluid dynamics (CFD) (Sun and Tang 2002), as well as using a mechanical insect wing model (Fry et al. 2005) show that COT in insect flight is high compared to larger animals, which can be explained by the fact that insect wings generate relatively high drag coefficients resulting from the low Re aerodynamics (Fry et al. 2005, Dickson et al. 2008; Fig. 13.4a, see also colour plate section).

COT_{mech} in flapping flight consists mostly of aerodynamic power and inertial power for accelerating the wings, and with respect to this, flying insects can be divided into two groups. For insects with a relatively low-frequency high-amplitude wingbeat, such as fruit flies and hawkmoths, COT_{mech} is governed by the aerodynamic forces, while for insects with high-frequency low-amplitude wingbeats, such as hoverflies and bees, wing inertia dominates flight energetics (Mao and Gang 2003). Insects can also reduce inertia power by elastically storing negative work performed during a wingbeat in the thorax and wings (Alexander and Bennet-Clark 1977, Dickinson and Lighton 1995). This would enable the group with relatively high inertia power to significantly reduce the total power required (up to 33% based on CFD data), while this has little to no effect on insects with high-amplitude wingbeats (Mao and Gang 2003).

The deformations responsible for elastic storage in insect flight are largely governed by inertia forces, at least for a large range of insect species (Daniel and Combes 2002), although it remains to be seen whether this is also valid for larger and more flexible structures as feathered bird wings and membranous bat wings. A CFD study showed that wing flexibility also has a significant effect on the aerodynamic performance of insect flight (Young et al. 2009), although a more recent study that combined a CFD model with a finite element (FE) model of the wing found only a low performance increase of 3.4% resulting from wing flexibility (Nakata and Liu 2012).

Insects have relatively smaller wings than birds and bats (Greenewalt 1962), and generation of an individual vortex ring for each wing requires higher lift coefficient (C_L) for insects compared to flying vertebrates, which results in a high induced drag. Interestingly, at the low Re that insects operate at, COT is minimized at the high C_L values found (Fig. 13.4a). Thus insect wings are expected to operate at higher post-stall angle of attack (α) throughout a larger flight range than vertebrates. This hypothesis is supported by a recent study showing that the hummingbird hawkmoths, *Macroglossum stellatarum*, use LEVs also at relatively high forward flight speed (Johansson et al. 2013). Although these trends suggest that insects minimize COT in forward flight, the overall combination of relatively low maximum flight speeds and high COT makes most insects less likely to seasonally migrate using active flight than birds and bats. Instead alternative migratory strategies have evolved where migration spans multiple generations, such as found in monarch butterflies (Brower 1996; Chapman et al., Chapter 2, Fig. 2.6), or strategies where insects rely on wind drift for migration. In fact, by selecting appropriate wind conditions, insects can reach effective migration speeds similar to those of small birds (Alerstam et al. 2011).

13.3.6 Morphology of wings and fins

One of the factors influencing the propulsive efficiency is the planform shape of the wing or fin. A common feature of the planform of biological foils at intermediate and high Re is what is known as a lunate shape, a foil with an arced, backward swept leading edge. The most well-known examples are the caudal fins of tunas, e.g. *Thunnus* spp., the wings of swifts and the swept handwings of large marine soaring birds. These foils have more gentile stall characteristics, and lower induced drag (van Dam 1987). The induced drag is inversely proportional to the aspect ratio (AR) and an alternative strategy to reduce the induced drag is found among albatrosses, which have very long, high aspect ratio wings (Anderson 2007). However, the long wings of albatrosses are associated with high inertial costs of flapping and these birds are therefore generally gliding. The microstructure of the foil may influence propulsive efficiency by controlling the flow over the foil. For example, the velvety upper surface of owl feathers have been suggested to improve the flow conditions on the wings by reducing separation (Winzen et al. 2013). Also feather porosity affects flow conditions, where air passing through the primary feathers can cause modulations to the boundary layer of the top surface of the feather, improving its lift to drag ratio (L/D) (Eder et al. 2011).

For animals moving in fluids, in addition to storing energy as mechanical deformations of the wings and fins similar to in locomotion across surfaces, the animals may reuse energy that has been deposited in the fluid as a wake of another animal or during a previous wing beat or tail stroke. We have already mentioned wake capture as a mechanism to improve performance in insects. Insects such as dragonflies and damselflies have four wings that can flap virtually independently from each other. The extra degrees-of-freedom in kinematics that results from having four wings instead of two has contributed to the fact that they are among the most agile and manoeuvrable insects. It has been suggested that dragonflies can change the phase shift between front and hind wings, altering the wing wake interactions, in order to rapidly switch between an energetically efficient flight kinematics and a kinematics that result in high force production used in fast flight manoeuvers (Wang and Russell 2007, Usherwood and Lehmann 2008).

Caudal fins and tail flukes in undulatory swimming animals rarely have a span that extends beyond the width/height of the body and fins ahead of the tail. The tails thus have a relatively lower span than the wings found in flying animals. One potential reason for this is that the body and fins ahead of the tail alter the flow meeting the tail in such a way that positive interactions occurs (Tytell 2006, Wu 2011). In addition, expanding the tail beyond the wake of the body and fins may result in a reduced efficiency due to the vortex interactions that occurs, similar to the interaction between the wake of the elytra and the wings in flying beetles (Fig. 13.8, and see also colour plate section; Johansson et al. 2012).

A great deal of attention has lately been focused on understanding the performance limitations to fish swimming (Tytell 2007, Borazjani and Sotiropoulos 2010, Tytell et al. 2010, Webb and Cotel 2010). This has been done by comparing the two extremes: anguilliform and thunniform swimming. Anguilliform swimming is exemplified by eels, *Anguilla* spp., which swim with body undulations with more than one wavelength present simultaneously along the body. The width of the body is narrow compared to the length and the head goes through substantial lateral displacement. Thunniform swimming is exemplified by tunas or mackerels, which swim with less than a wavelength present along the body simultaneously. The body is relatively high compared to the length and goes through little lateral displacement with the main lateral motion at the tail. A characteristic of these fishes is also a very narrow peduncle just ahead of the high aspect ratio caudal fin, which increases efficiency in this mode of swimming (Lighthill 1970). Interestingly, when comparing the velocity/power (i.e. the inverse of COT) eels and tunas turn out to be equally efficient swimmers, but with eels having an optimum at low swimming speed and tunas at high swimming speed (Schultz and Webb 2002). Comparison of wakes show that eels generate a wake without any significant flow induced in the swimming direction (thrust and drag cancel over the body), resulting in

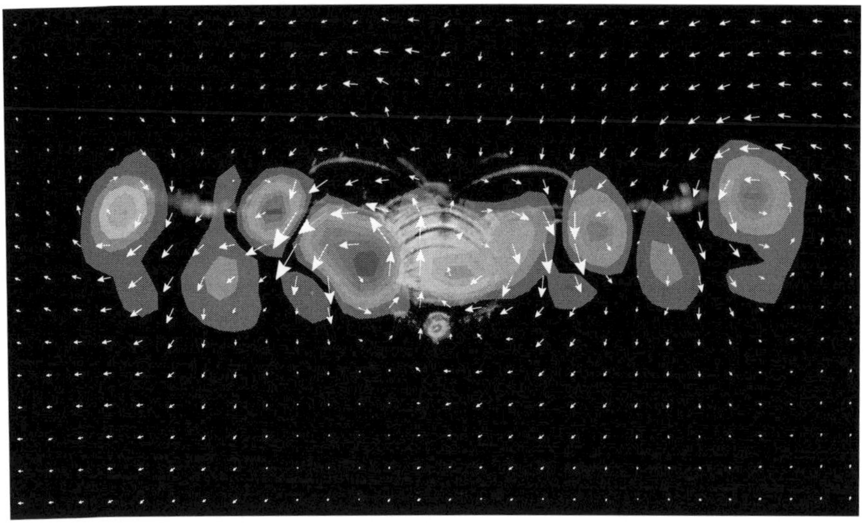

Figure 13.8 Velocity and vorticity measurements in the wake of a dung beetle during mid-downstroke as seen from behind. The wake contains multiple vortices along the wingspan. Going from the left wingtip to the centre of the body: wing tip vortex, interaction vortex, elytra tip vortex, and body or root vortex. The interaction vortex is the result of a negative interaction between the flow generated by the elytra and the wing, reducing the span efficiency. The strong body/root vortices show that the body does not generate significant lift forces. From Johansson et al. (2012). (See Plate 6).

a wake consisting of vortex rings with a laterally induced flow (Tytell and Lauder 2004). Thunniform swimmers, on the other hand, show an alternating thrust generating wake (Nauen and Lauder 2002), suggesting that the thrust generation is separated from the drag generated by the body. In swimming animals the induced power is called wasted wake power and is estimated from the kinetic energy deposited into the wake per unit time (Hall and Hall 1996). Since thrust and drag generation is both done by the body, in undulatory swimmers an estimation of the forces generated is difficult and the wake power coefficient has been suggested to be used as a proxy for the efficiency of swimming (Tytell 2007, Webb and Cotel 2010). The propulsive efficiency of swimming fish has traditionally been viewed as high, although recent analyses suggest that a large proportion of the energy is lost into the wake (Tytell et al. 2010, Webb and Cotel 2010). Interestingly, efficiency may actually be lower than, or in the same range as, the whole cycle efficiency of jet propulsion in squid (Anderson and Grosenbaugh 2005), which is generally thought to be a rather inefficient propulsive mechanism. Comparative studies of different swimming modes based on comparable quantitative measures of performance are thus called for.

13.4 Future perspectives

A recently emerging research field is the combined modelling of fluid dynamic forces and mechanical properties of fluid foils. By describing a basic motion of the foil, but allowing the foil to deform in response to the fluid dynamic and inertial forces, the effect of the mechanical properties of biological materials on the fluid dynamics can be examined. Initial studies in this field show that, at least for insects, wing flexibility may enhance force production significantly, but that it has little effect on aerodynamic efficiency (Nakata and Liu 2012). It still remains an open question what these effects are on more flexible and complex wing structures, such as feathered wings in birds, membranous wings in bats and fins in fish. In addition, our understanding is still limited regarding how active control of mechanical properties can modulate performance, for example with muscle control of tendon properties (Azizi and Roberts 2009). Closely linked to this is modelling of the parameter space of motions and morphologies that may bring us closer to an understanding of the relative performance of animals compared to a global optimum for performing a specific task. The results of these kinds of studies will increase our understanding of adaptations for reducing COT and allowing for control of the forces generated. Eventually this will provide inspiration to the field of robotics, where bio-inspired constructions are abundant. For example, the better performance of walking and running in a complex environment and higher turning performance compared to wheeled systems have lately inspired the construction of walking robots with multi-segmental legs with actuators that mimic muscles and springs to mimic tendons (Wyffels et al. 2010, Zhou and Bi 2012). Also fish-inspired swimming robots (Zhou and Low 2012) and flying robots using insect and bird inspired flapping flight (Gerdes et al. 2012, Ma et al. 2013) become increasingly complex with a higher and higher degree of autonomy. However, we are still far from constructing a machine that can match the performance of the bar-tailed godwit crossing the Pacific in one go.

References

Abourachid, A. (2003). A new way of analysing symmetrical and asymmetrical gaits in quadrupeds. *Comptes Rendus Biologies*, **326**, 625–30.

Ahlborn, B. K., Blake, R. W., and Chan, K. H. S. (2009). Optimal fineness ratio for minimum drag in large whales. *Canadian Journal of Zoology*, **87**, 124–31.

Alerstam, T., Chapman, J. W., Backman, J., et al. (2011). Convergent patterns of long-distance nocturnal migration in noctuid moths and passerine birds. *Proceedings of the Royal Society B: Biological Sciences*, **278**, 3074–80.

Alerstam, T., Hedenström, A., and Åkesson, S. (2003). Long-distance migration: evolution and determinants. *Oikos*, **103**, 247–60.

Alerstam, T., Rosén, M., Bäckman, J., Ericson, P. G. P., and Hellgren, O. (2007). Flight speeds among bird species: allometric and phylogenetic effects. *PLOS Biology*, **5**, 1656–62.

Alexander, D. E. (2004). *Nature's Flyers: Birds, Insects, and the Biomechanics of Flight*. Johns Hopkins University Press, Baltimore.

Alexander, R. M. (1988). Why mammals gallop. *American Zoology*, **28**, 237–45.

Alexander, R. M., and Bennet-Clark, H. C. (1977). Storage of elastic strain energy in muscle and other tissues. *Nature*, **265**, 114–17.

Altshuler, D. L., and Dudley, R. (2003). Kinematics of hovering hummingbird flight along simulated and natural elevation gradients. *Journal of Experimental Biology*, **206**, 3139–47.

Anderson, E. J., and Demont, M. E. (2000). The mechanics of locomotion in the squid loligo pealei: locomotory function and unsteady hydrodynamics of the jet and intramantle pressure. *Journal of Experimental Biology*, **203**, 2851–63.

Anderson, E. J., and Grosenbaugh, M. A. (2005). Jet flow in steadily swimming adult squid. *Journal of Experimental Biology*, **208**, 1125–46.

Anderson, J. D. (2007). *Fundamentals of Aerodynamics*. McGraw-Hill, New York.

Azizi, E., and Roberts, T. J. (2009). Biaxial strain and variable stiffness in aponeuroses. *Journal of Physiology*, **587**, 4309–18.

Barrett, D. S., Triantafyllou, M. S., Yue, D. K. P., Grosenbaugh, M. A., and Wolfgang, M. J. (1999). Drag reduction in fish-like locomotion. *Journal of Fluid Mechanics*, **392**, 183–212.

Bartol, I. K., Krueger, P. S., Stewart, W. J., and Thompson, J. T. (2009a). Hydrodynamics of pulsed jetting in juvenile and adult brief squid *Lolliguncula brevis*: evidence of multiple jet 'modes' and their implications for propulsive efficiency. *Journal of Experimental Biology*, **212**, 1889–1903.

Bartol, I. K., Krueger, P. S., Stewart, W. J., and Thompson, J. T. (2009b). Pulsed jet dynamics of squid hatchlings at intermediate Reynolds numbers. *Journal of Experimental Biology*, **212**, 1506–18.

Baudinette, R. V., Snyder, G. K., and Frappell, P. B. (1992). Energetic cost of locomotion in the tammar wallaby. *American Journal of Physiology*, **262**, R771–8.

Bechert, D. W., Bruse, M., Hage, W. and Meyer, R. (1997). Biological surfaces and their technological application—laboratory and flight experiments on drag reduction and separation control. *roAIAA*, Paper 97–1960, 1–34.

Bechert, D. W., Bruse, M., Hage, W., and Meyer, R. (2000). Fluid mechanics of biological surfaces and their technological application. *Naturwissenschaffen*, **87**, 157–71.

Bejan, A., and Marden, J. H. (2006). Unifying constructal theory for scale effects in running, swimming and flying. *Journal of Experimental Biology*, **209**, 238–48.

Bennett, L. (1970). Insect flight: lift and rate of change of incidence. *Science*, **167**, 177–9.

Bennett, M. B., and Taylor, G. C. (1995). Scaling of elastic strain energy in kangaroos and the benefits of being big. *Nature*, **378**, 56–9.

Biewener, A. A. (1989). Scaling body support in mammals: limb posture and muscle mechanics. *Science*, **245**, 45–8.

Biewener, A. A. (2003). *Animal Locomotion*. Oxford University Press, Oxford.

Birch, J. M., and Dickinson, M. H. (2001). Spanwise flow and the attachment of the leading-edge vortex on insect wings. *Nature*, **412**, 729–33.

Birch, J. M., and Dickinson, M. H. (2003). The influence of wing-wake interactions on the production of aerodynamic forces in flapping flight. *Journal of Experimental Biology*, **206**, 2257–72.

Birch, J. M., Dickson, W. B., and Dickinson, M. H. (2004). Force production and flow structure of the leading edge vortex on flapping wings at high and low Reynolds number. *Journal of Experimental Biology*, **207**, 1063–72.

Block, B. A., Booth, D., and Carey, F. G. (1992). Direct measurement of swimming speeds and depth of blue marlin. *Journal of Experimental Biology*, **166**, 267–84.

Bomphrey, R. J., Taylor, G. K., and Thomas, A. L. R. (2009). Smoke visualization of free-flying bumblebees indicates independent leading-edge vortices on each wing pair. *Experiments in Fluids*, **46**, 811–21.

Bonfil, R., Meÿer, M., Scholl, M. C., et al. (2005). Transoceanic migration, spatial dynamics and population linkages of white sharks. *Science*, **310**, 100–3.

Borazjani, I. and Sotiropoulos, F. (2010). On the role of form and kinematics on the hydrodynamics of self-propelled body/caudal fin swimming. *Journal of Experimental Biology*, **213**, 89–107.

Brennen, C., and Winet, H. (1977). Fluid mechanics of propulsion by cilia and flagella. *Annual Review of Fluid Mechanics*, **9**, 339–98.

Brower, L. (1996). Monarch butterfly orientation: missing pieces of a magnificent puzzle. *Journal of Experimental Biology*, **199**, 93–103.

Byquist, T. (1973). *Wave Making Resistance of a Series of Bodies of Revolution*, The Royal Institute of Technology.

Carsen, A. E., Hatcher, B. G., and Scheibling, R. E. (1996). Effects of the flow velocity and body size of swimming trajectories of sea scallops, *Placopecten magellanicus* (Gemlin): a comparison of laboratory and field measurements. *Journal of Experimental Marine Biology and Ecology*, **203**, 223–43.

Cavagna, G. A., and Kaneko, M. (1977). Mechanical work and efficiency in level walking and running. *Journal of Physiology*, **268**, 467–81.

Chan, B., Balmforth, N. J., and Hosoi, A. E. (2005). Building a better snail: lubrication and adhesive locomotion. *Physics of Fluids*, **17**, 113101.

Chen, P.-Y., Mckittrick, J., and Meyers, M. A. (2012). Biological materials: functional adaptations and bioinspired designs. *Progress in Materials Science*, **57**, 1492–1704.

Chia, F.-S., Buckland-Nicks, J., and Young, C. M. (1984). Locomotion of marine invertebrate larvae: a review. *Canadian Journal of Zoology*, **62**, 1205–22.

Clark, B. D., and Bemis, W. (1979). Kinematics of swimming of penguins at the Detroit zoo. *Journal of Zoology*, **188**, 411–28.

Combes, S. A., and Daniel, T. L. (2001). Shape, flapping and flexion: wing and fin design for forward flight. *Journal of Experimental Biology*, **204**, 2073–85.

Crandell, K. E., and Tobalske, B. W. (2011). Aerodynamics of tip-reversal upstroke in a revolving pigeon wing. *Journal of Experimental Biology*, **214**, 1867–73.

Dabiri, J. O. (2005). On the estimation of swimming and flying forces from wake measurements. *Journal of Experimental Biology*, **208**, 3519–32.

Dabiri, J. O., Colin, S. P., Katija, K., and Costello, J. H. (2010). A wake-based correlate of swimming performance and foraging behavior in seven co-occurring jellyfish species. *Journal of Experimental Biology*, **213**, 1217–25.

Dabiri, J. O., and Gharib, M. (2005). The role of optimal vortex formation in biological fluid transport. *Proceedings of the Royal Society, B: Biological Sciences*, **272**, 1557–60.

Daniel, T. L. (1984). Unsteady aspects of aquatic locomotion. *American Zoology*, **24**, 121–34.

Daniel, T. L., and Combes, S. A. (2002). Flexible wings and fins: bending by inertial or fluid-dynamic forces? *Integrative and Comparative Biology*, **42**, 1044–9.

Davenport, J. (1987). Locomotion in hatchling leatherback turtles *Dermochelys coriacea*. *Journal of Zoology*, **212**, 85–101.

Dickinson, M. H., Farley, C. T., Full, R. J., Koehl, M. a. R., Kram, R., and Lehman, S. (2000). How animals move: an integrative view. *Science*, **288**, 100–6.

Dickinson, M. H., and Götz, K. (1993). Unsteady aerodynamic performance of model wings at low Reynolds numbers. *Journal of Experimental Biology*, **174**, 45–64.

Dickinson, M. H., Lehmann, F.-O., and Sane, S. P. (1999). Wing rotation and the aerodynamic basis of insect flight. *Science*, **284**, 1954–60.

Dickinson, M. H., and Lighton, J. R. (1995). Muscle efficiency and elastic storage in the flight motor of drosophila. *Science*, **268**, 87–90.

Dickson, W. B., Straw, A. D., and Dickinson, M. H. (2008). Integrative model of *drosophila* flight. *AIAA Journal*, **46**, 2150–64.

Dillon, R. H., and Fauci, L. J. (2000). An integrative model of internal axoneme mechanics and external fluid dynamics in ciliary beating. *Journal of Theoretical Biology*, **207**, 415–30.

Dong, G.-J., and Lu, X.-Y. (2005). Numerical analysis on the propulsive performance and vortex shedding of fish-like travelling wavy plate. *International Journal for Numerical Methods in Fluids*, **48**, 1351–73.

Dudley, R. (2002). *The Biomechanics of Insect Flight: Form, Function, Evolution*. Princeton University Press, Princeton, NJ.

Eder, H., Fiedler, W., and Pascoe, X. (2011). Air-permeable hole-pattern and nose-droop control improve aerodynamic performance of primary feathers. *Journal of Comparative Physiology A: Neuroethology, Sensory, Neural, and Behavioral Physiology*, **197**, 109–17.

Edman, K. A. (1988). Double-hyperbolic force-velocity relation in frog muscle fibres. *Journal of Physiology*, **404**, 301–21.

Egevang, C., Stenhouse, I. J., Phillips, R. A., Petersen, A., Fox, J. W., and Silk, J. R. D. (2010). Tracking of arctic terns *Sterna paradisea* reveals longest animal migration. *Proceedings of the National Academy of Sciences USA*, **107**, 2078–81.

Ellington, C. P. (1984). The aerodynamics of hovering insect flight. IV. Aerodynamic mechanisms. *Philosophical Transactions of the Royal Society B: Biological Sciences*, **305**, 79–113.

Ellington, C. P., Van Den Berg, C., Willmott, A. P., and Thomas, A. L. R. (1996). Leading-edge vortices in insect flight. *Nature*, **384**, 626–30.

Feldkamp, S. D. (1987). Foreflipper propulsion in the California sea lion, *Zalophus californianus*. *Journal of Zoology*, **212**, 43–57.

Fish, F. E. (1984). Mechanics, power output and efficiency of the swimming muskrat (*Ondatra zibethicus*). *Journal of Experimental Biology*, **110**, 183–201.

Fricke, H., and Hissmann, K. (1992). Locomotion, fin coordination and body form of the living coelacanth *Latimeria chalumnae*. *Environmental Biology of Fishes*, **34**, 329–56.

Fry, S. N., Sayaman, R., and Dickinson, M. H. (2005). The aerodynamics of hovering flight in drosophila. *Journal of Experimental Biology*, **208**, 2303–18.

Gerdes, J. W., Gupta, S. K., and Wilkerson, S. A. (2012). A review of bird-inspired flapping wing miniature air vehicle designs. *Journal of Mechanisms and Robotics*, **4**, 02100321–03.

Geyer, H., Seyfarth, A., and Blickhan, R. (2006). Compliant leg behaviour explains basic dynamics of walking and running. *Proceedings of the Royal Society B: Biological Sciences*, **273**, 2861–7.

Gharib, M., Rambod, E., and Shariff, K. (1998). A universal time scale for vortex ring formation. *Journal of Fluid Mechanics*, **360**, 121–40.

Gill, R. E., Jr, Tibbitts, T. L., Douglas, D. C., et al. (2009). Extreme endurance flights by landbirds crossing the pacific ocean: Ecological corridor rather than barrier? *Proceedings of the Royal Society B: Biological Sciences*, **276**, 447–57.

Greenewalt, C. H. (1962). Dimentional relationships for flying animals. *Smithsonian Miscellaneous Collections*, **144**, 1–46.

Hall, K. C., and Hall, S. R. (1996). Minimum induced power requirements for flapping flight. *Journal of Fluid Mechanics*, **323**, 285–315.

Hedenström, A. (2003). Scaling migration speed in animals that run, swim and fly. *Journal of Zoology*, **259**, 155–60.

Hedenström, A. (2008). Power and metabolic scope of bird flight: a phylogenetic analysis of biomechanical predictions. *Journal of Comparative Physiology A: Neuroethology, Sensory, Neural, and Behavioral Physiology*, **194**, 685–91.

Hedenström, A. (2010). Extreme endurance migration: what is the limit to non-stop flight? *PLoS Biology*, **8**, e1000362.

Hedenström, A. H., Johansson, L. C., and Spedding, G. R. (2009). Bird or bat: comparing airframe design and flight performance. *Bioinspiration and Biomemetics*, **4**, 015001.

Hedenström, A., Johansson, L. C., Wolf, M., Von Busse, R., Winter, Y., and Spedding, G. R. (2007). Bat flight generates complex aerodynamic tracks. *Science*, **316**, 894–7.

Hedenström, A., Van Griethuijsen, L., Rosén, M., and Spedding, G. R. (2006). Vortex wakes of birds: recent developments using digital particle image velocimetry in a wind tunnel. *Animal Biology*, **56**, 535–49.

Henningsson, P., and Bomphrey, R. J. (2012). Time-varying span efficiency through the wingbeat of desert locusts. *Journal of the Royal Society, Interface*, **9**, 1177–86.

Henningsson, P., and Bomphrey, R. J. (2013). Span efficiency in hawkmoths. *Journal of the Royal Society, Interface*, **10**, 20130099.

Henningsson, P., Johansson, L. C., and Hedenström, A. (2010). How swift are swifts *Apus apus*? *Journal of Avian Biology*, **41**, 94–8.

Hildebrand, M. (1959). Motions of the running cheetah and horse. *Journal of Mammalogy*, **40**, 481–95.

Hoerner, S. F. (1965) *Fluid-Dynamic Drag*. Published by author.

Hubel, T. Y., Hristov, N. I., Swartz, S. M., and Breuer, K. S. (2009). Time-resolved wake structure and kinematics of bat flight. *Experiments in Fluids*, **46**, 933–43.

Hubel, T. Y., Hristov, N. I., Swartz, S. M., and Breuer, K. S. (2012). Changes in kinematics and aerodynamics over a range of speeds in *Tadarida brasiliensis*, the Brazilian free-tailed bat. *Journal of the Royal Society, Interface*, **9**, 1120–30.

Hubel, T. Y., Riskin, D. K., Swartz, S. M., and Breuer, K. S. (2010). Wake structure and wing kinematics: the flight of the lesser dog-faced fruit bat, *Cynopterus brachyotis*. *Journal of Experimental Biology*, **213**, 3427–40.

Hubel, T. Y., and Tropea, C. (2010). The importance of leading edge vortices under simplified flapping flight conditions at the size scale of birds. *Journal of Experimental Biology*, **213**, 1930–9.

Hudson, P. E., Corr, S. A., and Wilson, A. M. (2012). High speed galloping in the cheetah (*Acinonyx jubatus*) and the racing greyhound (*Canis familiaris*): spatio-temporal and kinetic characteristics. *Journal of Experimental Biology*, **215**, 2425–34.

Hughes, N. F. (2004). The wave-drag hypothesis: an explanation for size-based lateral segregation during the upstream migration of salmonids. *Canadian Journal of Fisheries and Aquatic Sciences*, **61**, 103–9.

Hui, C. A. (1988). Penguin swimming. I. Hydrodynamics. *Physiological Zoology*, **61**, 333–43.

Johansson, L. C., Engel, S., Baird, E., Dacke, M., Muijres, F. T., and Hedenström, A. (2012). Elytra boost lift, but reduce aerodynamic efficiency in flying beetles. *Journal of the Royal Society, Interface*, **9**, 2745–8.

Johansson, L. C., Engel, S., Kelber, A., Klein Heerenbrink, M., and Hedenström, A. (2013). Multiple leading edge vortices of unexpected strength in freely flying hawkmoth. *Scientific Reports*, 3, Paper 3264.

Johansson, L. C., and Hedenström, A. (2009). The vortex wake of blackcaps (*Sylvia atricapilla* L.) measured using high-speed digital particle image velocimetry (DPIV). *Journal of Experimental Biology*, **212**, 3365–3376.

Johansson, L. C., and Lauder, G. V. (2004). Hydrodynamics of surface swimming in leopard frogs (*Rana pipiens*). *Journal of Experimental Biology*, **207**, 3945–58.

Johansson, L. C., and Lindhe Norberg, U. M. (2000). Asymmetric toes aid underwater swimming. *Nature*, **407**, 582–3.

Johansson, L. C., and Lindhe Norberg, U. M. (2001). Lift-based paddling in diving grebe. *Journal of Experimental Biology*, **204**, 1687–96.

Johansson, L. C., and Norberg, R. Å. (2003). Delta-wing function of webbed feet gives hydrodynamic lift for swimming propulsion in birds. *Nature*, **424**, 65–8.

Johansson, L. C., and Wetterholm Aldrin, B. S. (2002). Kinematics of diving Atlantic puffins (*Fratercula arctica* L.): evidence for an active upstroke. *Journal of Experimental Biology*, **205**, 371–8.

Johansson, L. C., Wolf, M., Von Busse, R., Winter, Y., Spedding, G. R., and Hedenström, A. (2008). The near and far wake of Pallas' long tongued bat (*Glossophaga soricina*). *Journal of Experimental Biology*, **211**, 2909–18.

Koekkoek, G., Muijres, F. T., Johansson, L. C., Stuiver, M., Van Oudheusden, B. W., and Hedenström, A. (2012). Stroke plane angle controls leading edge vortex in a bat-inspired flapper. *Comptes Rendus Mécanique*, **340**, 95–106.

Kram, R., and Dawson, T. J. (1998). Energetics and biomechanics of locomotion by red kangaroos (*Macropus rufus*). *Comparative Biochemistry and Physiology. Part B: Biochemistry and Molecular Biology*, **120**, 41–9.

Kram, R., and Taylor, C. R. (1990). Energetics of running: a new perspective. *Nature*, **346**, 265–7.

Langman, V. A., Rowe, M. F., Roberts, T. J., Langman, N. V., and Taylor, C. R. (2012). Minimum cost of transport in Asian elephants: do we really need a bigger elephant? *Journal of Experimental Biology*, **215**, 1509–14.

Lee, D. V., Bertram, J. E. A., Anttonen, J. T., Ros, I. G., Harris, S. L., and Biewener, A. A. (2011). A collisional perspective on quadrupedal gait dynamics. *Journal of the Royal Society, Interface*, **8**, 1480–6.

Lehmann, F.-O. (2004). The mechanisms of lift enhancement in insect flight. *Naturwissenschaften*, **91**, 101–22.

Lehmann, F.-O., and Dickinson, M. H. (1997). The changes in power requirements and muscle efficiency during elevated force production in the fruit fly *Drosophila melanogaster*. *Journal of Experimental Biology*, **200**, 1133–43.

Lentink, D., and Dickinson, M. H. (2009). Rotational accelerations stabilize leading edge vortices on revolving fly wings. *Journal of Experimental Biology*, **212**, 2705–19.

Lentink, D., Muijres, F. T., Donker-Duyvis, F. J., and Van Leeuwen, J. L. (2008). Vortex-wake interactions of a flapping foil that models animal swimming and flight. *Journal of Experimental Biology*, **211**, 267–73.

Lentink, D., Müller, U. K., Stamhuis, E. J., et al. (2007). How swifts control their glide performance with morphing wings. *Nature*, **446**, 1082–5.

Lighthill, M. J. (1970). Aquatic animal propulsion of high hydromechanical efficiency. *Journal of Fluid Mechanics*, **44**, 265–301.

Lindstedt, S. L., Hokanson, J. F., Wells, D. J., Swain, S. D., Hoppeler, H., and Navarro, V. (1991). Running energetics in the pronghorn antelope. *Nature*, **353**, 748–50.

Lovvorn, J. R., Liggins, G. A., Borstad, M. H., Calisal, S. M., and Mikkelsen, J. (2001). Hydrodynamic drag of diving birds: effects of body size, body shape and feathers at steady speeds. *Journal of Experimental Biology*, **204**, 1547–57.

Lu, X. Y., and Yin, X. Z. (2005). Propulsive performance of a fish-like travelling wavy wall. *Acta Mechanica*, **175**, 197–215.

Ma, K. Y., Chirarattananon, P., Fuller, S. B., and Wood, R. J. (2013). Controlled flight of a biologically inspired, insect-scale robot. *Science*, **340**, 603–7.

Mao, S., and Gang, D. (2003). Lift and power requirements of hovering insect flight. *Acta Mechanica Sinica*, **19**, 458–69.

Martinez, M. M., Full, R. J., and Koehl, M. A. (1998). Underwater punting by an intertidal crab: a novel gait revealed by the kinematics of pedestrian locomotion in air versus water. *Journal of Experimental Biology*, **201**, 2609–23.

Mate, B., Mesecar, R., and Lagerquist, B. (2007). The evolution of satellite-monitored radio tags for large whales: one laboratory's experience. *Deep-Sea Research II*, **54**, 224–47.

Maxworthy, T. (1979). Experiments on the Weis-Fogh mechanism of lift generation by insects in hovering flight. Part 1. Dynamics of the 'fling'. *Journal of Fluid Mechanics*, **93**, 47–63.

Muijres, F. T., Bowlin, M. S., Johansson, L. C., and Hedenström, A. (2012a). Vortex wake, downwash distribution, aerodynamic performance and wingbeat kinematics in slow-flying pied flycatchers.. *Journal of the Royal Society, Interface*, **9**, 292–303.

Muijres, F. T., Johansson, L. C., Barfield, R., Wolf, M., Spedding, G. R., and Hedenström, A. H. (2008). Leading-edge vortex improves lift in slow-flying bats. *Science*, **319**, 1250–3.

Muijres, F. T., Johansson, L. C., Bowlin, M. S., Winter, Y., and Hedenström, A. (2012b). Comparing aerodynamic efficiency in birds and bats suggests better flight performance in birds. *PLOS One*, **7**, e37335.

Muijres, F. T., Johansson, L. C., and Hedenström, A. (2012c). Leading edge vortex in a slow-flying passerine. *Biology Letters*, **8**, 554–7.

Muijres, F. T., Johansson, L. C., Winter, Y., and Hedenström, A. (2011a). Comparative aerodynamic performance of flapping flight in two bat species using time-resolved wake visualization. *Journal of the Royal Society, Interface*, **8**, 1418–28.

Muijres, F. T., Spedding, G. R., Winter, Y., and Hedenström, A. (2011b). Actuator disk model and span efficiency of flapping flight in bats based on time-resolved PIV measurements. *Experiments in Fluids*, **51**, 511–25.

Myers, M. J., and Steudel, K. (1985). Effect of limb mass and its distribution on the energetic cost of running. *Journal of Experimental Biology*, **116**, 363–73.

Myers, M. J., and Steudel, K. (1997). Morphological conservation of limb natural pendular period in the domestic dog (*Canis familiaris*): implications for locomotor energetics. *Journal of Morphology*, **234**, 183–96.

Nachtigall, W. (1960). Über kinematik, dynamik und energetik des schwimmens einheimischer dytisciden. *Zeitschrift für vergleichende Physiologie*, **43**, 48–118.

Nakata, T. and Liu, H. (2012). Aerodynamic performance of a hovering hawkmoth with flexible wings: a computational approach. *Proceedings of the Royal Society B: Biological Sciences*, **279**, 722–31.

Nauen, J. C., and Lauder, G. V. (2002). Hydrodynamics of caudal fin locomotion by chub mackerel, *Scomber japonicus* (scombridae). *Journal of Experimental Biology*, **205**, 1709–24.

Norberg, U. M. (1975). Hovering flight in the pied flycatcher (*Ficedula hypolecua*). *Swimming and Flying in Nature*, eds Wu TT, Brokaw C, & Brennen C (Springer US), pp. 869–881.

Norberg, R. Å. (1985). Function of vane asymmetry and shaft curvature in bird flight feathers: inferences on flight ability of archaeopteryx. In M. K. Hecht (ed.), *The Beginnings of Birds: Proceedings of the International Archaeopteryx Conference, Eichstätt,1984*, pp. 303–318, Freunde des Jura-Museums Eichstätt, Germany.

Nudds, R. L., Folkow, L. P., Lees, J. J., Tickle, P. G., Stokkan, K.-A., and Codd, J. R. (2011). Evidence for energy savings from aerial running in the svalbard rock ptarmigan

(*Lagopus muta hyperborea*). *Proceedings of the Royal Society B: Biological Sciences*, **278**, 2654–61.

Pennycuick, C. J. (1968a). Power requirements for horizontal flight in the pigeon columba livia. *Journal of Experimental Biology*, **49**, 527–55.

Pennycuick, C. J. (1968b). A wind-tunnel study of gliding flight in the pigeon *Columba livia*. *Journal of Experimental Biology*, **49**, 509–26.

Pennycuick, C. J. (1971). Gliding flight of the dog-faced bat *Rousettus aegyptiacus* observed in a wind tunnel. *Journal of Experimental Biology*, **55**, 833–45.

Pennycuick, C. J. (1992). *Newton Rules Biology*. Oxford University Press, Oxford.

Pennycuick, C. J. (2008). *Modelling the Flying Bird*. Elsevier, London.

Pierce, S. E., Clack, J. A., and Hutchinson, J. R. (2012). Three-dimensional limb joint mobility in the early tetrapod ichthyostega. *Nature*, **486**, 523–6.

Poelma, C., Dickson, W. B., and Dickinsson, M. H. (2006). Time-resolved reconstruction of the full velocity field around a dynamically-scaled flapping wing. *Experiments in Fluids*, **41**, 213–25.

Rayner, J. M. V. (1979). A vortex theory of animal flight. Part 2. The forward flight of birds. *Journal of Fluid Mechanics*, **91**, 731–63.

Reilly, S. M., Mcelroy, E. J., and Biknevicius, A. R. (2007). Posture, gait and the ecological relevance of locomotor costs and energy-saving mechanisms in tetrapods. *Zoology*, **110**, 271–89.

Ren, L., Miller, C. E., Lair, R., and Hutchinson, J. R. (2010). Integration of biomechanical compliance, leverage, and power in elephant limbs. *Proceedings of the National Academy of Sciences USA*, **107**, 7078–82.

Roberts, T. J., and Azizi, E. (2011). Flexible mechanisms: the diverse roles of biological springs in vertebrate movement. *Journal of Experimental Biology*, **214**, 353–61.

Rubenson, J., Lloyd, D. G., Heliams, D. B., Besier, T. F., and Fournier, P. A. (2011). Adaptations for economical bipedal running: the effect of limb structure on three-dimensional joint mechanics. *Journal of the Royal Society, Interface*, **8**, 740–55.

Ruina, A., Bertram, J. E. A., and Srinivasan, M. (2005). A collisional model of the energetic cost of support work qualitatively explains leg sequencing in walking and galloping, pseudo-elastic leg behavior in running and the walk-to-run transition. *Journal of Theoretical Biology*, **237**, 170–92.

Saffman, P. G. (1995). *Vortex Dynamics*. Cambridge University Press, Cambridge.

Sane, S. P. (2003). The aerodynamics of insect flight. *Journal of Experimental Biology*, **206**, 4191–208.

Schmidt-Nielsen, K. (1972). Locomotion: energy cost of flying, swimming and running. *Science*, **177**, 222–8.

Schultz, W. W., and Webb, P. W. (2002). Power requirements of swimming: do new methods resolve old questions? *Integrative and Comparative Biology*, **42**, 1018–25.

Sharp, N. C. C. (1997). Timed running of a cheetah (*Acinonyx jubatus*). *Journal of Zoology*, **241**, 493–4.

Shen, L., Zhang, X., Yue, D. K. P., and Triantafyllou, M. S. (2003). Turbulent flow over a flexible wall undergoing a streamwise travelling wave motion. *Journal of Fluid Mechanics*, **484**, 197–221.

Spedding, G. R., and McArthur, J. (2006). Deducing aerodynamic mechanisms from near- and far-wake measurements of fixed and flapping wings at moderate Reynolds number. *Aerospace Sciences Meeting and Exhibit*, Reno, NV, USA.

Spedding, G. R., and McArthur, J. (2010). Span efficiencies of wings at low Reynolds numbers. *Journal of Aircraft*, **47**, 120–8.

Srinivasan, M. (2011). Fifteen observations on the structure of energy-minimizing gaits in many simple biped models. *Journal of the Royal Society, Interface*, **8**, 74–98.

Srinivasan, M., and Ruina, A. (2006). Computer optimization of a minimal biped model discovers walking and running. *Nature*, **439**, 72–5.

Stewart, W. J., Bartol, I. K., and Krueger, P. S. (2010). Hydrodynamic fin function of brief squid, *Lolliguncula brevis*. *Journal of Experimental Biology*, **213**, 2009–24.

Sun, M., and Tang, J. (2002). Lift and power requirements of hovering flight in *Drosophila virilis*. *Journal of Experimental Biology*, **205**, 2413–27.

Sutherland, K. R., and Madin, L. P. (2010). Comparative jet wake structure and swimming performance of salps. *Journal of Experimental Biology*, **213**, 2967–75.

Tokić, G., and Yue, D. K. P. (2012). Optimal shape and motion of undulatory swimming organisms. *Proceedings of the Royal Society B: Biological Sciences*, **279**, 3065–74.

Tytell, E. D. (2006). Median fin function in bluegill sunfish *Lepomis macrochirus*: streamwise vortex structure during steady swimming. *Journal of Experimental Biology*, **209**, 1516–34.

Tytell, E. D. (2007). Do trout swim better than eels? Challenges for estimating performance based on the wake of self-propelled bodies. *Experiments in Fluids*, **43**, 701.

Tytell, E. D., Borazjani, I., Sotiropoulos, F., Baker, T. V., Anderson, E. J., and Lauder, G. V. (2010). Disentangling the functional roles of morphology and motion in the swimming of fish. *Integrative and Comparative Biology*, **50**, 1140–54.

Tytell, E. D., and Lauder, G. V. (2004). The hydrodynamics of eel swimming I. Wake structure. *Journal of Experimental Biology*, **207**, 1825–41.

Usherwood, J. R., and Lehmann, F.-O. (2008). Phasing of dragonfly wings can improve aerodynamic efficiency by removing swirl. *Journal of the Royal Society, Interface*, **5**, 1303–7.

Van Dam, C. P. (1987). Efficiency characteristics of crescent-shaped wings and caudal fins. *Nature*, **325**, 435–7.

Van Den Berg, C., and Ellington, C. P. (1997). The three-dimensional leading-edge vortex of a 'hovering' model hawkmoth. *Philosophical Transactions of the Royal Society B: Biological Sciences*, **352**, 329–40.

Videler, J. J., Stamhuis, E. J., and Povel, G. D. E. (2004). Leading-edge vortex lifts swifts. *Science*, **306**, 1960–2.

Vogel, S. 1994. *Life in Moving Fluids*. Princeton University Press, Princeton.

Walker, J. A., and Westneat, M. W. (2002). Kinematics, dynamics, and energetics of rowing and flapping propulsion in fishes. *Integrative and Comparative Biology*, **42**, 1032–43.

Wang, Z., and Russell, D. (2007). Effect of forewing and hindwing interactions on aerodynamic forces and power in hovering dragonfly flight. *Physical Review Letters*, **99**, 1–4.

Warrick, D. R., Tobalske, B. W., and Powers, D. R. (2005). Aerodynamics of the hovering hummingbird. *Nature*, **435**, 1094–7.

Warrick, D. R., Tobalske, B. W., and Powers, D. R. (2009). Lift production in the hovering hummingbird. *Proceedings of the Royal Society B: Biological Sciences*, **276**, 3747–52.

Watanabe, Y., and Sato, K. (2008). Functional dorsoventral symmetry in relation to lift-based swimming in the ocean sunfish *Mola mola*. *PLoS One*, **3**, 1–7.

Watanabe, Y. Y., Sato, K., Watanuki, Y., et al. (2011). Scaling of swim speed in breath-hold divers. *Journal of Animal Ecology*, **80**, 57–68.

Watson, R. R., Rubenson, J., Coder, L., Hoyt, D. F., Propert, M. W. G., and Marsh, R. L. (2011). Gait-specific energetics contributes to economical walking and running in emus and ostriches. *Proceedings of the Royal Society B: Biological Sciences*, **278**, 2040–2046.

Webb, P. W., and Cotel, A. J. (2010). Turbulence: does vorticity affect the structure and shape of body and fin propulsors? *Integrative and Comparative Biology*, **50**, 1155–66.

Weis-Fogh, T. (1960). A rubber-like protein in insect cuticle. *Journal of Experimental Biology*, **37**, 889–907.

Weis-Fogh, T. (1973). Quick estimates of flight fitness in hovering animals, including novel mechanisms for lift production. *Journal of Experimental Biology*, **59**, 169–230.

Williams, T. M. (1999). The evolution of cost efficient swimming in marine mammals: limits to energetic optimization. *Philosophical Transactions of the Royal Society B: Biological Sciences*, **354**, 193–201.

Winzen, A., Klaas, M., and Schröder, W. (2013). High-speed piv measurements of the near-wall flow field over hairy surfaces. *Experiments in Fluids*, **54**, 1–14.

Wu, T. Y. (2011). Fish swimming and bird/insect flight. *Annual Review of Fluid Mechanics*, **43**, 25–58.

Wyffels, F., D'haene, M., Waegeman, T., Caluwaerts, K., Nunes, C., and Schrauwen, B. (2010). Realization of a passive compliant robot dog. *Biomedical Robotics and Biomechatronics (BioRob), 2010 3rd IEEE RAS and EMBS International Conference on 26–29 Sept*. 882–6.

Wyneken, J. 1997. Sea turtle locomotion: mechanisms, behavior, and energetics. In P. L. Lutz and J. A. Musick (eds), *The Biology of Sea Turtles*. CRC Press, Boca Raton.

Yao, Y., Lu, C.-J., Si, T., and Zhu, K. (2011). Water tunnel experimental investigation on the drag reduction characteristics of the traveling wavy wall. *Journal of Hydrodynamics, Series B*, **23**, 65–70.

Young, J., Walker, S. M., Bomphrey, R. J., Taylor, G. K., and Thomas, A. L. R. (2009). Details of insect wing design and deformation enhance aerodynamic function and flight efficiency. *Science*, **325**, 1549–52.

Zhou, C., and Low, K. H. (2012). Design and locomotion control of a biomimetic underwater vehicle with fin propulsion. *IEEE/ASME Transactions on Mechatronics*, **17**, 25–35.

Zhou, X., and Bi, S. (2012). A survey of bio-inspired compliant legged robot designs. *Bioinspiration and Biomemetics*, **7**, 041001.

CHAPTER 14

A synthesis of animal movement across scales

Lars-Anders Hansson, Susanne Åkesson, Christer Brönmark, Ben B. Chapman, Anders Hedenström, Johan Hollander, Christoffer Johansson, Miriam Liedvogel, Åke Lindström, Max Lundberg, Rachel Muheim, Jan-Åke Nilsson, Henrik G. Smith, Glenn P. Svensson, and Helena Westerdahl

As outlined in the Preface and Introduction, the aim of this book is not to provide a comprehensive review, but rather to cherry-pick interesting and exciting issues and examples from contemporary movement ecology research. In this last chapter we aim at synthesizing knowledge regarding animal movements in the context of spatial and temporal scales, based on the main conclusions presented in the previous chapters. We will also aim for some crystal ball-gazing by pointing out research paths that may be fruitful to pursue or are not well understood and need further attention. However, we will begin where we started in the Introduction, i.e. with the central trade-off between being sedentary and mobile.

14.1 Evolutionary compromises along the movement–sedentary gradient

Moving is not always an option, and as noted in the Introduction of this book, a sedentary strategy might be beneficial when resources are evenly distributed in space and time and easy to retrieve. For example, the beech tree, *Fagus* spp., discussed in the Introduction, has nutrients relatively evenly distributed in the soil and access to radiation from the sun for the production of photosynthetic energy. However, when conditions for generating energy and nutrient uptake are no longer suitable, such as during winter, a solution for a sedentary organism may be to hibernate or rest. For a beech tree, this means to let the leaves fall and survive the winter at a low metabolic rate. In contrast, when resources are patchy, difficult, or even impossible to acquire, a mobile strategy might be more beneficial. The pied flycatcher, *Ficedula hypoleuca*, which we also mentioned as an example in the Introduction, is able to actively hunt for patchy food resources, such as insects; an activity that, of course, is impossible for a completely sedentary organism. Besides taking on a hibernating strategy, such as in bears, hedgehogs, and bats, or changing diet like blue tits, *Cyanistes caeruleus*, mobile organisms also have the opportunity to move to geographically distant regions with more suitable conditions. Organisms that possess the ability to move are in the position to exploit seasonally favourable habitats during the entire annual cycle. Therefore, in many taxa migration is the rule rather than the exception (Chapman et al., Chapter 2; Liedvogel and Lundberg, Chapter 12). Further, less mobile organisms are more vulnerable to local processes than those being able to move, which might have considerable effects if the environment changes (Smith et al., Chapter 4).

Long journeys and movements away from the present site are not only motivated by the lack of

Animal Movement Across Scales. Edited by Lars-Anders Hansson and Susanne Åkesson.
© Oxford University Press 2014. Published 2014 by Oxford University Press.

resources or harsh environmental conditions, such as low temperatures, but, as Westerdahl et al. (Chapter 8) discuss, it is possible that pathogen-avoidance is one of the driving forces behind animal movements. Avoiding infections and parasites might actually provide 'fitness fuel' for making long journeys. Hence, the bar-tailed godwit, *Limosa lapponica*, for example, performs non-stop flights which take more than a week, from breeding areas in Alaska to wintering areas in New Zealand. Such outstanding work performances may be possible due to a very low exposure to parasites at both breeding and overwintering sites, allowing these shorebirds to make relatively small investments in the immune-machinery without having to pay the cost of an increased risk of contracting pathogens (Westerdahl et al., Chapter 8). Despite the risk of a non-stop flight over oceans, there could be a fitness gain in reducing costs of infections for the adult itself, and also for its offspring. Hence, animals may benefit from breeding in the far north where their offspring are less likely to encounter pathogens or parasites. An example of this is tundra swans, *Cygnus columbianus*, where those that breed in Alaska have a lower prevalence of blood parasites than swans breeding at more southerly sites. Thus, we may conclude that performing long migratory journeys and breeding at northern latitudes may not only provide migrants with an immense abundance of food resources and longer days, but could also provide a pause from disease vectors and parasites as discussed in Westerdahl et al. (Chapter 8). Such benefits might be very fruitful from a fitness point of view.

Organisms that travel between temperate and tropical areas, such as many songbirds, are exposed to more parasites than birds sedentary in temperate regions, particularly when wintering in tropical Africa. This implies that migrants need an immune system that can cope with a wider range of pathogens than less mobile organisms. It furthermore suggests that resistance against a broad range of parasites may be more common in migratory, than sedentary species, since they often encounter a wider diversity of parasites and often spend shorter time being exposed to parasites compared to sedentary species. In the long run this probably makes migratory species less able to co-evolve with each of their parasites (Westerdahl et al., Chapter 8).

14.2 Dispersal, gene flow, and assisted movements

Moving is not only migration (see Box 1.1), but being able to move also provides opportunities for the adult individual to disperse to other regions, and even zygotes of sedentary organisms can disperse with wind or water. The rate of dispersal and the exchange of gene flow among populations is thought to be a crucial component for the evolution of phenotypic plasticity (Hollander et al., Chapter 7). In populations that disperse there will be gene flow over a wide geographic area and, consequently, dispersing populations are exposed to a larger variation in abiotic and/or biotic conditions; they live in a more heterogeneous, fine-grained environment. As a result, individuals in a dispersing population experience a larger range of selection pressures, resulting in the evolution of different phenotypic adaptations and possibly a difference in the degree of phenotypic plasticity. Since the tendency to disperse is likely to be heritable, an interesting hypothetical effect might be that dispersal rates increase within a subpopulation of dispersers, e.g. by having different dispersal syndromes (Nilsson et al., Chapter 6), which in turn could lead to a range expansion of the species. Hence, phenotypic plasticity allows an organism to explore a larger array of environments, as the organism is able to switch between adaptive phenotypes (Hollander et al., Chapter 7).

Active migrations or dispersals (Box 1.1) are by far the most well-studied categories of movements and are very successful ways for an animal to optimize its performance. However, it should also be noted that undirected, passive dispersal, e.g. through wind or waves, might also be a very successful strategy. Although it obviously leads to a considerable loss of individuals, e.g. for those ending up drowning in the sea or drying out in a desert, passive, arbitrary dispersers may also reach places that are pristine and resource rich, which are unlikely to be found by active and directed dispersers or migrators following traditional paths. Moreover, considering the success of sessility among terrestrial plants and sedentary animals, it is not obvious that the energy or time spent on movements and migrations is always well spent. Hence, the selective forces behind migration, dispersal, and movements are manifold. Another consequence of dispersal or migration is that the organism is exposed to new types of infections and parasites, which, as discussed by Westerdahl et al. (Chapter 8), might affect their respective movement patterns. Hence, when studying animal movements it might be of importance to give a thought to why and when hosts move in relation to pathogen exposure. Accordingly, how pathogens move with their hosts, i.e. 'assisted movements/migrations', is an emerging field of research with potential implications for the distribution and abundance of host species, as well as the spread and emergence of diseases.

14.3 Effects of environmental change on movement and migration

Many species of migrating animals, including birds, mammals, and fish, have been reported to decline in numbers, and the rather alarming question 'Is animal migration disappearing?' has been raised (Wilcove and Wikelski 2008). The reasons for the declines are generally changes in migration barriers, overexploitation, climate change, or habitat destruction (Table 3.1). These processes may act both along the route and at the feeding or breeding habitat of the organism. As discussed in Smith et al. (Chapter 4), mobility has profound consequences for how habitat loss and land-use intensity affect animal communities. Clearly, the ability of organisms to persist in transformed human-dominated landscapes may depend on their ability to move and disperse between remaining semi-natural habitat fragments. However, it has become increasingly clear that a classical meta-population view on how mobile animals react to land-use change may represent a special case on how organisms utilize landscapes. Many transformed landscapes consist of a mosaic of habitats of different suitability for organisms, which are linked by organisms' daily or seasonal movements or by dispersal. While, for example, less mobile insect species spend most of their lives within a single patch and therefore may persist even in a fragmented landscape, more mobile ones can persist by moving between several habitat patches in a landscape and utilize geographically distant resources. In contrast, species with intermediate mobility need to move outside of their natal or home patch to find resources, but may not reach other patches within their mobility range if these are too spatially separated, which will be the case in a fragmented landscape. Therefore, species with intermediate mobility and foraging ranges may be most sensitive to habitat loss, fragmentation, and spatial scattering of resources. Some animals need to utilize different habitats for, e.g., nesting and foraging. When land-use change results in landscape simplification, habitats with such complementary resources may become spatially separated, indicating that only mobile animals may be able to cope with land-use change.

One suggested way to handle the negative effects of landscape fragmentation is to construct corridors between suitable habitats. However, although there are many reports on successful corridors, our perception of connectivity may sometimes reflect human perception rather than animal mobility, as discussed in Smith et al. (Chapter 4). Furthermore, if land-use change affects species through different ecological processes, at levels ranging from individuals to populations to communities, general recipes such as increasing connectivity by creating corridors is unlikely to provide a universal solution for all organisms affected by land-use change. Thus, if the main problem is the spatial separation of critical resources for nesting and foraging beyond the daily foraging limits of an organism, this will not be solved by corridors. Instead it is important to understand how species with different mobility utilize resources in, and disperse through, agricultural landscapes, and tailor conservation actions to the organism groups that are targeted (Smith et al., Chapter 4).

Another ongoing environmental alteration that evidently will, and already has, affected animal movement patterns is climate change. However, as discussed in Lindström et al. (Chapter 3), this large-scale alteration, just as for the landscape-scale changes discussed earlier, will affect different organisms in different ways. For example, an analysis of a long-term data set of amphibian migration showed that even highly related species may show opposing responses to climate change (Lindström et al., Chapter 3). Hence, studies at different scales illustrate how difficult it is to make firm predictions about what will happen with individual taxa. Therefore, although it is unlikely that climate change, or any other environmental alteration, will, as discussed earlier, make animal migration disappear, we should still expect considerable changes in movement and migration patterns among animals in the future. Actually, some species may even gain (i.e. increase in abundance) from environmental changes, whereas others may be unaffected. By discerning population trends we can identify the patterns and the risks with which dispersing or migrating taxa may go extinct or at least be gravely reduced in numbers as discussed in Lindström et al. (Chapter 3). In addition to analysing trends, we might use what is known about the rate of evolutionary processes. Based on this we should expect that more *K*-selected species along the *r*/*K*-selection continuum to be most at risk when the climate changes. Hence, since *K*-selected species are generally large, long-lived, and endothermic, which is the case for, e.g., most long-distance bird migrants, we should, as discussed in Lindström et al. (Chapter 3), expect pronounced future alterations in their migratory patterns. In addition to direct effects from environmental change, we might also expect indirect effects to further influence movement and migratory patterns. For example, the effects of habitat change on one migratory species can influence the fate of other migrants thousands of kilometres away (Lindström et al., Chapter 3), illustrating the intricate relationship between habitat change, climate change, and species interactions in shaping movement and migratory patterns.

14.4. Finding their way

It is surprising and fascinating how organisms find their way, especially during long journeys. As discussed in Åkesson et al. (Chapter 9), Muheim et al. (Chapter 10), and Svensson et al. (Chapter 11), animals often use several senses and cues for navigation, including landmarks, the Earth's magnetic field, stars, the sun, and chemical substances. For example, homing pigeons seem to combine information from compass information and memorized landmarks, suggesting that they combine information from several cues to perform their navigational tasks. Surprisingly, however, birds, for example, seem not to inherit detailed knowledge about the star pattern, but they are instead equipped with an inherited strategy to look for the centre of rotation and interpret this as 'polewards', providing a southward course during their first migration, as discussed in Åkesson et al. (Chapter 9). Much effort has been invested during the past decades to understand the detailed mechanisms underlying the different compasses used by animals for orientation (Muheim et al., Chapter 10). Research on the sensory mechanisms of animal orientation and navigation has become highly interdisciplinary, integrating the traditional observation-based study of the behavioural aspects of orientation and navigation with biophysical modelling, neuroanatomical, and molecular genetic tools. These new approaches have led to significant advances in our understanding of how the different sensory systems work that underlie animal orientation and navigation. One example is the magnetic sense in animals, where we have progressed from merely demonstrating in behavioural experiments that animals can perceive magnetic field information and use it for orientation and navigation, to having putative molecular mechanisms, receptor sites, and brain areas (Muheim et al., Chapter 10). Although our knowledge of animal navigation has improved considerably during the past decades, we are still far from fully understanding how animals can sense, for example, the magnetic field. Even the exact location of the magnetoreceptors and how the information is processed in the brain and transferred into a migration direction are still unknown. Equally little is understood about the star and sun compasses, and how animals perceive polarized light cues and use this information for compass orientation. To solve all of these mysteries will be the challenge of many more generations of researchers to come.

Many animals are known to perform extreme migrations across continents and oceans and to return to known locations along the route or at breeding or wintering destinations. To navigate during these long journeys they need one or several compasses, as well as knowledge of some sort of a map. Several major biological compasses have been described for animals, including geomagnetic and celestial (stellar, sun, skylight polarization pattern) information with surprising similarities between different animals (Åkesson et al., Chapter 9, and Muheim et al., Chapter 10). For instance, a general feature of the magnetic compass is its reliance on the angle of inclination rather than on polarity, which has been demonstrated in, e.g., birds, newts, and sea turtles. The widespread sun compass is equipped with a time-compensation mechanism and is used by, e.g., birds, insects, and fish, during long-distance migrations as well as during local movements within the home range, including hoarding. Further, the pattern of skylight polarization provides compass information and the possibility to detect the position of the sun, and is a primary cue for animals capable of detecting this pattern, such as insects, crustaceans, fish, and possibly birds. However, the fixed receptors in the compound eye provide an excellent mechanism for detecting this pattern by integrating information from several sensors, while it is still unknown what structure in the vertebrate eye can pick up this information as discussed in Muheim et al. (Chapter 10). To return to familiar places one will need more than a compass, and there are two major ways to detect locations. First, the animal might pay attention to its own movements relative to home during its travel, i.e. using path integration (measuring courses steered and compass courses taken), a strategy well known for walking insects, birds, and mammals (Åkesson et al., Chapter 9). Second, animals may use some sort of a map, and one extending over global distances is based on the combination of the angle of inclination and total field intensity of the Earth's magnetic field, forming a bi-coordinate map. This type of map has been extensively explored in experiments with sea turtles, lobsters, newts and birds, providing support that the geomagnetic field is central for navigation across scales. At more local scales landmarks and odours (Svensson et al. Chapter 11) may form useful maps. For several organisms for which we now have good understanding of their navigation mechanisms, we see adaptations involving a sequence or combination of strategies. For instance, when the newly hatched sea turtle is leaving its underground nest it will dig through the sand guided by gravity. Thereafter visual (brightness over open sea) cues will guide the turtle to the shoreline and well into water the turtle will steer consistently against the waves. In open water information from the geomagnetic field takes over and guides the turtle during many months to come, and by triggering adaptive course shifts when it is important to keep a safe route in warm waters (Chapman et al., Chapter 2). Another example involves the breeding migrations of salmonid fishes that spawn in small streams, migrate to sea as juveniles, and then return to their natal stream as adults. Recent studies suggest that they use geomagnetic cues to navigate in the open sea (Putnam et al. 2013) and then use olfactory cues to distinguish between different streams, allowing them to return to their natal stream and even the exact local site within the stream to reproduce (Brönmark and Hansson 2012, Brönmark et al. 2014). It is likely that we will see similar sequential navigation strategies like those we have just described when more animals are investigated in greater detail.

An additional navigational tool is odour cues which are used by many taxa when searching for important resources, such as mates or food, and they can often track such resources over very large spatial scales using their olfactory sense. Although the use of odour signals in the life of insects is well established and has been studied for decades, recent research on other animals has revealed that olfaction plays an important role in their lives as well, e.g. when seabirds forage over the open ocean, or when salmonids and sea lampreys search for suitable streams to spawn during their migration, which are behaviours that occur over scales of several hundreds or thousand of metres (Svensson et al., Chapter 11). The progress in identifying and synthesizing behaviourally active compounds for different species will clearly facilitate controlled laboratory and field experiments to increase our knowledge about long-range odour tracking in animals, and it may also provide new applications in pest control and management of invasive species (Brönmark and Hansson 2012).

14.5 Patterns

Observed patterns of movement are sometimes surprising although generally logical and based on the rules to minimize energy consumption, parasite exposure, predation exposure, or the time spent on migration (e.g. Hedenström and Lindström,

Chapter 5). For example, a nocturnal moth, using its olfactory sense when searching for a mate or food, often shows a Lévy flight pattern (a form of random walk) (Svensson et al., Chapter 11) to locate an odour plume. Similarly, migrating birds rarely take on a course with a fixed compass direction, which seems as the simplest solution for finding the target area. Instead they follow the great circle (e.g. Lindström et al., Chapter 3; Hedenström and Lindström, Chapter 5) to get to their target area by travelling the shortest possible distance. Other migrants may select a route intermediate between the direct compass and great circle routes. How they know that the Earth is a globe and thus that the great circle is the shortest way remains puzzling, but is beyond the scope of this book.

Migration is an evolutionary labile trait and it is not uncommon that also closely related species have markedly different migratory strategies. The evolutionary potential of migratory traits is strongly dependent on their genetic architecture (Liedvogel and Lundberg, Chapter 12). The threshold model of migration suggests that the expression of migratory behaviour in an individual requires that the effect of multiple genes reach a threshold. A consequence of the model is that genetic variation associated with migration could be hidden in a resident population and vice versa in a migrating population. This implies that selection could rapidly increase the frequency of individuals expressing the other ('hidden') strategy if environmental conditions were to change. This model also allows different importance of the environment in setting the threshold, which gives rise to a continuum of migratory strategies spanning from obligate to facultative migrants. Obligate migrants, such as many long-distance bird migrants, possess a mostly hardwired migratory behaviour that is minimally influenced by the environment. The decision to migrate in facultative migrants, on the other hand, is more sensitive to environmental factors, such as social interactions and resource abundances, as we will see later in, for example, red deer and zebras.

In some populations there are both migrating and resident individuals, a phenomenon called partial migration (Chapman et al., Chapter 2). The connection between general personality traits such as shyness–boldness and partial migration has recently received considerable attention (Chapman et al. 2011a, b; Nilsson et al., Chapter 6). Not all animals within a taxa or population behave in the same way, but some individuals are bolder than others and such individual differences may have considerable implications for dispersal, migration, and gene flow, as discussed by Nilsson et al. (Chapter 6). For example, in some species individuals with explorative, bold, asocial personalities have been shown to display a stronger tendency to disperse and migrate. In dispersal scenarios this may lead to a gradient of behavioural phenotypes across a landscape ('spatial sorting'; Shine et al. 2011), and this can have implications for invasion dynamics. Individual behavioural variation also provides a foundation for individual differences in migratory tendency in some animals (Chapman et al. 2011b), although few studies have been carried out in the field as yet. Hence, understanding the behavioural types of migrants can provide clues into the ecological basis of migratory behaviour. Furthermore, advances in tracking technologies and improved data capture have led to the insight that individual variation in migratory (and movement) behaviour is almost ubiquitous (Nilsson et al., Chapter 6). Explaining this variation is a major challenge in movement ecology and has significant implications for many important ecological and evolutionary processes such as metapopulation dynamics, gene flow, and disease transmission.

Many factors may affect the decision to move or migrate and thereby form the patterns we see in nature. By far the most common reason to move is a lack of resources at either a temporal (seasonal) or a spatial scale, or that resources are more plentiful at other sites, for example regarding seasonal autumn migration of birds due to low temperature and thereby a lack of food. However, also predation threat, or a trade-off between threat and feeding opportunities, may induce movements, such as in many lake fish, which migrate seasonally to refuges in rivers and streams, as a result of seasonal variations in resource levels and predation pressure (Chapman et al., Chapter 2; Nilsson et al., Chapter 6). Hence, the understanding of the ecological forces that shape movement behaviours is crucial for how to manage and protect important habitats and species (Smith et al., Chapter 4; Lindström et al., Chapter 3).

A migratory or dispersal journey generally occurs as a continuum; i.e. the animals move at a relatively constant speed (with fuelling stops; Hedenström and Lindström, Chapter 5). However, some animals, such as the red deer, *Cervus elaphus*, seem to track and follow resource quality or quantity, i.e. constantly striving at optimizing resource intake (Bischof et al. 2012). During spring migration the animals seem to 'jump' latitudinally and altitudinally from one vegetation boost to another as spring advances. This allows the migratory red deer to feed on high-quality vegetation, whereas resident deer are restricted to plants with lower food quality. Also zebras, *Hippotigris* spp., performing long seasonal migrations, are able to use local cues to predict food and water availability and can then use this information to delay their departure, or even reverse migration, in order to track suitable weather conditions and thereby food availability along their route (Bartlam-Brooks et al. 2013). Such 'optimized migration' patterns might not be the driving force for migration in deer and zebras only, but may be a common phenomenon in many other mammals, and even in many birds and fishes (Chapman et al., Chapter 2). Also temporal patterns exist, for example among many migrating bird species, which tend to move faster on their way to northern breeding sites in spring, than during the return flight in autumn, as shown in a recent meta-analysis (Nilsson et al. 2013). This seasonal pattern is likely caused by competition for breeding sites, as well as by daylight conditions.

14.6 Technical developments

According to the Popperian way of performing science (Popper 1959), research is driven by formulating and testing hypotheses. However, most of us are prone to admit that the development of new techniques solve many questions as well as open up new avenues for research, interestingly, and possibly sadly, often without bothering too much about the Popperian school. Many of these new technologies generate an endless amount of data that human brains, and even normal computers, have limited opportunities to grasp. However, a lot of very important data are generated by advanced technical instruments and equipment and their results nowadays fill up a major part of the pages of many well-renowned journals. Also within the research field of animal movement the technical advancements have been tremendous, especially concerning tracking and molecular techniques, as shown in, e.g., Nilsson et al., Chapter 6; Åkesson et al., Chapter 9; Liedvogel and Lundberg, Chapter 12; and Hedenström and Lindström, Chapter 5. Actually, this technical revolution is one of the main reasons behind the incredible advancement in animal movement research during the past decades. These new techniques give us an opportunity to address old questions that have previously been impossible to approach due to methodological constraints. Similarly, there were few possibilities to track individual movements of small, millimetre-scale, animals before the boom in nanotechnology made it possible to use 'nano-lamps' (quantum dots) to follow them and advanced computers to track their three-dimensional paths through the water (See Box 6.1 and Fig. 6.5). On land, there are also several new techniques available for tracking animal movement over large distances with high precision, for example harmonic radar (Box 11.2) and GPS loggers (Åkesson et al., Chapter 9) for satellite tracking of birds and mammals moving hundreds of kilometres (Muheim et al., Chapter 10; Åkesson et al., Chapter 9; Hedenström and Lindström, Chapter 5), which was impossible just a few decades ago. Those might also be used on small animals, such as insects, and may deliver unique insights into long-range olfactory search, and provide a basis for the development of general principles of odour tracking in insects on a broad landscape scale, with implications for both management of pest species and conservation of rare species, as discussed in Svensson et al. (Chapter 11). Another example is the development of high-throughput molecular techniques that have the potential to open up identification of 'the migration gene', i.e. the genetic code behind animal migration in birds, mammals, and also invertebrates, such as migratory direction in birds and the diel vertical migration of aquatic zooplankton (Liedvogel and Lundberg, Chapter 12; Chapman et al., Chapter 2). Further, quantitative measurements of the fluid dynamics of flying and swimming animals can now be performed at such a detail that we may address the differences in performance associated with differences in morphology and motion (Johansson et al., Chapter 13). Thereby we can get an improved understanding of one of the most important selection pressures shaping animal performance.

Hence, there is no doubt that technological developments have advanced research on animal movement although new instruments and technologies also provide the, less glamorous, opportunity to address questions that have never been asked, that is, providing answers lacking questions! Here is a potential risk in being carried away by the most recent instrument that produces data to fill up yet another terabyte hard drive. This humble warning is by no means an intention to halt future technical advancements, but rather to point out that in order to flourish, the research field of animal movement *also* requires bright hypotheses, hard field work, carefully designed behavioural experiments, and tedious laboratory work, as well as a sound theoretical foundation. Moreover, a rapidly expanding research field also requires syntheses of already retrieved results and conclusions in order to form the basis for future research. Thus, in addition to conferences and workshops, review papers, or special issues of journals, comprehensive and focal books are also crucial for a sound advancement of this scientific field. The overarching aim with this book is, indeed, to provide some assistance in this direction. Hence, in the last section of this book we will therefore take the opportunity to glance into the crystal ball and give our own personal view on which research paths that may be of interest to pursue in the future.

14.7 Future perspectives on animal movement research

Although animal movements have attracted interest since the days of Aristotle (Nussbaum 1978), actual research on animal movement has become a vibrant and rapidly advancing scientific field only during the past few decades, providing an enormous array of scientific papers, reviews, and even books. As is obvious from the chapters of this book, a main reason for this expansion is, as discussed earlier, the many technical advancements that have allowed scientists to address new questions and evaluate them from different viewpoints. There is no reason why the accelerating interest and advancements within research on animal movement should not continue in the future. It may, therefore, be of interest to make an attempt to identify which research questions and areas might be of importance to address in the future.

We may initially conclude that understanding the vast *intraspecific variation* in movement patterns that researchers have documented for a diverse range of taxa is a major and continuing challenge. Do different movement behaviours constitute different strategies? What mechanisms drive individual variation, and what are the consequences for such variety, for individuals, populations, species, and ecosystems? What are the fitness consequences of divergent movement behaviours? Further developing and refining an evolutionary ecology of animal movement is a tantalizing prospect for the coming years, as technological advances allow greater insights into the precise movements and fates of large numbers of mobile organisms.

A much neglected issue that is likely to become more important in the future is that conservation policy makers will have to recognize the importance of studying animal movements in the context of spreading parasites and in particular *pathogens* that may cause a potential hazard to the health of both humans and other animals. In addition, rapid changes of the ranges of hosts, vectors, and parasites caused by *global climate change* are likely to result in new opportunities for pathogens to invade naive hosts. However, we may also see that tolerance to infection becomes increasingly common in migratory, rather than in sedentary species, since mounting strong immune responses may be too costly for migratory species that already pay the cost of movement. Investigating different *mechanisms to deal with infections* with respect to sedentary and migratory organisms would indeed be an interesting avenue for future research.

Another emerging area of research within movement ecology is *individual variation* and repeatability in movement on all scales. Technical advancement has recently opened up this field for large animals, but very detailed movement patterns are still difficult to get for animals below 0.5 kg or so. One question that has been posed for a long time, but which has not been fully answered is how the dispersal process proceeds for individuals. When young animals try to get established in the population, be it in a group or on a piece of land, many search strategies are open, many decisions must be made and many fights must be won, but we know virtually nothing about the factors that are decisive for success or failure. As this part of an animal's life is often combined with high mortality rates, the lack of knowledge is unfortunate in hampering our predictions about population dynamics in a changing world. Moreover, our understanding of conservation in human-dominated landscapes is constrained by lack of knowledge of how animals utilize these landscapes through daily and seasonal movements. Since many organisms important for *ecosystem services* (pollination, biological control) or disservices (pests) are small, there is an enormous potential for the miniaturization of *animal tracking equipment*, and development of *landscape genetics* (Manel and Holdegger 2013) to inform the development of mitigation of biodiversity loss in these landscapes.

Our deeper knowledge regarding animal locomotion will, and has already, provided inspiration to the field of *robotics* and the design of autonomous micro-air vehicles, where bio-inspired constructions for movement on land and in water, as well as in air, will become abundant (Johansson et al, Chapter 13). Also navigating animals have inspired scientists in artificial intelligence when designing the algorithms that will guide the robot back to its starting point. However, despite our expanding biological and technical knowledge, we are still far from constructing a machine that can match the performance of, for example, the bar-tailed godwit crossing the Pacific in one go.

Hence, we must conclude that despite the recent very strong advancements in animal movement research, we are far from fully understanding the mechanisms, codes, senses, strategies, and the consequences of movement in nature. Although it is always sad to admit that we are far from understanding a subject, the positive side of the coin is that we will all have the opportunity to continue to be fascinated and surprised by animal movements also in the future. We hope that you as reader have been inspired by our book and that it can serve as a basis for your own thoughts, hypotheses, and field studies, and that we have encouraged your fascination and curiosity for animal movement at different scales.

References

Bartlam-Brooks, H. L. A., Beck, P. S. A., Bohrer, G., and Harris, S. (2013). In search of greener pastures: using satellite images to predict the effects of environmental change on zebra migration. *Journal of Geophysical Research: Biogeosciences*, **118**, 1427–1437.

Bischof, R., Loe, L. E., Meisingset, E. L., Zimmermann, B., Van Moorter, B., and Mysterud, A. (2012). A migratory northern ungulate in the pursuit of spring: jumping or surfing the green wave? *American Naturalist*, **180**, 407–424.

Brönmark, C., Chapman, B., Hansson, L.-A., Hulthén, K., Skov, C., and Nilsson, P. A. (2014). There and back again: Migration in freshwater fishes. *Canadian Journal of Zoology*, **92**, 467–479.

Brönmark, C., and Hansson, L.-A. (eds) (2012). *Chemical Ecology in Aquatic Systems*. Oxford University Press, Oxford.

Chapman, B. B., Brönmark, C., Nilsson, J. A., and Hansson, L.-A. (2011a). The ecology and evolution of partial migration. *Oikos*, **120**, 1764–75.

Chapman, B. B., Hulthen, K., Blomqvist, D. R., et al. (2011b). To boldly go: individual differences in boldness influence migratory tendency. *Ecology Letters*, **14**, 871–6.

Manel, S., and Holderegger, R. (2013). Ten years of landscape genetics. *Trends in Ecology and Evolution*, **28**, 614–621.

Nilsson, C., Klaassen, R., and Alerstam, T. (2013). Differences in speed and duration of bird migration between spring and autumn. *American Naturalist*, **181**, 837–45.

Nussbaum, M. C. (1978). *Aristotele´s de motu animalum*. Princeton University Press, Princeton, NJ.

Popper, K. 1959. *The Logic of Scientific Discovery*. Routledge, London.

Putnam, N. F., Lohmann, K. J., Putmane, M., Quinn, T. P., Klimley, A. P., and Noakes, D. L. G. (2013). Evidence for geomagnetic imprinting as a homing mechanism in pacific salmon. *Current Biology*, **23**, 312–16.

Shine, R., Brown, G. P., and Phillips, B. L. (2011). Reply to Lee: spatial sorting, assortative mating, and natural selection. *Proceedings of the National Academy of Sciences USA*, **108**, E348–E48.

Wilcove, D. S., and Wikelski, M. (2008). Going, going, gone: is animal migration disappearing? *Plos Biology*, **6**, 1361–4.

Glossary

Allothetic path integration Path integration based on external information.

Altitudinal migration Seasonal migration along an altitudinal gradient, i.e. from uplands to valleys.

Amphidromy A form of diadromous migration in aquatic organisms. Individuals spawn in either freshwater or marine habitats, and migrate to the alternative habitat to feed, returning to their natal habitat to spawn.

Amplified fragment length polymorphism (AFLP) A quick and inexpensive genetic fingerprinting method for obtaining anonymous multilocus genetic profiles. The main disadvantage of AFLP markers is the rather poor per-locus type of genetic information. AFLP is a dominant typing system, which means you can only score the absence or presence of the marker (scored as a band on a gel). Hence it does not tell you anything about the individual being homozygous or heterozygous.

Anadromy A form of diadromous migration in aquatic animals where individuals are born in freshwater habitats and migrate to marine habitats to feed, returning to reproduce in freshwater.

Angle of attack Angle between the chord line of a foil and the oncoming flow.

Aspect ratio Measure of the slenderness of the wing estimated as the ratio between the span and chord.

Azimuthal position of the sun in the sky The clockwise angle of the sun from North relative to local time, measured by their internal clock.

Beaconing The detection of an odour plume or chemical compound transported in air or in water by an animal, which then follows the odour source upstream to a goal.

Bi-coordinate gradient map Based on a combination of two gradients, such as, for example, the geomagnetic inclination and field strength, this map has a global extension and predicted variation across space. Each location in a bi-coordinate map will have its unique combinations of parameter values for the two gradients, and will provide site-specific knowledge to the animal using it.

Boundary layer Fluid closest to an object, affected by viscosity, which does not reach free stream velocity.

Camber Measure of the curvature of a foil relative to the chord.

Candidate gene approach Genes that have been shown to influence the expression of one focal phenotype in one or several organisms. These genes are candidates that potentially are influencing a similar phenotype in other organisms. Genetic variation within a candidate gene such as SNPs, insertions, or deletions could be investigated between individuals with different phenotypes. In the context of migration genetics a clear limitation of this approach is that very few genes are candidates for migratory traits.

Catadromy A type of diadromous migration in aquatic organisms where individuals spawn in marine habitats and migrate to freshwater to feed, migrating back to reproduce in saltwater.

Chain migration The maintaining of relative latitudinal position by migratory populations following migration. In this scenario, the more northerly populations in the breeding season remain more northerly in the non-breeding season following migration.

Circadian clock An endogenous rhythm oscillating with a period of approximately 24 h. The molecular genetic basis of the circadian clock is well understood. Conserved genes that regulate these biochemical oscillations have been described in a wide range of organisms.

Clock-and-compass model The initial migration of migratory naïve songbirds have been explained by a simple clock-and-compass model encoding time, direction, and distance to migrate. The endogenous migration programme and its components are inherited and expressed during solo migrations in birds.

Comparative genomics Across-species (or populations) analysis of the relationship of genome structure and function.

Compass orientation The ability to select a course by using one or several compasses. This course does not have to lead to a known goal, and is not selected based on landmarks, but rather on external compass (i.e. celestial or geomagnetic) information.

Conditioned responses Learned responses (or reflexive behaviour) to a conditioned stimulus.

Cost of transport (COT) Energy required moving one unit weight of an animal one unit of distance.

Diadromy A class of migratory behaviours in aquatic organisms, where animals migrate between marine and freshwater habitats (see also amphidromy, anadromy, and catadromy).

Dichroic chromophore Any light-harvesting molecule, like retinal, that differentially absorbs incoming light (i.e. is more sensitive to the e-vector aligned along one direction than to the perpendicular direction of polarized light).

Diel horizontal migration (DHM) A form of daily migration where animals migrate horizontally in aquatic systems, often from the shallow, littoral zone of lakes into the more open, pelagic water during the evenings.

Differential migration Within-population variation in migratory traits such as timing, destination, and migratory duration.

Differential reflection of polarized light Polarized light reflects more strongly at surfaces perpendicular than parallel to the axis of polarization. This may lead to differences in light intensity in experimental setups and thereby lead to optical artefacts.

Drag Fluid dynamic resistance originating from viscous resistance of moving an object through a fluid and from pressure differences between the front and rear of the object.

Drag coefficient Dimensionless coefficient describing drag of an object related to the shape relative to the direction of flow.

Effective muscle advantage (EMA) Relative moments exerted by limb extensor muscles compared to moments generated by ground reaction forces.

Epigenetic processes Processes that involves modifications of chromatin state (the packing of DNA and its accessibility to transcription) without changing the underlying DNA sequence. Epigenetic marks could lead to differences in gene expression underlying phenotypic differences and are mediated by the environment. These marks could be both reversible and transgenerational. The genome-wide collection of epigenetic marks is termed the epigenome and could, similarly to the transcriptome, be variable between different individuals, tissues, or developmental stages.

Expressed sequence tag (EST) Short unique DNA stretches within a coding region of a gene (derived from cDNA). ESTs are useful to identify and characterize full-length genes or can be used as markers for gene mapping.

Ferromagnetic materials have a large magnetic moment. They exhibit a long-range ordering phenomenon at the atomic level that causes the unpaired electron spins to line up parallel with each other in a region called a domain.

Fluid dynamics forces Forces (viscous drag, pressure forces, and added mass forces) generated by accelerating fluids (air and water).

Fluid foil A relatively flat, solid, object that may generate pressure forces when moved through a fluid. Biological foils include fins and wings.

Founder effect The loss of genetic variation in a newly established population that is founded by only very few individuals from a larger population.

Gait Discrete combinations of speed, duty factor (the time a foot is in contact with the ground), frequency, and stride length or time lags between limbs found in animals moving across surfaces.

Genetic correlation Quantified proportion of variability between two traits, which is shared due to genetic causes. This could be due to linkage disequilibrium between the genes or that the same set of genes are involved in both traits (pleiotropy).

Genotype Individual set of alleles that underlie the expression of a particular phenotypic trait or character, or an allelic profile for neutral loci not necessarily affecting the phenotype.

Geolocator A miniature light level data logger, collecting light intensity data by a sensor and store data about ambient light during long periods, typically a year or more. The data can be used to derive migration routes and wintering areas of migratory birds, on the basis of day-length (latitude) and variation in sunrise and sunset times (longitude).

Geomagnetic field The Earth's magnetic field can be considered as a giant bar magnet with magnetic field lines leaving the Earth's surface at the (geomagnetic) North Pole, curving around the globe, and re-entering the Earth at the South Pole. The vector of the magnetic field lines (i.e. inclination) and magnetic intensity vary with latitude. Intensity values are highest in areas near the poles (~68.000 nT) and decreases towards the equator (~25.000 nT).

Gradient map A map based on one or several gradients, varying in same or different geographical directions.

Heritability One of the central parameters in quantitative genetics. The most commonly used heritability, narrow sense heritability (h^2), is the proportion of phenotypic variation within populations explained by additive genetic variation. The strength of response of selection depends on the amount of additive genetic variation (assuming no genetic correlation and selection on correlated traits).

Homing How an animal returns to a known goal, e.g. its home.

Hybrid zone A geographic area where individuals from populations with different genomes meet and interbreed. Hybrid zones can span a migratory divide, and are well-suited scenarios to study processes involved in adaptive divergence and reproductive isolation.

Ideothetic path integration Path integration that relies on inertial signals, for instance generated by the vestibular system or by using proprioreceptors (i.e. 'step counting'), which is integrated throughout the movement path of an animal.

Inclination Inclination vector of the geomagnetic field lines is defined by the angle between the geomagnetic field lines and the Earth's surface (or gravity). Inclination changes with latitude such that the inclination vector points downwards in the northern hemisphere, and upwards in the southern hemisphere. It is perpendicular to the Earth's surface at the poles, and horizontal at the equator.

Inclination compass A compass that senses the axial alignment of the magnetic field and the inclination. Animals using an inclination compass do not distinguish north from south, but instead pole-wards from equator-wards. The intersection of the magnetic field lines with the horizon points always pole-wards, towards the closer of the two poles, in both the northern and southern hemispheres. The direction where the inclination angle and the horizon diverge always points towards the magnetic equator.

Incompressible fluid Fluid where density stays constant.

Invasive species Non-indigenous flora or fauna that influence the habitat they invade economically, environmentally, and/or ecologically.

Ladder wake Linked vortex structure with alternating signs of rotation of the vortices in each semi-stroke.

Leading edge vortex (LEV) Vortex in close proximity of the foil, generated by separation of the flow at the leading edge. The LEV enhances the lift production of the foil.

Leapfrog migration A population pattern of migration, where migratory populations 'leapfrog' other populations during the migratory period, leading to a switch in the relative latitudinal position of the populations between breeding and non-breeding seasons.

Lift Fluid dynamic force perpendicular to the motion of an object through a fluid, caused by pressure differences between the sides of the object.

Lift coefficient Dimensionless coefficient describing lift of an object related to the shape relative to the direction of flow.

Linkage disequilibrium A non-random association between alleles at different loci, which might be situated on the same (linked) or on different chromosomes (unlinked).

Linkage map A map of the distribution and order of genetic markers on different chromosomes (linkage groups). The map does not show physical distances (i.e. number of nucleotides) between markers but genetic distances. The genetic distances correspond to the recombination frequency between markers and are measured in centiMorgans (cM). In general, the closer two markers are on a chromosome, the more often they will be inherited together and the shorter the genetic distance will be.

Loxodrome (also known as the rhumb line) A line between two locations on the surface of a sphere (like the planet Earth) crossing all meridians of longitude at the same angle. This is the path of constant bearing relative to true north between the two locations.

Magnetic alignment Passive alignment relative to the magnetic field in a direction that does not lead to a specific goal. Often, animals are aligned along the cardinal direction of the magnetic field.

Magnetic compass orientation Use of directional magnetic information for orientation towards a specific direction.

Magnetic map/signpost sense A magnetic map sense usually implies that birds can sense magnetic intensity on the order of 10 nT, whereas a signpost sense assumes birds can recognize larger changes of magnetic intensity or inclination and use this information as a signpost to change one or more aspects of their behaviour (e.g. body fat deposition).

Magnetic moment of a magnet The torque exerted on a magnet when it is placed in a magnetic field and the force the magnet itself exerts on electric currents.

Map-and-compass The concept of navigating using a map and a compass was first defined by Gustav Kramer, who outlined three components: (1) navigation is a two-component process dependent on the ability to define the position and then use a compass independent on the position-fixing mechanism, (2) the map will provide the direction to go home, and (3) in the final step a compass is used to define the direction to move in order to reach home.

Microarray technology Comparative DNA hybridization technique that allows studying the expression of many genes at the same time. The target DNA (of your animal of interest) gets hybridized to short sequences (probes) designed from a large number of genes and that are attached to a solid support, a microarray chip. Complementary base pairings between the target DNA and the gene sequences on the chip produces light signal that can be quantified.

Microsatellite marker Commonly used molecular marker that consists of different numbers of usually short (two to six base pairs) sequence repeats and often show fast mutation rates and high genetic variation. Microsatellites could, for example, be used to study population genetic structure or be used for linkage mapping.

Migratory divide The boundary between two adjacent migratory populations with different migratory routes, i.e. migratory directions. The divide can, but does not need to, be associated with a physical feature, such as a mountain range.

Migratory restlessness ('Zugunruhe') Migratory songbirds in captivity are eager to migrate even in captivity, and show a characteristic behaviour during the migratory period. When expressing migratory restlessness they flap their wings and try to fly or jump into the side of the cage, pointing in the direction that corresponds to the migratory direction chosen by free-flying conspecifics.

Mitochondrial DNA (mtDNA) Mitochondrial markers are a classic tool for studying intraspecific population structure on a fine scale because of their considerable polymorphism and their strong sensitivity to founder events or population subdivision. The main reasons for this are the maternal mode of inheritance and the lack of recombination.

Mosaic map A map for which the spatial relationships of landscape features relative to home needs to be learned by the animal exploring the environment surrounding its home. The limit of a familiar area map or mosaic map is set by the limit of individual knowledge as a result of explorations.

Navigation How animals chart a course to a remote goal.

Next-generation sequencing (NGS) technology Massively parallel sequencing (determination of the nucleotides in a stretch of DNA) with a much reduced cost per sample compared to traditional sequencing. The two main technologies of NGS are Illumina and 454 sequencing.

Non-compass orientation The temporary deviation from an intended vector navigation course; e.g. a young animal deviates on its way towards its destination and move along a coastline or a line of trees.

Non-sense orientation Orientation along fixed geographic or magnetic courses recorded for free-moving animals or animals in cage experiments.

Null magnetic field Experimental condition where the Earth's magnetic field is cancelled, so that the animals perceive no magnetic field information at all.

Oceanodromy Migration in aquatic animals which takes place entirely within marine environments.

Orientation Directed movement.

Orthodrome (also known as the great circle) The intersection of a sphere (like the planet Earth) and a plane that passes through the centre (splitting the sphere into two equal halves). For any two points of the surface of the sphere there is a great circle through the two points, where the minor arc of the great circle is the shortest path connecting the two points.

Paramagnetic materials Materials that have small magnetic dipole moments that result from unpaired electrons. In the presence of an externally applied magnetic field they partly line up with the field, and thereby slightly increase it. Paramagnetic materials have small, constant, temperature-dependent positive susceptibilities, which are usually less than 1/1000 of the applied field.

Partial migration A widespread phenomenon where individuals from the same population vary in their migratory propensity; i.e. populations are composed of both migrants and residents.

Path integration (also termed dead-reckoning) An animal's constant recording of distances covered and angles steered and subsequent updating of its position relative to the starting location. From any position the animal will find the direct route back home by using its continuously updated path integrator.

Phenotype The observable characteristics or physical appearance (e.g. morphology, physiology, behaviour) of one or a set of focal traits. The individual's phenotype is shaped by its genotype, environmental factors, and their interactions.

Phenotypic plasticity The ability of a genotype to produce various phenotypes within an array of environments.

Philopatry The behaviour of inhabiting or returning to an individual's birthplace.

Piloting Goal-directed orientation based upon a strategy by which the animal is following a sequence of familiar landmarks, which could be of any type. Importantly, the landmarks do not have to define the position of the animal in space relative to the goal on a map, but rather provide a route to the goal.

Polarity compass A compass that senses the direction of the horizontal component of the magnetic field, thus pointing always towards magnetic North. It works similar to a human technical compass with a magnetic needle reliably pointing towards magnetic North.

Population genetics Analysis of allele frequency distribution and structure within populations or subpopulations, and their change under natural selection, genetic drift, mutation, and gene flow.

Potamodromy Migration in aquatic animals that takes place entirely within freshwater habitats.

Quantitative genetics Part of genetics involved in dissecting effect of environment and genetics on quantitative traits and explores response to selection and evolution of these traits.

Quantitative trait A continuous trait which is expected to be influenced by variation at many loci. Environmental conditions often have a large effect on these types of traits.

Quantitative trait locus (QTL) Specific regions in a genome that contain or are linked to the gene(s) underlying an identified quantitative trait.

Radical pair A pair of simultaneously produced and correlated transient (i.e. short-lived) radicals; for example, two correlated electron spins, one electron on each radical.

Reynolds number Ratio of inertial (pressure) and viscous forces.

RNAseq Sequencing of transcripts, typically those derived from coding genes (mRNA) using next-generation sequencing technology. This is a highly accurate and efficient approach for gene expression profiling.

Selection experiments Experimental breeding scheme in which individuals are selectively mated according to their phenotype.

Semelparous migration A form of migration when animals make a single return migratory journey to reproduce and then die.

Shear layer Layer with a velocity gradient resulting in viscous shear stresses.

Span efficiency Ratio of the minimum induced drag for a given force production to the actual induced drag of a foil. Minimum induced drag is achieved through a uniform induced velocity behind the foil. Any perturbation increases drag.

Stable isotope analyses Stable isotopes of carbon (C), nitrogen (N), sulphur (S), and hydrogen (H) from animal material (mostly applied for feather, claws, nail, fur) can be used as intrinsic markers in tracking a moving animal's whereabouts, e.g. wintering location, and thus can be used as a (rough) proxy for migratory behaviour. The method makes use of the fact that the isotopic signatures reflect the diet at the time they were synthesized, and thus can be related back to local food webs and geographic locations where the animal has been.

Stopover site A site where migrants temporarily interrupt their journey to replenish energy stores for further movement.

Superparamagnetic materials Materials that have a zero net magnetic moment, but in an external field they show an alignment of magnetic moments.

Telescopic migration A population pattern of migration which occurs when two populations coexist during one season and migrate to different latitudes during the migratory period.

Time-compensated sun compass A compass that works very similarly to human sun compasses. It takes into consideration the movement of the sun across the sky during the day, requiring the animals to measure time and to compensate for the azimuthal change of the sun with a time compensation mechanism.

Transcriptome The part of the genome that gets transcribed. As the epigenome, the transcriptome is much more dynamic than a genome and could vary for any given developmental stage or physiological condition.

Vector navigation The direction that a young animal selects when migrating to reach its population-specific wintering area.

Vortex Region of fluid with a spinning motion around an imaginary axis.

Vortex ring Vortical flow forming a closed loop (toroid structure).

Zugunruhe Migratory restlessness (expression of migratory flight in captive birds).

About the Centre for Animal Movement Research (CAnMove)

The Centre of Animal Movement Research (CAnMove), associated with the Department of Biology at Lund University, was funded by a 10-year grant from the Swedish Research Council and Lund University in 2008 and has been formed around the wide interest to study movement ecology across scales. CAnMove is contributing to the progress of animal movement research by implementing new technology to track small organisms such as *Daphnia* (nanotechnology) and insects and birds (LIDAR), but also songbirds (geolocators, radiotelemetry), seabirds (GPS loggers), raptors (satellite telemetry), fish (PIT-tags), and bats (radiotelemetry). However, movement is also investigated at all levels from genes encoding migration to effects of movements on population scales and across tropic levels. We also investigate how phenotypic, i.e. morphological and physiological, adaptations have evolved to meet a migratory as well as a sedentary lifestyle.

The primary goal of CAnMove is to understand the underlying ecological and evolutionary causes and consequences of animal movements, with an emphasis on the process(es) of migration across a large range of spatial scales. With a concerted effort we aim at reaching significant progress on an international scale. Output from this research will be instrumental when addressing problems about the evolution of migration and dispersal, climate change, and its effects on biodiversity, spread of diseases, and invasive species, all of which are fundamental issues to human societies from a global perspective.

Associated with CAnMove are more than 60 senior and junior scientists, postdocs, PhD students, and technical personnel. CAnMove regularly organize an international PhD student course on Ecology of Animal Movement, and symposia and topological meetings on movement ecology. Central to the CAnMove programme is to stimulate interactions and communication of migration phenomena to a world-wide audience. At the CAnMove home page (http://canmove.lu.se) we present recent research findings and programme activities. We welcome communications and research interactions on movement ecology.

Index

A
aggression 96–7
agricultural landscapes 51–64
  dispersal ability 59–60
  habitat fragmentation 52–7, 59–60, 261
  heterogeneity 58–9
  intensive farming 52, 54, 57
  landscape complementation 53, 61–2, 63
  landscape supplementation 53, 61
  metacommunities 53, 54
  metapopulations 53
  mosaic farmland 53–4
  multiple patch use 60–3
  resource separation 60–2
  source habitats and biological control 60
  spill-over 53, 57–9, 61–2
airfoil 241
albatross
  black-browed (*Thalassarche melanophrys*) 94, 104
  wandering (*Diomedea exulans*) 151, 168–9, 210
alewives (*Alosa pseudoharengus*) 39
Allee effect 47
allelochemicals 198
allomones 198
allothetic path integration 154, 269
altitudinal migration 26, 269
ambient noise 116
amphibian migration 14–15, 42
amphidromy 24, 269
amplified fragment length polymorphism (AFLP) 227, 269
anadromy 24, 269
anemotaxis 200
angle-of-attack 242, 269
anguilliform swimming 251–2
animal navigation, *see* navigation
animal personality 90–105
  behavioural consistency 90–1
  behavioural plasticity 93
  behavioural syndromes 91–3, 94, 98
  behavioural types 91–3
  dispersal 95–8
  fitness 95
  heritability 92–3
  invasive species 99–100
  local movements 93–5
  metapopulation dynamics 98–9
  migration 101–4, 264
  range expansion 99
  reintroduced species 100
antennules 205–6
ant navigation 1, 151–2, 157–8, 170–2
appetitive flight 203
arrival time hypothesis 18
arthropod migration patterns 15–16
aspect ratio 242, 250, 269
assisted movements 260–1
assortative mating 116
Atlantic salmon (*Salmo salar*) 207
avian influenza 140
avian malaria 131–2, 136, 137, 138–9
azimuthal position of sun in sky 156, 269

B
baculoviruses 139
barnacle geese (*Branta leucopsis*) 29–30
bar-tailed godwit (*Limosa lapponica*) 14, 17, 130, 232, 260
basal metabolic rate 233, 234
Batrachochytrium dendrobatidis 139
bats
  flight 248–9
  migration 13–14, 84–5
  pathogen spread 140
beaconing 154, 269
beetle banks 61
behavioural consistency 90–1
behavioural plasticity 93
behavioural syndromes 91–3, 94, 98
behavioural types 91–3
Bernoulli's principle 240
Bewick's swan (*Cygnus bewickii*) 84
bi-coordinate gradient map 156, 169–70, 269
biological compasses 156, 262–3
biological control 60
biotic interactions 44–8
bird flight 245–6
bird foraging 60–1
bird migration
  biotic interactions 44–5, 46
  circannual clock 223–4
  climate change 40–1, 42–4
  detours 82
  feeding for flight 83–4, 167–9
  fly-and-forage 79–80
  genetics 219–29
  habitat change 36, 38–9, 40
  immune response 129–30, 131–2, 133
  magnetic compass 163–4, 165–6, 181, 183–4, 187, 188
  multi-day stops 76–9
  nocturnal 23, 161–3
  non-stop flight capacity 75–6
  pathogen avoidance 135–7
  pathogen spread 138–9, 140
  patterns of 14, 25–6, 29–30
  repeated flights 76–9
  skylight polarization 160, 190
  soaring 80–2
  star compass 161–3
  strategies 74–84
  sun compass 158–9
  timing 223–6
  tracking devices 73, 78
  unexpected journeys 82–3
  vector navigation 166–9
bird odour tracking 209–10
black-browed albatross (*Thalassarche melanophrys*) 94, 104
blackcap (*Sylvia atricapilla*) 166, 220, 222, 223, 224, 225, 227
bluebirds (*Sialia mexicana*) 96–7, 98, 99
blue crab (*Callinectes sapidus*) 205, 206
blue mussel (*Mytilus* spp.) 118

blue petrel (*Helobaena caerulea*) 209–10
blue tit (*Cyanistes caeruleus*) 103, 134, 224
blue tit (*Parus caeruleus*) 44
boundary layer 240, 269
breeding migrations 14, 15, 19, 21, 24, 26
brown trout 207–8
bumblebees 60, 62, 63
butterflies
  landscape fragmentation 54–6
  migration 85, 158
  resource separation 60
  see also monarch butterfly

C
camber 242, 269
candidate gene approach 224, 227, 228, 269
casting flight 202, 203, 205, 208
catadromy 24, 269
caterpillars 139
cephalopod migration 15
cetacean migration 14
chain migration 16–17, 269
cheetah (*Acinonyx jubatus*) 235
chipmunk (*Tamias sibiricus*) 95
cilia 239, 241, 243, 248
circadian clock 156, 224, 269
circannual clock 223–4
climate change 28, 29, 40–4, 261–2
clock-and-compass model 166, 269
CLOCK gene 224
clock-shift experiments 157
Cluster N 188
cognitive maps 155
commensals 141
comparative genomics 269
compass navigation 156–61, 262–3
compass orientation 153, 188–9, 269
competitive release hypothesis 18, 105
conditioned response 270
convict cichlid (*Amatitlania nigrofasciata*) 94
corridors 56–7, 63, 261
cost of transport 232–4, 237–9, 247–50, 270
crane (*Grus grus*) 95
crustaceans
  migration patterns 15
  odour tracking 205–6
cryptochromes 186
cue conflict and cue calibration experiments 157

D
daily migration 19–21
dams 28, 39
Daphne magna, tracking 102
dead-reckoning 154
desert ant (*Cataglyphis fortis*) 151–2, 157–8, 170–2
desert locust (*Locusta migratoria*) 221
diadromy 14, 24–5, 270
dichroic chromophore 189, 270
diel horizontal migration 14, 20–1, 270
diel vertical migration 3, 19–20, 21
differential migration 19, 101, 270
differential reflection of polarized light 190, 270
dimethy sulphide 210
dinosaurs 12–13
dispersal
  behavioural syndrome 98–100, 128
  corridors 56–7
  definition 3
  gene flow 116–18, 260–1
  immune response 133–4
  landscape fragmentation 59–60
  of micro-organisms 141
  personality and 95–8
  phenotype immunocompetence hypothesis 128
  phenotypic plasticity 111–12, 113, 114, 115–16
dogs, odour tracking 211–12
drag 241–2, 250, 270
drag coefficient 242, 243, 247, 270
dragonflies 86, 250–1

E
ecological fitting 137
eels
  migration 223
  swimming style 251
effective muscle advantage 235, 270
electroantennography 197
elephant locomotion 235
emerging diseases 137–9
enemy release hypothesis 128, 129
epigenetic processes 226, 270
ethyl mercaptan 210
evolution
  compromises along movement–sedentary gradient 259–60
  migration 11–13, 264
  olfaction 199–200
  phenotypic plasticity 114–15
evolution of increased competitive ability hypothesis 129
expressed sequenced tag (EST) 221, 270

F
farmland, *see* agricultural landscapes
fasting endurance hypothesis 18
feeding for migration 83–4, 167–9
ferromagnetic materials 180, 270
ferromineral-based magnetoreception 180–4
fineness ratio 243
fins 241–2, 250–2
fish magnetoreception 180
fish migration
  biotic interactions 45
  genetics 223, 224
  habitat change 39
  patterns of 14, 22, 24–5
fish odour tracking 206–9
fish swimming styles 246, 251–2
fitness 95
flagella 239, 241, 243, 248
flight range curve 74
fluid dynamic forces 240, 270
fluid foil 241–2, 244, 248–50, 270
fly-and-forage 79–80
flying, physics of 74, 233–4, 239–52
foil 241
foraging
  individual differences 94
  movements 3
  optimal foraging theory 62
founder effect 270
fruit flies (*Drosophila melanogaster*) 202–3, 249
fuelling for migration 83–4, 167–9
fungal disease 139

G
gait 270
Gammarus pulex 112
garden warbler (*Sylvia borin*) 103
geese migration 40
gene expression profiling 221
gene flow 115–18, 260–1
genetic accommodation 121
genetic assimilation 121
genetic correlation 92, 270
genetic footprints 130
genetics
  immune response 130–2
  migration 219–29
  olfactory receptors 199–200
  personality traits 92–3
  phenotypic plasticity 115–16
genome-wide association studies 221
genomics 221
genotype 270
geolocator 78, 270
geomagnetic field 156, 167, 168, 180, 270
gliding flight 81
gradient map 156, 270
grain-size theory 111

great reed warblers (*Acrocephalus arundinaceus*) 226
great snipe (*Gallinago media*) 14, 83
great tit (*Parus major*) 92, 96, 134
green darner (*Anax junius*) 15, 86
green turtle navigation 154
grey squirrel (*Sciurus carolinensis*) 129

H
habitats
  changing 36–40
  fragmentation 52–7, 59–60, 261
  selection and phenotypic plasticity 113–14
Haidinger's brushes 189
harmonic radar 204–5
herbicides 57
heritability 270
  migratory traits 220, 222
  personality traits 92–3
homing 3, 153, 270
homing pigeons 159, 169–70, 172, 210
honeybees 157
hopping 235, 236, 239
house finches (*Carpodacus mexicanus*) 225
house sparrow (*Passer domesticus*) 99
humans
  dispersal 98
  hunting by 44
  malaria infection 130
  MHC 131
  olfaction 212
  pathogen transport 129, 137
  polarization vision 189
hybrid zone 118, 271

I
ideothetic path integration 154, 271
immune response 129–33
inclination 271
inclination compass 164, 165, 271
incompressible fluid 239, 271
individual differences 90, 94, 101, 104
infection, *see* pathogen exposure
insecticides 57
insects
  flight 244–5, 249–50
  migration 15–16, 22–3, 85–6
  navigation 171–2
  odour tracking 200–5
  sun compass 156–8
invasive species 271
  infection risk 129
  personality traits 99–100
  phenotypic plasticity 118–20

J
Japanese grenadier anchovy (*Coilia nasus*) 135
jet propulsion 241, 243–4, 248

K
kairomones 198
kangaroos 235, 239
K-selected species 46, 262

L
ladder wake 246, 271
lagena 183
landmark navigation 171–2
landscape barriers 26–8, 40
landscape complementation 53, 61–2, 63
landscape fragmentation 52–7, 59–60, 261
landscape supplementation 53, 61
latitudinal migration 25–6
leading-edge enemy release hypothesis 128, 129
leading edge vortex 245, 271
leapfrog migration 17, 271
lemming cycles 45
Lévy flight 204
lift 241–2, 271
lift coefficient 242, 246, 250, 271
lift-induced drag 241–2
lift to drag ratio 233
light-based magnetoreception 185–8
lightlogger 78
linkage disequilibrium 227, 271
linkage map 221, 271
lizards (*Lacerta vivipara*) 98, 99
lobster migration 15
local adaptation 111, 112, 115, 120–1
locomotion 232–52
  across surfaces 234–9
  cost of transport 232–4, 237–9, 247–50, 270
  power 232, 234
  speed 232, 234, 235–6, 242–7
  through fluids 239–52
loggerhead sea turtles (*Caretta caretta*) 169
loxodrome 76–7, 271

M
macrophages 183
magnetic alignment 163, 271
magnetic coil experiments 157
magnetic compass orientation 163–6, 179–88, 271
magnetic map 169, 181, 183, 271
magnetic moment 181, 271
mainland–island systems 53
main olfactory epithelium 210
major histocompatibility complex (MHC) genes 130–2, 209–10
malaria 130, 131
  avian 131–2, 136, 137, 138–9
mammals
  migration patterns 13–14
  odour tracking 210–12
map and compass 156–61, 271
map-based navigation 154–6
marsh harriers (*Circus aeruginosus*) 104
mass force 240
matrix 52, 54, 56, 57, 63
melatonin-induced immunoenhancement 133
Mendelian traits 222
metabolic rate 233, 234
metacommunities 53, 54
metapopulation 53, 98–9
mice, immune function 133
microarray technology 221, 271
microsatellite marker 221, 226, 271–2
migration
  costs and benefits 12
  definition 3
  evolution 11–13, 264
  extinct creatures 12–13
  genetics 219–29
  individual differences 104
  patterns 11–30
  speed 75, 232, 234
  strategies 73–86
  threshold model 224–6, 264
  timing 19–21, 23–4, 223–6
migratory divide 17, 272
migratory restlessness 158, 166, 219, 272
mismatch to novel parasites hypothesis 128
mitochondrial DNA 226, 272
mollusc migration patterns 15
monarch butterfly (*Danaus plexippus*) 16, 23, 85, 86, 135, 158, 220, 221
mosaic map 154–6, 272
mosquito (*Aedes aegypti*) 203
moths
  migration patterns 15–16
  odour tracking 201–2, 203, 204–5
movement, definition 3

N
nanotechnology 102
Nathusius' bat (*Pipistrellus nathusii*) 84, 85

navigation 151–72, 262–3, 272
bi-coordinate gradient map 156, 169–70, 269
biological compasses 156, 262–3
concepts and terminology 152–6
landmark use 171–2
magnetic compass 163–6, 179–88, 271
map and compass 156–61, 271
map-based 154–6
odour tracking 200–12
olfactory map theory 210
path integration 154, 170–1, 272
route-based 154
sensory mechanisms 179–90
skylight polarization 159–61, 189–90
star compass 161–3, 188–9
study methodologies 157
sun compass 156–9, 161, 188–9
vector-based 153, 166–9, 273
next-generation sequencing technology 221, 272
nocturnal migration 23–4, 161–3
non-compass orientation 153, 272
non-sense orientation 153, 272
novel weapon hypothesis 129
nucleus of basal optic root 188
null magnetic field 180, 272

O

ocean currents 28
oceanodromy 14, 272
olfaction 195–213, 263
evolution 199–200
odour plumes 196–9
odour tracking 200–12
olfactory map theory 210
olfactory receptor genes 199–200
optic tectum 188
optimal migration theory 74, 265
orientation 153, 262–3, 272
magnetic compass 163–6, 179–88, 271
skylight polarization 160, 189–90
star compass 161–3, 188–9
sun compass 156–9, 188–9
see also navigation
orthodrome 76, 82, 272
osprey (*Pandion haliaetus*) 80
ostrich (*Struthio camelus*) 239

P

Pacific oyster (*Crassostrea gigas*) 119
paddling 244, 246–7
painted lady butterfly (*Vanessa cardui*) 158
paramagnetic materials 185, 272
parasites, *see* pathogen exposure
partial migration 17–18, 101–4, 264, 272
path integration 154, 170–1, 272
pathogen exposure 126–42
avoidance by migration 134–7
cost of migration 126
dispersal 133–4
emerging diseases 137–9
immune response 129–33
micro-organisms influencing host movements 139–40
passive dispersal of micro-organisms 140–1
range expansion 128–9
transport of pathogens 137–9
penguin (*Spheniscus demersus*) 210
peregrine falcon (*Falco peregrinus*) 44
periwinkle (*Littorina saxatilis*) 111
personality traits, *see* animal personality
pesticides 29
p/g ratio 18, 19, 21
phenotype 272
phenotype–environment mismatch 114
phenotypic accommodation 121
phenotypic plasticity 110–11, 272
dispersal 111–12, 113, 114, 115–16
dynamics and evolution 114–15
environmental heterogeneity 111–13
genetics 115–16
habitat selection 113–14
invasive species 118–20
potential of speciation 120–1
pheromones 198, 199, 201–2, 204–5, 208
philopatry 113, 272
photoreceptors 185, 186, 188, 189–90
pied flycatcher (*Ficedula hypoleuca*) 42
pike (*Esox lucius*) 24, 94
piloting 153, 154, 272
pinch point 248
plaice (*Pleuronectes platessa*) 152
Plasmodium relictum 137
polarity compass 164–5, 272
polarized light sensitivity 159–61, 189–90
pollination 62–3
population genetics 226, 272
potamodromy 14, 272
potential flight range 74
power 74, 75, 232, 234
pressure forces 240
pronghorn antelope (*Antilocarpa americana*) 235
pulse remagnetization 182

Q

quantitative genetics 220, 222–3, 272
quantitative trait 222, 225, 272
quantitative trait locus 221, 272

R

radical-pair-based magnetoreception 185–8, 273
radio-frequency magnetic fields 182–3
rainbow trout (*Oncorhynchus mykiss*) 180, 221, 226
random walk 203–4
range expansion
dispersal behavioural syndromes 99
infection risk 128–9
ranging flight 203
rare species conservation 64
rats, odour tracking 211
reaction norms 110
red-backed shrike (*Lanius collurio*) 79
red bollworm moth (*Diparopsis castanea*) 154
red deer (*Cervus elaphus*) 265
red knot (*Calidris* spp.) 38–9, 133
red-spotted newt (*Nothophthalmus viridescens*) 188
reindeer (*Rangifer tarandus*) 135
reintroduced species 100
reptile migration patterns 15
resource separation 60–2
retinal photoreceptors 185, 186, 188
Reynold's number 240, 273
rheotaxis 200
r/K-selection continuum 46, 262
RNAseq 221, 273
roach (*Rutilus rutilus*) 101, 103, 104
roads 27
robins (*Erithacus rubecula*) 163–4
route-based navigation 154
rowing 244
ruff (*Philomachus pugnax*) 39
running, physics of 233, 234–9

S

salmon migration
climate change 41
genetics 224
habitat change 39
navigation 263
olfactory cues 207
patterns 22
swimming behaviour 247
sand cricket (*Gryllus firmus*) 226
Sardinian warbler (*Sylvia melanocephala*) 103
scanning harmonic radar 204–5

sea lamprey 208–9
seasonality
  immune function 133
  migration patterns 21
  partial migration 17
selection experiments 92, 220, 222, 224, 273
semelparous migration 22, 273
semi-natural grasslands 56, 62
semiochemicals 198
sensilla 201, 205
shark scales 243
sharp-tailed sandpiper (*Calidris acuminata*) 82
shear layer 248, 273
short-tailed shearwater (*Puffinus tenuirostris*) 82–3
Siberian chipmunk (*Tamias sibiricus*) 95
signature mixtures 198
silver-haired bat (*Lasionycteris noctivagans*) 85
silver Y moths (*Autographa gamma*) 16, 21, 85
single-sensillum recording 197
skylight polarization 159–61, 189–90
smoltification 22, 226
snail (*Radix balthica*) 112
snow geese 40, 44–5
soaring 80–2
sociality
  dispersal 97–8
  local movements 95
  migration 15, 47
source habitats 60
source–sink systems 53
span efficiency 248, 249, 273
spatial scale of migration 26
speed of movement 232, 234, 235–6, 242–7
spider dispersal 59
spill-over 53, 57–9, 61–2
squid locomotion 248
stable isotope analyses 12, 223, 273
stall 242
star compass 161–3, 188–9
starlings (*Sturnus vulgaris*) 153
star-nosed mole (*Condylura cristata*) 212
stopover site 36, 38–9, 42, 273
streamlining 243
sun board 161
sun compass 156–9, 161, 188–9
sunstone 161
superparamagnetic materials 181, 273
swimming, physics of 233, 234, 239–52
swordtail fish (*Xiphophorus* spp.) 115
synomones 198

T

taxis 200
telescopic migration 17, 273
tendons 238
thermals 81
thermal tolerance hypothesis 18
threshold model of migration 224–6, 264
thrush nightingales (*Luscinia luscinia*) 167–8
thrust 244
thuniform swimming 251–2
tiger snake (*Notechis scutatus*) 115
time-compensated sun compass 156–9, 273
timing
  of dispersal 114
  of migration 19–21, 23–4, 223–6
tobacco budworm (*Heliothis virescens*) 202
tolerance 132
toxoplasma 139
tracking devices 73, 78, 102, 265
transcriptome 221, 273
transgenerational migration 16, 22–3
treehoppers (*Enchenopa binotata*) 117
Trichomonas gallinae 139
trigeminal nerve 180, 183, 184
tropical wintering–parasite exposure hypothesis 133
trout migration 226
turkey vulture (*Cathartes aura*) 210

U

ungulate migration 13
urbanization 29, 40

V

vector navigation 153, 166–9, 273
Viking navigation 160–1
viscous forces 240
vomeronasal organ 211
vortex 240, 273
vortex ring 248, 273

W

wake capture 245, 250
walking, physics of 234–9
wandering albatross (*Diomedea exulans*) 151, 168–9, 210
wasted wake power 252
water shrew (*Sorex palustris*) 212
West Nile virus 140
wheatear (*Oenanthe oenanthe*) 76, 77–8
white stork (*Ciconia ciconia*) 80
wildebeest migration 44
willow warbler (*Phylloscopus trochilus*) 17, 227
wings
  morphing 246
  morphology 226, 250–2
  physics of 241–2
winter immunoenhancement 133
wood thrush (*Hylocichla mustelina*) 79

Y

yellow fever mosquito (*Aedes aegypti*) 203

Z

zebra 265
zoonotic diseases 140
Zugunruhe 158, 166, 219, 273

The manufacturer's authorised representative in the EU for product safety is Oxford University Press España S.A. of El Parque Empresarial San Fernando de Henares, Avenida de Castilla, 2 - 28830 Madrid (www.oup.es/en or product.safety@oup.com). OUP España S.A. also acts as importer into Spain of products made by the manufacturer.

Printed and bound by CPI Group (UK) Ltd, Croydon, CR0 4YY

11/08/2026

02193551-0001